Medical Importance of the
Normal Microflora

Medical Importance of the Normal Microflora

Edited by

Gerald W. Tannock

Department of Microbiology
University of Otago
Dunedin
New Zealand

KLUWER ACADEMIC PUBLISHERS

DORDRECHT / BOSTON / LONDON

A C.I.P. Catalogue record for this book is available from the Library of Congress

ISBN 0 412 79390 3

Published by Kluwer Academic Publishers,
P.O. Box 17, 3300 AA Dordrecht, The Netherlands.

Sold and distributed in North, Central and South America
by Kluwer Academic Publishers,
101 Philip Drive, Norwell, MA 02061, U.S.A..

In all other countries, sold and distributed
by Kluwer Academic Publishers Group,
P.O. Box 322, 3300 AH Dordrecht, The Netherlands

Printed in Great Britain

Contents

Contributors

Rodney D. Berg
Department of Microbiology and Immunology
Louisiana State University Medical School
Louisiana State University Medical Center – Shreveport
1501 Kings Highway
PO Box 33932
Shreveport, LA 71130
USA

John Birkbeck
InforMed System Ltd
PO Box 17
Waimauku 1250
New Zealand

S. P. Borriello
Central Public Health Laboratory
61 Colindale Avenue
London NW9 5HT
UK

Vinton S. Chadwick
Wakefield Gastroenterology Centre and Research Institute
Wakefield Hospital
PO Box 7168
Wellington
New Zealand

Casey Chen
Department of Periodontology
University of Southern California
Los Angeles, CA
USA

Wangxue Chen
Wakefield Gastroenterology Centre and Research Institute
Wakefield Hospital
PO Box 7168
Wellington
New Zealand

Anthony P. Corfield
Mucin Research Group
University Department of Medicine Laboratories
Bristol Royal Infirmary
Bristol BS2 8HW
UK

Paul L. Fidel Jr
Department of Microbiology, Immunology and Parasitology
Louisiana State University Medical Center
1901 Perdido Street
New Orleans, LA 70112-1393
USA

Marc Habash
Lawson Research Institute
268 Grosvenor Street
London, Ontario N6A 4V2
Canada

Christine Heinemann
Lawson Research Institute
268 Grosvenor Street
London, Ontario N6A 4V2
Canada

Howard F. Jenkinson
Department of Oral and Dental Science
University of Bristol
Lower Maudlin Street
Bristol BS1 2LY
UK

Jean O. Kim
Department of Pediatrics
Division of Infectious Diseases
Maricopa Medical Center
2601 East Roosevelt
Phoenix, AZ 85008
USA

Adrian Lee
School of Microbiology and Immunology
University of New South Wales
Sydney, NSW 2033
Australia

William C. Noble
St John's Institute of Dermatology
St Thomas' Hospital
Lambeth Palace Road
London SE1 7EH
UK

Andrew B. Onderdonk
Clinical Microbiology Laboratory
Brigham and Women's Hospital
75 Francis Street
Boston, MA 02115
USA

Gregor Reid
Lawson Research Institute
268 Grosvenor Street
London, Ontario N6A 4V2
Canada

Anthony M. Roberton
Molecular Genetics and Microbiology Research Group
School of Biological Sciences
University of Auckland
Private Bag 92019
Auckland
New Zealand

C. Roffe
Department of Geriatric Medicine
City General Hospital
Stoke-on-Trent
UK

Ian R. Rowland
Northern Ireland Centre for Diet and Health
School of Biomedical Sciences
University of Ulster
Coleraine BT52 ISA
UK

Jørgen Slots
Department of Periodontology
School of Dentistry
University of Southern California
Los Angeles, CA
USA

Gerald W. Tannock
Department of Microbiology
University of Otago
PO Box 56
Dunedin
New Zealand

Johan Van Eldere
Laboratorium Bacteriologie
Universitaire Ziekenhuizen Leuven
UZ Sint-Rafael
Kapucijnenvoer 33
3000 Leuven
Belgium

Jeffrey N. Weiser
Department of Microbiology
University of Pennsylvania School of Medicine
302B Johnson Pavilion
36th and Hamilton Walk
Philadelphia, PA 19104
USA

Preface

This book is a logical sequel to a small monograph entitled *Normal Microflora. An Introduction to Microbes Inhabiting the Human Body*, which was published by Chapman & Hall in 1995. This introductory text, concerning the collection of microbes that colonize the human body even in health, has been well received. *Medical Importance of the Normal Microflora* now develops an aspect of the host–microflora relationship that is often neglected in medical texts. It provides, in a single volume, a comprehensive survey of the association of the normal microflora with certain diseases. The normal microflora, although inhabiting the bodies of healthy humans can, when appropriate predisposing conditions exist, act as a source of aetiological agents of disease. Opportunistic pathogens are lurking on and within our bodies. The members of the normal microflora that may be involved, the predisposing conditions that occur and the diseases with which they are associated are described in this book. I am very grateful to the contributors of the chapters in this book. They are all busy investigators and expert in their field. They nevertheless took the time and effort to produce chapters that enable the reader to appreciate the complexity of the normal microflora–host relationship and its importance in the practice of medicine and dentistry. The study of the normal microflora, and the description in ecological terms of phenomena relating to it, has been greatly influenced by the work of Dwayne C. Savage. It has been my privilege to have enjoyed the friendship of Jean and Dwayne Savage over many years, and so it is my pleasure to respectfully dedicate this book to them.

G. W. Tannock
Dunedin, May 1998

1

The normal microflora: an introduction

Gerald W. Tannock

The 'normal microflora' is the term most commonly used when referring to the microbial collection that consistently inhabits the bodies of healthy animals. Other terms used are 'normal flora', 'commensals' and 'indigenous microbiota'. Of these, the strictly correct term is 'indigenous microbiota', since it refers to a collection of microscopic creatures that are native to the body. 'Flora' and 'microflora' have an unfortunate botanical connotation. Commensalism refers to an association between two organisms in which one partner benefits from the relationship but the other obtains neither benefit nor harm. The normal microflora–animal relationship is not one of commensalism, however, since each partner influences the other markedly. Many scientists would prefer the use of 'indigenous microbiota' since it is more correct than the alternatives. 'Normal microflora' has, however, been used extensively in the medical literature for many decades, has international recognition, is likely to remain in common usage, and is therefore used in this book.

The normal microflora is comprised of a diverse collection of microbial species, mostly bacterial. Some idea of the diversity of microbial types that can be present can be gained by reference to Tables 1.1–1.6. Some of these microbial species attain high population levels (Table 1.7) and it has been estimated that more microbial cells inhabit the human body than there are eukaryotic cells of which it is constituted (10^{14}:10^{13}; Luckey, 1972).

The study of the composition of the normal microflora has relied almost exclusively on the quantitative culture of microbes from samples

G. W. Tannock (ed.), *Medical Importance of the Normal Microflora*, 1–23.
© 1999 *Kluwer Academic Publishers. Printed in Great Britain.*

Table 1.1 Microbes commonly detected on human skin

Gram-positive cocci
  Staphylococcus aureus
  S. auricularis
  S. capitis
  S. cohnii
  S. epidermidis
  S. haemolyticus
  S. hominis
  S. saccharolyticus
  S. saprophyticus
  S. simulans
  S. warneri
  S. xylosus
  Micrococcus luteus
  M. lylae
  M. nishinomiyaensis
  M. kristinae
  M. sedentarius
  M. roseus
  M. varians
Gram-positive bacilli
  Corynebacterium jeikeium
  C. urealyticum
  C. minutissimum
  Propionibacterium acnes
  P. avidum
  P. granulosum
  Brevibacterium epidermidis
Gram-negative bacilli
  Acinetobacter johnsonii
Yeasts
  Malassezia furfur
Moulds
  Trichophyton mentagrophytes var. *interdigitale*
Mite
  Demodex folliculorum

collected from various body sites. For some body sites, e.g. the skin, total microscopic counts of microbial cells may match the viable cell counts obtained by culture methods. In other samples, notably from the ecosystems populated by fastidious anaerobes, culture results may comprise between 50% and 80% of the total microscopic count. Enumeration of particular microbial genera or species relies on the use of selective media. The inability to culture all of the microbes present in samples, and the use

Table 1.2 Bacteria commonly detected in the upper respiratory tract of humans

---

Bacteria isolated from the anterior nares
    Staphylococcus epidermidis
    S. aureus
    Corynebacterium species
Bacteria isolated from the nasopharynx
    As for the anterior nares plus:
    Moraxella catarrhalis
    Haemophilus influenzae
    Neisseria meningitidis
    N. mucosa
    N. sicca
    N. subflava
Bacteria isolated from the oropharynx
    As for the nasopharynx plus:
    Streptococcus anginosus
    S. constellatus
    S. intermedius
    S. sanguis
    S. oralis
    S. mitis
    S. acidominimus
    S. morbillorum
    S. salivarius
    S. uberis
    S. gordonii
    S. mutans
    S. cricetus
    S. rattus
    S. sobrinus
    S. crista
    S. pneumoniae
    S. pyogenes ($< 10\%$ of human population)
    Haemophilus parainfluenzae
    Mycoplasma salivarius
    M. orale

---

of a limited range of reliable selective media doubtless introduces bias into analyses of the composition of the normal microflora (Drasar and Barrow, 1985). Analysis of terrestrial and aquatic ecosystems in recent years has benefited from the use of molecular methods with which community profiles have been established (Schmidt, DeLong and Pace, 1991). The molecular methods involve the amplification by polymerase chain reactions of 16S ribosomal RNA genes (16S rDNA) from microbial DNA

Table 1.3 Gram-positive bacilli and filamentous bacteria commonly detected in the oral cavity of humans

Actinomyces israelii, A. viscosus, A. naeslundii
Eubacterium alactolyticum, E. saburreum
Lactobacillus casei
Bifidobacterium dentium
Corynebacterium matruchotii
Propionibacterium species
Rothia dentocariosa

Table 1.4 Gram-negative bacteria commonly detected in the oral cavity of humans

Prevotella melaninogenica, P. intermedia, P. loescheii, P. denticola
Porphyromonas gingivalis, P. assacharolytica, P. endodontalis
Fusobacterium nucleatum, F. naviforme, F. russii, F. peridonticum, F. alocis, F. sulci
Leptotrichia buccalis
Selenomonas sputigena, S. flueggei
Capnocytophaga ochracea, C. sputagena, C. gingivalis
Campylobacter rectus, C. curvus
Veillonella parvula, V. atypica, V. dispar

Table 1.5 Bacterial genera commonly detected in human faeces

Acidaminococcus
Bacteroides
Bifidobacterium
Clostridium
Coprococcus
Enterobacter
Enterococcus
Escherichia
Eubacterium
Fusobacterium
Klebsiella
Lactobacillus
Megamonas
Megasphaera
Methanobrevibacter
Methanosphaera
Peptostreptococcus
Proteus
Ruminococcus
Veillonella

Table 1.6 Microbial groups commonly detected in vaginal washings obtained from humans

Anaerobic Gram-positive cocci
Bacteroides
Candida
Corynebacterium
Eubacterium
Gardnerella
Lactobacillus
Mycoplasma
Propionibacterium
Staphylococcus
Streptococcus
Ureaplasma

Table 1.7 Approximate numbers of bacteria inhabiting various regions of the human body

Region	Size of total population
Skin	10^2 to 10^6 per square centimetre (numbers vary according to habitat)
Saliva	10^8 per millilitre
Dental plaque	10^{11} per gram
Buccal epithelium	25 bacteria per epithelial cell
Ileal contents	10^8 per millilitre
Colon (faeces)	10^{10} per gram
Vaginal washings	10^7 per millilitre

extracted from samples collected from particular habitats. The amplified 16S rDNA sequences, hopefully containing copies of the gene from all of the species represented in the sample, are cloned. The 16S rDNA clones are screened (some sequences will have been cloned more than once) and representative clones are sequenced. Since 16S rDNA sequences are the cornerstone of microbial taxonomy, alignment of the sequences with those stored in databanks permits the recognition of which species are represented in the habitat, including those that cannot be cultivated by conventional techniques. Species may subsequently be enumerated directly in samples by means of oligonucleotide probes based on the 16S rDNA sequences (*in situ* hybridization). These probe molecules are labelled with a fluorescent dye and the procedure is termed 'fluorescent *in situ* hybridization' (FISH; Welling *et al.*, 1997). Temperature-gradient gel electrophoresis (TGGE) is being developed as an additional molecular means of analysis of the intestinal microflora. In this technique, 16S rDNA is amplified by polymerase chain reactions from the DNA of microbial cells in a sample. The various molecular forms (from different microbial species) of 16S rDNA in the sample can be separated from one another using TGGE. A temperature gradient is established in a polyacrylamide gel (6% acrylamide, 0.1% bis-acrylamide, 8 mmol/l urea, 20% formamide, 2% glycerol) in parallel to the electric field. The DNA samples migrate through a gradient from low to high temperature. At the temperature of partial denaturation of the double-stranded DNA molecule, the nucleic acid is drastically retarded and molecules of the same size but of different thermal stability can be separated (Riesner *et al.*, 1992). 16S rDNAs from different species have different nucleotide base sequences in the variable regions and hence have differing thermal stability. Sequences differing in only one base substitution can be separated by this method. A pattern of DNA molecules in the gel can be observed, after application of an appropriate detection system (for example silver staining), which is characteristic of the microbial content of the sample. 16S rDNA bands can be eluted from the gel for further amplification by PCR and then sequenced in order to provide identification or characterization of the microbe from which it was amplified.

The application of these molecular analytical methods will doubtless add another dimension to studies of the normal microflora. They do not, however, solve the problems of bias and lack of sensitivity. Amplification of 16S rDNA requires that the microbial cells in the sample first be lysed for the extraction of DNA. There is a vast difference in the susceptibility of the cells of different microbial species to lytic procedures. As only one lytic method is applied to the sample, it is unlikely that template DNA for the polymerase chain reaction is extracted with equal success from all the species represented in the habitat. Additionally, polymerase chain reactions are known to amplify rDNA molecules from mixed populations with differing efficiency (Reysenbach *et al.*, 1992). These factors will bias the community profile obtained. Oligonucleotide probes must reach their target sequence which is inside the bacterial cell by passing through the cell wall. This is more easily achieved with some species compared to others. Gram-positive bacterial cells such as those of lactobacilli, for example, are more difficult to permeabolize than are others (Welling *et al.*, 1997). Currently, the lowest level of detection (microscopy) using FISH is 10^6 bacterial cells per gram. Additional concerns relate to the logistics of the use of oligonucleotide probes in analysis of complex microbial communities such as those encountered in body habitats. There are at least 30–40 numerically predominant species of bacteria in human faeces, for example. The application of such a large number of oligonucleotide probes to each sample does not seem feasible at present.

Interestingly, Wilson and Blitchington (1996) compared culture and 16S rDNA sequence analysis as methods of analysis of human faecal samples. They found that, overall, there was good agreement between the results obtained by the two methods as to the biodiversity of the samples. It is important to note that polymerase chain reaction and oligonucleotide probe methods cannot differentiate between strains belonging to the same microbial species. Probably the most practical use of the molecular methods will be in the detection and enumeration of particular groups of microbes, or microbes of specific medical importance, rather than attempts at comprehensive analysis of an ecosystem. A shining example of a medical application of this type of molecular technology was the detection by polymerase chain reaction, targeted at the 16S rRNA gene, of the non-cultivable microbe *Tropheryma whippelii* in small bowel lesions characteristic of Whipple's disease (Relman *et al.*, 1992).

The acquisition of the normal microflora begins at, or soon after, birth. Colonization of neonatal surfaces and internal cavities that have openings to the exterior of the body occurs within 24 hours following birth. Proliferation of microbial types in these sites appears to be initially unchecked, resulting in a heterogeneous collection of microbes. Regulatory mechanisms generated within habitats (autogenic factors) and by external forces (allogenic factors) permit the continuing presence of some microbial types in body ecosystems but the elimination of others. These qualitative and quantitative changes in the microbial communities provide examples of biological successions (Tannock, 1994). Eventually, the composition of the microflora, at least at the level of genera, becomes more stable and the adult normal microflora is attained (the climax communities). The different biochemical and physiological conditions prevailing in various regions of the body provide environments in which certain species of microbes can flourish while other microbial types, lacking appropriate properties, cannot. Containing unique molecular, structural and microbial characteristics, each of the body sites constitutes, by definition, an ecosystem. Since each ecosystem harbours a characteristic microflora, the normal microflora can be divided into the skin (cutaneous) microflora, the upper respiratory tract microflora, oral cavity microflora, gastrointestinal microflora and genital tract microflora.

The composition of the normal microflora of adults has generally been considered to be stable as long as the host is not subjected to stressful circumstances or to the administration of antimicrobial drugs. The impression that the composition of the normal microflora was stable resulted from observations involving experimental and farm animals. Certainly, in these animals, the same genera and species of bacteria can be cultivated at constant population levels from animal to animal of the same species. These animals are often genetically homogeneous and are maintained under constant conditions. Diet and environmental factors were similar for all of the animals in these studies. The situation is quite different in the case of humans: each individual is genetically distinct and major differences in lifestyle occur. Even two decades ago, person to person variation (as much as 100 000-fold) in the numbers of particular bacterial species inhabiting the large bowel of humans was reported (Holdeman, Good and Moore, 1976). An awareness that the composition of the microflora in terms of bacterial strains was also apparent at that time, because serotyping studies of *Escherichia coli* isolates from faecal samples showed that the collection of enterobacterial serotypes changed over a period of time (Mason and Richardson, 1981). The recent use of genetic fingerprinting methods (Table 1.8) to analyse the composition of particular bacterial populations that are part of the faecal microflora of humans has demonstrated the diversity of strains that colonize the intestinal tract of different humans.

Not only do differences occur between individuals but, in some cases, even between samples collected from the same individual. As reported by McCartney, Wang and Tannock (1996), examination of bifidobacterial and lactobacillus populations in monthly faecal samples collected over a 12-month period showed that there was a marked variation in the complexity and stability of these bacterial populations among human subjects. In this study, one subject harboured a relatively simple (five strains detected during the 12-month period) and stable collection of bifidobacteria but the other subject harboured 32 strains, some of which appeared, disappeared and sometimes reappeared during the course of the study. The collection of strains detected in each subject was unique to the individual in that a strain common to both subjects was not detected. Similarly, each subject harboured a characteristic lactobacillus strain that predominated throughout the 12-month period (McCartney, Wang and Tannock, 1996). This study was extended by analysis, using genetic fingerprinting, of the bifidobacterial and lactobacillus populations of a further ten healthy humans. Two faecal samples were obtained from each subject. About half the subjects harboured a relatively simple bifidobacterial population and the others harboured a more complex collection of these bacteria. Most subjects harboured a simple lactobacillus population, often composed of a single numerically predominant strain. Unique collections of bifidobacteria and lactobacilli that persisted throughout the study were detected in each subject (Kimura *et al.*, 1997). Similarly, in a quite limited study, human subjects have been observed to harbour between six and 13 strains of *Streptococcus mitis* in the oral cavity. The individuals tended to be colonized by unique collections of strains, and different strains predominated as inhabitants of the surface of the buccal compared to the oropharyngeal mucosa (Hohwy and Killian, 1995). It seems that not only does each body region harbour a characteristic microflora defined in terms of genera and species but, at least in the intestinal tract, individuals harbour unique collections of bacterial strains.

Awareness of the variation in composition of the normal microflora at the level of bacterial strains, its uniqueness for individuals, the

Table 1.8 Molecular typing methods commonly used for the differentiation of bacterial strains

Method	Description
Multilocus enzyme electrophoresis (MLEE)	Characterization of isolates by the relative electrophoretic mobilities of a large number of water-soluble enzymes. Net electrostatic charge and hence the rate of migration of a protein during electrophoresis are determined by its amino acid sequence. Variations in the sequences of the genetic determinants of the enzymes are reflected in the mobilities of the proteins.
Pulsed field gel electrophoresis (PFGE)	Characterization of isolates based on restriction fragment length polymorphisms (RFLP) of chromosomal DNA. The bacteria are embedded in agarose, lysed *in situ*, and the chromosomal DNA is digested with a restriction endonuclease that cuts the DNA infrequently. Slices of agarose are added to an agarose gel and the restriction fragments are resolved into a pattern of discrete bands in the gel by an electrophoretic apparatus that switches the direction of the current according to a predetermined programme.
Ribotyping	DNA extracted from bacterial isolates is digested with an appropriate restriction endonuclease, the resulting fragments are separated in an agarose electrophoretic gel, transferred to a hybridization membrane, and probed with a radiolabelled ribososmal RNA gene sequence. Because bacteria have multiple copies of rRNA operons in their chromosome, several fragments in the restriction digest hybridize the probe. The patterns thus produced provide a means of differentiating between bacterial strains.
Random amplification of polymorphic DNA (RAPD)	A primer of about ten nucleotides in length is arbitrarily selected and allowed to anneal to bacterial DNA under conditions of low stringency. The short primer molecules hybridize at random sites to initiate DNA polymerization in the polymerase chain reaction (PCR). The proximity, number and location of these priming sites varies between strains and the electrophoretic pattern of DNA fragments amplified by PCR provides a fingerprint characteristic of each bacterial strain.

distribution of the microflora, and the biological successions that occur in neonates is important in explaining the occurrence of diseases due to components of the microbial collection. Some strains may have attributes that permit them to spread and to establish in normally sterile areas of the body when certain predisposing conditions prevail (for example post-surgical sepsis due to *Bacteroides fragilis*; urinary tract infections due to *E. coli*). The size of certain populations, when permitted by autogenic or allochthonous forces to increase to abnormal levels, may produce disease (e.g. translocation of *E. coli* to the blood circulation or antibiotic-asso-ciated colitis due to *Clostridium difficile*). Acute purulent meningitis in human infants less than 1 month of age is commonly due to *E. coli*. The involvement of this bacterial species in these infections is understandable when it is considered that infants can be exposed to *E. coli* during passage through the vagina, and that they harbour large populations of *E. coli* in their intestinal tract during the first weeks of life. The tissues of the infant are thus exposed to large numbers of potentially pathogenic cells. Simi-larly, intestinal infections due to enterotoxigenic *E. coli* strains are com-mon in newborn farm animals and are a major cause of morbidity and mortality in human infants in developing countries. Mechanisms that maintain enterobacterial populations at low levels are not yet operating in the neonatal intestinal tract. Large populations of *E. coli* are present in the faeces, ensuring ease of transmission of infection between individuals if a pathogenic strain of *E. coli* is present. *Candida albicans* is a member of the normal microflora of the human oral cavity that causes pseudomem-branous candidiasis (oral thrush) in infants. In this case, the bacterial components of the microflora that suppress the replication of the yeast in adults are lacking. Coupled with the immunological immaturity of the infant, the yeast is thus able to proliferate and invade the oral mucosa.

Even in the absence of disease, in the young or in the adult, the presence of large numbers of viable, and hence presumably metabolizing, microbial cells must exert major influences on the animal host. The influences of the normal microflora on the properties of the healthy host must therefore also be considered. The microflora as a whole has marked influences on the animal host as has been observed in experiments in which the characteristics of germfree (absence of a microflora) and con-ventional (presence of a microflora) animals were made. These compar-isons showed that many biochemical, physiological and immunological characteristics of the animal host are strongly influenced by the presence of the normal microflora (Table 1.9; Gordon and Pesti, 1971; Luckey, 1963). It can be noted in passing, that biochemical assays of microflora-associated activities provide a suitable method of analysing the overall functioning of the intestinal microflora (Table 1.10).

Reference to Table 1.9 reveals that most of the microflora-associated influences relate to the intestinal tract. This is presumably because the gastrointestinal tract harbours the largest number, and most diverse collection, of bacteria inhabiting the animal body. Organs remote from the intestinal tract are nevertheless affected by the normal microflora.

Heart, lung and liver weights are lower and there is lower regional blood flow to these organs in germfree, compared to conventional, animals. Cardiac output and oxygen use are lower in germfrees. These differences in organ size and circulatory phenomena of germfree animals probably reflect the minimal stimulation of the reticuloendothelial system in the lungs and liver in the absence of normal microflora antigens. Additionally, there is a lower metabolic load in the case of the liver, because microbial products that require detoxification by this organ are absent from the germfree body (Gordon and Pesti, 1971).

Muramyl peptides originating from the cell walls of bacteria are among the chemicals that modify sleep patterns (sleep substances or sleep factors). Commonly referred to as Factor S, muramyl peptides accumulate in the brain, cerebrospinal fluid and urine of sleep-deprived animals, including humans. The administration of these peptides to experimental animals results in an increase in the amount of slow-wave sleep. Conversely, perturbation of the intestinal microflora by antibiotics has been reported to reduce the amount of slow-wave sleep of rats. Muramyl peptides originating in bacterial populations inhabiting the body thus apparently influence brain biochemistry (Brown *et al.*, 1990). This physiological effect and also the appetite-suppressing effect of peptidoglycan fragments are probably due to the release of the host's endogenous cytokines in response to the absorption of bacterial substances into the blood circulation (Biberstine and Rosenthal, 1994).

Wound healing has been reported to differ in comparisons of germfree and conventional guinea pigs. In germfree animals, regeneration of the epidermis preceded that of the dermis. The epidermis developed over a loose mesenchymal base rather than over granulation tissue as was the case in conventional animals. How the normal microflora influences tissue repair, and why the phenomenon observed in guinea pigs is not apparent in germfree mice, is unknown (Gordon and Pesti, 1971).

The impact of the normal microflora on the nutrition of the animal host is notable and is exemplified best by the microflora of the proximal digestive tract of ruminants. These animal species (for example sheep, cattle) rely on the normal microflora of the rumen–reticulum to digest plant-derived substances in their diet and for the provision of essential amino acids and vitamins. Plant structural materials (cellulose, hemicelluloses) form the major components of the diet of ruminants but the animals cannot synthesize digestive enzymes capable of catalysing hydrolysis of these plant materials. The rumen–reticulum accommodates a large collection of microbes capable of degrading the plant material. The short-chain fatty acids (acetic, propionic, butyric) resulting from the overall rumen fermentation carried out by these microbes are absorbed from the rumen–reticulum into the blood circulation. They provide the main source of energy for the ruminant. Microbes carried out of the rumen–

Table 1.9 Comparison of selected properties of germfree and conventional animals

Host characteristics	Conventional	Germfree
Bile acid metabolism	Deconjugation Dehydrogenation Dehydroxylation	Absence of deconjugation, dehydrogenation and dehydroxylation
Bilirubin metabolism	Deconjugation and reduction	Little deconjugation; absence of reduction
Cholesterol	Reduction to coprostanol	Absence of coprostanol
β-aspartylglycine	Absent	Present
Intestinal gases	Hydrogen, methane, carbon dioxide	Absence of hydrogen and methane; less carbon dioxide
Short chain fatty acids	Large amounts, several acids	Small amounts of a few acids
Tryptic activity	Little activity	High activity
Urease	Present	Absent
β-glucuronidase (pH 6.5)	Present	Absent
Organ weights: heart, lung, liver	Higher	Lower
Cardiac output and oxygen utilization	Higher	Lower
Mucin content of intestinal mucus	Higher	Lower
Extent of degradation of mucins	More	Less
Caecal size (rodents)	Smaller	Larger
Enzyme activities associated with duodenal enterocytes	Lower	Higher
Intestinal wall	Thicker	Thinner
Intestinal mucosal surface area	Greater	Smaller
Rate of enterocyte replacement	Faster	Slower
Peristaltic movement of contents through small bowel	Faster	Slower

Table 1.9 Continued

Host characteristics	Conventional	Germfree
Body temperature	Higher	Lower
Serum cholesterol concentration	Lower	Higher
Lymph nodes	Larger	Smaller
γ-globulin fraction in blood	More	Less

Table 1.10 Examples of biochemical assays that have been used in the analysis of the normal microflora of the intestinal tract

Assay	Reference
Azoreductase activity	McConnell and Tannock, 1991
β-glucosidase activity	Tannock *et al.*, 1988
β-glucuronidase activity	McConnell and Tannock, 1993
Short-chain fatty acids	Tannock *et al.*, 1988
Phenolic products (skatole, indole)	Jensen, Cox and Jensen, 1995
Bile salt hydrolase activity	Tannock, Dashkevicz and Feighner, 1989
Ratio of conjugated/unconjugated bile salts	Tannock *et al.*, 1994
pH	Tannock *et al.*, 1988
Production of methane, carbon dioxide and hydrogen	Jensen and Jorgensen, 1994
Urease activity	Brockett and Tannock, 1982
Mucin degradation	Gustafsson and Carlstedt-Duke, 1984
Proteolytic activity	Tannock *et al.*, 1988
Urobilinogen	Tannock *et al.*, 1988
Coprostanol	Carlstedt-Duke *et al.*, 1987
β-aspartyl glycine	Welling, 1982

reticulum in the digesta are digested by mammalian processes and provide the ruminant with most of its requirements for amino acids, vitamins and lipids. Ruminant fat is rich in odd-numbered saturated fatty acids that are microbial in origin (Hungate, 1966).

The contribution that members of the normal microflora make to satisfying the nutritional requirements of monogastric animals is less clear than in the case of ruminants. Monogastric animals and birds with well-developed caeca harbour a microflora capable of degrading complex plant materials. Microbial activity in the large bowel of rats, for example, degrades 39% of dietary cellulose. As in the rumen–reticulum, short-chain fatty acids resulting from the fermentation of substrates in the large bowel are absorbed into the blood circulation. Cellulose fermentation in the large bowel is thought to contribute 10–12% of the daily energy requirements for the rabbit, up to 5% for rats and no more than 2% for pigs (McBee, 1977).

The microflora of the large bowel of humans, like that inhabiting the distal regions of the digestive tract of other animals, encounters diet-derived substrates that the host has been unable to digest. It has been calculated that about 20 g of plant structural material is ingested per day by individuals consuming a 'Western diet'. Some 5–10 g of dietary fibre can be recovered from faeces. Since humans do not have enzymes for the digestion of plant structural substances, microbial activity must have digested the remainder. About 50% of ingested cellulose and 70–90% of hemicelluloses are fermented in the large bowel of humans. About 10–15% of starch in foods such as oats, white bread and potatoes escapes digestion in the small bowel and is fermented by microbes in the large bowel. If 30 g of carbohydrate were fermented daily in the large bowel, 200 mmol of fermentation products (mostly short-chain fatty acids) would be produced. As 5–20 mmol (mostly acetic, propionic and butyric acids) is actually detected in faeces, this means that most of the microbial fermentation products are absorbed by the host or are degraded by colonic bacteria. Short-chain fatty acids produced in the colon are not thought to contribute significantly to the daily energy requirements of humans. *In vitro* experiments suggest, however, that butyric acid is the predominant energy source for enterocytes comprising the large bowel epithelium (Roediger, 1980).

Perhaps one of the greatest enigmas concerning the normal microflora is the mechanism by which huge numbers of microbial cells can persist in intimate association with the mucosal surfaces (respiratory tract, alimentary canal, female genital tract) without inducing a marked inflammatory or immunological response on the part of the host. In general, it has been accepted that low titres of antibodies reactive with indigenous bacteria can be expected to be detected in the sera of healthy humans (Table 1.11). In recent work by Kimura and colleagues (1997), however, much higher titres of antibodies reactive with antigens associated with whole cells of lactobacilli and bifidobacteria isolated from the same human subjects from which the blood samples were obtained have been detected (Table 1.12).

The major differences between the earlier and more recent studies may relate to the source of the bacterial strains used in the investigations (heterologous *versus* homologous origins) and the sensitivity of the methods used to detect the antibodies (agglutination *versus* ELISA). IgM antibodies reactive with the cells of lactobacilli were detected in the sera

Table 1.11 Detection of antibodies in the sera of healthy humans reactive with members of the normal microflora

Observation	Reference
Precipitins that reacted with *E. coli, B. fragilis* and *Pseudomonas aeruginosa* were detected by immunoelectrophoresis	Hoiby and Hertz, 1979
Bactericidal antibodies reactive with fusobacteria, *Veillonella* and *E. coli* were detected	Evans, Spaeth and Mergenhagen, 1966
Agglutinating antibodies reactive with enterobacteria were detected at titres of between 1:40 and 1:1280.	Gillespie *et al.*, 1950
IgA, IgM and IgG antibodies reactive with *E. coli* were detected by immunofluorescence	Cohen and Norins, 1966
IgG and IgM antibodies that reacted with *E. coli*, enterococci, *Veillonella* and oral streptococci were detected using agglutination tests; IgA antibodies reactive with *E. coli* and oral streptococci were also detected	Sirisinha and Charupatana, 1971

Table 1.12 Concentrations of serum antibodies reactive with whole cells of lactobacilli and bifidobacteria

Bacteria	Antibody titre (ELISA*)		
	IgA	*IgM*	*IgG*
Lactobacilli	< 2–32 † (5/10) ‡	16–128 (10/10)	32–1024 (10/10)
Bifidobacteria	<2–<2 (0/10)	<2–64 (9/10)	<2–512 (9/10)

* Enzyme linked immunosorbent assay. Titre = reciprocal of the highest dilution of serum which gave an A_{490} in ELISA of > 0.10 following subtraction of the blank value.
† Range of titres obtained from ten healthy subjects.
‡ Number of subjects, out of total examined, with a detectable titre (> 2) of antibodies.

examined by Kimura and colleagues. The presence of this class of immunoglobulin suggests exposure of the host to small amounts of antigen, and that this exposure was relatively recent (Berg, 1983). The results published by Kimura and colleagues also suggested that the antibody reactivity (IgA, IgM, IgG) was genus- rather than species-specific. It would be interesting to investigate the nature of the cell surface-associated antigens with which the serum antibodies react. There may well be a degree of cross-reactivity between these 'natural antibodies' and various Gram-positive members of the normal microflora.

Other experiments support the hypothesis that at least some members of the normal microflora do not elicit as great a host immune response as do non-indigenous bacteria. For example, Berg and Savage (1972) injected conventional mice intraperitoneally with heat-killed cells of *E. coli* or *Bacteroides* species of murine origin. They compared the immune response to these strains with that engendered by inoculation with *E. coli* or *B. fragilis* obtained from human sources. Four days after inoculation, the immune response to the murine strains was less than that produced against the human strains. Similarly, germfree mice monoassociated (oral inoculation) with murine bacterial strains (*E. coli*, lactobacilli, bacteroides) did not show a detectable, systemic, immune response but responded immunologically to strains of human origin with which they were colonized. It is therefore likely that at least some members of the normal microflora resemble their host antigenically. The immunological environment of the host may provide a selective force for the evolution of the special relationship between host and microbes. Immunological selection of genetic variants less antigenically 'foreign' to the host may favour the establishment of certain microbes as members of the normal microflora. Some credence is given to this hypothesis by the observation that *Vibrio cholerae* cells of differing antigenicity are selected while the bacteria are inhabiting the intestinal tract of gnotobiotic mice (Miller *et al.*, 1972; Sack and Miller, 1969).

The unexplained host tolerance to the presence of the normal microflora provides a fascinating area of speculation and potential research. In a conceptual model, Gaskins (1997) has proposed that class Ib major histocompatibility complex (MHC) molecules may present peptides derived from normal microflora antigens to intraepithelial lymphocytes (IEL). These class Ib restricted IEL might respond by secretion of immunosuppressive cytokines that would prevent, or reduce the level of, the activation of immunological cells in the lamina propria. The host, however, continues to be responsive to the presence of pathogenic bacteria. Hence, the correct proportions of pro-inflammatory (IL-2, interferon-γ) and anti-inflammatory (IL-10, TGF-β) cytokines must somehow be maintained in subepithelial tissues. The mechanisms whereby this balance between tolerance and immune responsiveness is maintained need to be researched actively because disease results when the immunological balance between host and normal microflora is upset (Chapter 8).

Paradoxically, there is considerable evidence that the normal microflora has a stimulating effect on the development of the immune system of the host. This phenomenon is evidenced by the underdeveloped, less common and smaller lymph nodes associated with the intestinal tract of germfree animals compared to those of conventional animals. The reduction in the amount of lymphoid tissue in germfree animals extends to the thymus (bursa of Fabricius in chickens), which is reduced in size in gnotobiotes. Plasma cells are less numerous in the lamina propria of the intestinal tract and in the mesenteric lymph nodes of germfree animals. The gamma globulin fraction of serum is less in germfree chickens, rats, and mice compared to conventional animals. The stimulating effect of the microflora on the immunological tissues of the host is thought to enhance aspects of resistance that are important in the early stages of infection by pathogens (Gordon and Pesti, 1971).

Also of relevance to resistance to infectious diseases is the role of the ecological factors (competition, antagonism) that regulate the composition of the microflora and help to suppress the establishment of pathogenic bacteria. This phenomenon is commonly termed 'microbial interference' or 'colonization resistance'. Most attention in the study of the microbial interactions that are involved in microbial interference has centred on the intestinal tract and non-specific resistance to *Salmonella* species. In the 1960s, Bohnhoff and colleagues (Bohnhoff and Miller, 1962; Bohnhoff, Miller and Martin, 1964a, b) reported that the multiplication of *Salmonella enteritidis* was inhibited *in vitro* by buffered suspensions of contents from the large intestine of mice. Acetic and butyric acids were present in the intestinal contents at concentrations that inhibited *Salmonella* growth *in vitro*. The inhibitory effect of the intestinal contents was influenced by pH and was better under anaerobic conditions. Intestinal contents from mice that had been treated *per os* with streptomycin (an antibiotic that is not absorbed from the intestinal tract) 24 hours previously had a higher pH and a lower concentration of short-chain fatty acids than those of untreated mice. *Salmonella* multiplication was not prevented by suspensions of intestinal contents from streptomycin-treated mice. Moreover, mice administered streptomycin were more susceptible to infection by *Salmonella*, when challenged by the oral route, compared to untreated animals. Meynell (1963) also observed that the large bowel contents of mice contained short-chain fatty acids. The E_h (oxidation–reduction potential) of the caecal contents of mice was found to be about – 200 mV. Under these conditions (short-chain fatty acids and anaerobiosis) *in vitro*, the growth of *Salmonella typhimurium* was prevented. Treatment of mice with streptomycin lowered the short-chain fatty acid concentrations and produced a less anaerobic environment (E_h of +200 mV). These conditions did not inhibit the growth of *Salmonella*. Previous researchers, notably Bergeim and colleagues (1941), had shown that short-chain fatty acids such as acetic and butyric had an inhibitory effect on the growth of enterobacteria in the pH range 5–6. At these pHs, the short-chain fatty acid molecules were mostly in the undissociated form. Inhibition of *Salmonella* in the large bowel of mice was due, therefore, to undissociated short-chain fatty acids (especially butyric acid) produced as fermentation products by obligately anaerobic members of the normal microflora.

The contribution of the immune system to resistance to *Salmonella* infection in mice must not be overlooked. Tannock and Savage (1976) showed that germfree mice challenged by the oral route with *Salmonella typhimurium* had caeca that were abnormal in appearance and reduced in size (average 1.7% of body weight) compared to those of germfree controls (average 5% of body weight). Germfree mice injected with heat-killed *S. typhimurium* or gnotobiotes associated with three indigenous bacterial species (lactobacilli, bacteroides, clostridia), and subsequently challenged with *S. typhimurium* also had small caeca. By contrast, gnotobiotic mice that had been both injected with the heat-killed *S. typhimurium* and associated with the simple microflora before challenge with the pathogen had caeca similar in size and appearance to those of germfree mice. The results of these experiments suggested that synergism had occurred between the interference effect of the intestinal microflora and the immune system of the animal host.

Short-chain fatty acids cannot explain the regulation of all of the component populations of the normal microflora. One mechanism, however, does appear capable of explaining the observed balance among the numerous species of bacteria that inhabit any given body ecosystem: competition for nutrients that are in limited supply. As long as there is a growth-limiting nutrient for each of the microbial populations in the ecosystem, all the populations can coexist (Freter, 1988). If the competition for nutrients is considered to be the major regulatory factor, modulation of the effect of nutrient availability is probably provided by the production of inhibitory substances (hydrogen sulphide, short-chain fatty acids). The production of substances by one type of microbe that are inhibitory to other microbial types (antagonism) is indeed often cited as an important mechanism by which microbial communities are regulated (Tannock, 1981). Peptides, proteins or protein–carbohydrate complexes with relatively narrow spectra of activity are produced by many bacterial species and are generally referred to as 'bacteriocins' (Table 1.13).

Table 1.13 Examples of bacteriocins

Producer	Bacteriocin
Lactococcus lactis	Nisin
Pediococci	Pediocins
Lactobacillus johnsonii	Lactacin F
Escherichia coli	Colicins
Pseudomonas aeruginosa	Pyocins
Micrococci	Micrococcins
Staphylococci	Staphylococcins

These inhibitory substances are assumed to suppress the populations of other strains or species that are potential or actual competitors of the producing strain in a particular habitat. Experimental evidence that this assumption holds true, however, is lacking (Tannock, 1981). The derivation of isogenic strains of bacteria (bacteriocin-producing, bacteriocin-non-producing, bacteriocin-susceptible, bacteriocin-resistant) and the availability of suitable animal model systems with which to test the competitive capabilities of combinations of these isogenic strains would permit direct experimentation concerning the ecological significance of bacteriocin production by members of the normal microflora.

There is probably an element of redundancy in the regulatory systems in which specific factors become more important under certain prevailing conditions. The complexity of the regulatory system is suggested by the results of experiments conducted by Freter, Abrams and Aranki (1973). They found that a collection of 55 strains ('N' strains) reduced *E. coli* numbers in the large bowel of gnotobiotic mice by about 1000-fold. Altering the animals' diet from refined (L-356) to crude (L-485), however, decreased the interference effect. A collection of 100 strains was required to reduce the *E. coli* population size to that encountered in conventional mice when diet L-485 was utilized.

Competition for space is likely to be an important regulatory factor for microbes that associate with epithelial surfaces. This is exemplified by an interaction between lactobacilli and *Candida pintolopesii* observed by Savage (1969). These two microbes inhabit the stomach of animals belonging to certain mouse colonies. The lactobacilli colonize the surface of the non-secretory epithelium of the forestomach; the yeast colonizes the secretory epithelium of the corpus. In both cases, a layer of microbial cells on the epithelium can be observed in stained sections prepared from the stomach. There is a sharp delineation of the two microbial populations according to the distribution of the specific epithelial tissues. Savage observed that, when penicillin was administered to the mice in their drinking water, the lactobacilli were eliminated from the stomach. The yeast, unaffected by the penicillin, was no longer confined to the stomach corpus, but now also colonized the non-secretory epithelium of the forestomach. When penicillin administration was stopped, lactobacilli recolonized the forestomach and the yeast was displaced, from then on being again restricted to the surface of the secretory epithelium. The interference mechanism exerted by the lactobacilli is unknown, but might be due to the production of lactic acid, which is toxic to some yeasts (Young, Krasner and Yudkosky, 1956).

The presence of the normal microflora is not always of benefit to the animal host: germfree mice have been reported to live longer than their conventional counterparts (for example 723 days for germfree male Swiss mice, 480 days for conventional mice; Gordon and Pesti, 1971). While the non-specific resistance to infectious disease mediated by the normal microflora is widely promulgated in undergraduate textbooks and by commercial enterprises (Chapter 20), a more balanced attitude to the microbial populations that inhabit the human body should be adopted. 'Balance' is the key word. Throughout life, the body harbours an extensive collection of microbial cells. Mostly, the human host is unaware of the presence of the normal microflora because the microbes and the host have attained, during evolutionary time, a balanced relationship. When this balance is upset, some of the microbial components of the microflora become the aetiological agents of disease. These detrimental effects of the normal microflora and the factors that produce imbalance between the normal microflora and the host have been ably reviewed by the contributors of the chapters that follow in this book. Each author is an internationally recognized researcher on the topic with which s/he deals. This book thus contains a comprehensive overview of the importance of the normal microflora to medical science and to the practice of medicine.

1.1 REFERENCES

Berg, R. D. (1983) Host immune response to antigens of the indigenous intestinal flora, in *Human Intestinal Microflora in Health and Disease*, (ed. D. J. Hentges), Academic Press, New York, pp. 101–126.

Berg, R. D. and Savage, D. C. (1972) Immunological responses to microorganisms indigenous to the gastrointestinal tract. *American Journal of Clinical Nutrition*, **25**, 1364–1371.

Bergeim, O., Hanszen, A. H., Pincussen, L. and Weiss, E. (1941) Relation of volatile fatty acids and hydrogen sulphide to the intestinal flora. *Journal of Infectious Diseases*, **69**, 155–166.

Biberstine, K. J. and Rosenthal, R. S. (1994) Peptidoglycan fragments decrease food intake and body weight gain in rats. *Infection and Immunity*, **62**, 3276–3281.

Bohnhoff, M. and Miller, C. P. (1962) Enhanced susceptibility to *Salmonella* infection in streptomycin-treated mice. *Journal of Infectious Diseases*, **111**, 117–127.

Bohnhoff, M., Miller, C. P. and Martin, W. R. (1964a) Resistance of the mouse's intestinal tract to experimental *Salmonella* infection. I. Factors which interfere with the initiation of infection by oral inoculation. *Journal of Experimental Medicine*, **120**, 805–816

Bohnhoff, M., Miller, C. P. and Martin, W. R. (1964b) Resistance of the mouse's intestinal tract to experimental *Salmonella* infection. II. Factors responsible for its loss following streptomycin treatment. *Journal of Experimental Medicine*, **120**, 817–828.

Brockett, M. and Tannock, G. W. (1982) Dietary influence on microbial activities in the caecum of mice. *Canadian Journal of Microbiology*, **28**, 493–499.

Brown, R., Price, R. J., King, M. G. and Husband, A. J. (1990) Are antibiotic effects on sleep behavior in the rat due to modulation of gut bacteria? *Physiology and Behavior*, **48**, 561–565.

Carlstedt-Duke, B., Alm, L., Hoverstad, T. *et al.* (1987) Influence of clindamycin, administered together with or without lactobacilli, upon intestinal ecology in rats. *FEMS Microbiology Ecology*, **45**, 251–259.

Cohen, I. R. and Norins, L. C. (1966) Natural antibodies to gram-negative bacteria: immunoglobulins G, A, and M. *Science*, **152**, 1257–1259.

Drasar, B. S. and Barrow, P. A. (1985) *Intestinal Microbiology*, American Society for Microbiology, Washington, DC.

Evans, R. T., Spaeth, S. and Mergenhagen, S. E. (1966) Bactericidal antibody in mammalian serum to obligatorily anaerobic gram-negative bacteria. *Journal of Immunology*, **97**, 112–119.

Freter, R. (1988) Mechanisms of bacterial colonization of the mucosal surfaces of the gut, in *Virulence Mechanisms of Bacterial Pathogens*, (ed. J. A. Roth), American Society for Microbiology, Washington, DC, pp. 45–60.

Freter, R., Abrams, G. D. and Aranki, A. (1973) Patterns of interactions in gnotobiotic mice among bacteria of a synthetic 'normal' intestinal flora, in *Germfree Research*, (ed. J. B. Henegham), Academic Press, New York, pp. 429–433.

Gaskins, H. R. (1997) Immunological aspects of host/microbiota interactions at the intestinal epithelium, in *Gastrointestinal Microbes and Host Interactions*, (eds R. I. Mackie, B. A. White and R. E. Isaacson), Chapman & Hall, New York, pp. 537–587.

Gillespie, H. B., Steber, M. S., Scott, E. N. and Christ, Y. S. (1950) Serological relationships existing between bacterial parasites and their host. I. Antibodies in human blood serum for native intestinal bacteria. *Journal of Immunology*, **65**, 105–113.

Gordon, H. A. and Pesti, L. (1971) The gnotobiotic animal as a tool in the study of host–microbial relationships. *Bacteriological Reviews*, **35**, 390–429.

Gustafsson, B. E. and Carlstedt-Duke, B. (1984) Intestinal water-soluble mucins in germfree, exgermfree and conventional animals. *Acta Pathologica Microbiologica et Immunologica Scandinavica (Section B)*, **92**, 247–252.

Hohwy, J. and Kilian, M. (1995) Clonal diversity of the *Streptococcus mitis* biovar 1 population in the human oral cavity and pharynx. *Oral Microbiology and Immunology*, **10**, 19–25.

Hoiby, N. and Hertz, J. B. (1979) Precipitating antibodies against *Escherichia coli*, *Bacteroides fragilis* ss. *thetaiotaomicron* and *Pseudomonas aeruginosa* in serum from normal persons and cystic fibrosis patients, determined by means of crossed immunoelectrophoresis. *Acta Paediatrica Scandinavica*, **68**, 495–500.

Holdeman, L. V., Good, I. J. and Moore, W. E. C. (1976) Human fecal flora: variation in bacterial composition within individuals and a possible effect of emotional stress. *Applied and Environmental Microbiology*, **31**, 359–375.

Hungate, R. E. (1966) *The Rumen and its Microbes*, Academic Press, New York.

Jensen, B. B. and Jorgensen, H. (1994) Effect of dietary fiber on microbial activity and microbial gas production in various regions of the gastrointestinal tract. *Applied and Environmental Microbiology*, **60**, 1897–1904.

Jensen, M. T., Cox, R. F. and Jensen, B. B. (1995) 3-Methylindole (skatole) and indole production by mixed populations of pig fecal bacteria. *Applied and Environmental Microbiology*, **61**, 3180–3184.

Kimura, K., McCartney, A. L., McConnell, M. A. and Tannock, G. W. (1997) Analysis of fecal populations of bifidobacteria and lactobacilli and investigation of the immunological responses of their human hosts to the predominant strains. *Applied and Environmental Microbiology*, **63**, 3394–3398.

Luckey, T. D. (1963) *Germfree Life and Gnotobiology*, Academic Press, New York.

Luckey, T. D. (1972) Introduction to intestinal microecology. *American Journal of Clinical Nutrition*, **25**, 1292–1294.

McBee, R. H. (1977). Fermentation in the hind gut, in *Microbial Ecology of the Gut*, (eds R. T. J. Clarke and T. Bauchop), Academic Press, New York, pp. 184–222.

McCartney, A. L., Wang, W. and Tannock, G. W. (1996). Molecular analysis of the composition of the bifidobacterial and lactobacillus microflora of humans. *Applied and Environmental Microbiology*, **62**, 4608–4613.

McConnell, M. A. and Tannock, G. W. (1991) Lactobacilli and azoreductase activity in the murine cecum. *Applied and Environmental Microbiology*, **57**, 3664–3665.

McConnell, M. A. and Tannock, G. W. (1993) A note on lactobacilli and β-glucuronidase activity in the intestinal contents of mice. *Journal of Applied Bacteriology*, **74**, 649–651.

Mason, T. G. and Richardson, G. (1981) *Escherichia coli* and the human gut: some ecological considerations. *Journal of Applied Bacteriology*, **51**, 1–16.

Meynell, G. G. (1963) Antibacterial mechanisms of the mouse gut. II. The role of E_h and volatile fatty acids in the normal gut. *British Journal of Experimental Pathology*, **44**, 209–219.

Miller, C. E., Wong, K. H., Feely, J. C. and Forlines, M. E. (1972) Immunological conversion of *Vibrio cholerae* in gnotobiotic mice. *Infection and Immunity*, **6**, 739–742.

Relman, D. A., Schmidt, T. M., MacDermott, R. P. and Falkow, S. (1992) Identification of the uncultured bacillus of Whipple's Disease. *New England Journal of Medicine*, **327**, 293–301.

Reysenbach, A.-L., Giver, L. J., Wickham, G. S. and Pace, N. R. (1992) Differential amplification of rRNA genes by polymerase chain reaction. *Applied and Environmental Microbiology*, **58**, 3417–3418.

Riesner, D., Steger, G., Wiese, U. *et al.* (1992) Temperature-gradient gel electrophoresis for the detection of polymorphic DNA and for quantitative polymerase chain reaction. *Electrophoresis*, **13**, 632–636.

Roediger, W. E. W. (1980) Role of anaerobic bacteria in the metabolic welfare of the colonic mucosa in man. *Gut*, **21**, 793–798.

Sack, R. B. and Miller, C. E. (1969) Progressive changes of vibrio serotypes in germfree mice infected with *Vibrio cholerae*. *Journal of Bacteriology*, **99**, 688–695.

Savage, D. C. (1969) Microbial interference between indigenous yeast and lactobacilli in the rodent stomach. *Journal of Bacteriology*, **98**, 1278–1283.

Schmidt, T. M., DeLong, E. F. and Pace, N. R. (1991) Analysis of a marine picoplankton community by 16S rRNA gene cloning and sequencing. *Journal of Bacteriology*, **173**, 4371–4378.

Sirisinha, S. and Charupatana, C. (1971) Antibodies to indigenous bacteria in human serum, secretions, and urine. *Canadian Journal of Microbiology*, **17**, 1471–1473.

Tannock, G. W. (1981) Microbial interference in the gastrointestinal tract. *ASEAN Journal of Clinical Science*, **2**, 2–34.

Tannock, G. W. (1994) The acquisition of the normal microflora of the gastrointestinal tract, in *Human Health: the Contribution of Microorganisms*. (ed. S. A. W. Gibson), Springer-Verlag, London, pp. 1–16.

Tannock, G. W. and Savage, D. C. (1976) Indigenous microorganisms prevent reduction in cecal size induced by *Salmonella typhimurium* in vaccinated gnotobiotic mice. *Infection and Immunity*, **13**, 172–179.

Tannock, G. W., Crichton, C., Welling, G. W. *et al.* (1988) Reconstitution of the gastrointestinal microflora of lactobacillus-free mice. *Applied and Environmental Microbiology*, **54**, 2971–2975.

Tannock, G. W., Dashkevicz, M. P. and Feighner, S. D. (1989) Lactobacilli and bile salt hydrolase in the murine intestinal tract. *Applied and Environmental Microbiology*, **55**, 1848–1851.

Tannock, G. W., Tangerman, A., van Schaik, A. and McConnell, M. A. (1994) Deconjugation of bile acids by lactobacilli in the mouse small bowel. *Applied and Environmental Microbiology*, **60**, 3419–3420.

Welling, G. W. (1982) Comparison of methods for the determination of β-aspartylglycine in fecal supernatants of leukemic patients treated with antimicrobial agents. *Journal of Chromatography*, **232**, 55–62.

Welling, G. W., Elfferich, P., Raangs, G. C. *et al.* (1997) 16S ribosomal RNA-targeted oligonucleotide probes for monitoring of intestinal tract bacteria. *Scandinavian Journal of Gastroenterology*, **32**(Suppl. 222), 17–19.

Wilson, K. H. and Blitchington, R. B. (1996) Human colonic biota studied by ribosomal DNA sequence analysis. *Applied and Environmental Microbiology*, **62**, 2273–2278.

Young, G., Krasner, R. I. and Yudkosky, P. L. (1956) Interactions of oral strains of *Candida albicans* and lactobacilli. *Journal of Bacteriology*, **72**, 525–529.

2

The human skin microflora and disease

William C. Noble

2.1 SKIN AS A HABITAT

The human skin is a complex habitat. It comprises areas that are relatively dry, such as the forearm, and others that are moist, such as the perineum and toewebs. Some areas have abundant lipid from sebaceous glands, such as the face, scalp and upper chest and back, while some, such as the axillae, also have secretions from specialized apocrine glands. Areas with many eccrine sweat glands may become highly saline as watery sweat induced by heat or exercise evaporates. On balance, an average adult has a total skin surface area of about $1.5\,m^2$ with a pH of about 5.5 and a temperature between 30° and 37 °C with substantial local variations. The various skin secretions supply amino acids, metal ions, fatty acids, sugars and other nutrient materials although in dilute solution (Marples, 1965; Noble, 1981, 1993). It should therefore be no surprise that skin is inhabited by many different microorganisms, some of which may be fairly localized.

The site of microbial growth varies. Some organisms, such as the propionibacteria, have their headquarters, i.e. the site of reproduction, deep in the ducts of hair follicles, while the lipophilic yeasts are in the uppermost segment of these ducts. Both organisms are extruded, sometimes in very large numbers, on to the surface of the skin; for example, propionibacteria can reach population densities of well over $10^7\,cm^{-2}$ on the forehead. The staphylococci and the coryneforms also inhabit hair follicles but in addition form colonies on the skin surface that may contain

G. W. Tannock (ed.), *Medical Importance of the Normal Microflora*, 24–46.
© 1999 *Kluwer Academic Publishers. Printed in Great Britain.*

several thousand cells. The site of growth is important in respect of attempts to degerm the skin surface. Skin cannot be sterilized, at least, not by methods that do not damage it. Even the most efficient skin disinfectants do not remove more than 95% of the skin microflora because compounds that would penetrate the lipid and protein barrier to reach the depths of a hair follicle would be likely to damage the structure of the skin surface leaving it open to infection. Experimental infections on the skin of humans show, however, that even a 95% reduction can be useful in reducing the chance of infection (Singh, Marples and Kligman, 1971).

Although it has been suggested that skin lipids may act to keep out pathogens such as the streptococci and may have a selective role in allowing some staphylococcal species, but not others, to colonize the skin, there is no doubt that the single factor with the greatest effect is the availability of water. The effect of occluding skin, which prevents evaporative water loss, is to increase microbial populations manyfold (Fig. 2.1), although it must be recognized that under occlusion there are also changes in CO_2 concentration and in pH (Aly *et al.*, 1978); such changes also occur following the normal seasonal variations, although these are usually of lesser magnitude (Abe *et al.*, 1980).

Erosion of the habitat plays a part in controlling the skin microflora since the skin surface is renewed about every 5 days as the skin scales, or squames, are lost into the environment. Paradoxically, neither repetitive

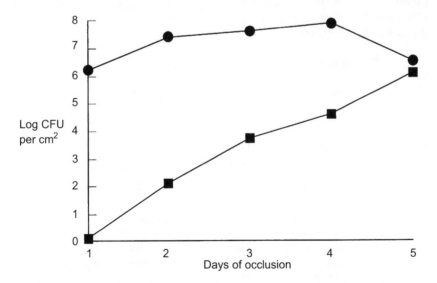

Figure 2.1 Effect of occlusion on components of the skin microflora: circles = the staphylococci. squares = lipophilic coryneforms. Source: based on Aly *et al.*, 1978.

washing nor not washing has much influence on the overall balance of the skin microflora. After washing, the microflora is replaced to its original level within a few hours, at which point substrate limitation and competition for nutrient probably control the increase.

2.2 COMPONENTS OF THE SKIN MICROFLORA

The normal skin microflora comprises four major groups of bacteria: these are the staphylococci, the coryneform bacteria, the micrococci and the Gram-negative bacilli, especially *Acinetobacter*, plus the yeast genus *Malassezia*. The taxonomy of each group will be described briefly, followed by the various infections of the skin and of other tissues that each is known to cause.

2.2.1 The genus *Staphylococcus*

There are now more than 30 species and subspecies of staphylococci known; those most common on human skin are shown in Table 2.1.

All are Gram-positive cocci and most are aerobes but relatively few are involved in disease. In humans the coagulase-positive species *S. aureus* is the chief pathogen, although *S. intermedius*, chiefly found on dogs and other animals, is also a rare commensal and pathogen of humans. The normal coagulase-negative microflora of the skin does not appear to be involved in skin disease, except perhaps in trivial localized lesions, but is involved in deep infection, especially where a catheter or other prosthesis has been inserted.

Table 2.1 Chief staphylococcal species of human skin

Species	Chief site of colonization
S. aureus	Nose, axillae, perineum, toewebs
S. auricularis	Ear
S. capitis	Head
S. cohnii	Feet
S. epidermidis	Head and thorax
S. haemolyticus	Thorax
S. hominis	Arms and legs
S. lugdunensis	No special niche determined
S. saccharolyticus	An anaerobe found on the face
S. saprophyticus	Feet
S. schleiferi	No special niche determined
S. simulans	No special niche determined
S. warneri	No special niche determined
S. xylosus	Feet

(a) Staphylococcus aureus

The anterior nares is the principal site of colonization; about 35% of the normal population carries *S. aureus* in this site and about two-thirds of these also carry on the turbinates. Carriage may be related to the HLA-Dr status of the individual, if this is so, it would account for reports of carriage in families and for differences between populations. The perineum is colonized in about 20% of the population and the axillae and toewebs in about 5–10% (Noble, Valkenburg and Wolters, 1967; Polakoff *et al.*, 1967). In persons without skin disease *S. aureus* is not found elsewhere on skin except as a contaminant from a carrier site. The classic epidemiological technique for distinguishing between strains of *S. aureus* is (bacterio)-phage typing, although this is being superseded by molecular methods such as restriction enzyme digests and PCR (Tenover *et al.*, 1994; Cuny and Witte, 1996). The phage types characteristically differ in different infec-tions: thus boils are most frequently caused by strains of phage group I, which includes the notorious phage type 80/81 and the probably clonal type 29 associated with toxic shock syndrome; phage type 71 in group II is most frequently found in impetigo and scalded skin syndrome; while the phage group III strains were formerly the most common in hospital sepsis with group I a close second. This picture has become blurred as a con-siderable proportion of the strains found in hospitals and the community are no longer susceptible to the international set of phages. In one com-munity survey of *S. aureus* carriage non-typable strains formed 34%, group III 20%, group I 15% and group II 14%, with the remaining 17% miscellaneous or mixed phage groups (Dancer and Noble, 1991).

Staphylococcus aureus produces a wide range of potential pathogenicity factors mostly not possessed by the coagulase-negative species (Table 2.2).

Table 2.2 Some pathogenicity factors of *Staphylococcus aureus*

Haemolysins ⎫ Leucocidins ⎬ Lipases ⎭	Cause non-specific effects by disrupting membranes*
Coagulase	Coagulates plasma
Enterotoxins	Described originally for their role in food poisoning these are important superantigens†
Toxic shock toxin	Another superantigen
Epidermolytic toxins	Split skin at stratum granulosum
Protein A	Induces a non-specific pseudo-allergic response

* Some have very profound effects; for example alpha haemolysin (alpha-toxin) may exhibit not only a haemolytic activity it can be cytotoxic, dermonecrotic and be lethal in sufficient doses in man and other animals.
† Superantigens are so-called because they activate T cells by a non-specific (Vβ) route and can activate about 25% of T cells rather than the more usual 0.001% for a specific antigen.

Folliculitis and boils (furuncles) are the most common endogenous skin infections to result more or less directly from nasal or skin carriage and are generally associated with minor trauma. Lesions of the head and neck are chiefly associated with nasal carriage, while lesions of the legs and buttocks are probably the result of perineal colonization (Kay, 1962). Local pyogenic skin infections are more common when both the temperature and humidity are at or near their peak (Fig. 2.2); tight clothing is often a contributing factor, those wearing Western-style clothing in hot humid climates suffering more infection than those wearing local dress (Selwyn, Verma and Vaishnav, 1967).

Nasal carriers of *S. aureus* also suffer more infection of catheter insertion sites in both haemodialysis and peritoneal dialysis. Davies *et al.* (1989) found that nasal carriers had a six times greater chance of peritoneal dialysis catheter site infection with *S. aureus* than did non-carriers. In haemodialysis, Yu and coworkers (1986) found that 93% of infections were with the same strain as was carried in the patient's nose. In these cases the route may be nose–hand–lesion; when this cycle is broken by reduction of nasal carriage infection rates drop significantly (Wenzel and Perl, 1995).

Impetigo contagiosa, as its name suggests, tends to occur in outbreaks of infection, most frequently in children, and is usually of exogenous origin although patients are frequently found to be temporary carriers of the infecting strain. Impetigo is a localized, skin-surface infection that

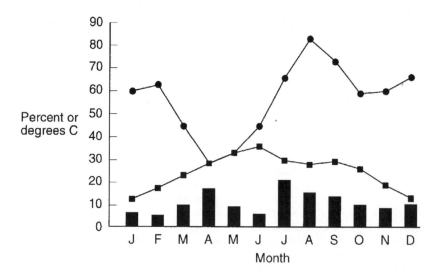

Figure 2.2 Effect of climate on the incidence of pyoderma in India: circles = humidity; squares = temperature; bars = percentage of outpatients with pyoderma. Source: based on Singh, 1973.

may also contain beta-haemolytic streptococci. In the USA impetigo was formerly chiefly associated with streptococci but there is evidence that this is changing (Barton and Friedman, 1987). There are two forms of impetigo: that characterized by 'honey-coloured stuck-on crusts' may be staphylococcal or streptococcal in origin while the blistering 'bullous' impetigo is always staphylococcal. Bullous impetigo is the mild part of a spectrum of disease, the other extreme being scalded skin syndrome, in which the skin peels away at the stratum granulosum. Epidemics of scalded skin syndrome may occur in maternity units when the source may be a staff member who has acquired a toxin-producing strain in the nose and transmitted this to the infants, perhaps resulting in colonization of the umbilical stump (Dancer *et al.*, 1988).

Secondary infection occurs in lesions of eczema, especially atopic dermatitis. The initial invading strain may originate from the patient or from a family member but frequent hospitalization and antibiotic therapy leads to acquisition of a hospital strain resistant to several antibiotics. When lesions are heavily colonized, frank infection is apparent, resembling impetigo, and the lesions are described as 'impetiginized'. The surrounding clinically uninvolved skin also acquires a rich microflora of *S. aureus* but at a density about 1/100 that of the lesional skin. The exact role of *S. aureus* in eczema is not known, although there is general agreement that it is secondary to the underlying disease. However, the production of superantigens such as the enterotoxins and toxic shock toxin has opened a new chapter in the study of inflammation since these superantigens can activate 25% or more of the T cells resulting in massive cytokine release (Schlievert, 1993). It is possible that superantigen production also exerts an effect even when only small numbers of staphylococci are present on an eczematous lesion, although this has not yet been established. Superantigens can penetrate intact normal skin and cause eczema-like inflammatory responses (Strange *et al.*, 1996).

(b) Coagulase-negative staphylococci

Among the coagulase-negative members of the skin's normal microflora one species, *S. epidermidis*, has become firmly associated with infection of catheters and other prostheses although both *S. haemolyticus* and *S. hominis* also contribute to these infections. These three species are the most common components of the skin's staphylococcal microflora, with *S. epidermidis* the most common numerically, and are acquired at or soon after birth (Fig. 2.3).

The chief species on the skin of the chest and back is *S. epidermidis*; *S. hominis* is dominant on the drier and less greasy arms and legs. It has long been known that these three species attach to and colonize catheters so that sedentary microbial growth occurs. The chief clinical problems

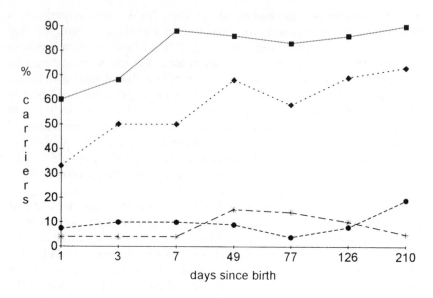

Figure 2.3 Colonization from birth with species of staphylococcus: squares = *S. epidermidis*; diamonds = *S. hominis*; circles = *S. aureus*; stars = *S. warneri*. *Source*: based on Carr and Kloos, 1977.

probably occur when these sedentary organisms break off to become planktonic in the blood stream and are able to settle and cause infection elsewhere. It had long been appreciated that the infecting organisms were frequently indistinguishable from those found on the skin around the catheter entry site (Fan *et al.*, 1988) but Ludlam and colleagues (1989) were able to show that the invaders were very often recently acquired members of the skin microflora, rather than long-established inhabitants. It is often necessary to remove the catheter in order to resolve the infection, despite the use of antibiotics, and a probable explanation for this is that adherent microbial colonies are much less susceptible to the antibiotics. Whether this is because they develop a glycocalyx that restricts the antibiotic access or because nutrient deprivation results in slow growth that is somehow less susceptible to antibiotic action is not known.

The distribution of coagulase-negative staphylococci in the blood stream of hospital patients from various countries is shown in Table 2.3.

The predominance of *S. epidermidis* is clear. *Staphylococcus epidermidis* is also the most common coagulase-negative staphylococcus in native valve infective endocarditis (Etienne and Eykyn, 1990). It is probable that these organisms gain access to the blood stream as a result of the transient bacteraemias that occur during everyday tasks such as tooth-brushing. Recently, minor members of the skin microflora such as *S. lugdunensis*

Table 2.3 Predominance of *Staphylococcus epidermidis* in
infection of the blood stream; percentage distribution
(*n* indicates number of positive cultures examined)

Species	Germany*	Greece†	Norway‡	USA§
S. epidermidis	57	48	78	69
S. hominis	26	11	12	16
S. haemolyticus	0	9	4	3
S. warneri	13	3	6	0
Others	4	29	0	11
n	23	35	68	36

* = Iwantscheff *et al.*, 1985; † = Papapetropoulos *et al.*, 1981; ‡ = Refsahl
and Anderson, 1992; § = Khatib *et al.*, 1995

have also been reported to cause very severe bacteraemias (Vandenesch *et al.*, 1993); while there is an impression that this is a new phenomenon, it is most probably in major part due to improvements in staphylococcal taxonomy, with the fairly recent recognition of new staphylococcal species.

A less obvious route of infection exists with urinary tract infection due to *S. saprophyticus*. Although the majority of infections are with *Escherichia coli*, infections with *S. saprophyticus* form between 5 and 10% of urinary tract infections in young women outside hospital, although they are rare in men. This sex difference is partly ascribable to the differing length of the urethra but Reuther and Noble (1993) have reported that between the years of 15 and 35 the flora of the feet in women, but not men, has a vastly increased proportion of *S. saprophyticus*, normally a minor component. The clinical significance of this has not been established.

(c) Interactions on skin

Interference mechanisms
Two aspects of interference between microorganisms on skin have excited interest in the past although they are not currently in vogue. One is the production of antibiotics by the skin microflora. The dermatophyte fungi are known to produce penicillins *in vivo* and to encourage the selection of penicillin-resistant staphylococci in the hedgehog and in humans, while the staphylococci themselves produce cyclic peptides that are active against other skin microflora. Attempts have been made to use this antibiotic production as a tool in therapy but without consistent success. In the past, during the pandemic staphylococcal infections of the 1950s and early 1960s, interference by a *S. aureus* strain known as 502A was very successful in preventing colonization and infection with the pandemic type 80/81 *S. aureus*. However, this was not mediated by

antibiotic production and most probably represented competitive adherence (Bibel *et al.*, 1983).

Gene transfer

Our understanding of the genetics of antimicrobial resistance in the staphylococci has increased in recent years. It is now well established that conjugative plasmids and conjugative transposons are able to transfer between most members of the skin staphylococcal microflora at frequencies of $1/10^4$ or greater when the strains are in stable cell-to-cell contact on a surface (Naidoo, 1984; Naidoo and Noble, 1978). Transfers also occur between enterococci and between enterococci and staphylococci under these conditions (Petts, Noble and Howell, 1997; Noble, Virani and Cree, 1992; Noble *et al.*, 1996). The skin of humans as well as experimental animals is one of the best surfaces for transfer although cotton dressings also function well (Townsend *et al.*, 1986); this should cause no surprise since the skin and mucous membranes are the natural habitat for these organisms. It is probable that the use of topical antibiotics has had a profound effect on the resistance patterns of the skin microflora and that gene transfer has played a considerable part in the spread of resistance.

There is much current concern of the possibility of staphylococci acquiring resistance to most or 'all' antibiotics and this concern has centred on the MRSA. Although strictly MRSA is the acronym for methicillin-resistant *Staphylococcus aureus*, the term **multiresistant** would generally be as appropriate. MRSA has attracted attention simply because of the fact of multiresistance and the problems this poses for therapy. There is no evidence that otherwise it is any more dangerous (i.e. more virulent) than other strains of *S. aureus*. The increased use of antibiotics in intensive care units, and the self-evident fact that patients in ICU are at much greater risk of serious infection, means that considerable efforts are made to maintain ICU free of MRSA or to eradicate MRSA if they are introduced. The aging population in the developed nations has meant that more elderly people are at risk of surgery and there has been concern that nursing homes for the elderly may become repositories for individuals carrying MRSA acquired during a hospital stay, perhaps in decubitus or other chronic ulcers (Bradley, 1997). These organisms may subsequently be reintroduced into the hospital environment, perhaps into the ICU; in one substantial study (Moreno *et al.*, 1995) more than half the isolations of MRSA in a hospital were from community cases. It is difficult to arrive at any consensus on nasal carrier rates for MRSA since these appear to vary greatly with the country of origin. In Japan, nasal carrier rates for staff in one hospital for ward staff was 91% for MRSA, with an overall rate of 8.5% for MRSA but 27.6% for total *S. aureus* (Saionji and Shitara, 1992). In Australia none of 808 medical students with varying degrees of exposure to the hospital environment carried MRSA (Stubbs *et al.*, 1994) while in England, Dancer and Noble (1991) found three carriers of MRSA in a population of 500 healthy women attending an antenatal clinic.

2.2.2 The coryneform microflora

Historically, the human skin was assumed to possess only one genus of coryneform, *Corynebacterium*, but improved taxonomy has revealed that at least four genera are found on the skin (Table 2.4); the taxonomy is incomplete, however, and it is known that new species exist, yet to be fully described.

The genera include *Corynebacterium* and *Propionibacterium*, both of which require lipid for successful growth on agar; because of the poor growth on non-lipid-containing agar these are called the lipophilic or 'small colony' coryneforms. Most *Propionibacterium* species are micro-aerophils and are best treated as anaerobes in the laboratory. The remaining genera, *Brevibacterium* and *Dermabacter*, do not require lipid, grow well on ordinary media and are sometimes called the 'large colony' coryneforms. The term coryneform refers to the club shape of these pleomorphic Gram-positive bacilli.

The old medical term 'diphtheroid' should be restricted to the two species which, in addition to *Corynebacterium diphtheriae*, are able to produce diphtheria toxin when lysogenized by *tox*⁺ phages, these are

Table 2.4 Genera and species of coryneform bacteria reported from human skin

Lipophilic species	
Corynebacterium	*Propionibacterium*
C. amycolatum	P. acnes
C. jeikeium	P. avidum
C. minutissimum	P. granulosum
C. striatum	P. innocuum*
C. urealyticum	
C. xerosis	
Non-lipophilic species	
Brevibacterium	*Dermabacter*
B. casei	D. hominis
B. epidermidis	
B. mcbrellneri	
B. otitidis	

* *P. innocuum* has now been transferred to the genus *Propioniferax*; it has not been recorded from infections

C. ulcerans and *C. pseudotuberculosis* (*C. ovis*). *Corynebacterium ulcerans* has been implicated in cases of clinical diphtheria. *Corynebacterium diphtheriae* is not a normal member of the skin microflora although patients who have suffered cutaneous diphtheria may carry the organism for long periods after resolution of infection (Belsey and LeBlanc, 1975).

General aspects of the coryneform bacteria in infectious disease were reported in a comprehensive review by Funke *et al.* (1997).

(a) Corynebacterium minutissimum

Erythrasma is the skin disease caused by *C. minutissimum*, which appears in the occluded areas of skin such the axillae, groin, toewebs and where skin contacts skin, such as the submammary area in obese individuals. It is characterized by erythematous scaly patches which fluoresce coral red under longwave ultraviolet light (a test frequently used in diagnosis) due to the production of porphyrins. Under normal lighting the lesions resemble fungal infections except that there is no healing behind the advancing edge.

Corynebacterium minutissimum is a member of the skin normal flora in about 20% of healthy individuals, where it occurs at densities of about $10^4\,\mathrm{cm}^{-2}$ in the axillae, groin, etc. When the characteristic lesions are present, however, the density of *C. minutissimum* exceeds $2 \times 10^5\,\mathrm{cm}^{-2}$ and treatments that reduce the density to its normal level, such as antibiotics, antibacterial soaps or even enthusiastic use of ordinary soap and water, resolve the lesions. Erythrasma appears, therefore, to represent the appearance of disease as a simple result of the overgrowth of the skin microflora, although the pathogenicity factors are not known.

(b) Trichomycosis and malodour

The terms trichomycosis axillaris or trichomycosis pubis refer to the growth of as yet unspeciated *Corynebacterium* species to form colonies on the shaft of the hairs in these regions. Several distinct organisms can be recovered from these colonies and the possibility that other genera such as the staphylococci are also involved cannot be excluded. Although there is damage to the hairs, which can only be detected by the use of an electron microscope, there would be little justification for interest in this affection if it were not frequently accompanied by profound malodour. This is apparently the result of microbial modification of testosterone secreted by the apocrine glands, which occur in these anatomical regions (Gower, Nixon and Mallet, 1988). It is probable that a mixture of volatile substances comprises the characteristic axillary malodour but interesting compounds, including a boar pheromone, are known to be produced. *In vitro*, mixtures of coryneforms produce more extensive modifications of testosterone than does each strain separately. Human infants are apparently able to recognize clothing worn by their mothers by the scent and the role of body odours may approach the pheromone concept more closely than is currently acknowledged.

(c) Deep infection due to cutaneous corynebacteria

Two minor components of the skin microflora are *C. jeikeium* and *C. urealyticum*. These are phenotypically very similar organisms and are both most frequently highly antibiotic-resistant. *Corynebacterium jeikeium* is found most frequently in males and is associated with bacteraemia in patients with leukaemia who are immunosuppressed in relation to a bone marrow transplant. In contrast, *C. urealyticum* is more common on the skin of females and is a very rare cause of kidney stones.

As might be anticipated, corynebacteria recognized as belonging to the skin microflora are occasionally recovered from deep infections. Changes in the taxonomy of the coryneforms, arising from the fact that they remain yet poorly described, may continue to cause problems, as in *C. diphtheriae* strains misidentified as low-virulence *Corynebacterium* species (Pennie, Malik and Wilcox, 1996). There may be confusion, for example, between *C. xerosis* and *C. striatum*; indeed, reference collection strains under these names may represent several different taxa – some, for example, may represent the species *C. amycolatum* (Coyle *et al.*, 1993). Recent examples of infection are here reported under the names assigned them by the authors: *Corynebacterium minutissimum* is reported to have caused recurrent breast abscess in an immunocompetent woman and to have infected dialysis patients (Berger *et al.*, 1984). Dialysis patients have also been reported to be infected with *C. striatum*; intensive-care patients suffered respiratory tract infection with a pigmented variant of this species in an outbreak of cross-infection (Leonard *et al.*, 1994). *Corynebacterium striatum* has also been recovered from pulmonary abscess (Batson *et al.*, 1996). Fatal infection has been reported due to *C. amycolatum* (Berner *et al.*, 1997). Weller, McLardy Smith and Crook (1994) have described osteomyelitis due to *C. jeikeium*, while *C. urealyticum* has been reported from a necrotic soft-tissue lesion (Saavedra *et al.*, 1996).

(d) Propionibacterium *species*

Three *Propionibacterium* species are found on normal human skin: these are *P. avidum*, found chiefly in the axilla and not implicated in skin disease although occasionally recovered from deep lesions (Estoppey *et al.*, 1997), plus *P. acnes* and *P. granulosum*, both implicated in acne vulgaris. *Propionibacterium acnes* and *P. granulosum* are found chiefly on the face and the upper chest and back, although *P. granulosum* is normally

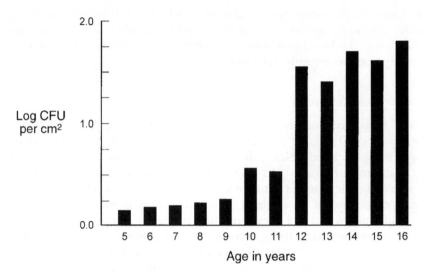

Figure 2.4 Colonization of children with *Propionibacterium acnes* in relation to age. Source: based on Matta, 1974.

present at densities about 1/100 those of *P. acnes*. Although typing schemes for *P. acnes* have not generally been successful, it has been shown using pyrolysis mass spectrometry that individuals carry distinct strains and that some persons carry more than one strain (Goodacre *et al.*, 1996). The skin population density of both *P. acnes* and *P. granulosum* increases markedly as puberty approaches (Fig. 2.4) and, since both are highly lipolytic and found to hydrolyse the triglyceride secreted as a major component of human sebum to release free fatty acids, it was initially assumed that this resulted in acne.

With time, the role assigned to the propionibacteria in acne has diminished and although a role in inflammation can be proposed, it is now believed that the organisms are not the **cause** of acne. However, erythromycin resistance in *P. acnes* impedes therapy with this antibiotic, indicating that some microbial role in inflammation exists.

Figure 2.5 shows pathways by which *P. acnes* could be implicated in acne.

A normal sebaceous gland becomes blocked as a result of changes in desquamation in the neck of the gland, sebum continues to be secreted and the propionibacteria that live deep in the ducts continue to grow (Leeming, Holland and Cunliffe, 1988). In the laboratory it can be seen that at sharp pH and pO_2 optima *P. acnes* will produce protease and hyaluronate lyase (Ingham *et al.*, 1983). These enzymes would make the gland walls leaky to the body defence mechanisms and the protease would activate complement by the alternative pathway; the cell wall of

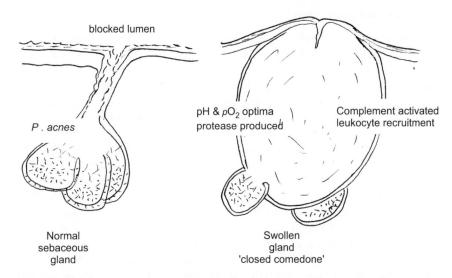

blocked lumen

P . acnes

pH & pO_2 optima
protease produced

Complement activated
leukocyte recruitment

Normal
sebaceous
gland

Swollen
gland
'closed comedone'

Figure 2.5 Complex aetiology of acne. Blockage of a normal sebaceous duct in which *P. acnes* continues to grow and sebum continues to be produced leads to a swollen gland (a closed comedone). At particular pH and pO_2 values protease is produced by *P. acnes*, making the cell wall 'leaky' to the body defence mechanism. Leukocytes are recruited, complement is activated by the protease or the cell wall of *P. acnes* and inflammation results.

P. acnes alone also has this action on complement (Massey, Mowbray and Noble, 1978; Webster *et al.*, 1978). Once complement is activated inflammation will occur, giving rise to the characteristic inflamed acne pustule (Tucker *et al.*, 1980). *Propionibacterium acnes* also produces porphyrins, which may oxidize squalene, a major component of the sebum, with resultant changes in the cellular structure of the follicle leading to blockage and the formation of a closed comedone (Saint-Leger *et al.*, 1986). Proinflammatory cytokines such as interleukins may also be induced by *P. acnes* products suggesting an alternative route for initiation of inflammation (Vowels, Yang and Leyden, 1995).

A fourth *Propionibacterium* species, *P. propionicum*, has been recovered from lachrymal gland infections (Brazier and Hall, 1993). This species has probably been misidentified in the past as *P. acnes* and its habitat is not known, although it may well be the skin.

Propionibacteria are also recovered from a variety of deep lesions (Brook, 1994) and are found as a constant, though small, component of infections following neurosurgery (Esteban *et al.*, 1995; Critchley and Strachan, 1996), where contamination from the scalp is most probable, but they are also found in other deep-seated infections such as those of hip prostheses and endocarditis (Stoll *et al.*, 1996; Armstrong and Wuerflein, 1996).

(e) Brevibacterium *species*

Organisms now described as *Brevibacterium epidermidis* isolated from the erosive, malodorous form of athlete's foot were found to resemble *B. linens*, an organism used in the dairy industry to impart flavour and odour to cheese by breaking down the protein releasing L-methionine and subsequently releasing methane thiol and other odorous gases. These organisms are now known to comprise part of the normal human skin microflora (Pitcher and Malnick, 1984). Other new species were to follow, including *B. casei*, which has been found chiefly in deep infections (Gruner *et al.*, 1994; Kaukoranta-Tolvanen *et al.*, 1995). In skin disease, the pathogenicity factors are probably closely related to the factors found desirable in cheese manufacture, i.e. powerful proteolytic enzymes. Most recently, *B. otitidis* has been described from the human ear (Pascual *et al.*, 1996) and *B. mcbrellneri* has been recovered from piedra, a growth of organisms on the hair of pubic regions (McBride *et al.*, 1993).

Some varieties of the malaria mosquito, *Anopheles gambiae*, have a propensity to bite preferentially around the foot and ankles of humans and it has been shown experimentally that they respond to a trap baited with Limburger cheese (Knols, 1996). In all probability this corresponds to the cheesy malodour recorded for feet colonized or infected with *Brevibacterium*. Disguising foot malodour may help to prevent mosquito bites!

(f) Dermabacter hominis

In 1988, Jones and Collins described a new monotypic genus, *Dermabacter hominis*, isolated from the skin of the normal human forearm. Later, strains conforming to this description and formerly assigned to CDC Groups 3 and 5 were described from 11 patients; five of the isolates were from blood; Funke *et al.* (1994) similarly reported 30 strains of *Dermabacter*, 21 of which had been recovered from blood. Eight of Funke's strains did not conform to *D. hominis*, although undoubtedly belonging to the genus, suggesting that further subdivision will be necessary.

2.2.3 *Micrococcus* species

In the past the *Micrococcus* species have been associated with the staphylococci, for they are Gram-positive cocci that tend to occur in packets, but they are taxonomically much closer to the coryneform bacteria. *Micrococcus* species are common on the skin, more especially in young females, being present at various sites in at least 60% of individuals (Noble, 1981). There are several species (Table 2.5) but only two (*M. luteus* and

Table 2.5 Species of *Micrococcus* recorded from human skin, with proposed new genera (based on Stackebrandt *et al.*, 1995)

M. (Kocuria) kristinae
M. (Micrococcus) luteus
M. (Micrococcus) lylae
M. (Dermacoccus) nishinomiyaensis
M. (Kocuria) roseus
M. (Kytococcus) sedentarius
M. (Kocuria) varians

M. varians) have been reported in deep infection such as bacteraemia, and even those are rare (Magee *et al.*, 1990).

One species, *M. sedentarius*, has been reported as a cause of pitted keratolysis (Nordstrom *et al.*, 1987), a disease in which painful erosions occur on the soles of the feet due to the action of proteases, although this is more frequently reported to be due to the actinomycete *Dermatophilus congolensis* (Woodgyer *et al.*, 1985). Radical revision of the generic status of the *Micrococcus* species has been suggested as indicated in Table 2.5 (Stackebrandt *et al.*, 1995).

2.2.4 Gram-negative bacilli

Probably because the skin is a fairly dry habitat, Gram-negative bacilli are, with one exception, rare on the skin in comparison with the Gram-positives. The exception is *Acinetobacter*, which has been described as an honorary Gram-positive. Early literature described these organisms as the genera *Mima* and *Herellea* and later these became *Acinetobacter lwoffi* and *A. anitratus*; and as such they were accorded a role in hospital infections. *Acinetobacter* spp. are found colonizing the moister areas of skin such axillae, groin, etc. and also the antecubital fossa, which is the site from which the vast majority of blood samples are obtained. Better taxonomy has shown that the species colonizing skin, *A. johnsonii* and *A. lwoffi*, are rarely involved in infection and that the most common pathogen in hospitals is *A. baumannii* (Grimont and Bouvet, 1991). Many other species are known from soil and other inanimate sources.

A number of genera such as *Enterobacter* can be found on the hands, which become temporarily colonized and constitute a source for cross-infection in hospital workers. However, *Pseudomonas aeruginosa* and *Proteus mirabilis* can be found in the toewebs of normal individuals and can be found as skin invaders in persons with very moist feet or those whose feet are wet as a result of employment. For example, Howell *et al.* (1988) found that 61% of coal miners with damaged toewebs were apparently infected with Gram-negative bacilli while those with clinically normal feet had carrier rates of 31%.

Infection of the skin, especially the toewebs but also the fingernails, by *P. aeruginosa* is often accompanied by the green pigment pyocyanin seen in laboratory culture of these organisms and forms a useful pathognomic feature.

2.2.5 *Malassezia* species

The only yeasts known to regularly inhabit human skin are members of the genus *Malassezia*. Formerly described as *Pityrosporum*, these yeasts are probably universal inhabitants of the head and thorax in adult humans but the fact that they are difficult to grow has resulted in a paucity of good data. The former *P. orbiculare* and *P. ovale* are lipid-requiring residents of human while *P. pachydermatis* is a non-lipid-requiring inhabitant of animal, especially dog, skin that is occasionally found colonizing human skin (Bandahaya, 1993). Recent advances in taxonomy have revealed that, while *M. (P.) pachydermatis* remains the sole non-lipid-requiring species, the lipid-requiring species now number six (Table 2.6).

Table 2.6 Species of *Malassezia* recorded from human skin (based on Gueho, Midgley and Guillot, 1996)

M. furfur
M. globosa
M. obtusa
M. pachydermatis
M. restricta
M. slooffiae
M. sympodialis

The taxonomy is too recent to afford much in the way of distributions or disease associations, but the former two species were known to be involved in the skin diseases seborrhoeic dermatitis and pityriasis versicolor and to play a part in the aetiology of severe dandruff (Schmidt, 1997). These yeasts may also play a part in exacerbation of atopic dermatitis (Mukai *et al.*, 1997). Occasional reports of deep infection in infants receiving total parenteral nutrition describe infection with lipid-requiring species (Marcon and Powell, 1992), where it is supposed that the high lipid content of the fluid plays a selective role, and with *P. pachydermatis* (Welbel *et al.*, 1994).

2.3 CONCLUSION

Several factors combine to ensure that the skin microflora will continue to fascinate ecologists, and medical and molecular microbiologists. It may seem curious that a habitat so apparently easily accessible as the skin surface should carry microbial species (and most probably genera) that remain to be described. Two factors contribute to this: one is that taxonomy is unfashionable and poorly funded, but the other is that molecular techniques are proving powerful with groups such as the coryneforms and yeasts that have proved difficult to study using phenotypic tools. Advances in medical technology have provided organisms new scope for colonization. The ability of *Staphylococcus epidermidis* to adhere to medical plastics and so evade, to some extent, both antibiotics and the body defence mechanism could not have been predicted, putting advances such as peritoneal dialysis or joint replacements at risk. Finally, in an era preoccupied with the possible appearance of 'superbugs' in the hospital environment, there is scope for studies of ecological genetics in nature. We have rushed into intensive studies of plasmids, transposons and insertion sequences with remarkably little attention to the habitat in which gene exchange actually occurs. We are in danger of knowing the DNA sequence without knowing the habitat whence it came.

2.4 REFERENCES

Abe, T., Mayuzumi, J., Kikuchi, N. and Arai, S. (1980) Seasonal variations in skin temperature, skin pH, evaporative water loss and skin surface lipid values on human skin. *Chemical Pharmaceutical Bulletin*, **8**, 387–392.

Aly, R., Shirley, C., Cunico, B. and Maibach, H. I. (1978) Effect of prolonged occlusion on the microbial flora, pH, carbon dioxide and transepidermal water loss on human skin. *Journal of Investigative Dermatology*, **71**, 378–381.

Armstrong, R. W. and Wuerflein, R. D. (1996) Endocarditis due to *Propionibacterium granulosum*. *Clinical Infectious Diseases*, **23**, 1178–1179.

Bandahaya, M. (1993) The distribution of *Malassezia furfur* and *Malassezia pachydermatis* on normal human skin. *Southeast Asian Journal of Tropical Medicine and Public Health*, **24**, 343–346.

Barton, L. L. and Friedman, A. D. (1987) Impetigo: a reassessment of etiology and therapy. *Pediatric Dermatology*, **4**, 185–188.

Batson, J. H., Mukkamala, R., Byrd, R. P. Jr and Roy, T. M. (1996) Pulmonary abscess due to *Corynebacterium striatum*. *Journal of the Tennessee Medical Association*, **89**, 115–116.

Belsey, M. A. and LeBlanc, D. R. (1975) Skin infections and the epidemiology of diphtheria: acquisition and persistence of *C. diphtheriae* infection. *American Journal of Epidemiology*, **102**, 179–184.

Berger, S. A., Gorea, A., Stadler J. *et al.* (1984) Recurrent breast abscesses caused by *Corynebacterium minutissimum*. *Journal of Clinical Microbiology*, **20**, 1219–1220.

Berner, R., Pelz, K., Wilhelm, C. *et al.* (1997) Fatal sepsis caused by *Corynebacterium amycolatum* in a premature infant. *Journal of Clinical Microbiology*, **35**, 1011–1012.

Bibel, D. J., Bayles, C., Strauss, W. G. *et al.* (1983) Competitive adherence as a mechanism of bacterial interference. *Canadian Journal of Microbiology*, **29**, 700–703.

Bradley, S. F. (1997) Methicillin-resistant *Staphylococcus aureus* in nursing homes. Epidemiology, prevention and management. *Drugs and Aging*, **10**, 185–198.

Brazier, J. S. and Hall, V. (1993) *Propionibacterium propionicum* and infections of the lachrymal apparatus. *Clinical Infectious Diseases* **17**, 892–893.

Brook, I. (1994) Infection caused by propionibacterium in children. *Clinical Pediatrics*, **33**, 485–490.

Carr, D. L. and Kloos, W. E. (1977) Temporal study of the staphylococci and micrococci of normal infant skin. *Applied and Environmental Microbiology*, **34**, 673–680.

Coyle, M. B., Leonard, R. B., Nowowiejski, D. J. *et al.* (1993) Evidence of multiple taxa within commercially available reference strains of *Corynebacterium xerosis*. *Journal of Clinical Microbiology*, **31**, 1788–1793.

Critchley, G. and Strachan, R. (1996) Postoperative subdural empyema caused by *Propionibacterium acnes* – a report of two cases. *British Journal of Neurosurgery*, **10**, 321–323.

Cuny, C. and Witte, W. (1996) Typing of *Staphylococcus aureus* by PCR for DNA sequences flanked by transposon Tn*916* target region and ribosomal binding site. *Journal of Clinical Microbiology*, **34**, 1502–1505.

Dancer, S. J. and Noble, W. C. (1991) Nasal, axillary and perineal carriage of *Staphylococcus aureus* among women: identification of strains producing epidermolytic toxin. *Journal of Clinical Pathology*, **44**, 681–684.

Dancer, S. J., Simmons, N. A., Poston, S. M. and Noble, W. C. (1988) Outbreak of staphylococcal scalded skin syndrome among neonates. *Journal of Infection*, **16**, 87–103.

Davies, S. J., Ogg, C. S., Cameron, J. S. *et al.* (1989) *Staphylococcus aureus* nasal carriage, exit-site infection and catheter loss in patients treated with continuous ambulatory peritoneal dialysis (CAPD). *Peritoneal Dialysis Bulletin*, **9**, 61–64.

Esteban, J., Ramos, R. M., Jiminez-Castillo, P. and Soriano, F. (1995) Surgical wound infections due to *Propionibacterium acnes*: a study of 10 cases. *Journal of Hospital Infection*, **30**, 229–232.

Estoppey, O., Rivier, G., Blane, C. H. *et al.* (1997) *Propionibacterium avidum* sacroiliitis and osteomyelitis. *Revue du Rhumatisme* (English edn), **64**, 54–56.

Etienne, J. and Eykyn, S. J. (1990) Increase in native valve endocarditis caused by coagulase-negative staphylococci: an Anglo-French clinical and microbiological study. *British Heart Journal*, **64**, 381–384.

Fan, S. T., Teoh-Chan, C. H., Lau, K. F. *et al.* (1988) Predictive value of surveillance skin and hub cultures in central venous catheter sepsis. *Journal of Hospital Infection*, **12**, 191–198.

Funke, G., Stubbs, S., Pfyffer, G. E. *et al.* (1994) Characteristics of CDC group 3 and group 5 coryneform bacteria isolated from clinical specimens and assignment to the genus *Dermabacter*. *Journal of Clinical Microbiology*, **32**, 1223–1228.

Funke, G., von Graevenitz, A., Clarridge, J. E. III and Bernard, K. A. (1997) Clinical microbiology of coryneform bacteria. *Clinical Microbiology Reviews*, **10**, 125–159.

Goodacre, R., Howell, S. A., Noble W. C. and Neal, M. J. (1996) Sub-species discrimination, using pyrolysis mass spectrometry and self-organinising neural networks, of *Propionibacterium acnes* isolated from normal human skin. *Zentralblatt für Bakteriologie. International Journal of Medical Microbiology, Virology, Parasitology and Infectious Diseases*, **284**, 501–515.

Gower, D. B., Nixon, A. and Mallet, A. I. (1988) The significance of odorous steroids in axillary odour, in *Perfumery: The Psychology and Biology of Fragrance*, (eds S. van Toller and G. H. Dodd), Chapman & Hall, London, pp. 47–85.

Grimont, P. A. D. and Bouvet, P. J. M. (1991) Taxonomy of acinetobacter, in *The Biology of Acinetobacter*, (eds K. J. Towner, E. Bergogne-Berezin and C. A. Fewson), Plenum Press, New York, pp. 25–36.

Gruner, E., Steigerwalt, A. G., Hollis, D. G. *et al.* (1994) Recognition of *Dermabacter hominis*, formerly CDC fermentative coryneform group 3 and group 5, as a potential human pathogen. *Journal of Clinical Microbiology*, **32**, 1918–1922.

Gueho, E., Midgley, G. and Guillot, J. (1996) The genus *Malassezia* with description of four new species. *Antonie van Leeuwenhoek*, **69**, 337–355.

Howell, S. A., Clayton, Y. M., Phan Q. G. and Noble, W. C. (1988) Tinea pedis: the relationship between symptoms, organisms and host characteristics. *Microbial Ecology in Health and Disease*, **1**, 131–133.

Ingham, E., Holland, K. T., Gowland, G. and Cunliffe, W. J. (1983) Studies of the intracellular proteolytic activity produced by *Propionibacterium acnes*. *Journal of Applied Bacteriology*, **54**, 263–271.

Iwantscheff, A., Kuhnen, E. and Brandis, H. (1985) Species distribution of coagulase-negative staphylococci isolated from clinical sources. *Zentralblatt für Bakteriologie, Mikrobiologie, und Hygiene, Serie A*, **260**, 41–50.

Jones, D. and Collins, M. D. (1988) Taxonomic studies on some human cutaneous coryneform bacteria: description of *Dermabacter hominis* gen. nov., sp. nov. *FEMS Microbiology Letters*, **51**, 51–56.

Kaukoranta-Tolvanen, S. S., Sivonen, A., Kostiala, A. A. *et al.* (1995) Bacteremia caused by *Brevibacterium* species in an immunocompromised patient. *European Journal of Clinical Microbiology and Infectious Diseases*, **14**, 801–804.

Kay, C. R. (1962) Sepsis in the home. *British Medical Journal*, **i**, 1048–1052.

Khatib, R., Riederer, K. M., Clark, J. A. *et al.* (1995) Coagulase-negative staphylococci in multiple blood cultures: strain relatedness and determinants of same-strain bacteremia. *Journal of Clinical Microbiology*, **33**, 816–820.

Knols, B. G. J. (1996) On human odour, malaria mosquitoes, and Limburger cheese. *Lancet*, **348**, 1322.

Leeming, J. P., Holland, K. T. and Cunliffe, W. J. (1988) The microbial colonization of inflamed acne vulgaris lesions. *British Journal of Dermatology*, **118**, 203–208.

Leonard, R. B., Nowowiejski, D. J., Warren, J. J. *et al.* (1994) Molecular evidence of person-to-person transmission of a pigmented strain of *Corynebacterium striatum* in intensive care units. *Journal of Clinical Microbiology*, **32**, 164–169.

Ludlam, H. A., Noble, W. C., Marples, R. R. *et al.* (1989) The epidemiology of peritonitis caused by coagulase-negative staphylococci in continuous ambulatory peritoneal dialysis. *Journal of Medical Microbiology*, **30**, 167–174.

McBride, M. E., Ellner, K. M., Black, H. S. *et al.* (1993) A new *Brevibacterium* sp. isolated from infected genital hair of patients with white piedra. *Journal of Medical Microbiology*, **39**, 255–261.

Magee, J. T., Burnett I. A., Hindmarsh, J. M. and Spencer, R. C. (1990) *Micrococcus* and *Stomatococcus* spp. from human infections. *Journal of Hospital Infection*, **16**, 67–73.

Marcon, M. J. and Powell, D. A. (1992) Human infections due to *Malassezia* spp. *Clinical Microbiology Reviews*, **5**, 101–119.

Marples, M. J. (1965) *The Ecology of Human Skin*, Charles C. Thomas, Springfield, IL.

Massey, A., Mowbray, J. F. and Noble, W. C. (1978) Complement activation by *Corynebacterium acnes*. *British Journal of Dermatology*, **98**, 583–584.

Matta, M. (1974) Carriage of *Corynebacterium acnes* in school children in relation to age and race. *British Journal of Dermatology*, **91**, 557–561.

Moreno, F., Crisp, C., Jorgensen, J. H. and Patterson, J. E. (1995) Methicillin-resistant *Staphylococcus aureus* as a community organism. *Clinical Infectious Diseases*, **21**, 1308–1312.

Mukai, H., Kaneko, S., Saito, N. *et al.* (1997) Clinical significance of *Malassezia furfur* specific IgE antibody in atopic dermatitis. *Arerugi – Japanese Journal of Allergology*, **46**, 26–33.

Naidoo, J. (1984) Interspecific co-transfer of antibiotic resistance plasmids in staphylococci *in vivo*. *Journal of Hygiene, Cambridge*, **93**, 59–66.

Naidoo, J. and Noble, W. C. (1978) Transfer of gentamicin resistance between strains of *Staphylococcus aureus* on skin. *Journal of General Microbiology*, **107**, 391–395.

Noble, W. C. (1981) *Microbiology of Human Skin*, Lloyd-Luke, London.

Noble, W. C. (ed.) (1993) *The Skin Microflora and Microbial Skin Disease*, Cambridge University Press, Cambridge.

Noble, W. C., Valkenburg, H. A. and Wolters, C. H. L. (1967) Carriage of *Staphylococcus aureus* in random samples of the normal population. *Journal of Hygiene, Cambridge*, **65**, 567–573.

Noble, W. C., Virani, Z. and Cree, R. G. A. (1992) Co-transfer of vancomycin and other resistance genes from *Enterococcus faecalis* NCTC12201 to *Staphylococcus aureus*. *FEMS Microbiology Letters*, **93**, 195–198.

Noble, W. C., Rahman, M., Karadec, T. and Schwarz, S. (1996) Gentamicin resistance gene transfer from *Enterococcus faecalis* and *E. faecium* to *Staphylococcus aureus*, *S. intermedius* and *S. hyicus*. *Veterinary Microbiology*, **52**, 143–152.

Nordstrom, K. M., McGinley, K. J., Cappriello, L. *et al.* (1987) Pitted keratolysis: the role of *Micrococcus sedentarius*. *Archives of Dermatology*, **123**, 1320–1325.

Papapetropoulos, M., Pappas, A., Papavassiliou, J. and Legakis, N. J. (1981) Distribution of coagulase-negative staphylococci in human infections. *Journal of Hospital Infection*, **2**, 145–153.

Pascual, C., Collins, M. D., Funke, G. and Pitcher, D. G. (1996) Phenotypic and genotypic characterisation of two brevibacterium strains from the human ear: description of *Brevibacterium otitidis* sp. nov. *Medical Microbiology Letters*, **5**, 113–123.

Pennie, R. A., Malik, A. S. and Wilcox, L. (1996) Misidentification of toxigenic *Corynebacterium diphtheriae* as a *Corynebacterium* species with low virulence in a child with endocarditis. *Journal of Clinical Microbiology*, **34**, 1275–1276.

Petts, D. N., Noble, W. C. and Howell, S. A. (1997) Potential for gene transfer among enterococci from a single patient and the possibility of confounding typing results. *Journal of Clinical Microbiology*, **35**, 1722–1727.

Pitcher, D. G. and Malnick, H. (1984) Identification of *Brevibacterium* from clinical sources. *Journal of Clinical Pathology*, **37**, 1395–1398.

Polakoff, S., Richards, I. D. G., Parker M. T. and Lidwell, O. M. (1967) Nasal and skin carriage of *Staphylococcus aureus* by patients undergoing surgical operation. *Journal of Hygiene, Cambridge*, **65**, 559–566.

Refsahl, K. and Andersen, B. M. (1992) Clinically significant coagulase-negative staphylococci: identification and resistance patterns. *Journal of Hospital Infection*, **22**, 19–31.

Reuther, J. W. A. and Noble, W. C. (1993) An ecological niche for *Staphylococcus saprophyticus*. *Microbial Ecology in Health and Disease*, **6**, 209–212.

Saavedra, J., Rodriguez, J. N., Fernandez-Jurado, A. *et al.* (1996) A necrotic soft-tissue lesion due to *Corynebacterium urealyticum* in a neutropenic child. *Clinical Infectious Diseases*, **22**, 851–852.

Saint-Leger, D., Bague, A., Cohen, E. and Chivot, M. (1986) A possible role for squalene in the pathogenesis of acne I. In vitro study of squalene oxidation. *British Journal of Dermatology*, **114**, 535–542.

Saionji, K. and Shitara, M. (1992) Carrier state and comparative study of methicillin-resistant *Staphylococcus aureus* from the staff and patients. *Japanese Journal of Clinical Medicine*, **50**, 998–1003.

Schlievert, P. M. (1993) Role of superantigens in human disease. *Journal of Infectious Diseases*, **167**, 997–1002.

Schmidt, A. (1997) *Malassezia furfur*: a fungus belonging to the physiological skin flora and its relevance in skin disorders. *Cutis*, **59**, 21–24.

Selwyn, S., Verma B. S. and Vaishnav, V. P. (1967) Factors in the bacterial colonization and infection of the human skin. *Indian Journal of Medical Research*, **55**, 652–656.

Singh, G. (1973) Heat, humidity and pyodermas. *Dermatologica*, **147**, 342–347.

Singh, G., Marples, R. R. and Kligman, A. M. (1971) Experimental *Staphylococcus aureus* infection in humans. *Journal of Investigative Dermatology*, **57**, 149–162.

Stackebrandt, E., Koch, C., Gvosdiak, O. and Schumann, P. (1995) Taxonomic dissection of the genus *Micrococcus*: *Kocuria* gen. nov., *Nesterkonia* gen. nov., *Kytococcus* gen. nov., *Dermacoccus* gen. nov. and *Micrococcus* Cohn 1872 gen. emend. *International Journal of Systematic Bacteriology*, **45**, 682–692.

Stoll, T., Stucki, G., Bruhlmann, P. *et al.* (1996) Infection of a total knee joint prosthesis by *Peptostreptococcus micros* and *Propionibacterium acnes* in an elderly RA patient: implant salvage with longterm antibiotics and needle aspiration/irrigation. *Clinical Rheumatology* **15**, 399–402.

Strange, P., Skov, L., Lisby, S. *et al.* (1996) Staphylococcal enterotoxin B applied on intact normal and intact atopic skin induces dermatitis. *Archives of Dermatology*, **132**, 27–33.

Stubbs, E., Pegler, M., Vickery A. and Harbour, C. (1994) Nasal carriage of *Staphylococcus aureus* in Australian (preclinical and clinical) medical students. *Journal of Hospital Infection*, **27**, 127–134.

Tenover, F. C., Arbeit, R., Archer, G. *et al.* (1994) Comparison of traditional and molecular methods of typing isolates of *Staphylococcus aureus*. *Journal of Clinical Microbiology*, **32**, 407–415.

Townsend, D. E., den Hollander, L., Bolton, S. and Grubb, W. B. (1986) Clinical isolates of staphylococci conjugate on contact with dry absorbent surfaces. *Medical Journal of Australia*, **144**, 166.

Tucker, S. B., Rogers, R. S. III, Winkelmann, R. K. *et al.* (1980) Inflammation in acne vulgaris: leukocyte attraction and cytotoxicity by comedonal material. *Journal of Investigative Dermatology*, **74**, 21–25.

Vandenesch, F., Etienne, J., Reverdy, M. E. and Eykyn, S. J. (1993) Endocarditis due to *Staphylococcus lugdunensis*: report of 11 cases and review. *Clinical Infectious Diseases*, **17**, 871–876.

Vowels, B. R., Yang, S. and Leyden, J. J. (1995) Induction of proinflammatory cytokines by a soluble factor of *Propionibacterium acnes*: implications for chronic inflammatory acne. *Infection and Immunity*, **63**, 3158–3165.

Webster, G. F., Leyden, J. J., Norman, M. E. and Nilsson, U. R. (1978) Complement activation in acne vulgaris: *in vitro* studies with *Propionibacterium acnes* and *Propionibacterium granulosum*. *Infection and Immunity*, **22**, 523–529.

Welbel, S. F., McNeil, M. M., Pramanik, A. *et al.* (1994) Nosocomial *Malassezia pachydermatis* bloodstream infections in a neonatal intensive care unit. *Pediatric Infectious Disease Journal*, **13**, 104–108.

Weller, T. M. A., McLardy Smith, P. and Crook, D. W. (1994) *Corynebacterium jeikeium* osteomyelitis following total hip joint replacement. *Journal of Infection*, **29**, 113–114.

Wenzel, R. P. and Perl, T. M. (1995) The significance of nasal carriage of *Staphylococcus aureus* and the incidence of postoperative wound infection. *Journal of Hospital Infection*, **31**, 13–24.

Woodgyer, A. J., Baxter, M., Rush-Munro, F. M. *et al.* (1985) Isolation of *Dermatophilus congolensis* from two New Zealand cases of pitted keratolysis. *Australasian Journal of Dermatology*, **26**, 29–35.

Yu, V. L., Goetz, A., Wagener, M. *et al.* (1986) *Staphylococcus aureus* nasal carriage and infection in patients on hemodialysis. *New England Journal of Medicine*, **315**, 91–96.

3

The respiratory tract microflora and disease

Jean O. Kim and Jeffrey N. Weiser

3.1 INTRODUCTION

The human respiratory tract harbors hundreds of different bacterial species. The purpose of the following chapter is to review some of the bacterial and host factors that contribute to the normal microflora and the infections that originate in this site. There is generally a peaceful state of coexistence between humans and the multitude of organisms that occupy the mucosal surfaces of the upper respiratory tract. Many bacterial species appear to be highly adapted to colonize this site, and in many cases humans are their only natural host. These are considered to be commensal organisms because in the absence of underlying mucosal damage or immunological dysfunction they live in the upper respiratory tract without causing disease. Several of these species are also common etiologic agents of disease both within the respiratory tract and at more distant host sites following hematogenous dissemination. The respiratory tract is also a common point of entry for many strict pathogens such as *Mycobacterium tuberculosis* and *Streptococcus pyogenes* (group A beta-hemolytic streptococcus) whose presence usually correlates with disease. These organisms are not a part of the normal microflora and will not be discussed in this chapter. Three species, *Streptococcus pneumoniae* (the pneumococcus), *Haemophilus influenzae* and *Neisseria meningitidis* (the meningococcus) will be reviewed in detail. These have been selected because they are members of the normal microflora as well as being etiologic agents of common respiratory tract and disseminated infections.

G. W. Tannock (ed.), *Medical Importance of the Normal Microflora*, 47–73.
© 1999 *Kluwer Academic Publishers. Printed in Great Britain.*

Many of the host and bacterial factors involved in the diseases caused by these organisms are understood and will be reviewed.

3.2 ANATOMY OF THE RESPIRATORY TRACT

The respiratory tract begins at the nasopharynx and includes the nose and superior pharyngeal areas. Extending laterally from the nasopharynx are self-contained cavities known as sinuses that are normally aerated but are not involved in the process of air movement; these areas become fully developed during early childhood. Included in the upper respiratory tract is the middle ear, although it, too, does not contribute to respiration. Beyond the nasopharynx inhaled air combines with that from the oro-pharynx, which joins the nasopharynx at the posterior pharynx. Inferiorly is the trachea, which serves as the marker between the upper and lower respiratory tracts. From there air moves into the mainstem bronchi, which divide into eight to ten segmental bronchi in each lung, and then through smaller branches called bronchioles until it reaches the distal sacs known as alveoli. At this level exchange of gases, primarily oxygen, carbon dioxide and nitrogen, takes place between the bloodstream and moving air. Adjacent to these airways are the pleurae, which enclose the lung.

3.2.2 The normal host state

In the normal state the human is colonized with several bacterial species and strains without the occurrence of disease (Tramont and Hoover, 1995; Table 3.1). These bacteria normally act as symbiotic partners on the mucosal surface, protecting the host from becoming colonized with potentially more pathogenic or antibiotic-resistant organisms.

The presence of particular organisms within specific sites is determined by various mechanical, chemical and immunological host factors that favor as well as hinder specific bacterial adhesion, growth, persistence and spread from host to host. The ready access to the outside environment as well as its moisture and warmth make the nasopharynx the most advantageous place for bacterial growth. However, the aerodynamic environment of the nasopharynx is turbulent because of the presence of nasal hairs as well as the constant movement of air against the nasal turbinates with each inhalation and exhalation, preventing attachment of less adaptable organisms. Large particles in this humid environment become trapped in the mucous secretions and are transported down the mucus stream to the posterior pharynx, where they are swallowed peri-odically and eliminated by the gastrointestinal system. Transiently increased production of mucus related to local inflammation, as occurs in viral illnesses, induces protective reflexes such as sneezing and coughing, which clear the airways of secretions that contain bacteria.

Table 3.1 Organisms found in nares and nasopharynx

	Organism	*Disease state commonly produced*
Aerobic, Gram-positive	*Streptococcus pyogenes*	Pharyngitis, pneumonia, impetigo
	Streptococcus pneumoniae	Otitis media, sinusitis, pneumonia, bronchitis, meningitis, bacteremia
	Staphylococcus aureus	Pneumonia
	Staphylococcus epidermidis	None
	Corynebacterium diphtheriae	Pharyngitis, diphtheria
Aerobic, Gram-negative	*Branhamella catarrhalis*	Otitis media, sinusitis, bronchitis
	Haemophilus parainfluenzae	None
	Haemophilus influenzae, non-encapsulated	Otitis media, sinusitis, pneumonia
	Haemophilus influenzae, type b	Epiglottitis, periorbital cellulitis, orbital cellulitis, meningitis, bacteremia
	Neisseria meningitidis	Meningococcemia, meningoencephalitis, meningitis, bacteremia
Anaerobic, Gram-positive	*Propionibacterium* spp.	None
	Micrococcus spp.	None

Few organisms remain in the airway beyond the trachea. The bronchioles and alveoli are generally considered sterile, i.e. without colonizing bacteria in the normal state. Particles that survive the journey to the alveoli are phagocytosed by alveolar macrophages. The sinuses and middle ear spaces, which connect to the oropharynx through the eustachian tubes, are also considered sterile sites.

On the mucosal surface normal microflora must survive various immunological factors, including nasal secretions containing immunoglobulins, which consist predominantly of secretory immunoglobulin A_1 (IgA_1), complement, lysozyme and phagocytic cells present in the nasopharynx. The normal microflora appears to change as specific immunity develops. Carriage of a particular strain or serotype will often last for a matter of weeks to months until host clearance mechanisms prevail. In addition, it has recently been recognized that the respiratory tract is capable of an innate immunity as demonstrated by the production of antimicrobial peptides (Diamond *et al.*, 1991).

Table 3.2 Virulence factors

Organism	Virulence Determinant	Biological Function(s)
Streptococcus pneumoniae	Capsular polysaccharide (90 types)	Inhibition of phagocytosis
	IgA$_1$ protease	Inactivation of immunoglobulin A$_1$
	PspA	Inhibition of phagocytosis?
	CbpA	Adhesin
	Pneumolysin	Cytolytic toxin, complement activation
	Neuraminadase	Cleaves sialic acid on the mucosal surface
	Teichoic acid contains choline	Adherence?
Haemophilus influenzae	Capsular polysaccharide (six types*)	Serum resistance
	IgA$_1$ protease	Inactivation of immunoglobulin A$_1$
	Fimbriae	Adhesin
	Lipopolysaccharide containing choline	Endotoxin, antigenic variation, serum resistance
Neisseria meningitidis	Capsular polysaccharide (13 types†)	Serum resistance
	IgA$_1$ protease	Inactivation of immunoglobulin A$_1$
	Fimbriae	Adhesin
	Opacity proteins	Adhesins
	PIII protein	Blocking antibody
	Lipopolysaccharide containing sialic acid	Endotoxin, antigenic variation, serum resistance

* Non-typable strains lacking capsule are common cause of localized disease confined to the respiratory tract
†Unencapsulated strains are not commonly associated with disease

Factors that assist the bacteria in attaching to the epithelial surface or mucus enhance colonization. The specific protein or carbohydrate host receptors involved in these interactions are known in only a few instances (Table 3.2).

3.2.3 Alteration of normal microflora

There are several different conditions that determine which bacteria will exist as normal microflora in the respiratory tract. If these conditions are altered, the make-up of the normal microflora changes. One such factor is the age of the host. Infants initially become colonized within the first week of life; even as young as 1 day old the oropharynx of the infant is colonized with *Streptococcus salivarius* or *S. mitis* and later with *Lactobacillus*. Immediately following tooth eruption *S. mutans*, *S. sanguis* and several anaerobes become members of the oral microflora. Spirochetes, such as *Actinomyces* spp. and *Bacteroides melaninogenicus*, become more prevalent in the oropharynx during puberty (Cooperstock, 1992). Host environments with high rates of transmittence such as daycare centers may affect the host's microflora. It has been shown that frequent and prolonged contact with carriers who harbor antibiotic-resistant isolates increases the chance of being colonized with these strains. Exposure to systemic antimicrobial therapy, which is common in some situations, such as children with frequent bouts of otitis media, may change the normal microflora by eliminating susceptible organisms and replacing them with antibiotic-resistant organisms.

3.2.4 Alteration in host status

The remarkable balance between the human host and resident organisms may be upset by a number of host factors. The disease state usually occurs following disruption of the mucosal barrier. Factors that may contribute to this disruption include local viral infection, e.g. the common cold or influenza, trauma, surgery or any cause of significant inflammation or damage to the mucosa. When bacteria are able to occupy sites within the respiratory tract that are normally sterile, the ability of the host to clear organisms may concede to the ability of bacteria to replicate and trigger an inflammatory response. Conditions that predispose to the disease state include aspiration of pharyngeal contents into the lower respiratory tract causing pneumonia, obstruction of sinus passages causing sinusitis, or blockage of eustachian tubes resulting in otitis media. Tissue destruction can allow for further breakdown of barriers, which may lead to invasion with some bacterial species capable of surviving the full array of clearance mechanisms available to the healthy host. These organisms can then enter into the blood stream of the host, leading to a completely different spectrum of disease.

Organisms within the blood stream may be readily cleared via the reticuloendothelial system found in the liver and spleen. Immunological factors that assist in the clearance of bacteria from the blood stream include complement and antibodies (particularly IgG), which may have been induced from previous exposure or immunization. Other important factors include circulating neutrophils and macrophages, which phagocytose the bacteria–complement–antibody complexes. The subsequent release of cytokines and other inflammatory modulators induces systemic host responses, including neutrophil chemotaxis, leukocytosis and fever, which all act in concert to clear invading organisms from normally sterile sites.

Conditions that alter normal immunological defenses, such as immature or waning immunity as occurs during the first years of life and in the elderly, respectively, assist the invading bacteria in establishing infection either within the respiratory tract or elsewhere via the blood stream. For example, a deficiency of IgA at the site of mucosal defense may allow growth of organisms that would otherwise be kept in check. A defect in the ability to phagocytose bacteria would allow organisms not normally existing as commensals to persist and perhaps become invasive. Other defects in the immune system such as functional or anatomic asplenia, complement or antibody deficiencies, neutropenia or chemotaxis defects would pose a serious risk to the host for developing disseminated infection.

3.3 COMMON RESPIRATORY TRACT SYNDROMES

Infections within the respiratory tract are characterized by their distinct clinical presentations. The following are brief descriptions of the common respiratory tract syndromes.

3.3.1 Rhinitis

Rhinitis, characterized by inflammation of the mucosal surfaces of the nasopharynx, results in rhinorrhea or discharge from the nose with few other systemic symptoms. The etiology is most often viral (rhinoviruses, adenoviruses or enteroviruses). Allergy-induced rhinitis is also common in predisposed individuals. Bacteria do not often cause rhinitis and nasopharyngeal mucosal surface cultures are not useful.

3.3.2 Sinusitis

This syndrome may be divided into two categories, acute and chronic, based on duration and characteristics of presentation. Acute sinusitis usually occurs in the setting of a recent viral rhinitis. Common complaints are rhinorrhea or purulent nasal discharge, sneezing and nasal obstruction, together with facial tenderness or pressure, headache or fever. Cough and foul-smelling breath are also frequent complaints. More regional symptoms may occur, depending on the specific sinuses involved. For example, in sphenoid sinusitis severe headache radiating to the occiput and altered esthesia of fifth cranial nerve dermatomes may be present (Gwaltney, 1996). Pathogenesis of sinusitis usually involves accumulation of fluid in the enclosed cavities as a result either of viral infection or allergic inflammation and subsequent obstruction of normal drainage through the ostia. There may also be a destructive component involved in the pathogenesis with ciliary dysfunction and/or epithelial damage.

This creates an environment suitable for bacterial superinfection from nearby organisms. Members of the normal microflora most commonly identified in acute sinusitis are *Streptococcus pneumoniae, Haemophilus influenzae* and *Branhamella catarrhalis.* Complications from local infections include cavernous sinus or cortical vein thromboses related to sphenoid sinusitis.

Chronic sinusitis implies a duration of illness beyond that of acute infection, i.e. longer than 3 weeks of symptoms. Complaints may be of chronic nasal congestion, nasal discharge or cough. In addition, patients may be infected with organisms that differ from the pathogens associated with acute sinusitis.

Diagnosis of either condition is determined primarily by history and physical findings. However, diagnosis is aided by radiographs, either X-ray (including the Water's view) or computed tomography (CT) showing air–fluid levels or mucosal thickening, implying inflammation. Under certain circumstances, such as persistent sinusitis without improvement on appropriate therapy or sinusitis in immunocompromised patients, microbial cultures are indicated in order to optimize antimicrobial therapy. Sinus biopsy, however, is an invasive procedure and is rarely undertaken. Improvement is generally slow, even in cases of early diagnosis and adequate therapy. Resolution on sequential X-rays may take months. Adjunctive therapy such as nasal decongestants and steroids has not shown consistent benefit.

3.3.3 Otitis media

The middle ear cavity is normally a sterile space. In the presence of fluid and obstruction, however, organisms proliferate, causing either sympathetic effusion or purulent exudate in the enclosed cavity. Typically this disease occurs in the 6 months to 2 years age group in association with viral rhinitis and is among the most common problems requiring medical intervention. Many ear infections are viral in etiology, but without culturing the middle ear exudate these cannot be differentiated from those of bacterial etiology. The bacteria causing otitis media are similar to those associated with sinusitis (Klein, 1994). Infants may present with non-specific symptoms such as irritability, mild cold symptoms, low-grade fever and tugging of the affected ear. The older patient may complain of ear pain or decreased hearing. On physical examination the tympanic membrane is often bulging and erythematous or dull, with abnormal or no light reflex and decreased mobility to insufflation. The difficulty in diagnosis comes from the variety of presentations on examination and the non-specific nature of tympanic membrane erythema. Tympanocentesis – insertion of a sterile needle into the middle ear and obtaining of cultures – is rarely undertaken because of its traumatic nature.

As a result, otitis media is generally treated empirically as if it is bacterial in origin. Resolution of symptoms is usually rapid, although the tympanic membrane may remain abnormal for up to 2–3 weeks despite appropriate therapy and resolution of symptoms. Improvement usually continues over the following weeks without continuation of antibiotics.

3.3.4 Pharyngitis

Pharyngitis is a common diagnosis in all age groups, beginning in early childhood. Etiologies include Epstein–Barr virus, enteroviruses and adenoviruses, as well as bacteria, chiefly *Streptococcus pyogenes*, *Neisseria gonorrhoeae* and toxin-producing *Corynebacterium diphtheriae*. Because of the presence of normal microflora in the oropharynx, it is important to identify those patients who may harbor treatable pathogens as the cause for their sore throat. The classic presentation of group A streptococcal pharyngitis includes the presence of fever and sore throat with an erythematous posterior pharynx, palatal petechiae, mild exudate and cervical lymphadenopathy. This etiology is easily established by throat culture or rapid antigenic detection of this organism. *S. pyogenes* remains uniformly sensitive to penicillin, the drug of choice for therapy. Therapy is important for the prevention of serious post-infectious complications, acute glomerulonephritis and rheumatic fever. Pharyngitis caused by *N. gonorrhoeae* follows oral sexual contact with an infected individual. *C. diphtheriae*, i.e. diphtheria, causes a distinct disease characterized by a thick, adherent, gray exudate on the posterior pharynx.

3.3.5 Tracheitis

Bacterial tracheitis is an uncommon disease in the normal host. Prior to the development and widespread use of *Haemophilus influenzae* type b (Hib) vaccine this infection occurred primarily in the pediatric population. Bacterial tracheitis is still common in patients with a tracheostomy for prolonged mechanical ventilatory or anatomic abnormality. In these cases the patient presents with increase in production or change in quality of tracheal secretions and increased oxygen or ventilatory needs. Common organisms causing tracheitis include *Staphylococcus aureus* but may vary depending on the organisms colonizing the individual patient. Accurate diagnosis may be difficult, particularly in determining the causative pathogen as the trachea is often heavily colonized in these circumstances with multiple species, including many that could act as pathogens. If one organism is predominant, it may be prudent to direct therapy towards that organism, assuming there is overgrowth causing inflammation. In addition, Gram-staining of secretions may assist in affirming the diagnosis, since the presence of many neutrophils indicates inflammation and active disease. A short course of antibiotics may be indicated with follow-up evaluation based on tracheal cultures in conjunction with clinical signs and symptoms.

3.3.6 Bronchitis

Bronchitis is inflammation limited to the bronchi and occurs primarily in adolescents and adults. In infants and children the bronchial anatomy is immature and the smaller airways, the bronchioles, are more often affected by viral infections, particularly respiratory syncytial virus and rhinoviruses. Common presenting symptoms for bronchitis are fever, cough productive of sputum, and shortness of breath, without localizing signs on physical examination and chest X-ray. In contrast, patients with bronchiolitis present with predominantly obstructive airway symptoms, i.e. wheezing. Treatment of bronchitis is usually empiric based on the probable colonizing flora associated with this infection. The most common complication of this disease is pneumonia.

3.3.7 Pneumonia

Disease of the lower respiratory tract, especially when it involves the alveolar spaces, may be more threatening to the host because of its potential to interfere with gas exchange in the lung. Pneumonia remains the sixth leading cause of death in the developed world. It is defined as lobar consolidation of lung tissue with organisms and inflammatory exudate and subsequent collapse of that area. Onset of disease may be subacute in cases with prior viral infection or acute in primary bacterial infection. Hallmarks of presentation are fever, productive cough and dyspnea. Physical examination typically reveals dullness to percussion, diminished aeration, rales on auscultation and egophony over the affected area. Patients may also have a supplemental oxygen requirement, depending on the severity of consolidation within the lung parenchyma. Diagnosis is confirmed by sputum examination and chest radiographs revealing focal areas of opacity. Common etiologic agents of acute pneumonia include *S. pneumoniae*, *S. aureus*, *H. influenzae*, Enterobacteriaceae, *P. aeruginosa*, *Legionella* spp. and oral anaerobes (usually associated with aspiration). Therapy is directed towards the most common pathogens known to cause lower respiratory tract infections, Gramstaining and culture of sputum if these are available. Complications include pleural effusion, abscess formation and bacteremia leading to sepsis, depending on the bacterial etiology.

3.4 EXAMPLES OF ORGANISMS AND THEIR DISEASES

3.4.1 *Streptococcus pneumoniae*

The pneumococcus was first identified separately by George Sternberg and Louis Pasteur about 1880. Since that time the pneumococcus, labeled by Sir William Osler 'the captain of the men of death', has continued to be a major pathogen of the respiratory tract and cause of invasive infection.

(a) Bacterial characteristics

The pneumococcus is a Gram-positive coccus that grows in chains and lancet-shaped pairs and is identified by the following characteristics: lack of catalase production, alpha-hemolysis on blood agar, bile salt solubility and sensitivity to optochin. The organism is generally grown on blood agar, which provides a source of catalase. As is the case with many upper respiratory tract organisms capable of causing invasive disease, the pneumococcus expresses a capsular polysaccharide, which is also an important antigen and the basis for serotyping. Currently 90 serotypes, each with a structurally distinct capsular polysaccharide, are recognized (van Dam *et al.*, 1990). The presence of capsule may be detected by a Quellung reaction in which the organism's capsular polysaccharide interacts with a type-specific antiserum, making the cell appear swollen as the surrounding capsule becomes visible. Within the capsule is another layer of polysaccharide referred to as the C-polysaccharide, which is the peptidoglycan-linked teichoic acid of the pneumococcus (Tomasz, 1981). The surface glycolipid or lipoteichoic acid, also known as the F or Forssman antigen, is anchored in the membrane but is otherwise structurally identical to the C-polysaccharide. The cell-wall-associated and lipoteichoic acids are also unique in the pneumococcus because they have a ribitol rather than a glycerol backbone and because of the presence of choline phosphate or phosphorylcholine (ChoP; Fischer *et al.*, 1993). Choline, a major component of eukaryotic membrane lipids, has an uncommon bacterial structure. Choline, which is acquired from the growth medium, serves as a non-covalent anchor for several cell-surface proteins. One of these proteins is LytA, or autolysin, a murein hydrolase responsible for the marked autolytic behavior characteristic of this species.

(b) Epidemiology of carriage

Nasopharyngeal carriage rates range from 5–10% in healthy adults and 20–40% in healthy children (Austrian, 1986). In developing countries and populations in which there is close contact, such as military quarters and daycare centers, rates of carriage as high as 76% have been documented. Acquisition of pneumococcus in normal microflora begins in infancy and each newly acquired strain may be carried 1–18 months. The simultaneous carriage of as many as five different strains has been reported.

(c) Bacterial factors in carriage

The capsular polysaccharide appears to be important in colonization since unencapsulated isolates are seldom obtained from the nasopharynx. The capsule has been shown to be antiphagocytic but its exact role in promoting carriage remains unclear. Recent discoveries have suggested that the attachment factors for this organism may include both proteins and cell wall components (Geelen *et al.*, 1993; Cundell *et al.*, 1995). The choline moiety appears to be critical to binding proteins acting in attachment, such as choline binding protein A (CbpA), and possibly interacting directly with host receptors (Rosenow *et al.*, 1997). The pneumococcus undergoes phase variation between an avirulent phenotype that colonizes efficiently and a more virulent phenotype that colonizes poorly (Weiser *et al.*, 1994). These phenotypes, distinguished by differences in colony morphology, correlate with variation in the surface expression of ChoP, CbpA and other choline binding proteins, including PspA, a major surface protein of unknown function. In addition, the pneumococcus produces a number of secreted enzymes that alter the host environment and may contribute to colonization. One of these proteins is an IgA_1 protease (a zinc-metalloprotease) that cleaves human IgA_1 at its hinge region, removing the F_c fragment, the portion that contributes to triggering of an inflammatory response (Wani *et al.*, 1996). The precise role of the hyaluronidase and neuraminadase in colonization is less clear-cut.

(d) Host factors in carriage

The molecular details of the interaction with the host mucosal surface are still poorly understood. The pneumococcus appears to bind to multiple different cell-surface receptors. These include a GalNAβ1-4Gal structure found on pharyngeal epithelial cells and several other carbohydrate structures that are not completely characterized (Andersson *et al.*, 1983). There is also evidence of attachment via the receptor for platelet activating factor which is up-regulated during inflammation, as may occur following viral infection (Cundell *et al.*, 1995). Since the natural ligand for this receptor, platelet activating factor, also contains choline phosphate, the pneumococcus may be mimicking this structure to utilize its host cell-surface receptor.

(e) Epidemiology of disease

The pneumococcus is a leading pathogen of the respiratory tract and the most common bacterial cause of otitis media, sinusitis, community-acquired pneumonia and meningitis. There is a strong seasonal variation, which corresponds to peak transmission of respiratory-tract viruses, with the highest incidence in the winter season. The occurrence of bacteremic infection, which carries a considerably higher mortality rate, is predominantly in the first 2 years and last decade of life.

(f) Pathogenesis

Considering the prevalence of pneumococcal carriage, disease is a relatively unusual outcome. What appear to distinguish this species from other members of the microflora of the respiratory tract are the unique characteristics that enable it to survive normal host defenses (Tuomanen *et al.*, 1995). Its major pathogenic mechanism is its ability to trigger an inflammatory response yet escape from phagocytosis. The single known toxin, pneumolysin, does not appear to be an absolute requirement for local or disseminated infection (Berry *et al.*, 1989). The importance of humoral immunity in defense from this organism is demonstrated by the predisposition to pneumococcal infection in hosts with deficiency in this arm of the immune system. Protection from pneumococcal disease correlates with the level of type-specific antibody generated in days during infection or in weeks to months following carriage. In addition, a serum protein, referred to as C-reactive protein (CRP) because it binds the C-polysaccharide, appears to contribute to opsonization and clearance (Horowitz *et al.*, 1987).

  Infants become bacteremic primarily through direct invasion of organisms directly from the nasopharynx (occult bacteremia), whereas the elderly population usually becomes bacteremic as a complication of pneumonia. The route of invasion beyond the mucosal surface is felt to be via the lymphatics, although it is unclear whether this is the only route for dissemination. From the blood stream the pneumococcus is normally cleared by the spleen. Lack of splenic function results in predisposition to rapid and overwhelming pneumococcal sepsis.

(g) Clinical syndromes

Otitis media, sinusitis and bronchitis involve inflammation primarily restricted to the mucosal surfaces. The most common predisposing condition to developing pneumococcal infection of these sites is antecedent or concurrent viral infection. The pneumococcus is responsible for about 50% of acute upper respiratory tract infection of a bacterial etiology (Klein, 1994).

  Although not the most severe form of disease caused by the pneumococcus, pneumonia is the most significant because of its high prevalence and rate of life-threatening complications. Pneumococcal pneumonia is characterized by alveolar consolidation, i.e. local collection of pus and exudate in a lobar area of the lung. Isolation of *Streptococcus pneumoniae* from a sample of respiratory tract secretions is not necessarily diagnostic because of high rates of carriage in asymptomatic hosts. Cultures and Gram-staining, therefore, must be interpreted carefully. Adequate specimens contain purulent material, i.e. sputum that contains fewer than 10 epithelial cells per high power field are most likely to be helpful in diagnosis. A pure culture of pneumococcus from an acceptable sputum sample or a blood culture positive for the same organism would be highly suggestive of pneumococcal infection. An alternative method for obtaining culture sample would be to perform a bronchoalveolar lavage via bronchoscopy, although this more invasive procedure requires endotracheal intubation, making it unsuitable except in complicated or severe cases.

Once it has invaded the bloodstream, either through the upper respiratory tract or via a lower respiratory tract infection, the organism has the ability to establish infection in normally sterile sites such as the meninges. With the decline in the incidence of meningitis due to *Haemophilus influenzae* type b following the introduction of vaccination against this organism, the pneumococcus has become the leading cause of bacterial meningitis in the pediatric as well as the adult population. The classic presentation of bacterial meningitis in the adult patient includes a constellation of symptoms: fever, headache, photophobia and vomiting associated with physical findings such as nuchal rigidity and positive Brudzinski and Kernig's signs. However, in infants, who have the highest incidence of this disease, the presentation may be very subtle, with non-specific symptoms such as fever and irritability or lethargy. Clinical signs in this population may include bulging of the anterior fontanel but may or may not include nuchal rigidity from meningeal irritation. Diagnosis of pneumococcal meningitis relies on a cerebrospinal fluid sample with pneumococcus grown in culture. Other more rapid tests using antigen detection have been developed to attempt earlier diagnosis, but these techniques are less specific and less sensitive. Of the common forms of bacterial meningitis, pneumococcal meningitis has the highest rate of complications such as hearing loss, seizures, hydrocephalus, brain damage and death.

A variety of other sites may be infected following bacteremic dissemination. Like the meninges, these sites are often in areas where immune clearance is relatively deficient. Infections involving the joint fluid or bone may occur in normal host whereas the heart valves or peritoneal fluid may be involved in patients predisposed to seeding at these sites.

(h) Treatment

Treatment of pneumococcal infections was once considered straightforward because the organism was uniformly susceptible to penicillin.

In recent years, however, there has been acquisition and widespread dissemination of penicillin- and multi-antibiotic-resistant strains that has considerably complicated treatment (Austrian, 1994). Resistance to penicillin has occurred through mutation in penicillin binding proteins leading to stepwise increases in amounts of beta-lactam antibiotics required for killing (Tomasz, 1995). Isolates with intermediate levels of resistance (MIC from 0.1–2 µg/ml) represent up to 70% of strains in certain populations of some regions. There is an association with overuse of oral antibiotics and the prevalence of resistant strains. Of greater concern are the highly resistant isolates (MIC $\geqslant 2.0$ µg/ml), which are less common but would be difficult to treat with achievable doses of penicillin. Many of the penicillin-resistant strains are resistant to the cephalosporins as well as to antibiotics unrelated to penicillins, such as erythromycin and the sulfa agents. In treating infection outside the central nervous system, except for highly resistant strains, penicillin and cephalosporins remain the drugs of choice. For meningitis, the difficulty in delivering sufficient drug past the blood–brain barrier makes it necessary to include vancomycin in initial therapy until susceptibility testing is completed. Vancomycin is the only available agent to which resistance has not yet been reported, but the potential for the development of resistance as has occurred in the enterococcus is worrying. The spread of antibiotic resistance has increased the need for accurate diagnosis based on culture and sensitivity testing. Specific treatment recommendations are evolving and should reflect local resistance patterns. Long-term penicillin prophylaxis has been used for patients at high risk of overwhelming infection (sickle cell disease, asplenia) but its usefulness in the era of widespread resistance is no longer clear. The use of corticosteroids in pneumococcal meningitis to interrupt the inflammatory cascade is controversial.

(i) Prevention

Given the growing problem of multidrug resistance, immunization against pneumococcus would be desirable in populations at greatest risk (the young, the elderly, the immunosuppressed). A vaccine consisting of purified capsular polysaccharide of the 23 most common types has been shown to be of benefit in protecting against pneumococcal bacteremia (Butler *et al.*, 1993). However, actual success with the 23-valent pneumococcal vaccines has been less than optimal for two reasons: despite the recommendation for universal vaccination at age 65, acceptance of the vaccine has been poor. In addition, the population in which immunization would be most beneficial, i.e. children under the age of 2 years, does not mount an antibody response to polysaccharide antigens. Efforts to conjugate the most common capsular polysaccharides to protein carriers to induce immunity in this population as was done for the Hib vaccine are now in progress.

3.4.2 *Haemophilus influenzae*

(a) Bacterial characteristics

Haemophilus influenzae is a Gram-negative, pleomorphic rod found only in humans, which, when first isolated, was felt to be the causative agent of influenza. It requires two factors acquired from blood for aerobic growth: X factor, which contains iron-containing pigments providing protoporphyrins for specific enzyme activity, and V factor, which is a coenzyme in the form of nicotinamide adenine dinucleotide (NAD), nicotinamide adenine dinucleotide phosphate (NADP) or nicotinamide nucleoside. Optimal growth occurs on chocolate agar, which is blood agar media with lysed or digested red blood cells to provide these nutrients. *Haemophilus* exists in both encapsulated (types a–f) and non-encapsulated or non-typable forms.

(b) Epidemiology of nasopharyngeal carriage

Colonization with this organism occurs early in life, with each strain being carried for days to months as in the case of the pneumococcus. Non-typable *Haemophilus* are carried by 50–80% of individuals, whereas encapsulated strains colonize only 3–5% of hosts.

(c) Bacterial factors in carriage

The vast majority of *Haemophilus* isolates are non-typable. The capsular polysaccharide renders the organism resistant to phagocytosis and the bactericidal activity of complement and antibody, but it is not required for colonization of humans. About 75% of non-typable strains express one of two high-molecular-weight proteins (HMW1 and HMW2) that function in adherence to epithelial cells (Barenkamp and St Geme, 1996). All strains appear capable of expressing additional adhesins, which include fimbriae with binding specificity for the Anton blood group antigen. The ability to undergo phenotypic or antigenic variation appears to be particularly important for this organism (Hood *et al.*, 1996). *Haemophilus* synthesizes a rough lipopolysaccharide with marked intra- and interstrain structural heterogeneity in its composition due to differences in the linkage of saccharide units in the outer core (Weiser, 1993). The LPS is decorated with phosphorylcholine, which is antigenically indistinguishable from the ChoP on the pneumococcal teichoic acids. ChoP has also been found on the surface glycolipid of *Mycoplasma* spp., suggesting that

Haemophilus, pneumococcus and *Mycoplasma* may have a common mechanism involving this host-like structure in their pathogenesis. Structural variation of the LPS is generated by multiple translational switches created by highly repetitive DNA sequences in multiple genes involved in expression of the LPS (Weiser *et al.*, 1989). The cell-surface expression of fimbriae also displays phase variation; however, the molecular switch responsible is transcriptional (van Ham *et al.*, 1993). Complete sequencing of the *H. influenzae* genome has shown that highly repetitive sequences which generate molecular switches are a common feature of many genes, especially those involved in the expression of cell surface components and acquisition of iron (Fleischmann *et al.*, 1995). Phase variation confers a versatility on the organism that appears to be important for its adaptation to different host environments or evasion of host defenses. Like the pneumococcus, *H. influenzae* expresses an IgA_1 protease, which cleaves human IgA_1 in its hinge region (Plaut, 1983). The *H. influenzae* enzyme is a serine-type endopeptidase, suggesting that this species and *S. pneumoniae* have evolved distinct mechanisms for equivalent functions.

(d) Host factors in colonization

The organism appears to attach preferentially to the mucous layer, which may promote local dissemination (Moxon and Wilson, 1991). Organ culture studies suggest that the organism preferentially adheres to areas on the mucosal surface where there has been interruption of the epithelial cell layer. Bacterial factors other than LPS that promote an inflammatory response and local damage are not well understood.

(e) Epidemiology of disease

Historically, infection by encapsulated *Haemophilus*, chiefly caused by organisms with the polyribose phosphate (PRP) type b capsular polysaccharide, has been especially common in young children. Prior to the introduction of the *Haemophilus influenzae* type b (Hib) vaccine, this organism was the leading cause of bacterial meningitis, epiglottitis, septic arthritis and orbital/periorbital cellulitis. The incidence of meningitis was 1.24 per 100 000 population per year in the USA. Since widespread immunization with PRP, and in particular conjugated forms of PRP, the incidence of invasive infection caused by *H. influenzae* type b has declined precipitously. One of the unexpected consequences of immunization has been a drastic reduction in Hib carriage. Success in preventing Hib infection, however, has not altered the incidence or spectrum of disease due to non-typable *H. influenzae*, which is still a major factor in disease confined to the upper respiratory tract and in chronic bronchitis. Indeed non-typable *H. influenzae* now accounts for about 40% of otitis media cases of bacterial origin. Since the dramatic decrease in Hib-related infections, there has been a notable rise in incidence of Hib-like infections such as septic arthritis and orbital cellulitis, caused by other types of *Haemophilus influenzae* including types a, c and f. Accurate diagnosis of *H. influenzae* infection in the upper respiratory tract is challenging because of the frequency of carriage in normal hosts.

(f) Pathogenesis

Disease results from contiguous spread within the respiratory tract in susceptible hosts. Host factors that predispose to infection such as antecedent viral disease, aspiration or eustachian tube dysfunction are similar to those described for the pneumococcus. Specific antibody is a key element in host defense. Individuals with deficient antibody production are at increased risk of non-typable and Hib infection. In the unvaccinated host, susceptibility to Hib infection wanes later in childhood as a result of acquisition of anti-PRP antibody. Persons with decreased splenic function are at risk for overwhelming invasive Hib infection as is the case with other encapsulated respiratory pathogens. Disease outside the respiratory tract is generally a result of bacteremic spread and is unusual for non-typable isolates.

(g) Clinical syndromes

The spectrum of disease caused by *Haemophilus influenzae* is determined by the presence of type b capsule. Hib is a potentially more invasive organism whereas non-typable strains may cause generally less severe infections confined to the respiratory tract. There are, however, reports of unencapsulated strains capable of causing disseminated infection.

Like other commensals of the nasopharynx, *Haemophilus influenzae* has access to nearby structures in the middle ear and sinuses, particularly when there is mucosal compromise due to previous viral infection and accumulation of fluid and inflammation preventing adequate drainage. In the lower respiratory tract non-typable strains are a relatively common cause of chronic bronchitis in older individuals and, should clearance be impaired, pneumonia may develop. The onset of symptoms is usually insidious, in contrast to pneumococcal pneumonia. Although abscess formation is not common, bacteremia with *Haemophilus influenzae* pneumonia is common only for Hib.

Cellulitis of the periorbital and orbital regions is typically caused by *Haemophilus influenzae* type b. These syndromes are distinguished by clinical presentation significant for exophthalmos and impaired extraocular movement. Local invasion may lead to central nervous system involvement.

The type b strains have a predilection for setting up inflammation and infection of the epiglottis for reasons that remain unclear. The incidence of this disease has decreased dramatically since the implementation of routine Hib vaccination. The classic presentation of epiglottitis includes an anxious-appearing child, usually 2 years of age or older, who is acutely ill, febrile, sitting upright and leaning forward, with the mouth partly open, drooling to avoid swallowing against an inflamed epiglottis. This constitutes a medical emergency warranting swift attention for intubation or tracheostomy to be performed before complete obstruction of the airway occurs. The differential diagnosis of this presentation includes processes such as retropharyngeal cellulitis or peritonsillar abscess and laryngotracheobronchitis (croup).

Once beyond normal clearing mechanisms of the respiratory tract, Hib may gain access to the blood stream, causing bacteremia, sepsis and disseminated infection such as meningitis. The clinical presentation of meningitis is indistinguishable from that caused by other bacteria. Sequelae and complications of this disease are well known and have been the major impetus for developing an effective Hib vaccine in recent years. Other infections complicating Hib bacteremia include septic arthritis and osteomyelitis.

(h) Treatment

Infections of the upper respiratory tract are often self-limiting. Antimicrobial therapy, however, remains the mainstay of therapy for disease due to *H. influenzae*. In the past ampicillin has been an excellent choice for treating *Haemophilus* infections. Unfortunately, about one in four strains is now resistant to ampicillin because of inactivation by beta-lactamases. Also, fear of the idiosyncratic reaction (aplastic anemia) associated with the use of chloramphenicol has led to decrease in the use of this agent. Currently, the drug of choice in empiric therapy of seriously ill patients is a second- or third-generation cephalosporin. These are not affected by the beta-lactamases of *H. influenzae*. Of these only the third-generation cephalosporins are able to penetrate the blood–brain barrier adequately to treat meningitis. Once the organism has been cultured and *in vitro* sensitivity testing performed, more specific antibiotic coverage may be selected. Culture results and sensitivity testing are seldom available in selecting agents for therapy of upper respiratory tract infections. For involvement of these sites the sulfa drugs and the newer macrolides are suitable alternatives for non-parenteral therapy.

Extensive studies have been performed in the pediatric population to examine the prior or concomitant use of steroids to block the CNS damage from the inflammatory cascade that could follow bacterial lysis following antibiotic administration. In the case of Hib meningitis steroid use has been shown to decrease incidence of hearing loss, the most common sequela associated with this infection. This has led to the recommendation that steroids be administered prior to or at the time of giving the first dose of antibiotics in situations of suspected Hib meningitis.

(i) Prevention strategies

As previously mentioned, immunization with conjugated Hib vaccine, which began about 1989, has dramatically reduced the incidence of Hib infection in vaccinated populations. There are several different preparations of the vaccine, each based on different carrier proteins, given in several doses to children beginning about 2 months of age. The addition of the protein carrier renders the PRP immunogenic in the pediatric population, which is at highest risk for serious infection and fails to respond to polysaccharide antigen alone. In cases of invasive Hib infection, administration of prophylaxis, which effectively eliminates carriage and further transmission of the organism, is indicated for close contacts (including family members and day-care contacts) in situations where susceptible hosts may be exposed. There are currently no measures available for prevention of non-typable *H. influenzae* disease.

3.4.3 *Neisseria meningitidis*

N. meningitidis, or the meningococcus, is rarely recognized as a cause of disease within the respiratory tract. This organism, however, is responsible for severe forms of invasive infection, including epidemic disease, following dissemination from its site of residence in the nasopharynx.

(a) Characteristics

This organism is a Gram-negative diplococcus considered to be fastidious in its *in vitro* growth requirements. It grows aerobically, although it favors 5–8% carbon dioxide to support ion transport in its moist environment. Media that allow the meningococcus to grow include blood agar, supplemented chocolate agar, tryptic soy and Mueller–Hinton agar. Its pattern of utilization of carbohydrates distinguishes it from non-pathogenic species of the same genus as it produces acid from glucose and maltose but not sucrose. As is the case with the pneumococcus, the meningococcus undergoes autolysis through production of autolysin. In addition it exists in an encapsulated form, which can be detected by the Quellung reaction using appropriate capsule-specific antiserum. There are 13 different capsular polysaccharides, serotypes A, B, C, D, X, Y, Z, E, W-135, H, I, K and L, of which only eight commonly cause disease in humans. The capsules are distinct in repeating units of polymers of carbohydrate structures such as N-acetyl neuraminic acid with 2,8 linkage in serogroup B and O,N-acetyl neuraminic acid with 2,9 linkage in serogroup C.

(b) Epidemiology of colonization

Like the other residents of the respiratory tract, the meningococcus exists in the nasopharynx, whence it is transmitted through respiratory secretions from host to host. Most persons colonized with the organism harbor non-groupable meningococci that are not pathogenic. There is a subset of the population that carries one of the more pathogenic, encapsulated strains of meningococcus; however, the vast majority of hosts remain completely asymptomatic for the duration of colonization. The duration of carriage is usually months. Some studies have shown endemic meningococcal disease to be due to the wide variability of serotypes whereas epidemic disease is usually associated with a single serotype.

(c) Bacterial factors in colonization

Both unencapsulated and encapsulated forms are found in the human nasopharynx, but only the encapsulated organisms have the capacity for invasive infection. This is because of the role of the capsular polysaccharide in preventing clearance, as has been discussed for the pneumococcus and capsulated *H. influenzae* (Broud, Griffiss and Baker, 1979). Beneath the capsular layer is an outer membrane made up of various proteins and LPS or lipo-oligosaccharide (LOS). The LOS undergoes structural or antigenic variation and is sialylated, a characteristic that has been shown to contribute to serum resistance (Mandrell *et al.*, 1991; Gotschlich, 1994). The meningococcus expresses fimbrial adhesins, which also display both on–off and antigenic variations (Virji *et al.*, 1993). Other structural proteins of the meningococcus include multiple opacity proteins (Opa), which exhibit on–off phase variation and act as virulence factors by mediating adhesion to the epithelium by binding to the carcinoembryonic antigen, CD66a, molecule (Virji *et al.*, 1996). Another protein called PIII is present on the surface of the *Neisseria*. It serves as a protein to which specific antibody may bind and paradoxically inhibit killing of the organism by the immune system (Rice *et al.*, 1986). The meningococcus synthesizes an IgA_1 protease, which resembles that of *H. influenzae* (Koomey and Falkow, 1984).

(d) Epidemiology of disease

The rapid and fulminant nature of some forms of meningococcal disease together with its often epidemic spread are the cause of considerable anxiety. The highest attack rates tend to occur in the late winter and early spring. Epidemics continue to occur worldwide each year in both developed and developing countries, although individual serotypes that cause these epidemics may vary depending on the sanitary conditions and population density of the locale. Non-epidemic meningococcal infections occur in developed countries at an incidence of 1 per 100 000 population per year. Baseline mortality rates, apart from concurrent epidemics in developed countries, range from 7% for meningitis to 19% for meningococcemia (Broud, Griffiss and Baker, 1979).

In endemic infection disease most often occurs in children below the age of 5 years. During epidemics the group most affected is slightly older, 5–19 years, because of increased transmission rates in this group. Close contact in day-care and nursery school situations provide a favorable environment for transmission of respiratory droplets from asymptomatic carriers to susceptible, naive hosts. Also, the household attack rate of about 1% is much higher than that of the general population. Persons living in close quarters such as the military or college dormitories/halls of residence may be at higher risk of acquiring pathogenic strains because of their close proximity. Disease, although an unusual outcome, will most often occur through a host acquiring a serotype to which s/he has not previously been exposed. Serotypes that most commonly cause epidemics are serogroups A, B, and C. Patterns of attack rate and studies of susceptible populations have shown that type A epidemics occur in less developed nations, type B occur in developed countries and type C occur in both settings (Broud, Griffiss and Baker, 1979). Attack rates in type A and C epidemics are five to ten times those of type B epidemics, which may range from 50–100 cases per 100 000 persons.

(e) Pathogenesis

The meningococcus has the ability to colonize mucosa with little apparent effect on the host, but once within the bloodstream its behavior can be extremely virulent, even to the immunocompetent individual. Bacterial structures related to colonization and evasion of host defenses have been noted. Factors other than endotoxin that contribute to the fulminant nature of invasive infection and the picture of hemodynamic compromise, capillary leakage and disseminated intravascular coagulopathy are not well understood.

Protection against invasive infection correlates with the level of serum antibody. These antibodies are not present in the infant over 6 months of age, after maternal antibody has waned, until about 2 years of age, when adequate antibody response to polysaccharide antigen may be mounted. This scenario provides a window of opportunity for an organism such as the meningococcus to invade and survive unchallenged by a mature immune response. Normally, antibodies formed against a combination of capsular polysaccharides and cell wall antigens will opsonize the bacteria and either complex with complement to destroy the organism or lead to phagocytosis by circulating neutrophils and macrophages. However, should inadequate amounts of antibody and/or complement be present, the bacteria's growth remains unchecked. Deficiency in terminal components of the complement cascade, in particular, is associated with the development of chronic meningococcemia and invasive disease (Densen, 1991). The carrier state has been shown to be an immunizing process. There appears to be some degree of cross-reactivity, with antibodies to non-pathogenic strains of meningococcus and non-meningococcal species eliciting a bactericidal effect against encapsulated meningococcal strains. Those individuals recently acquiring pathogenic strains yet lacking appropriate capsule-specific antibodies in their serum are at highest risk of meningococcemia (Goldschneider, Gotschlich and Artenstein, 1969). Infection often occurs in the setting of a preceding viral syndrome involving the upper respiratory tract.

(f) Clinical syndromes

There are at least four distinct clinical syndromes associated with meningococcal infection. Although the presentations of the different syndromes appear similar, the natural course and outcome of each process is distinctive.

Meningococcal bacteremia without signs of sepsis is the least dramatic presentation. Patients usually have problems such as upper respiratory tract symptoms or fever and are found to have meningococcus growing in blood culture. This process may resolve spontaneously without antibiotic therapy.

Meningococcemia is a syndrome that often portends a grave prognosis, as onset is usually very acute and often dramatic. Patients have bacteremia but are also septic, i.e. they have hypotension and general malaise. Examination of the skin is of critical importance in this infection as one of the earlier clues to the disease is the presence of petechiae and purpura on the trunk, lower body and points of pressure. When these lesions coalesce, they appear to be ecchymotic. Embolic lesions, in addition, may appear secondary to a state of disseminated intravascular coagulopathy. Hemodynamic instability causes poor perfusion of peripheral extremities, which may become severe with compromise of the viability of the fingers and toes and eventual vascular collapse.

Meningococcal meningitis usually presents without signs of meningococcemia. The key differences are the symptoms and clinical signs attributable to central nervous system infection (headache, fever and signs of meningeal irritation) associated with cerebrospinal fluid findings consistent with meningitis (elevated white cell count, low glucose and high protein). Focal neurological signs are not common. Meningoencephalitis due to meningococcus is the culmination of the above presentation of meningitis with an altered mental status or obtundation. These patients also have abnormal deep tendon reflexes and may have loss of suppression of primitive reflexes, the so-called pathologic reflexes. The prognosis in meningococcal meningitis is considerably better than in meningococcemia and other forms of bacterial meningitis discussed in this chapter.

In addition to these presentations a more chronic form of infection, called chronic meningococcemia, may occur in which symptoms are as non-specific as intermittent or low-grade fever and rash that lasts for weeks to months. In this setting other more focal infections such as septic arthritis or soft-tissue abscesses may occur in the absence of an acute fulminant illness.

(g) Treatment

Because of the potentially serious nature of meningococcal infections, early initiation of therapy, both antimicrobial and supportive care, is essential to preserving a good outcome. Although the bactericidal effect may be adequate with appropriate therapy, fulminant meningococcemia is still associated with a poor prognosis, even with appropriate and timely care. To date there have been only limited reports of penicillin-resistant meningococci worldwide. Therefore, intravenous penicillin, for a duration of 5–7 days, remains the treatment of choice for meningococcal infections. Alternative antibiotics include third-generation cephalosporins such as cefotaxime or ceftriaxone, or chloramphenicol. The use of steroids in meningococcal meningitis has not been established. Prophylaxis of intimate and household contacts is important to prevent secondary cases. The goal is to eliminate nasopharyngeal carriage of the organism from the index case as well as close contacts. Recommended antibiotics for prophylaxis include rifampicin, ceftriaxone or ciprofloxacin. Since secondary cases may occur for many weeks, immunoprophylaxis with polyvalent meningococcal vaccine may be indicated in outbreaks, depending on the serogroup.

(h) Prevention strategies

The current vaccine consists of purified polysaccharide of serogroups A, C, Y and W-135. There is no currently available vaccine for serogroup B disease as the O,N-acetyl neuraminic acid is poorly immunogenic. Immunization with the meningococcal vaccine is indicated in high-risk groups such as asplenic patients and those with deficiency in terminal complement components. Also, persons traveling to areas with high rates of epidemic or endemic disease should receive the vaccine prior to departure.

Rates of infection are not considered to be high enough to justify wide-spread immunization in most developed countries. As with the pneumococcus, conjugate vaccines capable of targeting the population at greatest risk are now in clinical trials.

3.4.4 Other respiratory tract pathogens

The organisms discussed above are among the most pathogenic organisms that also exist commonly as colonizers of the upper respiratory tract. However, this list should not be considered exhaustive; there are many bacteria that colonize the nasopharynx but cause disease less commonly or infection that is generally less severe. Among these is *Branhamella catarrhalis*, a Gram-negative organism that is often implicated in the etiology of otitis media, sinusitis and pneumonia. It may colonize the upper respiratory tract, but it rarely leads to invasive disease. The Gram-positive organism *Staphylococcus aureus* is frequently a member of the normal microflora and has the capacity to elaborate a number of toxins. When it causes disease such as pneumonia, it can be a complicated, necrotic process, often with pleural effusion and abscess formation. Infection with *S. aureus* is also problematic because this organism is often resistant to the antibiotics commonly used to treat other respiratory tract infections.

Mycoplasma pneumoniae is a pathogen, not considered a member of the normal microflora, which is responsible for a large proportion of cases of community-acquired or 'walking pneumonia'. *Chlamydia pneumoniae*, an obligate intracellular bacterium, is also a more recently recognized pathogen causing community-acquired pneumonia. The most well known pathogen of the oropharynx, *Streptococcus pyogenes*, is not a part of the normal microflora and its presence often correlates with disease in the pharynx. Long-term persistence of the organism in the posterior pharynx despite appropriate antibiotic treatment has been ascribed to the concomitant presence of beta-lactamase producing organisms such as *S. aureus*. In such cases, therapy should include beta-lactamase-resistant antibiotics.

There are some transient inhabitants of the nasopharynx that are associated with distinct disease syndromes, such as *Bordetella pertussis*, which causes pertussis or whooping cough, and *Corynebacterium diphtheriae*, which causes diphtheria. As a result of widespread vaccination based on the toxins secreted by these organisms, the overall incidence of these infections is now low, although transmission of these organisms continues in the developed world. *B. pertussis* is frequently transmitted by adults, who may suffer from a prolonged but mild coughing illness, to susceptible infants and children who develop the well-known syndrome of whooping cough. Diphtheria has re-emerged in parts of the world where rates of immunization are no longer adequate and the adult population has become susceptible.

3.5 SUMMARY

The nasopharynx provides a home for many different bacteria that live in harmony with the human host under normal circumstances. Because of a variety of host factors this relationship may be threatened and mucosal barriers breached. Among the many residents of the upper respiratory tract only a few species, which share the ability to evade the humoral immune defense system, are particularly adept at taking advantage of disturbances in the host. For these organisms there is a delicate balance between coexistence as a member of the microflora and infection of the respiratory tract and beyond.

3.6 REFERENCES

Andersson, B., Dahmen, J. *et al.* (1983) Identification of an active disaccharide unit of a glycoconjugate receptor for pneumococci attaching to human pharyngeal epithelial cells. *Journal of Experimental Medicine*, **158**, 559–570.

Austrian, R. (1986) Some aspects of the pneumococcal carrier state. *Journal of Antimicrobial Chemotherapy*, **18**(Suppl. A), 35–45.

Austrian, R. (1994) Confronting drug-resistant pneumococci. *Annals of Internal Medicine*, **121**, 807–809.

Barenkamp, S. J. and St Geme, J. W. (1996) Identification of a second family of high-molecular weight adhesion proteins expressed by non-typable *Haemophilus influenzae*. *Molecular Microbiology*, **19**(6), 1215–1223.

Berry, A. M., Yother, J. *et al.* (1989) Reduced virulence of a defined pneumolysin-negative mutant of *Streptococcus pneumoniae*. *Infection and Immunity*, **57**(7), 2037–2042.

Broud D. D., Griffiss, J. M. and Baker C. J. (1979) Heterogeneity of serotypes of *Neisseria meningitidis* that cause endemic disease. *Journal of Infectious Diseases*, **140**, 465–470.

Butler, J. C. *et al.*(1993) Pneumococcal polysaccharide vaccine efficacy. *Journal of the American Medical Association*, **270**(15), 1826–1831.

Cooperstock, M. (1992) Indigenous flora in host economy and pathogenesis, in *Textbook of Pediatric Infectious Diseases*, 3rd edn, (eds R. D. Feigin and J. D. Cherry), W. B. Saunders, Philadelphia, PA, vol. 1, pp. 91–119.

Cundell, D. R., Gerard, N. P. *et al.* (1995) *Streptococcus pneumoniae* anchor to activated human cells by the receptor for platelet-activating factor. *Nature*, **377**, 435–438.

Densen, P. (1991) Complement deficiencies and meningococcal disease. *Clinical and Experimental Immunology*, **86**(Suppl. 1), 57–62.

Diamond, G., Zasloff, M. *et al.* (1991) Tracheal antimicrobial peptide, a cysteine-rich peptide from mammalian tracheal mucosa: peptide isolation and cloning of a cDNA. *Proceedings of the National Academy of Sciences of the USA*, **88**, 3952–3956.

Fischer, W., Behr, T. *et al.* (1993) Teichoic acid and lipoteichoic acid of *Streptococcus pneumoniae* possess identical structures: investigation of teichoic acid (C polysaccharide). *Biochemistry*, **215**, 851–857.

Fleischmann, R. D. *et al.* (1995) Whole-genome random sequencing and assembly of *Haemophilus influenzae* Rd. *Science*, **269**, 497–512.

Geelen, S., Bhattacharyya, C. *et al.* (1993) The cell wall mediates pneumococcal attachment to and cytopathology in human endothelial cells. *Infection and Immunity*, **61**(4), 1538–1543.

Goldschneider I., Gotschlich, E. C. and Artenstein, M. S. (1969) Human immunity to the meningococcus. I. The role of the humoral antibody. *Journal of Experimental Medicine*, **129**, 1307–1326.

Gotschlich, E. C. (1994) Genetic locus for the biosynthesis of the variable portion of *Neisseria gonorrhoeae* lipooligosaccharide. *Journal of Experimental Medicine*, **180**, 2181–2190.

Gwaltney, J. (1996) Acute community-acquired sinusitis. *Clinical Infectious Diseases*, **23**, 1209–1225.

Hood, D. W., Deadman, M. E. *et al.* (1996) DNA repeats identify novel virulence genes in *Haemophilus influenzae*. *Proceedings of the National Academy of Sciences of the USA*, **93**, 11121–11125.

Horowitz, J., Volanakis, J. E. *et al.* (1987) Blood clearance of *Streptococcus pneumoniae* by C-reactive protein. *Journal of Immunology*, **138**(8), 2598–2603.

Klein, J. O. (1994) Otitis media. *Clinical Infectious Diseases*, **19**, 823–833.

Koomey, J. M. and Falkow, S. S. (1984) Nucleotide sequence homology between the immunoglobulin A1 protease genes of *Neisseria gonorrhoeae*, *N. meningitidis* and *Haemophilus influenzae*. *Infection and Immunity*, **43**, 101–107.

Mandrell, R. E., Kim, J. J. *et al.* (1991) Endogenous sialylation of the lipooligosaccharides of *Neisseria meningitidis*. *Journal of Bacteriology*, **173**, 2823–2832.

Moxon, E. R. and Wilson, R. (1991) The role of *Haemophilus influenzae* in the pathogenesis of pneumonia. *Reviews in Infectious Diseases*, **13**(6), S518–S527.

Plaut, A. G. (1983) The IgA1 proteases of pathogenic bacteria. *Annual Review of Microbiology*, **37**, 603–622.

Rice, P. A., Vayo, H. E., Tam, M. R. and Blake, M. S. (1986) Immunoglobulin G antibodies directed against protein III block killing of serum resistant *Neisseria gonorrhoeae* by immune sera. *Journal of Experimental Medicine*, **164**, 1735–1748.

Rosenow, C., Ryan, P. *et al.* (1997) Contribution of novel choline-binding proteins to adherence, colonization and immunogenicity of *Streptococcus pneumoniae*. *Molecular Microbiology*, **25**, 819–829.

Tomasz, A. (1981) Surface components of *Streptococcus pneumoniae*. *Reviews of Infectious Diseases*, **3**(2), 190–210.

Tomasz, A. (1995) Pneumococcus at the gates. *New England Journal of Medicine*, **333**(8), 514–515.

Tramont, E. C. and Hoover, D. L. (1995) Host defense mechanisms, in *Mandell, Douglas and Bennett's Principles and Practice of Infectious Diseases*, 4th edn, (eds G. L. Mandell, J. E. Bennett and R. Dolin), Churchill Livingstone, New York, vol. 1, pp. 30–35.

Tuomanen, E. I., Austrian, R. *et al.* (1995) Pathogenesis of pneumococcal infection. *New England Journal of Medicine*, **332**, 1280–1284.

van Dam, J. E. G., Fleer, A. *et al.* (1990) Immunogenicity and immunochemistry of *Streptococcus pneumoniae* capsular polysaccharides. *Antonie van Leeuwenhoek*, **58**, 1–47.

van Ham, S. M., van Alphen, L. *et al.* (1993) Phase variation of *H. influenzae* fimbriae: transcriptional control of two different genes through a variable combined promoter region. *Cell*, **73**, 1187–1196.

Virji, M., Saunders, J. R. *et al.* (1993) Pilus-facilitated adherence of *Neisseria meningitidis* to human epithelial and endothelial cells: modulation of adherence phenotype occurs concurrently with changes in primary amino acid sequence and the glycosylation status of pilin. *Molecular Microbiology*, **10**(5), 1013–1028.

Virji, M., Watt, S., Barker, S. *et al.* (1996) The N-domain of the human CD66a adhesion molecule is a target for Opa proteins of *Neisseria meningitidis* and *Neisseria gonorrhoeae*. *Molecular Microbiology*, **22**(5), 929–939.

Wani, J., Gilbert, J. *et al.* (1996) Identification, cloning and sequencing of the immunoglobulin A1 protease gene of *Streptococcus pneumoniae*. *Infection and Immunity*, **64**, 3967–3974.

Weiser, J. N. (1993) The oligosaccharide of *Haemophilus influenzae*. *Microbial Pathogenesis*, **13**, 335–342.

Weiser, J. N., Love, J. M. *et al.* (1989) The molecular mechanism of phase variation of *H. influenzae* lipopolysaccharide. *Cell*, **59**, 657–665.

Weiser, J. N., Austrian, R. *et al.* (1994) Phase variation in pneumococcal opacity: relationship between colonial morphology and nasopharyngeal colonization. *Infection and Immunity*, **62**(6), 2582–2589.

Youmans, G. (1986) Host–bacteria interactions: external defense mechanisms, in *The Biological and Clinical Basis of Infectious Diseases*, 3rd edn, W. B. Saunders, Philadelphia, PA. pp. 8–17.

4

Dental caries

Howard F. Jenkinson

4.1 INTRODUCTION

The human oral cavity contains several types of epithelial surfaces and is the only site in the body that contains hard non-shedding natural surfaces (on teeth) for microbial colonization. Teeth are composed of four tissues: pulp, dentine, cementum and enamel. Dentine is the major component of the tooth: it supports the outer highly calcified enamel and protects the inner pulp, which contains nerve cells and blood supply. Cementum covers the roots of the tooth. Teeth possess a number of distinct surfaces that may be colonized by microorganisms. Pits and fissures on the biting (occlusal) surfaces frequently carry large populations of bacteria. Smooth surfaces are more exposed to environmental cleansing forces and tend to be colonized by a more limited number of bacteria. When recession of the gingival (gum) tissues occurs, as it does with natural aging, the cementum may become exposed and then colonized by microorganisms.

Dental plaque is the complex microbial community found on oral surfaces embedded in a matrix of host salivary components and bacterial extracellular products. The association of dental plaque with dental caries (tooth decay) and periodontal disease has prompted much research into the mechanisms responsible for bacterial adhesion to oral surfaces. The initial stages of adhesion involve interactions between the surfaces of the bacteria and the components of the acquired pellicle adsorbed to enamel. The acquired pellicle contains salivary glycoproteins, antibodies, serous proteins originating as an exudate in gingival crevicular fluid, bacterial products, dietary components and other compounds entering saliva from

G. W. Tannock (ed.), *Medical Importance of the Normal Microflora*, 74–100.

gastric or respiratory reflux. Pioneer bacteria, identified as chiefly *Streptococcus, Neisseria, Haemophilus, Actinomyces* and *Capnocytophaga*, interact directly with the acquired pellicle. There then follows a gradual increase in the complexity of the microflora with coadhesion of bacteria leading to film formation, and eventually to a climax community of high species diversity containing many filamentous organisms. Dental caries is caused by acid, produced by the metabolism of plaque and associated bacteria, attacking the tooth enamel. Caries is probably initiated by the combined activities of a number of bacterial species, but there is strong evidence that progression of the disease is promoted by *Streptococcus mutans*, and by other mutans group streptococci, e.g. *Streptococcus sobrinus*.

4.2 THE MANY FACETS OF CARIES AETIOLOGY

Fitzgerald and Keyes, working at the National Institutes of Health, Bethesda, MD in the late 1950s, were the first to demonstrate, in carefully controlled studies using rodents, that dental caries was an infectious disease caused by streptococci. It was not until 1968, however, that the caries-inducing streptococci were recognized as *Streptococcus mutans*, the same bacterium described by Clarke over 40 years earlier in the UK (Tanzer, 1992). The organism *S. mutans*, of which there are three serotypes, designated c, e and f, is a member of the mutans group streptococci. Species in this group share common phenotypic traits and include *Streptococcus cricetus* (serotype a), *Streptococcus rattus* (serotype b), *Streptococcus macacae* and *Streptococcus ferus* (serotype c), *Streptococcus sobrinus* (serotypes d or g) and *Streptococcus downei* (serotype h). The serotyping scheme is based on cell-wall polysaccharide antigens, and the mutans group streptococci are differentiated from other oral streptococci by their ability to ferment sorbitol and mannitol.

A possible genetic predisposition to caries in humans has been extensively investigated and hotly debated. Since the disease results from the interaction of exogenous (bacterial) and endogenous (host) traits it is likely that tooth decay is related to factors under genetic control. Caries attack rate in humans is associated with at least five factors: the physicochemical properties and integrity of dental enamel; salivary flow rate and composition; topical or communal water fluoridation; nutrition and diet; dental hygiene and the oral microflora. Clearly the first two factors are under genetic influence: there is significant heritability for salivary flow rate and salivary pH, and a marked genetic component to dentate status and dental caries experience has been convincingly demonstrated from twins studies (Hassell and Harris, 1995). Research has also shown that children acquire *S. mutans* strains from their mothers, and that female infants acquire streptococcal genotypes identical to their mother's with more fidelity than do male infants. On the other hand, male children harbouring the same genotype as their mothers had a 13 times greater likelihood of developing caries than female children who acquired their mother's strains (Li and Caufield, 1995).

While caries outcome may be determined by both source and genotype of mutans group streptococci, the idea of 'more cariogenic' and 'less cariogenic' types has been hard to reconcile. Several studies on African populations have reported high prevalences of mutans group streptococci. These findings are unexpected because, unlike in Europe and North America, caries and sugar intake are both low on the African continent. In fact the widespread occurrence of mutans group streptococci and low caries prevalence has led to suggestions that these African populations may harbour less-cariogenic types of streptococcus. However, differences in caries experience of children in Africa, Europe and North America cannot be explained by the prevailing mutans streptococcal species but should be attributed to the differences in the cariogenicity of various diets (van Palenstein Helderman *et al.*, 1996).

4.3 DEVELOPMENT OF THE NORMAL ORAL MICROFLORA

Streptococci are dominant in the mouth throughout the first year of life, comprising 98% of the cultivable bacteria within 2 days of birth and 70% after 12 months. *Streptococcus mitis*, *Streptococcus oralis* and *Streptococcus salivarius* make up 55% of pioneer species in neonates, while *Streptococcus anginosus*, *Streptococcus gordonii* and *Streptococcus sanguis* collectively account for about 11% (Pearce *et al.*, 1995). Almost half the streptococci colonizing the oral mucosa of neonates produce immunoglobulin A1 (IgA1) proteases, which cleave specifically human IgA1 at the hinge region. Secretory IgA agglutinates bacteria and inhibits bacterial adhesion, so destruction of these functions may provide a colonization advantage to bacteria, especially those establishing in the oral cavity and pharynx under exposure of IgA antibodies in mother's milk (Kilian *et al.*, 1996). Another factor that may be relevant to primary colonization by *S. oralis* and related species is the ability of these bacteria to secrete a wide range of hydrolytic enzymes that degrade host glycoproteins. Neuraminidase and hexosaminidase, in particular, may destroy host cell recognition functions, as well as providing for nutrition and growth of bacteria on released sugars (Homer *et al.*, 1996).

Following tooth eruption there is an increase in the isolation frequencies of *S. sanguis*, mutans group streptococci, *Actinomyces naeslundii*, *Fusobacterium nucleatum* and obligate anaerobes. In adults, certain streptococcal species predominate at different sites: for example, *S. oralis* is found on teeth, *S. sanguis* on buccal mucosa, and *S. mitis* and *S. salivarius* on the dorsum of the tongue and in the pharynx (Frandsen, Pedrazzoli and Kilian, 1991). Site specificity of adhesion is believed to account, at least in part, for these distributions and in turn provides an explanation for the relationship of specific bacterial groups or species with disease states.

The mutans group streptococci comprise just a few of the 20 or more species of streptococcus found in the human oral cavity and represent a minor component of the adult normal microflora consisting of at least 300 different bacterial species. It is important to emphasize, therefore, that, while mutans group streptococci are most strongly implicated in dental caries, the disease is initiated primarily by the metabolic activities of dental plaque, in which mutans group streptococci do not predominate or may even be undetectable (Frandsen, Pedrazzoli and Kilian, 1991). If conditions favour active progression of dental caries then there is a concomitant enrichment for mutans streptococci and other bacteria, such as lactobacilli, that are highly acid-tolerant.

4.4 RELATIONSHIP OF PLAQUE AND DISEASE

Development of caries is associated with dental plaque on the root and smooth (coronal) surfaces of teeth and in pits and fissures. Early lesions of the smooth surfaces appear as white spots on enamel. These may then progress, through early cavitation and collapse of enamel, to gross cavitation (Fig. 4.1).

Root surface lesions appear as soft areas of the cementum or dentine, but not as white spots because roots are not covered by enamel. Dental plaque generally originates on surfaces at the gingival margin or in gingival crevices. At these sites the primary colonizing species of bacteria bind to the acquired pellicle. The organisms that are most consistently isolated from clean tooth surfaces and have the highest affinities for binding to experimental pellicles include *S. oralis*, *S. sanguis*, *S. mitis*, *S. gordonii*, *A. naeslundii* and *Capnocytophaga* (Fig. 4.1). Binding of these organisms paves the way for 'secondary colonizers' to become established in the complex microbial community. The secondary colonizers adhere to the primary colonizers, either directly or by bridging of bacterial cells by intermediary molecules. The latter may include, for example, bacterial polysaccharides or salivary components that become adsorbed to the surface of bacterial cells and expose ligands to which other bacteria may adhere (Fig. 4.1). The secondary colonizers arguably include the mutans group organisms *S. mutans* and *S. sobrinus*, which have been shown generally not to have a high affinity for binding to experimental salivary pellicles. Other secondary colonizers include *F. nucleatum*, which makes up a large component of mature plaque, and *Porphyromonas* and *Prevotella* species (Fig. 4.1) that are implicated as primary aetiological agents in several periodontal disease states. For an explicit and extensive description of the interactions occurring between microorganisms in the

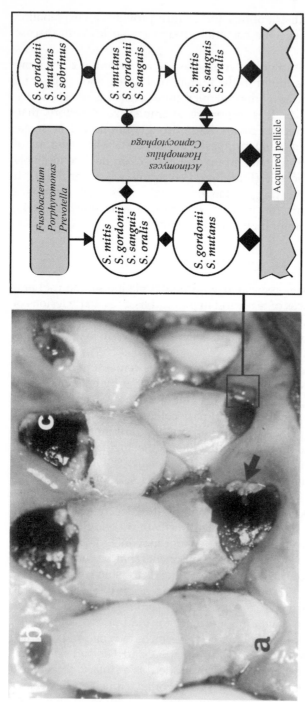

Figure 4.1 Carious lesions of human teeth and the microflora of dental plaque. The photograph shows smooth-surface coronal lesions in various stages, from white spots (a) through primary cavitation and enamel collapse (b) to gross cavitation (c). Plaque accumulations are visible on smooth surfaces and in particular at the gum (gingival) margins (arrowed). (Photograph kindly provided by J. C. Rodda.) The diagram on the right depicts the bacterial species likely to be found within such plaque and some of the adhesive interactions known to occur between the various organisms. Primary colonizers adhere to salivary components (filled diamonds) within the acquired pellicle. Adhesins (arrows) expressed on the surfaces of bacterial cells recognize receptors on the surfaces of other cells, resulting in coadhesion. Interbacterial binding is also promoted by salivary components that include proteins and glycoproteins (filled diamonds) and polysaccharides, especially glucans (filled circles). Source: redrawn from Jenkinson, 1994a with permission from Elsevier Science Ltd.

build-up of dental plaque the reader is referred to the recent review of Whittaker, Klier and Kolenbrander (1996).

4.5 BACTERIAL CELL ADHESION

There is much interest in defining the cell surface components of the primary colonizers that are responsible both for initial adhesion to acquired pellicle and for interbacterial adhesion. The most widespread interbacterial adhesion reactions involve streptococci. Components mediating interbacterial binding include cell surface proteins and polysaccharides. The best characterized at the molecular level are the linear cell wall phosphopolysaccharides produced by *S. oralis* and *S. sanguis*, which act as receptors for adhesion of bacteria such as *A. naeslundii*, *Veillonella atypica*, *Haemophilus parainfluenzae* and *Prevotella loescheii* (Whittaker, Klier and Kolenbrander, 1996). Streptococcal proteins involved directly in binding other oral microorganisms include the antigen I/II polypeptides SspA and SspB in *S. gordonii* (discussed below) and CshA, one of the largest (259 kDa) cell-wall-associated streptococcal proteins characterized, which is found in *S. gordonii*, *S. sanguis*, *S. oralis* and *S. mitis*. CshA was first identified by gene inactivation experiments (McNab and Jenkinson, 1992) as conferring the property of cell surface hydrophobicity in *S. gordonii*. The polypeptide comprises 2508 amino acid (aa) residues and contains an extensive aa repeat block region (residues 879–2417) and a non-repetitive amino (N)-terminal region spanning 836 aa residues (McNab *et al.*, 1994). Repeated blocks of aa residues are a characteristic of cell-wall-anchored proteins in Gram-positive bacteria, and in many instances the repeat blocks function in substrate binding. While the 13 repeat blocks of 101 aa residues present within CshA are implicated in determining cell surface hydrophobicity, they have not yet been attributed a ligand binding function. Instead, the non-repetitive N-terminal region appears to mediate adhesion of streptococcal cells to a variety of different substrates that include *A. naeslundii* and human fibronectin (McNab *et al.*, 1996). Isogenic *CshA* mutants of *S. gordonii* are deficient in binding to these substrates, but they are unaffected in their ability to adhere to experimental salivary pellicles. In addition, *CshA* mutants are unable to colonize the oral cavity of mice (McNab *et al.*, 1994). Evidently, therefore, CshA is an essential colonization factor for *S. gordonii*; however the physiological function of this protein *in vivo* is not yet known.

A number of salivary components present in the acquired pellicle have been shown to function as receptors for microbial cell adhesion. These include proline-rich proteins (PRPs), proline-rich glycoproteins (PRGs), secretory immunoglobulin A (sIgA), lysozyme, mucin-like salivary agglutinin glycoprotein (SAG) and α-amylase (see Scannapieco, 1994).

Table 4.1 Specific components of acquired pellicle that support adhesion of dental plaque bacteria, compiled from information within the following references: Ligtenberg *et al.*, 1992; Murray *et al.*, 1992; Scannapieco, 1994; Whittaker, Klier and Kolenbrander, 1996)

Acquired pellicle component	*Adhering bacterial species*
Mucin glycoproteins (MG1 and MG2)	*Capnocytophaga ochracea, Eikenella corrodens, Fusobacterium nucleatum, Streptococcus oralis*
Mucin MG2 (sialic-acid-rich)	*Streptococcus gordonii, S. oralis, Streptococcus sanguis*
α-amylase	*Streptococcus crista, S. gordonii, Streptococcus mitis, Streptococcus parasanguis, Streptococcus salivarius*
Proline-rich proteins (PRPs)	*Actinomyces naeslundii, Porphyromonas gingivalis, Prevotella loescheii, S. gordonii, Streptococcus mutans*
Proline-rich glycoproteins (PRGs)	*A. naeslundii, F. nucleatum, S. gordonii, S. oralis, S. sanguis*
Statherin	*A. naeslundii, F. nucleatum*
Lysozyme	*Capnocytophaga gingivalis, S. mutans, S. sanguis*
Fibronectin	*P. gingivalis, S. gordonii, Treponema denticola*
Bacterial glucans	*Streptococcus cricetus, S. gordonii, S. mutans, Streptococcus sobrinus*

All the primary colonizers bind to one or more of these salivary molecules (Table 4.1).

Other components present within acquired pellicle that may act as receptors for microbial adhesion include fibronectin and bacterially derived polymers such as cell-wall polysaccharides and extracellular glucan polymers (Table 4.1). However, despite intense research over several decades, relatively little is known at the molecular level about the bacterial molecules or adhesins that recognize these pellicle receptors. A large number of distinct salivary adhesins have been described, but many studies have used different strains of streptococci and thus the range of adhesins possessed by individual organisms remains to be determined. Only relatively recently have genetic methods of analysis been developed for studying oral streptococcal adhesion (Jenkinson, 1995). The generation of isogenic bacterial mutants has enabled the identification of functional adhesins and has confirmed that streptococci express multiple adhesins, sometimes recognizing the same receptors. The elaboration of multiple cell surface adhesins presumably enhances the colonization potential of bacteria *in vivo* where multiple binding interactions will increase the strength of adhesion and increase the availability of potential colonization sites.

Oral bacteria, in addition to adhering to deposited salivary components, are aggregated by fluid phase salivary components. This is an essential mechanism in the natural bacterial clearance processes.

Thus, oral bacteria must be able to adhere to deposited salivary components (receptors) despite there being present an excess of fluid-phase receptors. One explanation for this ability is that salivary components deposited on a surface may reveal different receptor conformations for adhesion from when they are in fluid phase. Accordingly, bacteria may have acquired adhesins that recognize different forms of the receptors. Furthermore, the deposition of salivary components on to an adhered organism will provide new receptors for subsequent microbial adhesion. Salivary component binding is therefore relevant both to initiation of colonization and to subsequent microbial cell proliferation and accumulation.

A large proportion (40%) of total parotid salivary protein is made up of PRPs. These proteins bind Ca^{2+} ions and are selectively adsorbed from saliva by hydroxylapatite (the mineral component of dental enamel). In this state they are potent inhibitors of surface-induced mineral deposition. Binding of acidic PRPs to hydroxylapatite occurs via a 30 aa residue N-terminal segment. Concomitantly with binding, a major conformational change is induced and the bound PRPs are then able to support the adhesion of bacteria such as *A. naeslundii* and *S. gordonii* (Gibbons and Hay, 1988; Gibbons, Hay and Schlesinger, 1991). Neither of these bacterial species is able to bind acidic PRPs in fluid phase, indicating that the bacterial adhesins recognize only the immobilized forms of PRPs. A C-terminal decapeptide of PRP-1 linked to agarose beads supports adhesion of *S. gordonii* (Gibbons, Hay and Schlesinger, 1991), suggesting that the positively charged C-terminal regions of bound PRPs are the receptors for bacterial adhesion.

The bacterial molecules mediating cell adhesion to PRPs have not yet been characterized. Adhesion of *A. naeslundii* cells to PRP-1 might be associated with the production of surface fimbriae since strains lacking type 1 fimbrial structures are deficient in binding to PRP-1-coated surfaces. These strains are also deficient in binding to experimental salivary pellicle, suggesting that PRPs may be major pellicle receptors (Clark *et al.*, 1989). Isogenic mutants of *A. naeslundii* in which the fimbrial type 1 subunit gene (*fimP*) is inactivated are deficient in binding to immobilized PRPs (Yeung, 1995). However, antibodies raised to the type 1 fimbrial protein subunit (molecular mass approximately 60 kDa) did not inhibit cell adhesion to salivary pellicle (Cisar *et al.*, 1991), and independently isolated adhesion-defective mutant cells of *A. naeslundii* T14V reacted normally with type 1 fimbrial subunit antibodies (Nesbitt *et al.*, 1992). These observations are taken to indicate that the PRP adhesin, while being intimately associated with the production and/or structure of type 1 fimbriae, is not the fimbrial structure subunit.

Some *A. naeslundii* strains also produce type 2 fimbriae. The structural subunit is a 54 kDa protein that is 34% identical in aa sequence to FimP (type 1 subunit protein). Production of type 2 fimbriae is associated with adhesion of *A. naeslundii* to glycosidic receptors present on epithelial cells, polymorphonuclear leukocytes and streptococcal cells. Lectin-like adhesion of *A. naeslundii* to these substrates is inhibited by galactose (Gal) and N-acetylgalactosamine (GalNAc; Whittaker, Klier and Kolenbrander, 1996) and strains lacking type 2 fimbriae are deficient in binding galactoside-containing receptors (Cisar *et al.*, 1984). The GalNAcβ receptor specificity of the type 2 fimbrial lectin is subject to strain variation and affects the tissue specificity of adhesion of *Actinomyces* (Strömberg and Borén, 1992). Recent evidence suggests that the type 2 fimbrial lectin adhesin (95 kDa), like the type 1 fimbrial adhesin, is distinct from the fimbrial structural subunit (Klier *et al.*, 1997).

A. *naeslundii* cells bind to type I and type III collagens, and this adherence property may be associated also with the production of type 1 fimbriae (Liu *et al.*, 1991). Adhesion of cells to collagen-coated hydroxylapatite is inhibited by free collagen but not by saliva (Hawkins, Cannon and Jenkinson, 1993). Lactobacilli (McGrady *et al.*, 1995) and streptococci (Switalski *et al.*, 1993) have also been shown to bind collagen. A collagen binding protein (90 kDa) from *S. mutans* has sequence similarity to the collagen binding protein (Cba) from *Staphylococcus aureus*. Interestingly, antibodies to the *S. mutans* protein block adhesion of both *S. mutans* and *A. naeslundii* cells to collagen (Lawry and Switalski, 1996), suggesting that the collagen binding domain sequences might be conserved in these organisms. The ability of bacteria to bind collagen may be relevant to the initiation of root surface caries since collagen type I is a predominant component of dentine. In this respect, it may be significant that the levels of *S. mutans*, *Actinomyces* and lactobacilli are consistently higher in root carious lesions (Bowden *et al.*, 1990; Beighton and Lynch, 1995).

PRPs are just one of many potential pellicle receptors for streptococci (Table 4.1). Different streptococcal species, and even different strains within species, appear to have widely differing abilities to adhere to salivary components. For example, α-amylase, the most abundant protein in saliva, is bound by only a few species of streptococci. Binding of α-amylase to *S. gordonii* cells is associated with the activities of a group of cell surface proteins, notably a 20 kDa polypeptide (Scannapieco, 1994). Amylase may serve directly as a pellicle receptor for streptococci. In addition, the presence of starch may enhance the adhesion of amylase binding streptococci to amylase (Scannapieco, Torres and Levine, 1995).

Cell surface proteins with highly conserved primary sequences, denoted the LraI family (Jenkinson, 1994b), are present in streptococci, pneumococci and enterococci and are implicated in streptococcal adhesion to salivary pellicle and *A. naeslundii*. The LraI family polypeptides (approximate molecular mass 35 kDa) are lipoproteins, covalently modified post-translationally with lipid at the N-terminal cysteine (Jenkinson,

1992), and associated with the outer face of the cytoplasmic membrane. The available evidence, based on gene structure and protein sequence homologies, suggests that these lipoproteins are substrate (solute) binding protein components of an ATP binding cassette (ABC) uptake system (traffic ATPase) that is highly conserved across the streptococci. It has been difficult to reconcile transporter function directly with streptococcal cell adhesion. However, there is presently no evidence to discount the possibility that the binding-protein components of traffic ATPases might dually function as adhesins (Jenkinson, 1992). Indeed, the extracellular component of a putative ABC-type transport system in *Lactobacillus reuteri* has recently been shown to bind collagen (Roos *et al.*, 1996).

The mechanisms used by oral bacteria for adhesion are diverse and complex. Different complements of adhesins may be used by bacteria to adhere to the same substrate. In addition, homologous adhesins may perform different functions in different bacterial species. The generation and analysis of isogenic mutants of *S. mutans* and *S. gordonii* has helped define more clearly the activities of adhesins and virulence factors in these organisms (Russell, 1993; Jenkinson, 1995). New methods have recently been developed for allelic replacement in *S. sobrinus* and *A. naeslundii*, which will enable better understanding of the adhesion mechanisms used by these cariogenic organisms. Of all the multiple surface components of streptococci that appear to be involved in the adhesion of cells to salivary components, none has been so thoroughly characterized as the antigen I/II family of polypeptides. These proteins mediate adhesion of streptococcal cells to salivary glycoproteins, and in particular to a high-molecular-mass mucin-like molecule present in saliva, designated salivary agglutinin glycoprotein (SAG).

4.6 CELL-SURFACE PROTEINS OF MUTANS GROUP STREPTOCOCCI

Properties of *S. mutans* associated with cariogenicity include the ability of organisms to adhere to salivary pellicle and to produce extracellular polysaccharides, and the ability to grow and survive within dental plaque under acidic conditions. Adherence and extracellular polysaccharide formation are determined by the activities of secreted proteins. The characterization of *S. mutans* secreted proteins was entwined initially within caries protection studies in various animal models. Immunization of primates with whole cells of *S. mutans* was found to stimulate production of antibodies that provided effective protection against caries (Lehner *et al.*, 1975). These experiments were extended to identify specific protein components of *S. mutans* that might induce protective immunity, with the view to producing an anticaries acellular vaccine. Russell and Lehner (1978) identified a series of protein antigens, denoted I, II, III and IV, from *S. mutans*, while Russell *et al.* (1983) independently described four antigens designated A, B, C and D. Antigen I (identical to antigen B) was a cell-wall protein of apparent molecular mass 185 kDa, and was renamed antigen I/II on the evidence that antigen II was a breakdown product of antigen I. Antigen I/II from *S. mutans* was as effective as whole cells in inducing immunological protection against caries in the Rhesus monkey (Lehner, Russell and Caldwell, 1980). Antigen A (WapA, 45 kDa) and antigen III are now known to be identical (Russell, Harrington and Russell, 1995) and may be involved in the retention of lipoteichoic acid on the cell surface (Harrington and Russell, 1993). Conflicting data have been obtained on the function of WapA, and on its ability to elicit protective antibodies in animal models of caries (Russell, Harrington and Russell, 1995). Lastly, antigen D has been characterized as a phosphocarrier protein (HPr) involved in the phosphotransferase system (PTS)-mediated uptake of sugars (Sutcliffe, Hogg and Russell, 1993), but the significance of the cell-surface-associated form of HPr is unknown.

4.7 ROLE OF ANTIGEN I/II FAMILY POLYPEPTIDES IN STREPTOCOCCAL COLONIZATION

Polypeptides immunologically cross-reactive with *S. mutans* antigen B (antigen I/II) have been detected in a wide range of streptococcal species indigenous to the human oral cavity. These polypeptides have been associated with the ability of bacteria to bind salivary glycoproteins. There is no evidence for similar proteins in *Streptococcus pyogenes* (group A streptococci), group B streptococci or *Streptococcus pneumoniae*, nor for immunologically cross-reactive antigens in other oral bacterial genera. The cloning of eight antigen I/II polypeptide genes from different species and serotypes of oral streptococci has confirmed the widespread distribution and structural similarities of this family of polypeptides (Table 4.2).

The mature proteins (after removal of the leader (signal) peptide) are all between 1500 and 1566 aa residues in length and they are conveniently divided into six regions based on their primary sequence characteristics (Fig. 4.2).

The sequences recognized as functionally and immunologically significant are present within the A, P and C regions. The alanine-rich repetitive region (A) is comprised of three 82 aa residue repeat blocks of amino acids and a fourth partial repeat (Fig. 4.2). The region contains a periodic distribution of hydrophobic aa residues forming a heptad repeat characteristic of α-helical coiled-coil structure. The sequence (D) residing between the A region and proline rich repeats (P) exhibits the highest degree of variability among the antigen I/II polypeptides from different species (Fig. 4.2). The P region is usually composed of three copies of a 39 aa residue repeat and is highly conserved. The N-terminal end of this

Table 4.2 Secreted proteins of *S. mutans* and related proteins in other streptococci

Secreted protein (mol. mass)	Function or activity	Related proteins in other streptococcal species
Antigen I/II (includes antigen B, SpaP, P1, PAc, Sr) (c. 165 kDa)	Binding to salivary proteins and glycoproteins	SpaA, PAg (*S. sobrinus*); SspA, SspB (*S. gordonii*); SoaA (*S. oralis*); *S. intermedius, S. sanguis*
Antigen III, antigen A, WapA (c. 45 kDa)	Binding of cell-surface lipoteichoic acid	Not determined
Antigen D (HPr) (c. 8 kDa)	Phosphocarrier protein	HPr (*S. gordonii, S. pyogenes, S. salivarius*)
GtfB (162 kDa), GtfC (149 kDa)	Water-insoluble (predominantly α-1,3-linked) glucan synthesis	GtfI (*S. downei*); GtfI (*S. sobrinus*); GtfL (*S. salivarius*)*; GtfJ (*S. salivarius*)*
GtfD (155 kDa)	Water-soluble (predominantly α-1,6-linked) glucan synthesis	GtfM (*S. salivarius*); GtfT (*S. sobrinus*)†; GtfG (*S. gordonii*); GtfK (*S. salivarius*); GtfS (*S. downei*)†
Ftf (84 kDa)	Fructan synthesis	Ftf (*S. salivarius*)
Gbp74 (59 kDa)	Binding to α-1,6-linked glucan	Gbp87 (*S. sobrinus*)
Gbp59 (c. 59 kDa)	Binding to α-1,6-linked glucan	Not determined
MsmE (c. 45 kDa)	Binding of melibiose, raffinose, isomaltosaccharides	Not determined

* Phylogenetically GtfB, GtfC, GtfI and GtfI are the most closely related, while GtfL is more closely related to GtfD and GtfM, and GtfI is more closely related to GtfK and GtfT (Simpson, Giffard and Jacques, 1995)
† Phylogenetically, GtfI is more closely related to GtfJ and GtfK, while GtfS is more distantly related

Dental caries

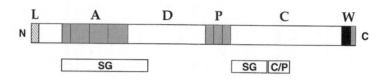

Figure 4.2 Structural features and functional regions of streptococcal antigen I/II polypeptides. Structural regions based on primary sequence and found in all known members of the polypeptide family are shown. L = leader (signal) peptide; A = alanine rich repeats; D = divergent region; P = proline rich repeats; C = highly conserved C-terminal region; WA = wall anchorage region. The characteristics of these regions are discussed in the text. Specific sequences within *S. mutans* polypeptides involved in binding salivary glycoproteins (SG), and within *S. gordonii* polypeptides binding Ca^{2+} and *Porphyromonas gingivalis* (C/P), are depicted.

region appears to elicit a strong immune response in humans. The C-terminal 500 aa residues of antigen I/II polypeptides are also highly conserved (62% identical aa residues). Secondary structural analysis suggests that this may resemble a globular domain, and functional analyses have shown that many of the binding activities of the antigen I/II polypeptides reside in this region. The C-terminus of all antigen I/II proteins contains sequences that are proposed to be involved in covalent linkage of the proteins to the Gram-positive bacterial cell wall (Schneewind, Fowler and Faull, 1995).

Insertional inactivation of the *S. mutans* spaP gene was shown to decrease binding of *S. mutans* cells to SAG-coated hydroxylapatite (Lee *et al.*, 1989). Inactivation of the *pac* gene of *S. mutans* (Koga *et al.*, 1990), and of the gene encoding antigen B in *S. mutans* LT11 (Harrington and Russell, 1993), also generated mutants deficient in adhesion to experimental salivary pellicles. Conversely, inactivation of the *sspA* and *sspB* genes, which encode two separate antigen I/II polypeptides in *S. gordonii*, did not affect adhesion of cells to salivary pellicle, although adhesion to immobilized SAG was reduced by 50% (Demuth *et al.*, 1996). Therefore it appears that, while the overall structure and sequence of antigen I/II polypeptides are well conserved across streptococci, their relative functional roles in adhesion of the different species may be different.

Streptococci, as well as binding to salivary glycoproteins in pellicle, also bind salivary glycoproteins that are present in fluid phase, Antigen I/II polypeptides exhibit multiple binding sites for salivary glycoproteins, and it appears that discrete regions within the polypeptides may discriminate between soluble and immobilized forms of salivary glycoproteins. Antibodies reacting with epitopes within the A or P regions of SpaP specifically block binding of cells to fluid-phase SAG, whereas antibodies to epitopes within the C-terminal regions more effectively inhibit adherence to immobilized SAG (Brady *et al.*, 1992). Recombinant peptides comprising segments of the *S. mutans* A region of SpaP are able to bind salivary glycoproteins, and functional sequences of up to 20 aa residues have been delineated (Jenkinson and Demuth, 1997). Saliva and serum samples from naturally sensitized human subjects contain high antibody titres to epitopes within the A region (Matsushita *et al.*, 1994). The P repeats also carry immunodominant B-cell epitopes and may elicit antibodies that are cross-reactive with human IgG (Moisset *et al.*, 1994). The C-terminal region contains only minor B-cell epitopes and sequences that interact with immobilized salivary glycoproteins. Synthetic peptides corresponding to regions encompassing aa residues 1005–1044 and 1085–1114 of *S. mutans* antigen I/II polypeptide are more than 90% inhibitory to adhesion of *S. mutans* cells to salivary glycoproteins (Kelly *et al.*, 1995). Comparison of these putative adhesion-mediating sequences of *S. mutans* antigen I/II polypeptide with those of other antigen I/II polypeptides reveals that they are 65% identical across eight antigen I/II polypeptides, and that the variable residues within these sequences are species-specific (Jenkinson and Demuth, 1997). Thus the structural and functional analyses of the antigen I/II polypeptides suggest how they may be relevant both for adhesion of streptococci to salivary pellicle and for salivary-glycoprotein-mediated interbacterial adhesion. This may be shown experimentally in that binding of SAG from fluid phase to the surface of one *Streptococcus* species promotes adhesion of cells of other species expressing antigen I/II polypeptides (Lamont *et al.*, 1991).

The pivotal role of antigen I/II polypeptide function in the formation of dental plaque is highlighted by the revelation of other binding functions for these polypeptides. For example, the antigen I/II polypeptides of *S. gordonii* are involved in the interactions of streptococcal cells with *P. gingivalis* (Lamont *et al.*, 1994) and with *A. naeslundii* (Demuth *et al.*, 1996). The binding sequence for *P. gingivalis* is located in the C-terminal region just downstream from the salivary glycoprotein adhesion sequences (Jenkinson and Demuth, 1997), and is close to a high-affinity Ca^{2+} binding site (Duan *et al.*, 1994; Fig. 4.2). The clustering of these binding sites indicates that Ca^{2+} may play an important role in antigen I/II polypeptide binding function, and indeed it is well documented that interactions of streptococci with salivary glycoproteins and other oral bacteria exhibit dependency for Ca^{2+}.

Research on the antigen I/II polypeptides has had a major impact on the development of immunization strategies against dental caries. However, one factor that has complicated the use of antigen I/II protein as a vaccine component is that highly related polypeptides are present on many streptococci, including species that are not strongly cariogenic. To overcome this it may be possible to exploit the species-specificity of antigen I/II function in order to delineate a region capable of eliciting a mutans-streptococcus-specific response. Further work is needed to dissect and define in more detail the species-specific active regions of these polypeptides if they are to be used as subunit vaccines.

4.8 ROLE OF POLYSACCHARIDE METABOLISM IN DENTAL CARIES

The ability to synthesize extracellular glucans is a property of many oral streptococcal species, and of *A. naeslundii*. Glucans are thought to play a key role in the formation of plaque because they adhere to smooth surfaces and mediate the coadhesion of bacterial cells (Fig. 4.1). The glucan-mediated accumulation of mutans group streptococci in dental plaque is enhanced through the activities of cell-surface-associated glucan binding proteins (GBPs). Furthermore, water-soluble forms of extracellular glucans and fructans provide a source of storage polymer that may be degraded by extracellular enzymes such as dextranases and fructanases for energy supply under conditions of nutrient deprivation. This in turn leads to acid production and demineralization of enamel. Thus the synthesis, binding and degradation of extracellular carbohydrate polymers by streptococci is a critical reaction complex in the aetiology of dental caries (Kuramitsu, 1993).

Glucosyltransferases (GTFs) and fructosyltransferases (FTFs) are secreted enzymes responsible for the synthesis of glucans and fructans respectively from sucrose. In *S. mutans*, GTFs are encoded by three genes denoted *gtfB*, *gtfC* and *gtfD*. The GtfB enzyme (162 kDa) synthesizes a water-insoluble glucan made up of α-1,3-linked glucose residues; GtfC (149 kDa) synthesizes a low-molecular-mass partly water-soluble glucan; and GtfD (155 kDa) produces a water-soluble α-1,6-linked glucan. Sucrose induces expression of *gtfB* (Hudson and Curtiss, 1990) and this is repressed by glucose, fructose and other PTS-sugars (Wexler, Hudson and Burne, 1993). Enzymes with high sequence similarities to GtfB and synthesizing water-insoluble glucans are produced by *S. sobrinus* (GtfI) and *S. downei* (GtfI) (Table 4.2). Enzymes similar to GtfD in that they synthesize water-soluble glucans, but not phylogenetically related, are produced by *S. sobrinus* (GtfT) and *S. downei* (GtfS). Some of these enzymes (GtfD, GtfI, GtfI) are primer-stimulated or -dependent whereas others (GtfB, GtfC, GtfT and GtfS) do not require a glucan primer for polysaccharide synthesis. There is no clear relationship, therefore, between primer requirement of an enzyme and the water-solubility of the product synthesized. However, it is thought that the dextran products of GtfT and GtfS may serve as primers for the corresponding GtfI (primer-dependent) enzymes. Most human isolates of *S. mutans*, but not *S. sobrinus*, also express one or more antigenically distinct FTFs synthesizing fructans with inulin-like β-2,1-linkages. The FTF enzyme (83 kDa) from *S. mutans* GS5 has considerable sequence similarity to the *Bacillus subtilis* levansucrase and expression of FTF is increased in the presence of sucrose (Kiska and Macrina, 1994).

The importance of extracellular polysaccharide production to cariogenesis has been documented through the analysis of isogenic mutants of *S. mutans* in rodent models of experimental caries. Collectively, the data from these studies show that colonization and caries production by mutans group streptococci, in rodents fed high-sucrose diets, is dependent upon the synthesis by bacteria of water-insoluble glucans (Kuramitsu, 1993). However, it is now suggested that the products of all four genes, *gtfB, gtfC, gtfD* and *ftf*, in *S. mutans* are involved in eliciting experimental caries, with differing and site-specific roles for water-insoluble (non-degradable) and water-soluble (degradable) polymers. The products of the *gtfB* and *gtfC* genes contribute to the severity of smooth-surface carious lesions and mutants able to make only soluble glucan or fructan are significantly reduced in ability to cause buccal (smooth-surface) lesions. Interestingly, however, *gtfB* mutants, which are deficient in water-insoluble glucan synthesis, cause significantly elevated sulcal (pit and fissure) lesions (Munro, Michalek and Macrina, 1995). An explanation for this is that in the absence of GtfB more degradable polymer is synthesized through the activities of GtfD and Ftf, thus enhancing the amount available for hydrolysis and ensuing acid production. These data imply that the ability of streptococci to cause caries depends upon the amount and type of polysaccharide produced, as well as upon the rate of degradation of polysaccharides. In particular, a major role for water-soluble polysaccharide production and degradation is now evoked. This may explain the reduced cariogenicity of streptococcal mutants deficient in endodextranase activity (Tanzer, 1992), even though dextranase (*dexA*) mutants of *S. mutans* are more adherent to smooth surfaces (Colby *et al.*, 1995).

The GTFs secreted by oral streptococci contain a common four-domain structure as follows (from the N-terminus): a signal sequence (36–38 aa residues); a poorly conserved region of about 200 aa residues; the catalytic domain (800 aa residues), which is highly conserved; a C-terminal domain of approximately 500 aa residues, which binds glucan molecules (Fig. 4.3).

The catalytic domain contains the active site with an essential Asp residue, as well as residues important in determining the chemical structure of the glucan synthesized (Shimamura *et al.*, 1994). The glucan binding domains are composed of a series of repeated aa residue sequence blocks originally designated A, B or C repeats (Ferretti, Gilpin and Russell, 1987; Banas, Russell and Ferretti, 1990). All the GTFs contain A repeats (38–72% identical aa residues) and some contain shorter (15–20 aa residues) C repeats downstream and separated from the A repeats by 3–5, or more, aa residues. The AC repeat cycle occurs five times in GtfI

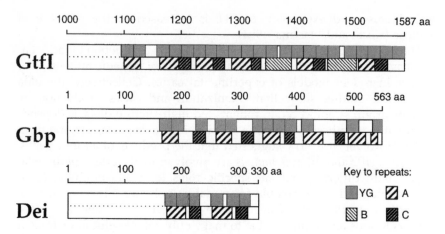

Figure 4.3 Comparison of the arrangement of aa residue repeat block sequences within the C-terminal glucan binding domains of GtfI (*S. downei*), Gbp$_{74}$ (*S. mutans*) and Dei (*S. sobrinus*). All these proteins contain A repeats (each approximately 34 aa residues) and C repeats (each approximately 20 aa residues), while GtfI is unique in containing two B repeats (48 aa residues each). The regular arrays of YG repeats (usually 21 aa residues) and their relationship with the A and C repeats are shown. The scales represent the amino acid number from the start of the polypeptide precursor. Note that, although the repeat blocks contain varying degrees of sequence homology, the sequences of the C-terminal glucan binding domains of the GTFs and GBPs do not align because the repeat units have undergone a number of independent duplications and rearrangements.

(Fig. 4.3) and GtfB, four times in GtfS (*S. downei*) and twice in GtfC (Gilmore, Russell and Ferretti, 1990). The AC repeats within GtfS contain 60–80% identical aa residues to a consensus AC repeat sequence of 53 aa residues. Only GtfI contains B repeats (Fig. 4.3), consisting of two identical blocks of 48 aa residues. The functional significance (if any) of the different repeat block structures is not known, although deletion analysis has shown that the C-terminal three A and two B repeats present in GtfI (Fig. 4.3) are not essential for enzymic activity (Ferretti, Gilpin and Russell, 1987).

A detailed comparison of all the sequenced GTFs has further suggested the presence of a fundamental repeated conserved 'YG motif' running through the A, B and C repeats within the C-terminal regions (Giffard and Jacques, 1994). The YG repeat is a 21 aa residue sequence containing one (or more) aromatic residues (usually tyrosine, Y) followed by glycine (G), usually four residues downstream. The YG repeats are present as highly regular arrays within GtfI (Fig. 4.3) and other GTFs, GBPs and the C-terminal domains of *Clostridium difficile* toxin A, *Streptococcus pneumoniae* autolysin and *S. pneumoniae* surface protein PspA (Giffard and

Jacques, 1994). The wide range of similarity indicates that the repeats have arisen from multiple duplication events, with evidence for functional selection of conserved residues for ligand (carbohydrate) binding. Since antibodies to the repeat regions are able to block substrate binding it is possible that immune surveillance has resulted in selection pressure for sequence variations within the substrate binding domain.

Other GBPs produced by oral streptococci (Table 4.2) are not known to have any enzymic activity. The GBPs are proposed to act as cellular receptors for glucans formed by bacterial cells during the accumulation phase of plaque formation or present in acquired pellicle. The GBP of *S. mutans* Ingbritt is a 563 aa residue protein (apparent molecular mass 74 kDa on SDS-gel electrophoresis) with considerable sequence identity (up to 60%) over 400 aa residues with the YG repeat regions of GtfB, GtfC and GtfI (Banas, Russell and Ferretti, 1990; Fig. 4.3). Structurally and antigenically related GBPs have been described for *S. sobrinus* (Kuramitsu, 1993). Dei (31 kDa) is a special GBP that is an inhibitor of dextranase produced by *S. sobrinus* (Sun *et al.*, 1994). *S. mutans* Gbp contains five A repeats (and a sixth partial A repeat) and four C repeats, while Dei contains two A and two C repeats (Fig. 4.3); the YG repeats may be discerned within the A and C repeat regions of both these proteins (Fig. 4.3). Despite extensive sequence information on the glucan binding domains of GTFs and GBPs, it is not known if a relationship exists between the number and sequence of repeat blocks and the specificity of substrate binding. This is an important issue for future research, as will be determining the three-dimensional structure of the substrate binding site.

Most work on glucan metabolism and its role in the initiation of caries has concentrated on the mutans group streptococci. It is important to note that other streptococci such as *S. gordonii* and *S. salivarius* also produce extracellular glucans. *S. gordonii* contains a single gene (*gtfG*) encoding a GTF (Sulavik and Clewell, 1996) that synthesizes a mixed-linked glucan that is partly water-soluble and promotes accumulation of *S. gordonii* cells on surfaces (Vickerman, Clewell and Jones, 1991). On the other hand, *S. salivarius* produces at least four enzymic activities one of which (GtfL) is primer-independent and synthesizes a water-insoluble glucan (Simpson, Giffard and Jacques, 1995). One of the main sources of GTFs in saliva may be from *S. salivarius* and, when incorporated into pellicle, GTFs synthesize glucans to which other streptococci adhere. *S. salivarius* is generally regarded as a poor colonizer of the human tooth surface, possibly because it exhibits low affinity of binding to salivary pellicle. It is interesting to note, however, that animal studies have suggested that some strains of *S. salivarius* may be cariogenic (Drucker, Shakespeare and Green, 1984), so general conceptions of species being either cariogenic or non-cariogenic may not be entirely tenable.

4.9 SUGAR METABOLISM AND PLAQUE ECOLOGY

It is estimated that less than 10% of sucrose available to mutans group streptococcal cells is converted into hexose polymers, most being transported into the cell and used directly as a carbon and energy source. Evidence suggests that there are at least four transport systems for sucrose (LeBlanc, 1994). A high-affinity sucrose-specific PTS system, which might function as a scavenger for sucrose, and a lower-affinity PTS system (Poy and Jacobson, 1990) are possibly major routes of sugar transport by *S. mutans*. A third non-PTS system (Slee and Tanzer, 1982) is active at fast growth rates (Ellwood and Hamilton, 1982). Under these conditions, cells exhibit homolactic fermentation producing much larger amounts of lactic acid than at slower growth rates, when the cells are heterofermentative. Sucrose transport occurs also via the multiple sugar metabolism (Msm) system, which is a traffic ATPase (ABC uptake system) transporting several sugars, including melibiose and isomaltosaccharides released from the breakdown of bacterial polysaccharides (Tao *et al.*, 1993). Transport of sugars via Msm requires a functional PTS system (Cvitkovitch, Boyd and Hamilton, 1995). Precisely how the activities of the various sucrose uptake systems are integrated under different environmental and physiological conditions remains to be fully determined.

The frequent consumption of fermentable dietary carbohydrates is associated with increased risk of dental caries. As already discussed, increased proportions of mutans group streptococci and lactobacilli occur in active carious lesions, with correspondingly reduced proportions of other streptococcal species. Mixed-culture chemostat studies have been crucial to understanding the basis of this shift in microflora associated with caries. In brief, a nine species consortium of plaque bacteria (including *Streptococcus* species, *Actinomyces*, *Lactobacillus*, *Neisseria*, *Veillonella*, *Fusobacterium* and *Prevotella*) could be maintained relatively stably in continuous culture and was only disrupted when the pH was allowed to fall below pH 5.0. Under these conditions the predominant species always became *S. mutans*, *Lactobacillus casei* and *Veillonella dispar*, suggesting that the selection of more cariogenic species following regular sugar consumption is probably a consequence of their ability to more successfully compete at lower pH (Marsh, 1994). These observations, taken in conjunction with clinical observations over many decades, have led to the proposal of the 'ecological plaque hypothesis'. In this, it is suggested that a change in a key environmental factor triggers a shift in the balance of the normal microflora. Under conditions that prevail in health, potentially pathogenic species (such as mutans streptococci) may be only weakly competitive and therefore constitute a minor proportion of the flora. If, for example, the frequency of dietary fermentable carbohydrate were to increase, plaque pH would be lower for longer periods, and streptococci and lactobacilli might proliferate at the expense of less acid-tolerant species (Marsh, 1994). Shifts in the microflora predisposing to caries are also caused by interrupted salivary flow, immunological defects and broad-spectrum antibiotic treatment.

4.10 PREVENTION OF DENTAL CARIES

It is well established that avoidance between meals of food or drinks containing fermentable sugars, the stimulation of saliva flow after meals and high regard to oral hygiene are the major strategies in the day-to-day prevention of dental caries. Also, fluoride has been shown to increase the resistance of enamel to demineralization and to have antimicrobial (anti-glycolytic) properties that are enhanced at lower pH. However, following observations now made over 20 years ago, that immunization with whole cells of *S. mutans* prevents colonization of the tooth surface by this microorganism in animal models, much effort has been expended to develop a vaccine. Although a safe and effective vaccine is not yet available, considerable information has been obtained on the immunological properties of specific surface antigens of oral streptococci.

Components of mutans streptococcal cells important to adhesion and accumulation of the bacteria have been assessed for vaccine suitability, including antigen I/II polypeptide, glucosyltransferase and GBPs. All these antigens are able to induce protection against caries in various primate and rodent experimental models (Lehner, Challacombe and Caldwell, 1980; Smith, Taubman and Ebersole, 1978; Smith and Taubman, 1996). As more becomes known about the functional regions within these proteins, immunization protocols may be developed that use synthetic peptides containing relevant epitopes to evoke protective immunity. These have the advantage that undesirable antibodies, e.g. tissue cross-reactive, are not elicited. This is also an advantage of using passive immunization, and promising results have been obtained by passively immunizing human subjects orally with monoclonal antibodies to antigen I/II polypeptide (Ma *et al.*, 1990). The inhibitory effect of these antibodies on *S. mutans* colonization was long-lasting and persisted after the antibodies had been eliminated from the oral cavity. This mode of passive immunization may therefore induce long-term changes in the oral microflora that are maintained without the need for continual presence of antibodies.

For the development of synthetic peptide vaccines, the peptide antigen must contain both B-cell and T-cell epitopes that induce cross-protective antibodies against the native target. Recent interest in the prospect of developing an antigen I/II polypeptide-directed vaccine has led to the identification, by epitope scanning analysis, of the major antigenic determinants within antigen I/II polypeptide (Matsushita *et al.*, 1994). These are multiple and scattered throughout the molecule; however the A region (Fig. 4.2) does appear to be strongly antigenic in humans and in mice. Intranasal immunization of mice with a 19 aa residue peptide from the A region, coupled to cholera toxin B subunit, suppressed colonization of mice by *S. mutans* (Takahashi *et al.*, 1991). More recently a unique peptide TYEAALKQYEADL from the A region has been suggested as a suitable vaccinal immunogen, since it corresponds to the sequence of a B-cell epitope and contains its own T-cell epitope (for mice; Senpuku *et al.*, 1995). However, there are two broad concerns with respect to developing anti-antigen-I/II-based strategies. The first is that, assuming that both systemic and mucosal antibodies are desirable, salivary IgA antibody titres to antigen I/II peptides have been shown to not necessarily correlate with titres of serum IgG antibodies to the peptides (Challacombe, 1980; Matsushita *et al.*, 1994). The second is that it may be difficult to produce a specific anti-*S.-mutans* response because of the widespread distribution of antigen I/II polypeptides across the oral streptococci. Indeed, evoking species-cross-reacting antibodies might upset the delicate balance between less or non-cariogenic streptococci and the host, with adverse consequences.

Synthetic peptides representing the glucan binding domain and catalytic domains of mutans group streptococcal GTFs are immunogenic in rodents and antibodies interfere with GTF activity. A 22 aa residue peptide TGAQTIKGQKLYFKANGQQVKG, part of a 'YG' sequence from the C-terminal substrate binding region of the *S. downei* GtfI, was synthesized on a core matrix of three lysines to yield a complex of four identical 22-mer peptides per molecule (Taubman, Holmberg and Smith, 1995). Rats immunized with the peptide construct showed significantly reduced sulcal caries after infection with *S. sobrinus* or *S. mutans*. The effectiveness of this peptide as an immunogen may be enhanced because the sequence also shows 50% homology to the *S. mutans* Gbp. Thus a combination of antibody-mediated inhibition of GTF activity and GBP-mediated inter-bacterial adhesion might more effectively interfere with mutans group streptococcal cell accumulation. A second peptide construct comprising a catalytic region sequence identical in GtfI (*S. downei*) and GtfB (*S. mutans*; Russell *et al.*, 1988), and synthesized on a core matrix as above, was also found to elicit antibodies protective against caries induced by *S. sobrinus* and *S. mutans* (Taubman, Holmberg and Smith, 1995). These experiments pave the way for the construction of possibly more effective peptide vaccines that incorporate epitopes from separate regions of the same antigen molecule (GTF) or additional epitopes from another target (e.g. antigen I/II).

There are many other potential ways of preventing mutans group streptococcal colonization beyond manipulation of host immune defences. As more information about the structure of adhesins and the molecular basis of their activities becomes available it may be possible to synthesize agonists that block adhesion. Promising results have been reported from using various receptor analogues to prevent bacterial adhesion to respiratory and gastrointestinal tract epithelia (Ofek, Kahane and Sharon, 1996). Replacement therapy is another possible strategy for controlling *S. mutans*. One example might be to engineer a strain of *S. mutans* deficient in lactic acid production and producing bacteriocin active against other (cariogenic) mutans group streptococci, and to introduce this into the oral cavity where it might grow and survive at the expense of other mutans group streptococci (Hillman, Chen and Snoep, 1996).

Many of these antistreptococcal strategies have far-reaching implications for devising new means of controlling or preventing colonization of humans by other microbial pathogens. Nevertheless, research aimed at developing new therapies for dental caries is no less compelling now than it was two decades ago. There is a worldwide need for improvement in existing preventive products and techniques (Mandel, 1996). Caries, and its sequelae, are the primary cause of tooth loss in the aging population of the USA, and caries is on the increase in many countries in South and Central America, Asia, Africa and the Middle East, and in several European countries.

4.11 REFERENCES

Banas, J. A., Russell, R. R. B. and Ferretti, J. J. (1990) Sequence analysis of the gene for the glucan-binding protein of *Streptococcus mutans* Ingbritt. *Infection and Immunity*, **58**, 667–673.

Beighton, D. and Lynch, E. (1995) Comparison of selected microflora of plaque and underlying carious dentine associated with primary root caries lesions. *Caries Research*, **29**, 154–158.

Bowden, G. H., Ekstrand, J., McNaughton, B. *et al.* (1990) The association of selected bacteria with the lesions of root surface caries. *Oral Microbiology and Immunology*, **5**, 346–351.

Brady, L. J., Piacentini, D. A., Crowley, P. J. *et al.* (1992) Differentiation of salivary agglutinin-mediated adherence and aggregation of mutans streptococci by use of monoclonal antibodies against the major surface adhesin P1. *Infection and Immunity*, **60**, 1008–1017.

Challacombe, S. J. (1980) Serum and salivary antibodies to *Streptococcus mutans* in relation to the development and treatment of human dental caries. *Archives of Oral Biology*, **25**, 495–502.

Cisar, J. O., David, W. A., Curl, S. H. *et al.* (1984) Exclusive presence of lactose-sensitive fimbriae on a typical strain (WVU45) of *Actinomyces naeslundii*. *Infection and Immunity*, **46**, 453–458.

Cisar, J. O., Barsumian, E. L., Siraganian, R. P. *et al.* (1991) Immunochemical and functional studies of *Actinomyces viscosus* T14V type 1 fimbriae with monoclonal and polyclonal antibodies directed against the fimbrial subunit. *Journal of General Microbiology*, **137**, 1971–1979.

Clark, W. B., Beem, J. E., Nesbitt, W. E. *et al.* (1989) Pellicle receptors for *Actinomyces viscosus* type 1 fimbriae *in vitro*. *Infection and Immunity*, **57**, 3003–3008.

Colby, S. M., Whiting, G. C., Tao, L. *et al.* (1995) Insertional inactivation of the *Streptococcus mutans dexA* (dextranase) gene results in altered adherence and dextran catabolism. *Microbiology*, **141**, 2929–2936.

Cvitkovitch, D. G., Boyd, D. A. and Hamilton, I. R. (1995) Regulation of sugar uptake via the multiple sugar metabolism operon by the phosphoenolpyruvate-dependent sugar phosphotransferase transport system of *Streptococcus mutans*, in *Genetics of Streptococci, Enterococci and Lactococci*, (eds J. J. Ferretti, M. S. Gilmore, T. R. Klaenhammer and F. Brown), Developments in Biological Standardization, vol. 85, S. Karger, Basel, pp. 351–356.

Demuth, D. R., Duan, Y., Brooks, W. *et al.* (1996) Tandem genes encode cell-surface polypeptides SspA and SspB which mediate adhesion of the oral bacterium *Streptococcus gordonii* to human and bacterial receptors. *Molecular Microbiology*, **20**, 403–413.

Drucker D. B., Shakespeare, A. P. and Green, R. M. (1984) The production of dental plaque and caries by the bacterium *Streptococcus salivarius* in gnotobiotic WAG/RIJ rats. *Archives of Oral Biology*, **6**, 437–443.

Duan, Y., Fisher, E., Malamud, D. *et al.* (1994) Calcium-binding properties of SSP-5, the *Streptococcus gordonii* M5 receptor for salivary agglutinin. *Infection and Immunity*, **62**, 5220–5226.

Ellwood, D. C. and Hamilton, I. R. (1982) Properties of *Streptococcus mutans* Ingbritt growing on limited sucrose in a chemostat: repression of the phosphoenolpyruvate phosphotransferase transport system. *Infection and Immunity*, **36**, 576–581.

Ferretti, J. J., Gilpin, M. L. and Russell, R. R. B. (1987) Nucleotide sequence of a glucosyltransferase gene from *Streptococcus sobrinus* MFe28. *Journal of Bacteriology*, **169**, 4271–4278.

Frandsen, E. V. G., Pedrazzoli, V. and Kilian, M. (1991) Ecology of viridans streptococci in the oral cavity and pharynx. *Oral Microbiology and Immunology*, **6**, 129–133.

Gibbons, R. J. and Hay, D. I. (1988) Human salivary acidic proline-rich proteins and statherin promote the attachment of *Actinomyces viscosus* LY7 to apatitic surfaces. *Infection and Immunity*, **56**, 439–445.

Gibbons, R. J., Hay, D. I. and Schlesinger, D. H. (1991) Delineation of a segment of adsorbed salivary acidic proline-rich proteins which promotes adhesion of *Streptococcus gordonii* to apatitic surfaces. *Infection and Immunity*, **59**, 2948–2954.

Giffard, P. M. and Jacques, N. A. (1994) Definition of a fundamental repeating unit in streptococcal glucosyltransferase glucan-binding regions and related sequences. *Journal of Dental Research*, **73**, 1133–1141.

Gilmore, K. S., Russell, R. R. B. and Ferretti, J. J. (1990) Analysis of the *Streptococcus downei gtfS* gene, which specifies a glucosyltransferase that synthesizes soluble glucans. *Infection and Immunity*, **58**, 2452–2458.

Harrington, D. J. and Russell, R. R. B. (1993) Multiple changes in cell wall antigens of isogenic mutants of *Streptococcus mutans*. *Journal of Bacteriology*, **175**, 5925–5933.

Hassell, T. M. and Harris, E. L. (1995) Genetic influences in caries and periodontal diseases. *Critical Reviews in Oral Biology and Medicine*, **6**, 319–342.

Hawkins, B. W., Cannon, R. D. and Jenkinson, H. F. (1993) Interactions of *Actinomyces naeslundii* strains T14V and ATCC 12104 with saliva, collagen and fibrinogen. *Archives of Oral Biology*, **38**, 533–535.

Hillman, J. D., Chen, A. and Snoep, J. L. (1996) Genetic and physiological analysis of the lethal effect of L-(+)-lactate dehydrogenase deficiency in *Streptococcus mutans*: complementation by alcohol dehydrogenase from *Zymomonas mobilis*. *Infection and Immunity*, **64**, 4319–4323.

Homer, K. A., Kelley, S., Hawkes, J. *et al.* (1996) Metabolism of glycoprotein-derived sialic acid and N-acetylglucosamine by *Streptococcus oralis*. *Microbiology*, **142**, 1221–1230.

Hudson, M. C. and Curtiss III, R. (1990) Regulation of expression of *Streptococcus mutans* genes important to virulence. *Infection and Immunity*, **58**, 464–470.

Jenkinson, H. F. (1992) Adherence, coaggregation, and hydrophobicity of *Streptococcus gordonii* associated with expression of cell surface lipoproteins. *Infection and Immunity*, **60**, 1225–1228.

Jenkinson, H. F. (1994a) Adherence and accumulation of oral streptococci. *Trends in Microbiology*, **2**, 209–212.

Jenkinson, H. F. (1994b) Cell surface protein receptors in oral streptococci. *FEMS Microbiology Letters*, **121**, 133–140.

Jenkinson, H. F. (1995) Genetic analysis of adherence by oral streptococci. *Journal of Industrial Microbiology*, **13**, 186–192.

Jenkinson, H. F. and Demuth, D. R. (1997) Structure, function and immunogenicity of streptococcal antigen I/II polypeptides. *Molecular Microbiology*, **23**, 183–190.

Kelly, C. G., Todryk, S., Kendal, H. L. *et al.* (1995) T-cell, adhesion, and B-cell epitopes of the cell surface *Streptococcus mutans* protein antigen I/II. *Infection and Immunity*, **63**, 3649–3658.

Kilian, M., Reinholdt J., Lomholt H., *et al.* (1996) Biological significance of IgA1 proteases in bacterial colonization and pathogenesis: critical evaluation of experimental evidence. *Acta Pathologica Microbiologica et Immunologica Scandinavica*, **104**, 321–338.

Kiska, D. L. and Macrina, F. L. (1994) Genetic regulation of fructosyltransferase in *Streptococcus mutans*. *Infection and Immunity*, **62**, 1241–1251.

Klier, C. M., Kolenbrander, P. E., Roble, A. G. *et al.* (1997) Identification of a 95-kDa putative adhesin from *Actinomyces* serovar WVA963 strain PK1259 that is distinct from type 2 fimbrial subunits. *Microbiology*, **143**, 835–846.

Koga, T., Okahashi, N., Takahashi, I. *et al.* (1990) Surface hydrophobicity, adherence, and aggregation of cell surface protein antigen mutants of *Streptococcus mutans* serotype c. *Infection and Immunity*, **58**, 289–296.

Kuramitsu, H. J. (1993) Virulence factors of mutans streptococci: role of molecular genetics. *Critical Reviews in Oral Biology and Medicine*, **4**, 159–176.

Lamont, R. J., Demuth, D. R., David, C. A., *et al.* (1991) Salivary-agglutinin-mediated adherence of *Streptococcus mutans* to early plaque bacteria. *Infection and Immunity*, **59**, 3446–3450.

Lamont, R. J., Gil, S., Demuth, D. R. *et al.* (1994) Molecules of *Streptococcus gordonii* that bind to *Porphyromonas gingivalis*. *Microbiology*, 140, 867–872.

Lawry, J. and Switalski, L. M. (1996) Cloning and expression of collagen adhesin of *Streptococcus mutans*. *Journal of Dental Research (IADR Abstracts)*, **75**, 96, abstr. 626.

LeBlanc, D. J. (1994) Role of sucrose metabolism in the cariogenicity of the mutans streptococci, in *Molecular Genetics of Bacterial Pathogenesis*, (eds V. L. Miller, J. B. Kaper, D. A. Portnoy and R. R. Isberg), American Society for Microbiology, Washington, DC, pp. 465–477.

Lee, S. F., Progulske-Fox, A., Erdos, G. W. *et al.* (1989) Construction and characterization of isogenic mutants of *Streptococcus mutans* deficient in major surface protein antigen P1 (antigen I/II). *Infection and Immunity*, **57**, 3306–3313.

Lehner, T., Challacombe, S. J. and Caldwell, J. (1975) Immunological and bacteriological basis for vaccination against dental caries in rhesus monkeys. *Nature (London)*, **254**, 517–520.

Lehner, T., Russell, M. W. and Caldwell J. (1980) Immunisation with a purified protein from *Streptococcus mutans* protects against dental caries in rhesus monkeys. *Lancet*, **i**, 995–996.

Li, Y. and Caufield, P. W. (1995) The fidelity of initial acquisition of mutans streptococci by infants from their mothers. *Journal of Dental Research*, **74**, 681–685.

Ligtenberg, A. J. M., Walgreen-Weterings, E., Veerman, E. C. I. *et al.* (1992) Influence of saliva on aggregation and adherence of *Streptococcus gordonii* HG 222. *Infection and Immunity*, **60**, 3878–3884.

Liu, T., Gibbons, R. J., Hay, D. I. *et al.* (1991) Binding of *Actinomyces viscosus* to collagen: association with the type 1 fimbrial adhesin. *Oral Microbiology and Immunology*, **6**, 1–5.

Ma, J. K.-C., Hunjan, M., Smith, R. *et al.* (1990) An investigation into the mechanism of protection by local passive immunization with monoclonal antibodies against *Streptococcus mutans*. *Infection and Immunity*, **58**, 3407–3414.

McGrady, J. A., Butcher, W. G., Beighton, D. *et al.* (1995) Specific and charge interactions mediate collagen recognition by oral lactobacilli. *Journal of Dental Research*, **74**, 649–657.

McNab, R. and Jenkinson, H. F. (1992) Gene disruption identifies a 290 kDa cell-surface polypeptide conferring hydrophobicity and coaggregation properties in *Streptococcus gordonii*. *Molecular Microbiology*, **6**, 2939–2949.

McNab, R., Holmes, A. R., Clarke, J. M. *et al.* (1996) Cell surface polypeptide CshA mediates binding of *Streptococcus gordonii* to other oral bacteria and to immobilized fibronectin. *Infection and Immunity*, **64**, 4204–4210.

McNab, R., Jenkinson, H. F., Loach, D. M. *et al.* (1994) Cell-surface-associated polypeptides CshA and CshB of high molecular mass are colonization determinants in the oral bacterium *Streptococcus gordonii*. *Molecular Microbiology*, **14**, 743–754.

Mandel, I. D. (1996) Caries prevention: current strategies, new directions. *Journal of the American Dental Association*, **127**, 1477–1488.

Marsh, P. D. (1994) Microbial ecology of dental plaque and its significance in health and disease. *Advances in Dental Research*, **8**, 263–271.

Matsushita, K., Nisizawa, T., Nagaoka, S. *et al.* (1994) Identification of antigenic epitopes in a surface protein antigen of *Streptococcus mutans* in humans. *Infection and Immunity*, **62**, 4034–4042.

Moisset, A., Schatz, N., Lepoivre, Y. *et al.* (1994) Conservation of salivary glyco-protein-interacting and human immunoglobulin G-cross-reactive domains of antigen I/II in oral streptococci. *Infection and Immunity*, **62**, 184–193.

Munro, C. L., Michalek, S. M. and Macrina, F. L. (1995) Sucrose-derived exopolymers have site-dependent roles in *Streptococcus mutans*-promoted dental decay. *FEMS Microbiology Letters*, **128**, 327–332.

Murray, P. A., Prakobphol, A., Lee, T. *et al.* (1992) Adherence of oral streptococci to salivary glycoproteins. *Infection and Immunity*, **60**, 31–38.

Nesbitt, W. E., Beem, J. E., Leung, K.-P. *et al.* (1992) Isolation and characterization of *Actinomyces viscosus* mutants defective in binding salivary proline-rich proteins. *Infection and Immunity*, **60**, 1095–1100.

Ofek, I., Kahane, I. and Sharon, N. (1996) Toward anti-adhesion therapy for microbial diseases. *Trends in Microbiology*, **4**, 297–299.

Pearce, C., Bowden, G. H., Evans, M. *et al.* (1995) Identification of pioneer viridans streptococci in the oral cavity of human neonates. *Journal of Medical Microbiology*, **42**, 67–72.

Poy, F. and Jacobson, G. R. (1990) Evidence that a low-affinity sucrose phospho-transferase activity in *Streptococcus mutans* GS-5 is a high-affinity trehalose uptake system. *Infection and Immunity*, **58**, 1479–1480.

Roos, S., Aleljung, P., Robert, N. *et al.* (1996) A collagen binding protein from *Lactobacillus reuteri* is part of an ABC transporter system. *FEMS Microbiology Letters*, **144**, 33–38.

Russell, R. R. B. (1993) The application of molecular genetics to the microbiology of dental caries. *Caries Research*, **28**, 69–82.

Russell, M. W., Harrington, D. J. and Russell, R. R. B. (1995) Identity of *Streptococcus mutans* surface protein antigen III and wall-associated protein antigen A. *Infection and Immunity*, **63**, 733–735.

Russell, M. W. and Lehner, T. (1978) Characterisation of antigens extracted from cells and culture fluids of *Streptococcus mutans* serotype c. *Archives of Oral Biology*, **23**, 7–15.

Russell, R. R. B., Peach, S. L., Colman, G. *et al.* (1983) Antibody responses to antigens of *Streptococcus mutans* in monkeys (*Macaca fascicularis*) immunized against dental caries. *Journal of General Microbiology*, **129**, 865–875.

Russell, R. R. B., Shiroza, T., Kuramitsu, H. K. *et al.* (1988) Homology of glucosyl-transferase gene and protein sequences from *Streptococcus sobrinus* and *Streptococcus mutans*. *Journal of Dental Research*, **67**, 543–547.

Scannapieco, F. A. (1994) Saliva-bacterium interactions in oral microbial ecology. *Critical Reviews in Oral Biology and Medicine*, **5**, 203–248.

Scannapieco, F. A., Torres, G. I. and Levine, M. J. (1995) Salivary amylase promotes adhesion of oral streptococci to hydroxyapatite. *Journal of Dental Research*, **74**, 1360–1366.

Schneewind, O., Fowler, A. and Faull, K. F. (1995) Structure of the cell wall anchor of surface proteins in *Staphylococcus aureus*. *Science*, **268**, 103–106.

Senpuku, H., Miyauchi, T., Hanada, N. *et al.* (1995) An antigenic peptide inducing cross-reactive antibodies inhibiting the interaction of *Streptococcus mutans* PAc with human salivary components. *Infection and Immunity*, **63**, 4695–4703.

Shimamura, A., Nakano, Y. J., Mukasa, H. *et al.* (1994) Identification of amino residues in *Streptococcus mutans* glucosyltransferases influencing the structure of the glucan product. *Journal of Bacteriology*, **176**, 4845–4850.

Simpson, C. L., Giffard, P. M. and Jacques, N. A. (1995) *Streptococcus salivarius* ATCC 25975 possesses at least two genes coding for primer-independent glucosyltransferases. *Infection and Immunity*, **63**, 609–621.

Slee, A. M. and Tanzer, J. M. (1982) Sucrose transport by *Streptococcus mutans*: evidence for multiple transport systems. *Biochimica et Biophysica Acta*, **692**, 415–424.

Smith, D. J. and Taubman, M. A. (1996) Experimental immunization of rats with a *Streptococcus mutans* 59-kilodalton glucan-binding protein protects against dental caries. *Infection and Immunity*, **64**, 3069–3073.

Smith, D. J., Taubman, M. A. and Ebersole, J. L. (1978) Effects of local immunization with GTF fractions from *Streptococcus mutans* on dental caries in hamsters caused by homologous and heterologous serotypes of *Streptococcus mutans*. *Infection and Immunity*, **21**, 833–841.

Strömberg, N. and Borén, T. (1992) *Actinomyces* tissue specificity may depend on differences in receptor specificity for GalNAcβ-containing glycoconjugates. *Infection and Immunity*, **60**, 3268–3277.

Sulavik, M. C. and Clewell, D. B. (1996) Rgg is a positive regulator of the *Streptococcus gordonii gtfG* gene. *Journal of Bacteriology*, **178**, 5826–5830.

Sun, J.-W., Wanda, S.-Y., Camilli, A. *et al.* (1994) Cloning and DNA sequencing of the dextranase inhibitor gene (*dei*) from *Streptococcus sobrinus*. *Journal of Bacteriology*, **176**, 7213–7222.

Sutcliffe, I. C., Hogg, S. D. and Russell, R. R. B. (1993) Identification of *Streptococcus mutans* antigen D as the HPr component of the sugar-phosphotransferase transport system. *FEMS Microbiology Letters*, **107**, 67–70.

Switalski, L. M., Butcher, W. G., Caufield, P. C. *et al.* (1993) Collagen mediates adhesion of *Streptococcus mutans* to human dentin. *Infection and Immunity*, **61**, 4119–4125.

Takahashi, I., Okahashi, N., Matsushita, K. *et al.* (1991) Immunogenicity and protective effect against oral colonization by *Streptococcus mutans* of synthetic peptides of a streptococcal surface protein antigen. *Journal of Immunology*, **146**, 332–336.

Tanzer, J. M. (1992). Microbiology of dental caries, in *Contemporary Oral Microbiology and Immunology*, (eds J. Slots and M. A. Taubman), Mosby/Year Book, St Louis, MO, pp. 377–424.

Tao, L., Sutcliffe, I. C., Russell, R. R. B. *et al.* (1993) Transport of sugars, including sucrose, by the *msm* transport system of *Streptococcus mutans*. *Journal of Dental Research*, **72**, 1386–1390.

Taubman, M. A., Holmberg, C. J. and Smith, D. J. (1995) Immunization of rats with synthetic peptide constructs from the glucan-binding or catalytic region of mutans streptococcal glucosyltransferase protects against dental caries. *Infection and Immunity*, **63**, 3088–3093.

van Palenstein Helderman, W. H., Matee, M. I. N., van der Hoeven, J. S. *et al.* (1996) Cariogenicity depends more on diet than the prevailing mutans streptococcal species. *Journal of Dental Research*, **75**, 535–545.

Vickerman, M. M., Clewell, D. B. and Jones, G. W. (1991) Ecological implications of glucosyltransferase phase variation in *Streptococcus gordonii*. *Applied and Environmental Microbiology*, **57**, 3648–3651.

Wexler, D. L., Hudson, M. C. and Burne, R. A. (1993) *Streptococcus mutans* fructosyltransferase (*ftf*) and glucosyltransferase (*gtfBC*) operon fusion strains in continuous culture. *Infection and Immunity*, **61**, 1259–1267.

Whittaker, C. J., Klier, C. M. and Kolenbrander, P. E. (1996) Mechanisms of adhesion by oral bacteria. *Annual Review of Microbiology*, **50**, 513–552.

Yeung, M. K. (1995) Construction and use of integration plasmids to generate site-specific mutations in the *Actinomyces viscosus* T14V chromosome. *Infection and Immunity*, **63**, 2924–2930.

5

The oral microflora and human periodontal disease

Jørgen Slots and Casey Chen

5.1 INTRODUCTION

The periodontium consists of gingiva, periodontal ligament, root cementum and alveolar bone (Fig. 5. 1).

Periodontal diseases can be grouped broadly into gingivitis and periodontitis, and each can be further divided according to disease activity and severity, age of onset, related systemic disorders and other factors. Gingivitis connotes inflammation of gingiva, which does not affect the attachment apparatus of teeth. Periodontitis denotes destruction of connective tissue attachment and adjacent alveolar bone. The gingival crevice (periodontal pocket) in healthy periodontium measures 1–3 mm in depth. It increases in depth in periodontitis with loss of the connective tissue attachment and apical migration of the junctional epithelium. Nearly every adult experiences gingivitis or some degree of periodontitis. The advanced forms of periodontitis with extensive loss of tooth-supporting connective tissue and bone occur in approximately 7–15% of dentate adults (Brown and Loe, 1993). Bacteria are involved in virtually all forms of inflammatory periodontal disease.

Oral and periodontal microbiology occupies an important place in the history of microbiology, since the first bacteria described by Antonie van Leeuwenhoek in 1683 originated from dental plaque. Leeuwenhoek noted that persons with poor oral hygiene had particularly large numbers of periodontal bacteria. In 1890, Willoughby D. Miller, an

G. W. Tannock (ed.), *Medical Importance of the Normal Microflora*, 101–127.
© 1999 *Kluwer Academic Publishers. Printed in Great Britain.*

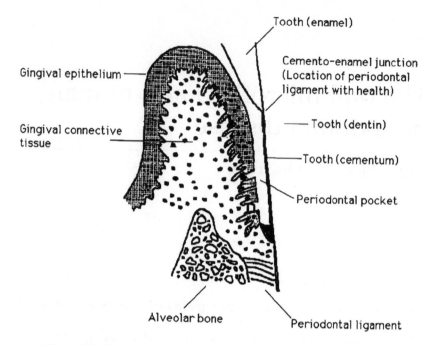

Tooth (enamel)

Gingival epithelium

Cemento-enamel junction
(Location of periodontal
ligament with health)

Gingival connective
tissue

—— Tooth (dentin)

Tooth (cementum)

Periodontal pocket

Alveolar bone

Periodontal ligament

Figure 5.1 Illustration of the development of periodontal disease.

American dentist working with Robert Koch in Berlin, initiated the study of bacteria in the etiology of periodontal disease. Miller believed that 'pyorrhoea alveolaris is not caused by a specific bacterium'. However, microbiological techniques of Miller's era preceded the discovery of anaerobic culturing and microbial taxonomy was in its infancy.

A great preponderance of bacteria colonize the gingival crevice, in which more than 500 microbial taxa can be encountered (Moore and Moore, 1994). In addition, the periodontal pocket may harbor *Entamoeba*, trichomonads and *Candida*. Cytomegalovirus, Epstein–Barr virus and other herpesviruses have also been identified in periodontal pocket specimens (Contreras and Slots, 1996; Parra and Slots, 1996).

Gingivitis is an antecedent to, but does not necessarily lead to, destructive periodontitis. Many bacterial species initiate gingival inflammation if present in high numbers in gingival sites. In contrast, a limited number of species appear to be responsible for the conversion of a gingivitis lesion to a periodontitis lesion. Chronic gingivitis probably constitutes a microbiologically non-specific infection and destructive periodontitis a specific infection. In medically compromised patients, microorganisms unusual for the periodontal pocket microflora may also give rise to periodontal tissue destruction.

Some organisms occur in high proportions in gingivitis as well as in periodontitis (e.g. *Fusobacterium nucleatum, Prevotella intermedia*). The organisms' presence in periodontitis lesions may indicate a causal relationship with periodontitis or, alternatively, a mere subgingival overgrowth resulting from a favorable ecological environment in deep periodontal pockets. Perhaps distinct subspecies, genotypes or serotypes within some species may play different roles in gingivitis and periodontitis. Other periodontitis organisms are rarely detected (e.g. *Porphyromonas gingivalis*) or occur in very low numbers (e.g. *Actinobacillus actinomycetemcomitans*) in long-standing gingivitis. Such organisms seem to constitute more obvious candidates for periodontal pathogens. However, because of the unavailability of an appropriate animal model for human destructive periodontal disease, Koch's postulates to delineate etiological significance have not been fulfilled for periodontal microorganisms.

Socransky's modifications of Koch's postulates are presently used to incriminate an organism in periodontitis (Socransky, 1979). To be a potential periodontitis pathogen, an organism must fulfill (or fit) the following criteria.

- The organism must occur at higher numbers at disease-active than at disease-inactive sites.
- The elimination of the organism should arrest disease progression.
- The organism should possess virulence factors relevant to initiation and progression of periodontitis.
- The cellular or humoral immune responses to the organism should reflect its unique role in the disease.
- Animal pathogenicity testing should imply potential for human periodontitis development.

Figure 5.2 shows that, with increasing severity of disease, the proportions of anaerobic, Gram-negative and motile organisms increase significantly. Table 5.1 lists specific organisms that have been associated with various types of periodontal disease.

5.2 PERIODONTAL HEALTH AND GINGIVITIS

5.2.1 Health

The healthy gingival sulcus harbors a scant microflora dominated by Gram-positive organisms (85%) and facultatively anaerobic species (75%). Spirochetes and motile rods make up less than 5% of the healthy microflora. *Actinomyces* and *Streptococcus* species each account for about 40% of total isolates. The Gram-negative organisms include low levels of *Fusobacterium, Prevotella* and *Veillonella* species

Table 5.1 Microbial species associated with various clinical forms of periodontitis (for review, see Socransky and Haffajee, 1994)

Species	Localized juvenile periodontitis	Early-onset periodontitis	Adult periodontitis	Refractory periodontitis
Actinobacillus actinomycetemcomitans	+++	++	++	++
Porphyromonas gingivalis	±	+++	+++	++
Prevotella intermedia/ nigrescens	++	+++	+++	+++
Bacteroides forsythus	±	++	+++	++
Fusobacterium species	+	++	+++	++
Peptostreptococcus micros	±	++	+++	++
Eubacterium species	−	+	++	+
Campylobacter rectus	+	++	++	+
Treponema species	++	+++	+++	++
Enteric rods and pseudomonads	−	−	±	+
Beta-hemolytic streptococci	?	++	++	++
Candida species	−	−	−	±

− = not elevated in comparison to health; ± = occasionally isolated; + = < 10% of the patients positive; ++ = < 50% of the patients positive; + + + = > 50% of the patients positive

5.2.2 Chronic gingivitis

The withdrawal of oral hygiene measures in subjects with clinical healthy gingivae will result in dental plaque formation and development of gingivitis. Chronic gingivitis is considered to be an inevitable antecedent to loss of periodontal attachment; however, some patients can exhibit gingivitis for extended periods without converting to periodontitis. In direct microscopic examination, motile rods and spirochetes each make up about 20% of total microorganisms. Culture studies show that the microflora in chronic gingivitis is dominated by Gram-positive and facultative organisms (about 55% each) but Gram-negative and anaerobic organisms are almost as abundant (about 45% each). Predominant Gram-positive organisms include *Actinomyces viscosus*, *A. naeslundii*, *Streptococcus sanguis*, *S. mitis* and *Peptostreptococcus micros*. Common Gram-negative organisms include *Fusobacterium nucleatum* subspecies, *P. intermedia*, *Veillonella parvula*, *Campylobacter* and *Haemophilus* species.

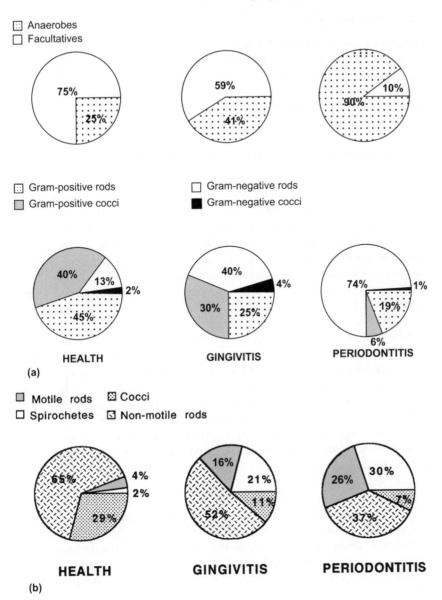

Figure 5.2 Subgingival microflora in periodontal health and disease. (a) Cultivable subgingival microbiota. (b) Subgingival morphotypes by direct microscopic examination.

5.2.3 Pregnancy gingivitis

Pregnancy gingivitis appears in some women from the second to eighth month of pregnancy. The gingiva is fiery red and enlarged. The disease has been associated with *P. intermedia*, which may 'bloom' in the periodontal pocket as a result of increased levels of the growth factors estradiol and progesterone in gingival crevicular fluid.

5.2.4 Acute necrotizing ulcerative gingivitis (ANUG)

ANUG represents one of the few acute infections of the marginal periodontium. Young people under psychosocial stress and, especially, HIV-infected individuals constitute high-risk groups (Dougherty and Slots, 1993; Goldhaber and Giddon, 1964). ANUG is associated with pain, bleeding, *fetor ex ore* and, occasionally, fever and malaise. The disease starts as necrotic lesions of one of more interdental papillae and progresses to its maximum extent within a few days. In malnourished individuals in certain Third-World countries ANUG-like lesions, termed cancrum oris (NOMA), can expand considerably beyond the gingiva and give rise to life-threatening infections.

ANUG lesions display a characteristic histopathological picture. Four zones have been described, starting from the outer surface of the ANUG lesion. The bacterial zone contains a variety of bacteria and may be rather similar to the subgingival microflora of periodontal lesions. The neutrophil zone is rich in leukocytes and underscores the acute state of the disease. The necrotic zone contains cell debris as well as spirochetes and Gram-negative rods. The invasion zone shows infiltration of medium-sized and large spirochetes into apparently normal gingival connective tissue. Selective spirochetal invasion of underlying connective tissue is unique to ANUG lesions

ANUG lesions harbor large numbers of spirochetes and *P. intermedia* (Loesche *et al.*, 1982). Direct microscopic examination of ANUG lesions from earlier studies identified high levels of fusobacteria ('fusospirochetal' disease). However, recent culture and immunofluorescence studies detected only small proportions of *Fusobacterium* species and found most of the Gram-negative rods to be members of the *P. intermedia* species.

Serum antibody reactivity in ANUG patients also points to spirochetes and *P. intermedia* as major pathogens (Chung *et al.*, 1983). Significantly higher IgG and IgM antibody levels to intermediate-sized spirochetes and higher IgG levels to *P. intermedia* have been found in ANUG patients in the acute clinical phase than in subjects with gingivitis and healthy control groups. ANUG patients revealed normal levels of serum antibodies against *F. nucleatum*, *P. gingivalis* and various Gram-positive rods.

Because the production of significant IgG antibody levels may take weeks, the increased antibody against intermediate-sized spirochetes and *P. intermedia* suggests that these bacteria proliferate above normal levels prior to overt disease and that it does not occur merely as a result of the pathological changes in ANUG lesions. In individuals at risk for ANUG, the suppression of spirochetes and *P. intermedia* may prevent the initiation of the disease.

Recent studies have associated ANUG lesions with Cytomegalovirus, Epstein–Barr virus and other herpesviruses (Contreras *et al.*, 1997). Conceivably, herpesviral infection may decrease the cellular host defense and permit overgrowth of pathogenic bacteria.

5.3 LOCALIZED JUVENILE PERIODONTITIS

Localized juvenile periodontitis displays distinct clinical, immunological and microbiological features. The disease may serve as a model for understanding the pathogenesis of human periodontitis. Localized juvenile periodontitis afflicts 0.1–1% of subjects in US and European populations and occurs with higher prevalence in African-Americans and some Third-World country populations (Brown and Loe, 1993). The disease has a familial predisposition, involves alveolar bone destruction around permanent incisors and first molars in otherwise healthy individuals, has its onset during the circumpubertal period (10 and 14 years of age), and is frequently associated with relatively little dental plaque, gingival inflammation and little or no approximal dental caries (Dougherty and Slots, 1993). The disease progresses rapidly during the early stages, but may later arrest or 'burn out', even in the absence of treatment. Progressing localized juvenile periodontitis is often refractory to mechanical periodontal debridement.

5.3.1 Association with *A. actinomycetemcomitans*

A. actinomycetemcomitans seems to be the primary pathogen in localized juvenile periodontitis (Slots, Reynolds and Genco, 1980; Slots and Schonfeld, 1991). Numerous studies have shown that:

- the organism is present subgingivally in nearly all localized juvenile periodontitis patients but is relatively uncommon in healthy subjects;
- the subgingival level of *A. actinomycetemcomitans* increases several months prior to the conversion from health to disease;
- localized juvenile periodontitis patients exhibit a strong immune response against the organism;
- successful treatment of localized juvenile periodontitis often depends on elimination of the organism from periodontal pockets.

African-Americans exhibit a relatively high prevalence of oral *A. actinomycetemcomitans*. In part, *A. actinomycetemcomitans* infections among African-Americans may represent carriage of the organism from populations in Africa with high prevalence of juvenile periodontitis. The familial distribution of localized juvenile periodontitis may also be due to the transmission of *A. actinomycetemcomitans* between family members rather than a consequence of the inheritance of genetic risk factors.

5.3.2 *A. actinomycetemcomitans* virulence factors

Fimbriae constitute the major adhesin of *A. actinomycetemcomitans*. Freshly isolated *A. actinomycetemcomitans* strains invariably possess fimbriae and attach better to oral surfaces than do non-fimbriated laboratory strains (Rosan *et al.*, 1988). However, non-fimbriated variants may not attach to granulocytes, thereby avoiding phagocytosis and allowing invasion of periodontal tissues. Other cell surface structures such as extracellular vesicles and extracellular amorphous material may also mediate tissue adherence (Meyer and Fives-Taylor, 1994).

Following adherence to host tissue or pre-existing microbial plaque *A. actinomycetemcomitans* may enter host tissues either extracellularly or intracellularly (Fives-Taylor, Meyer and Mintz, 1995; Christersson, Mbini and Zambon, 1987). Periodontal *A. actinomycetemcomitans* infections are particularly insidious because of the organism's ability to invade gingival tissue.

Polymorphonuclear leukocytes, which are the major defense cells in periodontal disease, can be impaired *in vitro* by the *A. actinomycetemcomitans* leukotoxin, chemotaxis-inhibitory products and antiphagocytic factors (Zambon, Christersson and Slots, 1983). Immune responses may be inhibited by a proteinaceous T-suppressor cell stimulator and a lipopolysaccharide capable of killing macrophages (Zambon, Christersson and Slots, 1983). The limited gingival inflammation characteristic of early localized juvenile periodontitis may be caused by *A. actinomycetemcomitans* toxins, which may impair the influx of inflammatory cells to the site of infection. Consistent with this hypothesis, the appearance of an increased gingival inflammation in the later stages of localized juvenile periodontitis is associated with elevated levels of neutralizing antibodies against the *A. actinomycetemcomitans* toxins.

5.3.3 Variation in virulence among *A. actinomycetemcomitans* strains

In the USA 20–25% of adolescents harbor *A. actinomycetemcomitans* in gingival crevice but fewer than 1% develop localized juvenile periodontitis. The high carrier rate may be due to the requirement of certain rare host susceptibility factors for disease development. Alternatively, certain *A. actinomycetemcomitans* strains may exert particularly high periodontopathic potential.

The *A. actinomycetemcomitans* species include at least five serotypes (a–e) based on carbohydrate surface antigens. *A. actinomycetemcomitans* serotype b strains predominate in periodontitis patients and serotype c strains in healthy individuals (Asikainen *et al.*, 1991). Serotype b antigen is defined by the O-antigen of lipopolysaccharide and consists of a polymer of trisaccharide repeating units composed of D-Fuc, L-Rha and D-GalNac residues (Perry *et al.*, 1996). In non-oral *A. actinomycetemcomitans* infection, serotype c seems to be the predominant pathogen.

Clonal analysis of *A. actinomycetemcomitans* by restriction fragment length polymorphism (RFLP) suggests a distinction between isolates with high and low virulence potential. A RFLP type B seemed to be particularly virulent, as indicated from its common presence in subjects who convert from health to localized juvenile periodontitis (DiRienzo and Slots, 1990). RFLP type B isolates are generally not recovered from healthy periodontal sites infected by *A. actinomycetemcomitans* (DiRienzo and Slots, 1990).

A. actinomycetemcomitans strain-to-strain variation in virulence may be due to varying ability to produce leukotoxin (Zambon *et al.*, 1996). Strongly leukotoxic *A. actinomycetemcomitans* strains have been detected in higher prevalence in infected periodontitis lesions than in infected healthy periodontal sites. Also, higher proportions of leukotoxin-producing strains have been recovered from younger patients (presumably having more aggressive type of disease) than from older localized juvenile periodontitis patients.

5.3.4 Bacterial interactions

Common oral streptococci and *Actinomyces* species inhibit the *in vitro* growth of *A. actinomycetemcomitans*. *A. actinomycetemcomitans* also displays bacteriocin properties capable of inhibiting oral streptococci and *Actinomyces*. This antagonistic mechanism may operate *in vivo* as well, as inferred from a frequent recovery of inhibitory organisms from periodontal sites free of *A. actinomycetemcomitans* and a much lower occurrence of these organisms in periodontitis lesions infected (Hillman, Socransky and Shrivers, 1985) with *A. actinomycetemcomitans*. If protective organisms predominate, they may prevent the establishment of *A. actinomycetemcomitans* and the development of disease. If *A. actinomycetemcomitans* establishes a sustained colonization first, it may suppress the protective organism and little or no supragingival plaque, gingivitis and dental caries occurs.

5.3.5 Host susceptibility

The familial pattern of localized juvenile periodontitis may be associated with a genetically determined susceptibility, in addition to the intrafamilial transmission of *A. actinomycetemcomitans* (Hart, 1996). The heredity of juvenile periodontitis appears to be complex. Several modes of inheritance of the disease have been suggested, including X-linked dominant, autosomal recessive and autosomal dominant. Various HLA (human leukocyte antigen) profiles have been associated with localized juvenile periodontitis.

Important susceptibility factors in localized juvenile periodontitis may encompass defects in polymorphonuclear leukocyte or monocyte chemotaxis and possibly phagocytosis. These defects can have a genetic or environmental basis. However, defective granulocytes are detected in only about 75% of patients with localized juvenile periodontitis, which suggests that additional susceptibility factors are important in disease development.

Puberty seems to play a role in the initiation of localized juvenile periodontitis. Onset of disease occurs at 10–14 years of age, and females experience the disease 6 months to 1 year earlier than males. Sex hormones serve as growth factors for *A. actinomycetemcomitans* and increased hormonal levels in the gingival crevicular fluid may stimulate an overgrowth of the organism in periodontal sites.

5.3.6 Antibody immune response against *A. actinomycetemcomitans*

Localized juvenile periodontitis patients exhibit markedly elevated serum IgG antibodies against *A. actinomycetemcomitans*. A large body of evidence indicates that the anti-*A.*-*actinomycetemcomitans* antibody is protective (Tew *et al.*, 1996). For example, relatively high levels of serum IgG_2 antibody against serotype type b antigen have been found in localized juvenile periodontitis patients with limited destruction, compared to patients with the generalized form of the disease. Antibody may facilitate phagocytic killing of *A. actinomycetemcomitans* by polymorphonuclear leukocytes and may neutralize the *A. actinomycetemcomitans* leukotoxin.

The pathognomonic molar–incisor lesions in localized juvenile periodontitis may be explained in the context of an effective serum IgG immune response against *A. actinomycetemcomitans* antigens. *A. actinomycetemcomitans* colonizes the oral cavity during early childhood and initiates periodontal destruction on the first permanent teeth to erupt, the incisors and first molars. Anti-*A.*-*actinomycetemcomitans* antibody response after initial periodontal destruction may block the seeding of *A. actinomycetemcomitans* cells to surfaces of later erupting teeth, thereby preventing periodontal breakdown of those teeth. When the antibody levels become high enough to neutralize the insult of the large number of *A. actinomycetemcomitans* cells in the deep original lesions, the disease may be contained and clinically appear as 'burned out'.

The effectiveness of serum IgG_2 immune response against *A. actinomycetemcomitans* infection may be related to the ability of polymorphonuclear leukocytes to bind bacteria via F_c surface receptors. Polymorphonuclear leukocytes bearing an F_c-receptor allotype H131 exhibit more efficient binding and killing of serum IgG_2-coated *A. actinomycetemcomitans* than leukocytes with the alternative F_c-receptor allotype R131 (Wilson and Kalmar, 1996). Hosts homozygous for the R131 gene may be at higher risk for early-onset periodontitis.

5.3.7 Generalized juvenile periodontitis

Generalized juvenile periodontitis may develop in persons with insufficient levels of protective anti-*A.-actinomycetemcomitans* antibodies. Fewer teeth are periodontally involved in patients with precipitating anti-*A.-actinomycetemcomitans* antibodies than in patients lacking such antibodies. However, generalized juvenile periodontitis can also result from combined infections of *P. gingivalis* and other periodontopathic organisms, and *A. actinomycetemcomitans*.

5.4 PREPUBERTAL PERIODONTITIS

Prepubertal periodontitis afflicts primary teeth and occurs in a localized and a generalized form (Dougherty and Slots, 1993). The disease affects about 1% of American children and appears to have a familial predisposition. Prepubertal periodontitis may be followed by severe periodontitis of the permanent teeth or by a normal permanent dentition.

In localized prepubertal periodontitis, the onset of disease is at the age of 4 years or before. The gingival tissue manifests only minor inflammation, dental plaque is minimal and disease progression can be halted by mechanical debridement combined with antibiotic therapy. Many individuals with localized prepubertal periodontitis demonstrate defective chemotaxis of either peripheral polymorphonuclear leukocytes of monocytes.

Localized prepubertal periodontitis is associated with *P. intermedia*, *A. actinomycetemcomitans*, *Eikenella corrodens*, *Fusobacterium* species, *P. gingivalis* and other organisms. Treatment should not include systemic tetracycline because of the young age of the patients; systemic metronidazole–amoxycillin (250 mg of each two to three times daily for 8 days) may be considered in older children.

In generalized prepubertal periodontitis, disease onset is at the time of tooth eruption. Severe gingival inflammation is a common feature;

the disease may be refractory to most treatments and affected children may exhibit otitis media and other recurrent infections. Most individuals with generalized prepubertal periodontitis show profound abnormalities of both peripheral polymorphonuclear leukocytes and monocytes. The microbiology of generalized prepubertal periodontitis has not been determined.

5.5 ADULT PERIODONTITIS

5.5.1 Clinical features

The term adult periodontitis is an umbrella appellation for a wide range of clinical and infectious entities. Clinically, few or all teeth in a dentition can be involved, individual teeth can show vertical or horizontal alveolar bone loss, the associated gingivitis can vary from very slight inflammatory changes to severe bleeding with pus formation, disease progression can be rapid or slow, supragingival plaque levels can show little or no relationship to the degree of periodontal breakdown, and periodontitis lesions can respond favorably or be recalcitrant to mechanical debridement. Differences in the disease-producing microflora account for some of the clinical variation. Also, as discussed in section 5.3, host responses against periodontal pathogens may modify considerably the clinical course of the disease.

5.5.2 Subgingival microflora

Adult periodontitis lesions contain high proportions of anaerobes (90%), Gram-negative organisms (75%), and spirochetes (30%). Many cases of adult periodontitis show greatly elevated proportions of a limited number of microorganisms. However, the composition of the periodontopathic microflora can differ markedly from patient to patient and from pocket to pocket in a given patient. *P. gingivalis*, *P. intermedia* and *A. actinomycetemcomitans* became recognized in the early 1980s as major pathogens in adult periodontitis. *P. gingivalis* is closely associated with advanced adult periodontitis and numerically is one of the most important pathogens in the disease. *P. gingivalis* possesses some of the highest virulence potential of any oral organism tested so far. *P. intermedia* and *P. nigrescens* are associated with periodontitis as well as gingivitis, complicating the exact assignment of periodontopathic significance. *A. actinomycetemcomitans* occurs in about one-third of severe adult periodontitis cases. The periodontopathic potential of this organism is detailed in section 5.3

Periodontitis progresses in spurts interspersed with quiescent phases in which little tissue destruction occurs. It is imperative to distinguish disease-active and disease-inactive periodontal lesions in order to differentiate between bacterial species that produce periodontal destruction and those that are merely secondary colonizers of deep periodontal pockets. Some of the species associated with active lesions include *Bacteroides forsythus*, *P. gingivalis*, *Peptostreptococcus micros*, *A. actinomycetemcomitans*, *Campylobacter rectus*, *P. intermedia*, *F. nucleatum* ssp. *nucleatum*, *Treponema denticola*, *Selenomonas noxia* and various *Eubacterium* species.

Certain subgingival microbial complexes show a particularly strong relationship with active periodontitis, underscoring the mixed infectious nature of the disease. Species that occur together may produce additive or synergistic damage to the periodontal tissues.

5.5.3 Periodontal superinfections with non-oral bacteria

Periodontal superinfections with non-oral bacteria occur more commonly in medically compromised patients, patients receiving prolonged chemotherapy against cancer or infectious diseases, or geriatric patients. Non-oral organisms including 45 species of enteric rods, pseudomonads and *Actinetobacter* have been recovered from deep periodontal pockets. The most frequently isolated species are *Enterobacter cloacae*, *Klebsiella pneumoniae*, *Pseudomonas aeruginosa* and *Klebsiella oxytoca*. Periodontal superinfections can also involve *Staphylococcus epidermidis*, *S. aureus*, enterococci and *Candida albicans*. The presence of superinfecting non-oral organisms can significantly influence the selection of a proper antibiotic therapy.

5.5.4 Viral infections and adult periodontitis

With the exception of destructive periodontitis in HIV-positive patients (section 5.6.1), the possible involvement of human viruses in periodontitis has still to be elucidated. Recent studies found human Cytomegalovirus and Epstein–Barr virus to be more frequent in subgingival specimens from advanced periodontitis than from gingivitis sites (Contreras and Slots, 1996; Parra and Slots, 1996). Several possible mechanisms of viral potentiation of periodontitis have been suggested. Viruses may impair the local immune defense and allow bacterial superinfections, promote microbial colonization of the gingival crevice by altering the expression of host cell surface receptor for bacteria, exert direct cytopathic effects or cause tissue destruction via immunopathological mechanisms.

5.5.5 Refractory periodontitis

The term is used to describe the approximately 10% of adult periodontitis patients who experience continued loss of periodontal attachment despite comprehensive mechanical periodontal therapy and systematic maintenance care. The major reason for unsuccessful periodontal therapy is inability to control subgingival pathogenic organisms because of their location in gingival tissue or root structures inaccessible to mechanical instrumentation. Some refractory patients may have weakened periodontal defense, which allows even small numbers of pathogens to overwhelm the periodontium. Supplemental usage of systemic antimicrobial agents can help eliminate or markedly suppress subgingival pathogens.

5.5.6 Peri-implantitis

Tooth-form implants are being used in humans to replace missing teeth. Most dental implants osseointegrate without complications. Successful implants exhibit scarce plaque accumulation, comprising mainly streptococci and *Actinomyces* species and a few spirochetes.

Implant may be lost because of shortcomings in surgical techniques or use of excessive occlusal loading forces (traumatic failures). Implant failure caused by trauma is associated with few or no periodontopathic organisms and spirochetes. Implant may also be lost as a result of microbial infections (infectious failures). Infectious failures demonstrate a complex microflora of conventional oral pathogens and of organisms unusual in the subgingival flora (Table 5.2; Alcoforado *et al.*, 1991).

A failing implant can be associated with as many as six species of periodontal pathogen. The infecting organisms originate from natural teeth with periodontal disease. Overgrowth of staphylococci, enteric rods, pseudomonads and yeasts may occur after prolonged use of systemic or topical antimicrobial agents.

Table 5.2 Pathogenic microflora in failing dental implants

Species	No. of implants (total = 18)	Percent of total microflora in infected implants
Peptostreptococcus micros	6	24
Campylobacter rectus	6	6
Fusobacterium spp.	5	11
Candida albicans	5	5
Prevotella intermedia	4	2
Capnocytophaga spp.	3	7
Pseudomonas aeruginosa	3	28
Staphylococcus spp.	2	2
Klebsiella pneumoniae	2	27
Escherichia coli	2	19
Actinobacillus actinomycetemcomitans	1	6
Non-pigmented Prevotella spp.	1	1
Eubacterium aerofaciens	1	2
Enterobacter cloacae	1	4

5.6 PERIODONTITIS IN MEDICALLY COMPROMISED PATIENTS

5.6.1 HIV

Periodontitis in HIV-infected individuals often exhibits exceedingly rapid onset and progression, and can be associated with ANUG and acute oral pain. Exposed alveolar bone may occur in areas of severe gingival tissue necrosis (Murray, 1994).

HIV periodontitis lesions reveal spirochetes, *Fusobacterium* species, *A. actinomycetemcomitans*, *C. rectus*, *P. micros* and *P. intermedia*, each averaging 5–20% of the subgingival microflora in positive patients. Some HIV periodontitis lesions yield organisms not usually recovered from common forms of adult periodontitis, including *Bacteroides fragilis, Fusobacterium necrophorum, Eubacterium aerofaciens, Clostridium* species, enterococci, *P. aeruginosa*, various enteric rods and *C. albicans*. The unusual organisms may contribute to the particular pathoetiological characteristics of HIV periodontitis. They may also constitute the source of some cases of life-threatening disseminated candidiasis and of systemic infections associated with *Pseudomonadaceae* and *Enterobacteriaceae* (Rams *et al.*, 1991; Zambon, Reynolds and Genco, 1990).

5.6.2 Diabetes mellitus

Type I (insulin-dependent) diabetes mellitus is associated with a higher frequency of periodontitis compared to age-matched non-diabetic controls. However, patients with well-controlled type I diabetes who practice good oral hygiene may not necessarily demonstrate increased prevalence of severe periodontitis. *Capnocytophaga* species and *P. intermedia* have been related to the initial breakdown in diabetic periodontitis. *C. rectus*, *A. actinomycetemcomitans* and *P. gingivalis* may also play roles in some forms of type I diabetic periodontitis. Type II (non-insulin-dependent) diabetes does not seem to constitute a risk factor for severe periodontitis.

5.6.3 Myelosuppressed cancer patients

Leukemia patients on myelosuppressive therapy are at great risk of developing progressive periodontitis and periodontal abscesses, especially during periods of pronounced granulocytopenia. Cytotoxic drug therapy may facilitate bacterial invasion and colonization of subepithelial gingival tissues. In contrast, renal transplant patients receiving non-cytotoxic azathioprine or corticosteroids do not demonstrate a higher incidence of periodontitis.

The combined use of immunosuppressing agents and broad-spectrum antibiotics gives rise to periodontal pocket colonization by opportunistic pathogens, including enteric rods, pseudomonads, staphylococci and yeasts. It is important to realize that in patients with acute leukemia these opportunistic pathogens might seed from periodontal lesions into the blood stream and induce life-threatening septicemia. Also, many superinfecting enteric rods and pseudomonads are resistant to chlorhexidine and other oral antiseptics and may overgrow in the oral cavity after prolonged use of mouthrinses.

5.6.4 Neutropenia

Patients with chronic or cyclical neutropenia of genetic origin frequently show advanced generalized periodontitis. Neutropenia may enhance subgingival microbial colonization and proliferation. *A. actinomycetemcomitans*, black-pigmented *Porphyromonas/Prevotella* species, *Fusobacterium* species and spirochetes occur in high levels in periodontitis lesions of neutropenia patients.

5.7 TRANSMISSION OF PERIODONTAL PATHOGENS

The source of most if not all human oral bacteria appears to be other humans. Although the subgingival microflora in adults appears to be relatively stable, changes in microflora may occur throughout life. Genotype analysis of *A. actinomycetemcomitans* and *P. gingivalis* isolates from family members show evidence of vertical and horizontal transmission (Saarela *et al.*, 1993; Alaluusua *et al.*, 1993). For *A. actinomycetemcomitans* and *P. gingivalis*, horizontal transmission between spouses occurs in 20–35% of families (Asikainen, Chen and Slots, 1996).

Bacteria-induced periodontitis behaves as a transmissible infection modified by host immune response, oral hygiene measures, cigarette smoking and other risk factors. Not surprisingly, destructive periodontal disease may follow the acquisition of periodontal pathogens from cohabitants. Subjects married to severe periodontitis patients harbor more periodontal pathogens and exhibit more severe periodontal tissue destruction than spouses of periodontally healthy or gingivitis subjects (Von Troil-Lindén *et al.*, 1995). Prevention of initial colonization by periodontal pathogens comprises a key strategy in prevention of periodontal disease.

5.8 VIRULENCE OF PERIODONTOPATHIC ORGANISMS

Virulence of an organism is a multifactorial property influenced by the inherent pathogenic potential of the organism, the habitat of the organism and a number of host determinants. Also, although a certain microbial property may be necessary for virulence, the mere possession of the property is not sufficient to make an organism virulent. In order to produce periodontitis, an organism must:

- establish close proximity to periodontal tissues;
- avoid being swept away by saliva or gingival crevice fluid;
- acquire essential nutrients for growth;
- resist bacterial antagonism and local host defenses;
- be able to induce periodontal tissue destruction.

Table 5.3 lists putative microbial virulence factors in destructive periodontal disease.

5.8.1 Proximity to periodontal tissues

Different types of bacteria–host interaction are recognized. 'Association' implies a weak, reversible attachment and may be characteristic of the subgingival colonization of spirochetes. 'Adhesion' describes a relatively stable attachment mediated by bacterial adhesins to host surface receptors; an example may be colonization of *P. micros* of gingival crevice epithelial cells. 'Invasion' describes the translocation of bacteria into tissue or inside host cells; an example may be the tissue invasion and intracellular invasion of *A. actinomycetemcomitans* in localized juvenile periodontitis lesions.

Table 5.3 Selected microbial virulence determinants in periodontal disease; specific organisms may possess a few or several of the virulence factors listed below (Slots and Genco, 1984; Holt and Bramanti, 1991)

Colonization	Evasion of host defense	Tissue destruction
Fimbriae	Leukoaggressins:	Collagenase
Capsule	Chemotaxis inhibitors	Hyaluronidase
Lipopolysaccharide	Phagocytosis inhibitors	Sulfur compounds
Microbial antagonism/	Killing inhibitors	Endotoxin-mediated
synergism	Proteases against:	bone resorption
	Immunoglobulins	Acid phosphatases
	Complement components	Epithelial cell toxin
	Plasma proteinase cascade	Fibroblast inhibitors
	components:	Endothelial cell toxin
	Fibrinolysin	
	Proteinase inhibitors	
	Siderophores	
	Cytotoxins	
	T-suppressor cell stimulation	
	Antigenic shift	
	Surface F_c-receptor	

5.8.2 Overcoming host defenses

After gaining access to the periodontal area, a microbe must overcome humoral and cellular defense mechanisms. Host defense mechanisms may be evaded by rendering them impotent (e.g. *P. gingivalis* proteases destroying immunoglobulins and complement proteins), by avoiding their effect (e.g. *Capnocytophaga* releasing products that inhibit chemotaxis of polymorphonuclear leukocytes) or by withstanding their action (e.g. *A. actinomycetemcomitans* exerting resistance to lysis by complement).

(a) Humoral host defenses

Immunoglobulins (Ig) and complement play important roles in the host defense against periodontal pathogens. Secretory IgA is the major antibody class in saliva. Its main action is to restrict bacterial adhesion. IgG is the dominant antibody class in gingival crevice fluid and gingival tissues. Its antibacterial functions include opsonization, IgG complement-mediated killing, and neutralization of bacterial enzymes and toxins.

Periodontal organisms may evade the effects of immunoglobulins by:

- degrading the immunoglobulin molecule with trypsinlike enzyme or specific immunoglobulin proteases (*P. gingivalis*, *B. forsythus*, *T. denticola* and some strains of *Capnocytophaga*);
- possessing F_c receptors for IgG (*Staphylococcus aureus*, *A. actinomycetemcomitans* and preventing proper display of the F_c portion of IgG on the bacterial surface critical for attachment of phagocytic cells;
- altering the antigens against which the immunoglobulins are directed (antigenic shift in streptococci).

(b) Complement

Bacterial resistance to complement can be achieved by enzymatic degradation of the complement proteins (*P. gingivalis*) or by forming bacterial cell envelope components (capsule, outer membrane proteins, lipopolysaccharide) exhibiting little or no ability to activate complement.

(c) Other serum factors

Additional host-protective serum factors include: lysozyme, which has a lytic effect especially against Gram-positive cells; transferrin and lactoferrin, which bind iron necessary for microbial growth; fibronectin, which can bind to bacteria and exert opsonic activity for phagocytic cells; and C-reactive protein, which appears in high levels during the acute phase of an inflammatory response and has opsonic and complement-activating potential. Microbes may evade these antimicrobial proteins by enzymatic degradation (*P. gingivalis*). They may obtain essential iron by producing high-affinity iron-chelating agents (siderophores) that compete directly with transferrin and lactoferrin or by degrading iron-binding and heme-containing proteins (haptoglobin, hemopexin, albumin).

(d) Cellular host defenses

An effective cellular antimicrobial defense involves:

- directed migration of resident or blood-borne phagocytes;
- attachment and ingestion of microbe;
- stimulation of oxidative burst;
- phagosome–lysosome fusion;
- killing of microbe by reduced pH, microbiocidal oxygen intermediates (hydroxyl radical, superoxide anion, hydrogen peroxide) or micro-biocidal cationic peptides;
- activation of specific immune response;
- phagocyte response to cytokines;
- removal of microbial and cellular debris.

In periodontal disease, the polymorphonuclear leukocyte is the key protective cell, as inferred from the occurrence of severe periodontitis in patients with defective cells. As an example, killing of *A. actinomycetemcomitans* requires polymorphonuclear leukocytes and opsonins, and may be executed by either oxidative or non-oxidative mechanisms.

(e) Innate cellular defense

On the one hand, leukocyte locomotion can be stimulated by N-formy-lated tripeptides released from the organism itself or by serum-derived chemoattractants such as C5a. On the other hand, microbes (*Capnocytophaga, A. actinomycetemcomitans*) may defend themselves by releasing toxins and other substances that can impede the directed migration of leukocytes to various chemoattractants. To avoid phagocytosis, a microbe may produce antiphagocytic surface components (either polysaccharide or protein in nature) that prevent leukocyte attachment and internalization. Phagocytosis of encapsulated bacteria may be avoided as a result of decreased binding of opsonic IgG and C3b, unavailability of opsonic products or decreased hydrophobicity of the microbial cell surface.

Microbial strategies to overcome oxygen-dependent and oxygen-independent killing mechanisms of leukocytes include little or no stimulation of the oxidative burst after phagocytosis (*A. actinomycetemcomitans*) or neutralization of toxic oxygen species by the microbial enzymes superoxide dismutase or catalase. In addition to the functional impairments described above, considerable attention has been given to the leukotoxin of *A. actinomycetemcomitans*. Apparently this leukotoxin shares biological properties with and is genetically similar to leukotoxins of other (non-oral) species in the family *Pasteurellaceae* well as with *Escherichia coli* α-hemolysin. Superinfecting *S. aureus* and *P. aeruginosa* and perhaps *C. rectus* also produce leukotoxins. By killing or impairing leukocytes entering the periodontal tissue and pocket, microbial leukotoxins may constitute important virulence determinants, especially in the absence of neutralizing antibodies.

(f) Immunological cellular defense

Microbes may circumvent immunologically regulated cellular defense mechanisms by changing antigenic composition (antigenic shift) or by inducing immunosuppression (*A. actinomycetemcomitans*-mediated activation of suppressor T cells). Periodontal infections may also activate macrophages to release prostaglandin E_2 and perhaps other factors suppressive of lymphocyte proliferation.

5.8.3 Periodontal tissue destruction

The development of periodontitis involves breakdown of the supportive periodontal ligament and alveolar bone. Periodontal tissues may be destroyed either by direct release of damaging bacterial products or by microbially induced destructive host reactions.

(a) Direct

Microbial enzymes, toxins and metabolites can directly harm the host. Some periodontal species produce collagenase capable of degrading native collagen (*P. gingivalis*) or trypsin-like enzyme active against altered collagen (*P. gingivalis, T. denticola, B. forsythus*, certain *Bacteroides, Prevotella* and *Capnocytophaga* strains). Also, *P. gingivalis, A. actinomycetemcomitans* and other organisms may activate collagenase from polymorphonuclear leukocytes, fibroblasts and macrophages. Acid phosphatases produced by many periodontal bacteria may participate in bone breakdown, especially when tissue-invading organisms reside in close proximity to bone.

Fibroblast-inhibiting toxins produced by several periodontal organisms (*P. gingivalis, P. intermedia, A. actinomycetemcomitans, Capnocytophaga sputigena*) may interfere with synthesis and turnover of collagen, resulting in a net loss of collagen. Lipopolysaccharides (endotoxins) from many organisms are able to induce bone resorption in *in vitro* bone systems.

Moreover, *A. actinomycetemcomitans* releases a specific bone-resorption-inducing toxin distinct from the endotoxin.

Several suspected periodontal pathogens release volatile sulfides (methylmercaptan, hydrogen sulfide, dimethyl sulfide) that are inhibitory to collagen and non-collagenous protein synthesis.

(b) Indirect

The periodontal host defense system is a two-edged sword. On the one hand, it protects against microbial infections and tumor development; on the other hand, it is capable of causing severe tissue injury. Lipopolysaccharide of *P. gingivalis* and other bacteria stimulates *in vitro* the release of prostaglandin E_2 and interleukin-1β (previously termed osteoclast activating factor) from macrophages and fibroblasts, and tumor necrotic factor from macrophages, all of which demonstrate bone-resorbing potential and inflammation-inducing activities. Butyric and other short-chain fatty acids may also stimulate interleukin-1β production, as well as suppressing the production of T lymphocytes. A number of immunopathological processes may also play roles in periodontal tissue destruction.

5.9 MICROBIAL DIAGNOSIS

Bacterial specificity in periodontal disease has important implications in clinical management of periodontitis patients, especially when antibiotic therapy is contemplated (Slots and Rams, 1996). Since different patients may harbor vastly different subgingival microflora, microbial analysis can be crucial in treatment of periodontal disease.

Culture methods can provide a determination of the relative occurrence of specific pathogens among total cultivable bacteria and crucial antibiotic sensitivity profiles of the pathogens. Oral microbiologists have developed selective and non-selective isolation media for periodontal organisms. Classification schemes for rapid identification of suspected periodontal pathogens have also been developed (Slots, 1986; Tanner, Lai and Maiden, 1992; Maiden, Lai and Tanner, 1992).

Subgingival microflora exhibits a predictable shift from predominantly cocci in health to mainly rods and motile organisms in disease. Direct examination of subgingival plaque by phase contrast and dark-field microscopy provides information about the size, shape and the motility of subgingival bacteria, and is an excellent method of screening for subgingival spirochetes.

Quantitative and semi-quantitative immunological detection methods have been developed to detect periodontal pathogens. Immunofluorescence assays provide a quantitative measurement of the target organisms

(Bonta *et al.*, 1985; Zambon *et al.*, 1985). In dot blot assays, bacterial specimens are immobilized on nitrocellulose membranes and allowed to react with immune sera. Following treatment with an enzyme-conjugated secondary antibody, the presence of the target organism is revealed with the addition of enzyme substrate. In enzyme-linked immunosorbent assay (ELISA; Dzink *et al.*, 1983), primary antibody is immobilized to the wells of a microtiter plate. Bacterial specimens are allowed to react with the primary antibody and the target bacteria are revealed with enzyme-conjugated primary or secondary antibody and enzyme substrate. Latex agglutination assays provide a fast and easy chair-side method for microbial identification (Nisengard *et al.*, 1992). The primary antibody is bound to polystyrene latex beads and the addition of target organisms in the specimens induces agglutination reactions visible to the unaided eye.

DNA probes have been developed for many periodontal pathogens. Available DNA probe methodologies can detect as few as 1000 cells of the target species (Tay *et al.*, 1992). However, whole cell genomic DNA contains sequences common to many bacteria and may yield unacceptable cross-reactions. Short DNA probes comprising 18–20 bases provide a more specific reaction but generate weaker signals, which reduces the sensitivity of the method. PCR detection methods appear to satisfy the requirement of high specificity and sensitivity for an ideal periodontal bacterial detection system. Optimized PCR methods can detect as few as 50 cells of the target species. Presently, eight periodontal pathogens can be reliably detected by PCR (Ashimoto *et al.*, 1996).

5.10 THERAPY

Modern periodontal therapy has a strong focus on suppression or eradication of specific periodontal pathogens (Slots and Rams, 1996). Periodontal prophylaxis and therapy seek to reduce or eliminate bacterial dental plaque and plaque-derived products and to aid in repair or regeneration of tissue. Conventional periodontal therapy encompasses oral hygiene and plaque control, scaling/root planing and surgery. Periodontal surgery helps to get access to the root surface for thorough cleaning and to promote re-establishment of periodontal connective tissue attachment. Conventional periodontal treatment is effective in curing gingivitis and many types of periodontitis. However, about 10% of patients with advanced periodontitis continue to experience further breakdown despite diligent mechanical debridement. These refractory cases occupy a large portion of the time and efforts of a periodontal specialist practice.

Treatments differ in their ability to eradicate periodontal pathogens. Non-surgical subgingival scaling and root planing can remove subgingival *C. rectus* but is mostly ineffective against *P. gingivalis*, *P. intermedia*,

B. forsythus, staphylococci and enteric rods, and virtually does not affect *P. micros* and *A. actinomycetemcomitans*. Debridement may fail to remove organisms because of colonization of subepithelial gingival tissues (*A. actinomycetemcomitans*), crevicular epithelial cells (*P. micros*, *P. intermedia*, *B. forsythus*), cementum and radicular dentinal tubuli, or furcation areas inaccessible to instrumentation.

Subgingival irrigation with antiseptic agents (e.g. chlorhexidine, iodine) or placement of slow-release antibiotic fibers in conjunction with mechanical debridement may enhance suppression of some periodontal organisms. However, topical antimicrobials are generally ineffective in reducing periodontal *A. actinomycetemcomitans*, *P. micros* and enteric rods/pseudomonads. Topical tetracycline therapy can even result in higher subgingival *A. actinomycetemcomitans* counts, possibly because of suppression of inhibitory organisms. Lack of effectiveness of topical antimicrobial therapy probably stems from inability of the agents to reach the target microorganisms. Possibly, topical antimicrobial agents may be valuable in controlling small numbers of periodontal pathogens in maintenance care.

To eradicate periodontal infections, systemic antimicrobial therapy is commonly used. Systemic antimicrobial medication enters the periodontal tissues and pocket via serum and can affect organisms outside the reach of cleaning instruments and topical antiseptic therapy. Systemic antimicrobial therapy is required for eradication of *A. actinomycetemcomitans* and some other bacterial species in the subgingival microflora. It is important to note that systemic antimicrobials aim solely to control specific periodontal pathogens and are not to be regarded as substitutes for conventional plaque control measures.

Because several microorganisms with differing antimicrobial susceptibility can produce destructive periodontal disease, the choice of antimicrobial therapy should ideally be based on a microbiological analysis including *in vitro* antimicrobial susceptibility testing. Other important therapeutic considerations are the medical status of the patient, potential side-effects of the antibiotics and possible drug interactions. Single drug therapies with penicillins, tetracyclines or metronidazole have been used frequently in periodontal practice. Since the realization that several pathogens often inhabit a given periodontal lesion, drug combination therapies have gained increasing importance. Metronidazole–amoxycillin (250 mg of each three times a day for 8 days) is effective in eradicating subgingival *A. actinomycetemcomitans* and other periodontopathic organisms. Metronidazole–ciprofloxacin (500 mg of each twice a day for 8 days) constitutes an interesting drug combination therapy for periodontal lesions harboring anaerobes and enteric rods. After systemic metronidazole–ciprofloxacin therapy, resistant streptococci predominate in the periodontal pocket microbiota. Because several viridans streptococcal strains inhibit several periodontal pathogens and are themselves not periodontopathogens, an overgrowth of viridans streptococci in a periodontal pocket constitutes a desirable therapeutic outcome.

5.11 SUMMARY

Relatively specific microflora are associated with various types of periodontal conditions. The healthy gingival sulcus harbors relatively few cells, usually of the genera *Streptococcus* and *Actinomyces*. The development of gingivitis is accompanied by a substantial increase in Gram-negative organisms, including *Fusobacterium nucleatum* subspecies, *Prevotella intermedia* and numerous spirochetes and motile rods. The subgingival microflora in adult periodontitis consists predominantly of *Porphyromonas gingivalis*, *Prevotella intermedia*, *Bacteroides forsythus*, *Fusobacterium nucleatum* subspecies, *Campylobacter rectus*, *Selenomonas* species, *Actinobacillus actinomycetemcomitans*, *Treponema denticola*, *Peptostreptococcus micros* and several other microbial species.

In localized juvenile periodontitis, *A. actinomycetemcomitans* seems to be a key pathogen. It appears that several unique clinical features of localized juvenile periodontitis result from a combination of periodontal *A. actinomycetemcomitans* infection, specific local and systemic immune factors and time of tooth eruption. Periodontitis in medically compromised patients may be associated with various enteric rods, pseudomonads, staphylococci, yeasts and other unusual periodontal organisms. Peri-implantitis lesions can show a great variety of infecting microorganisms.

The virulence potential of some major suspected periodontal pathogens have been identified. Subgingival microbial colonization may be mediated by fimbriae or other surface components and may be influenced by other microorganisms. Microbial mechanisms to overcome the periodontal defense may include leukotoxins and other microbial products capable of impairing polymorphonuclear leukocytes. Microbes may induce periodontal tissue destruction directly by releasing collagenases and bone-destroying toxins or indirectly by stimulating damaging host reactions.

The concept of bacterial specificity in human periodontitis forms the basis for treatment strategies that focus upon the suppression or elimination of selected periodontal pathogens. Some organisms can be removed by conventional periodontal therapy whereas other organisms require the adjunctive use of systemic antimicrobial agents. Metronidazole–amoxycillin and metronidazole–ciprofloxacin constitute promising combination therapies in many mixed periodontal infections. Microbial monitoring of the periodontal microflora can have major value in treatment planning and in assessing treatment efficacy.

5.12 REFERENCES

Alaluusua, S., Saarela, M., Jousimies Somer, H. and Asikainen, S. (1993) Ribotyping shows intrafamilial similarity in *Actinobacillus actinomycetemcomitans* isolates. *Oral Microbiology and Immunology*, **8**, 225–229.

Alcoforado, G. A., Rams, T. E., Feik, D. and Slots, J. (1991) Microbial aspects of failing osseointegrated dental implants in humans. *Journal de Pariodontologie*, **10**, 11–18.

Ashimoto, A., Chen, C., Bakker, I. and Slots, J. (1996) Polymerase chain reaction detection of 8 putative periodontal pathogens in subgingival plaque of gingivitis and advanced periodontitis lesions. *Oral Microbiology and Immunology*, **11**, 266–273.

Asikainen, S., Chen, C. and Slots, J. (1996) Likelihood of transmitting *Actinobacillus actinomycetemcomitans* and *Porphyromonas gingivalis* in families with periodontitis. *Oral Microbiology and Immunology*, **11**, 387–394.

Asikainen, S., Lai, C. H., Alaluusua, S. and Slots, J. (1991) Distribution of *Actinobacillus actinomycetemcomitans* serotypes in periodontal health and disease. *Oral Microbiology and Immunology*, **6**, 115–118.

Bonta, Y., Zambon, J. J., Genco, R. J. and Neiders, M. E. (1985) Rapid identification of periodontal pathogens in subgingival plaque: comparison of indirect immunofluorescence microscopy with bacterial culture for detection of *Actinobacillus actinomycetemcomitans*. *Journal of Dental Research*, **64**, 793–798.

Brown, L. J. and Loe, H. (1993) Prevalence, extent, severity and progression of periodontal disease. *Periodontology 2000*, **2**, 57–71.

Christersson, L. A., Mbini, B., Zambon, J. (1987) Tissue localization of *Actinobacillus actinomycetemcomitans* in human periodontitis. 1. Light, immunofluorescence and electron microscopic studies. *Journal of Periodontology*, **58**, 529–539.

Chung, C. P., Nisengard, R. J., Slots, J. and Genco, R. J. (1983) Bacterial IgG and IgM antibody titers in acute necrotizing ulcerative gingivitis. *Journal of Periodontology*, **54**, 557–562.

Contreras, A. and Slots, J. (1996) Mammalian viruses in human periodontitis. *Oral Microbiology and Immunology*, **11**, 381–386.

Contreras, A., Falkler W. A. Jr, Enwonwu, C. O. *et al.* (1997) Human Herpesviridae in acute necrotizing ulcerative gingivitis in Nigerian children. *Oral Microbiology and Immunology*, **12**, 259–265.

DiRienzo, J. M. and Slots, J. (1990) Genetic approach to the study of epidemiology and pathogenesis of *Actinobacillus actinomycetemcomitans* in localized juvenile periodontitis. *Archives of Oral Biology*, **35**(Suppl.), 79S–84S.

Dougherty, M. A. and Slots, J. (1993) Periodontal diseases in young individuals. *Journal of the California Dental Association*, **21**, 55–69.

Dzink, J. L., Socransky, S. S., Ebersole, J. L. and Frey, D. E. (1983) ELISA and conventional techniques for identification of black-pigmented *Bacteroides* isolated from periodontal pockets. *Journal of Periodontology Research*, **18**, 369–374.

Fives-Taylor, P., Meyer, D. and Mintz, K. (1995) Characteristics of *Actinobacillus actinomycetemcomitans* invasion of and adhesion to cultured epithelial cells. *Advances in Dental Research*, **9**, 55–62.

Goldhaber, P. and Giddon, D. B. (1964) Present concepts concerning the etiology and treatment of acute necrotizing ulcerative gingivitis. *International Dental Journal*, **14**, 468–496.

Hart, T. C. (1996) Genetic risks factors for early-onset periodontitis. *Journal of Periodontology*, **67**, 355–366.

Hillman, J. D., Socransky, S. S. and Shivers, M. (1985) The relationships between streptococcal species and periodontopathic bacteria in human dental plaque. *Archives of Oral Biology*, **30**, 791–795.

Holt, S. C. and Bramanti, T. E. (1991) Factors in virulence expression and their role in periodontal disease pathogenesis. *Critical Reviews of Oral Biology and Medicine*, **2**, 177–281.

Loesche, W. J., Syed, S. A., Laughon, B. E. and Stoll, J. (1982) The bacteriology of acute necrotizing ulcerative gingivitis. *Journal of Periodontology*, **53**, 223–230.

Maiden, M. F. J., Lai, C.-H. and Tanner, A. (1992) Characteristics of oral gram-positive bacteria, in *Contemporary Oral Microbiology and Immunology*, (eds J. Slots and M. A. Taubman), C. V. Mosby, St Louis, MO, pp. 342–372.

Meyer, D. H. and Fives-Taylor, P. M. (1994) Characteristics of adherence of *Actinobacillus actinomycetemcomitans* to epithelial cells. *Infection and Immunity*, **62**, 928–935.

Moore, W. E. C. and Moore, L V. H. (1994) The bacteria of periodontal diseases. *Periodontology 2000*, **5**, 66–77.

Murray, P. A. (1994) Periodontal diseases in patients infected by human immuno-deficiency virus. *Periodontology 2000*, **6**, 50–67.

Nisengard, R. J., Mikulski, L., McDuffie, D. and Bronson, P. (1992) Development of a rapid latex agglutination test for periodontal pathogens. *Journal of Periodontology*, **63**, 611–617.

Parra, B. and Slots, J. (1996) Detection of human viruses in periodontal pockets using polymerase chain reaction. *Oral Microbiology and Immunology*, **11**, 289–293.

Perry, M. B., MacLean, L. L., Gmur, R. and Wilson, M. E. (1996) Characterization of the O-polysaccharide structure of lipopolysaccharide from *Actinobacillus actinomycetemcomitans* serotype b. *Infection and Immunity*, **64**, 1215–1219.

Rams, T. E., Andriolo, M. Jr, Feik, D. *et al.* (1991) Microbiological study of HIV-related periodontitis. *Journal of Periodontology*, **62**, 74–81.

Rosan, B., Slots, J., Lamont, R. J. *et al.* (1988) *Actinobacillus actinomycetemcomitans* fimbriae. *Oral Microbiology and Immunology*, **3**, 58–63.

Saarela, M., von Troil-Lindén, B., Torkko, H. *et al.* (1993) Transmission of oral bacterial species between spouses. *Oral Microbiology and Immunology*, **8**, 349–354.

Slots, J. (1986) Rapid identification of important periodontal microorganisms by cultivation. *Oral Microbiology and Immunology*, **1**, 48–57.

Slots, J. and Genco, R. J. (1984) Black-pigmented *Bacteroides* species, *Capnocytophaga* species, and *Actinobacillus actinomycetemcomitans* in human periodontal disease: virulence factors in colonization, survival, and tissue destruction. *Journal of Dental Research*, **63**, 412–421.

Slots, J. and Rams, T. E. (1996) Systemic and topical antimicrobial therapy in periodontics. *Periodontology 2000*, **10**, 5–159.

Slots, J. and Schonfeld, S. E. (1991) *Actinobacillus actinomycetemcomitans* in localized juvenile periodontitis, in *Periodontal Disease: Pathogens and Host Immune Response*, (eds S. Hamada, S. C. Holt and J. R. McGhee), Quintessence, Tokyo, pp. 53–64.

Slots, J., Reynolds, H. S. and Genco, R. J. (1980) *Actinobacillus actinomycetemcomitans* in human periodontal disease: a cross-sectional microbiological investigation. *Infection and Immunity*, **29**, 1013–1020.

Socransky, S. S. (1979) Criteria for the infectious agents in dental caries and periodontal disease. *Journal of Clinical Periodontology*, **6**, 16–21.

Socransky, S. S. and Haffajee, A. D. (1994) Microbiology and immunology of periodontal diseases. *Periodontology 2000*, **5**, 7–168.

Tanner, A., Lai, C.-H. and Maiden, M. (1992) Characteristics of oral gram-negative species, in *Contemporary Oral Microbiology and Immunology*, (eds J. Slots and M. A. Taubman), C. V. Mosby, St Louis, MO, pp. 299–341.

Tay, F., Liu, Y. B., Flynn, M. J. and Slots, J. (1992) Evaluation of a non-radioactive DNA probe for detecting *Porphyromonas gingivalis* in subgingival specimens. *Oral Microbiology and Immunology*, 7, 344–348.

Tew, J. G., Zhang, J.-B., Quinn, S., *et al.* (1996) Antibodies of the IgG_2 subclass, *Actinobacillus actinomycetemcomitans*, and early-onset periodontitis. *Journal of Periodontology*, **67**, 317–322.

Von Troil-Lindén, B., Torkko, H., Alaluusua, S. *et al.* (1995) Periodontal findings in spouses. A clinical, radiographic and microbiological study. *Journal of Clinical Periodontology*, **22**, 93–99.

Wilson, M. E. and Kalmar, J. R. (1996) $F_c\gamma RIIa$ (CD32): a potential marker defining susceptibility to localized juvenile periodontitis. *Journal of Periodontology*, **67**, 323–331.

Zambon, J. J., Christersson, L. A. and Slots, J. (1983) *Actinobacillus actinomycetemcomitans* in human periodontal disease. Prevalence in patient groups and distribution of biotypes and serotypes within families. *Journal of Periodontology*, **54**, 707–711.

Zambon, J. J., Reynolds, H. S. and Genco, R. J. (1990) Studies of the subgingival microflora in patients with acquired immunodeficiency syndrome. *Journal of Periodontology*, **61**, 699–704.

Zambon, J. J., Reynolds, H. S., Chen, P. and Genco, R. J. (1985) Rapid identification of periodontal pathogens in subgingival dental plaque. Comparison of indirect immunofluorescence microscopy with bacterial culture for detection of *Bacteroides gingivalis*. *Journal of Periodontology*, **56**, 32–40.

Zambon, J. J., Haraszthy, V. I., Hariharan, G. *et al.* (1996) The microbiology of early-onset periodontitis: association of highly toxic *Actinobacillus actinomycetemcomitans* strains with localized juvenile periodontitis. *Journal of Periodontology*, **67**, 282–290.

6

Helicobacter pylori: opportunistic member of the normal microflora or agent of communicable disease?

Adrian Lee

6.1 INTRODUCTION

This is a book about the normal microflora. It concerns those bacteria that have, over the millennia, evolved to inhabit the multitude of ecological niches provided by the nooks and crannies of the human body. In certain circumstances these organisms may induce disease, especially if the balance of the microflora changes or the bacteria move into a different niche. The evolved pathogens that specifically cause major diseases such as diphtheria or tuberculosis are not our concern. Inclusion of the bacterium *Helicobacter pylori* is justified by the title to this chapter; there are those who still believe this bacterium is not a true pathogen, rather it is an opportunistic colonizer that inhabits already damaged gastric mucosa (Graham, 1995). As is seen below, the answer to the question is unequivocal. *H. pylori* is indeed a major gastroduodenal pathogen responsible for millions of deaths per year and for great discomfort to many of the world's population. Thus, we could have deleted this chapter and claimed that it sits more properly within a text on infectious disease. However, *H. pylori* shares many of the characteristics of the normal microflora. This is a bacterium that has evolved over the millennia to inhabit its niche of the gastric mucosa. The hypothesis suggested in the conclusion is that in previous times *H. pylori* was indeed normal microflora and it is only the consequences of human development that

G. W. Tannock (ed.), Medical Importance of the Normal Microflora, 128–163.
© 1999 Kluwer Academic Publishers. Printed in Great Britain.

have produced circumstances in which the bacterium is pathogenic. This is the reason that we consider that a chapter on this fascinating bacterium can greatly contribute to this treatise on the normal microflora and its medical importance.

6.1.1 The intestinal surface as a microbial habitat

As the first primitive fishes evolved and the amphibians crawled out of the primordial slime, they developed specialized intestinal tracts to benefit from the digestion of edible substrates. Constant exposure to ingested primitive prokaryotes resulted in colonization of the intestinal tracts by vast populations of bacteria protected by the casing of the gut and also benefiting by the available nutrient in the intestinal lumen. The gut microflora have developed into the most complex of the microbial ecosystems and, as is shown elsewhere in this book, play a major beneficial role in animal and human health (Lee, 1985).

6.1.2 Intestinal mucus an ecological niche for helical/spiral microflora

Intestinal surfaces became covered with mucus, which protected the tissue and aided the flow of intestinal chyme propelled by peristalsis. This nutrient-rich mucus layer provides an excellent niche for a range of bacteria well adapted to a viscous environment. Phase contrast examination of scrapings of intestinal mucus from different areas of the gut of most animal species reveals large numbers of highly motile bacteria darting along the tracts provided by the mucus threads (Phillips and Lee, 1983). Most of these organisms have a spiral/helical morphology, which gives them more torque as they 'bore' through their viscous habitat and thus they are at a selective advantage (Ferrero and Lee, 1988). Some bacteria have further adapted by acquiring specific adhesion mechanisms that allow them to survive in mucus by means of firm attachment to the gut surface. However, the organisms of interest here are the spiral-shaped bacteria. The conventional rodent is the best animal in which to observe them. Morphologically different species can be seen to inhabit the surfaces of the different regions of the gastrointestinal tract as is shown in Fig.6.1.

As is discussed later, many of these organisms are probably *Helicobacter* spp. and are important to the current discussion. Some belong to the genus *Campylobacter* and it is of relevance that the human pathogen *Campylobacter jejuni* can colonize mucus surfaces of animals without adhesion to the gut wall, demonstrating the effectiveness of this specialized motility in intestinal colonization (Lee, 1985).

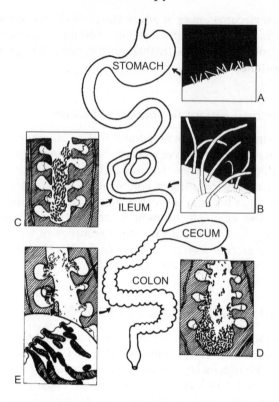

Figure 6.1 Diagramatic representation of the surface-associated microbiota of the rodent gastrointestinal tract.

Source: reproduced with permission from Lee, 1985.

6.1.3 The spiral inhabitants of gastric mucus

In the heyday of observational microbiology, similar bacteria were seen to colonize the gastric surfaces of many animal species. Thus as early as 1892, bacteria were reported in the stomachs of cats and dogs (Bizzozero, 1892). Non-human primates were seen to have the bacteria and there were even early reports of organisms in the stomachs of human patients (Doenges, 1939). Significantly, these bacteria all had a spiral morphology. They are illustrated in Fig. 6.2. Apart from some bizarre experiments in 1896 when the stomachs of cats and dogs were fed to rodents and the gastric bacteria were shown to colonize mice in high numbers (Salomon, 1896), there was little interest in these bacteria until the momentous discovery described in the next section.

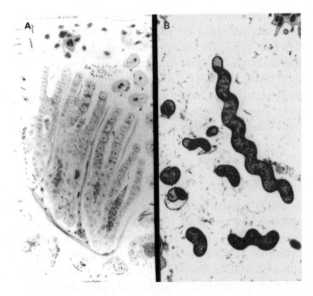

Figure 6.2 (A) The original drawing by Salomon of spiral shaped bacteria in the gastric mucosa of mice (Salomon, 1896). **(B)** Transmission electron micrograph (TEM) showing the same spiral-shaped bacteria colonizing the gastric mucosa of mice fed mucus scrapings obtained from a non-human primate (× 6000).

6.2 THE DISCOVERY OF *HELICOBACTER PYLORI* – A PARADIGM SHIFT IN THE MANAGEMENT OF GASTRODUODENAL DISEASE

Gastroenterologists are well aware of the importance of the normal microflora. They appreciate the beneficial role of the bacteria and are witness to the disastrous effects of diseases such as pseudomembranous colitis that occur when the balance of the microflora is disturbed. However, the stomach was a different story. They knew gastric contents have a very low pH of about 2 and that bacteria were killed at this pH and so the stomach was essentially 'sterile'. Bacteria had been grown from the stomach of persons with hypochlorhydria or from gastric contents buffered by food but these were not important and certainly were not normal. In pre-HIV days, they did not even bother to wear gloves for endoscopy even though their hands would get liberally coated with gastric mucus. This all changed in 1983, when Marshall and Warren published in *The Lancet* their observations of a spiral-shaped bacterium in the gastric mucosa that seemed to be associated with gastritis or inflammation of the gastric mucosa (Marshall, 1983; Warren, 1983). Warren, a histopathologist in Perth, Western Australia, had previously seen these organisms in 1979 and was convinced they had a role in gastritis.

However, it was not until he convinced a young medical registrar, Barry Marshall, to work with him on these organisms that a bacterium was finally cultured in 1982. This followed the accidental overincubation of culture plates from gastric biopsies due to the length of the Australian Easter holiday weekend (Marshall *et al.*, 1985b). After a period of taxonomic confusion during which the bacterium was variously called CLO (a *Campylobacter*-like organism), *Campylobacter pyloridis* and *Campylobacter pylori*, the final name was settled on, i.e. *Helicobacter pylori* (Goodwin *et al.*, 1989).

6.2.1 The early skepticism – *H. pylori* as opportunistic colonizer

In the early 1980s gastroenterologists were highly skeptical as to the importance of this discovery as it went against all their prior concepts of the gastric milieu. Thus, they were surprised that a bacterium was even able to live there let alone be involved in gastric pathology. The hypothesis of Warren and Marshall was flatly rejected and explained away as a simple opportunistic colonizer of an already damaged mucosa.

The author first became interested in these bacteria, together with Stuart Hazell, because of their similarity to many of the organisms that colonized the mucus of the lower bowel. We attempted to help put this organism in an acceptable framework for the gastroenterologist by proposing *H. pylori* to indeed be a bacterium highly evolved to the ecological niche provided by gastric mucus and introduced the term 'almost normal flora' (Lee and Hazell, 1988).

6.2.2 The proof of *H. pylori* as a pathogen

The early attempts to convince others that this bacterium could cause gastritis were two heroic antipodean self-administration experiments in which Barry Marshall himself and the New Zealander Arthur Morris swallowed living cultures of the organism (Marshall *et al.*, 1985a; Morris and Nicholson, 1987). Both developed gastric symptoms and biopsies showed that gastritis did develop. However, an $n = 2$ did not convince the skeptics and the gastroenterologist still did not believe. The inability of Arthur Morris to rid himself of the bacterium and the suggestion of an involvement in gastric cancer meant that further human studies were unethical. The first real evidence came as patients were treated with antimicrobials that cleared the stomach of infection (Borody, Carrick and Hazell, 1987; McNulty *et al.*, 1986; Rauws *et al.*, 1988). Loss of *H. pylori* was seen to correlate with regression of inflammation. The early therapies were poor and so failed and the bacterium came back. So too did the gastritis. By this time, it had been seriously proposed that the *H. pylori* infection actually caused peptic ulcers, a proposal that inflamed the gastroenterologist even more. Once again, early treatment studies showed that if the infection was cured the ulcers did not come back (Coghlan *et al.*, 1987). Peptic ulcers at this time were easily treated by the very effective H_2 antisecretory drugs cimitidine and ranitidine. However, once the antacid therapy was stopped up to 80% of ulcers recurred within 1 year. Successful *H. pylori* eradication resulted in no recurrence of the ulcer. Once again the skeptics remained unconvinced and claimed that the effects were due to the presence of mucosal protectants such as the bismuth compounds that were in the early therapies. The seminal publication was by Hentschel and colleagues in 1993, who showed that, using a therapeutic regimen of antibiotics alone, ulcer recurrence was prevented in those successfully cured of infection (Hentschel *et al.*, 1993). Subsequently many good trials have repeated the finding and in some studies ulcers have not recurred up to 10 years post-treatment (Bell and Powell, 1996). The publication by a consensus panel of the United States National Institutes of Health was the turning point. They recommended that all *H.-pylori*-positive ulcers should be treated with anti-*H.-pylori* therapy (Yamada *et al.*, 1994). The evidence is now unassailable. Cure the infection and you cure the ulcer. Only those who refuse to acknowledge the evidence continue to sound the trumpets of ignorance and skepticism (Graham, 1995).

6.2.3 The diseases caused by *H. pylori*

(a) Gastritis

The underlying lesion of all *H. pylori* infections is gastritis. This is important with respect to the concept of *H. pylori* as normal microflora. All *H. pylori* infections are accompanied by histological gastritis. This will vary in severity in different regions of the stomach but even if the bacterium is seen in some part of the gastric epithelium with no associated inflammation, somewhere else in the gastric mucosa of that same patient the bacterium will be seen together with inflammation. However, it is also important to appreciate that, in the majority of *H. pylori* infections, the gastritis will be symptom-free. More than half the world's population is infected with this bacterium and in some countries, e.g. Colombia and Estonia, more than 90% of the population are infected but the majority have no symptomatic disease (Maaroos, 1995). This is the fact that made the concept of 'almost normal flora' attractive (Lee and Hazell, 1988). The type of gastritis caused by *H. pylori* was previously called type B gastritis but the discovery led to a new classification system called the 'Sydney System', which has recently been upgraded (Dixon *et al.*, 1996; Price, 1991). *H.-pylori*-associated gastritis comprises about 90% of the gastritis found in patients world wide.

(b) Duodenal ulcer

Peptic ulcers are areas of damaged mucosa where the muscularis mucosae is penetrated. The damage is caused by acid as they heal quickly on acid-suppressive therapy. *H. pylori* causes the ulcers by damaging the gastric epithelium such that the acid directly impinges on to the tissue and a focus of ulceration results. Ulcers may be found in the stomach proper, i.e. a gastric ulcer, or in the duodenal bulb, i.e. a duodenal ulcer. There are three main complications of peptic ulcer: gastrointestinal bleeding, perforation and pyloric stenosis (gastric stenosis). These complications are not only distressing to the patient but in some cases they may be life-threatening. Prior to the introduction of anti-*H.-pylori* therapy, up to 10% of the population of countries in the developed world could be expected to suffer from an ulcer at some stage of their life. Currently up to 95% of duodenal ulcers have been shown to be caused by *H. pylori* (Dooley and Cohen, 1988). As the number of ulcers decrease as a result of the introduction of *H. pylori* eradication therapies this proportion will drop. Other causes include the ingestion of non-steroidal anti-inflammatory drugs (NSAIDs).

(c) Gastric ulcer

Up to 80% of gastric ulcers are thought to be due to *H. pylori*; however this will vary from country to country as ulcers are a common consequence of the long-term ingestion of NSAIDs such as aspirin (Ormand and Talley, 1990). These drugs result in a weakened mucosa. The *H.-pylori*-induced gastric ulcer tends to be found at the transitional zone between the non-acid-secreting antrum of the stomach and the parietal-cell-containing body mucosa. The location of the pathology is important as it is likely to be a consequence of the local environmental conditions in the mucosa and the behavior of the bacterium. This is discussed more fully below.

(d) Gastric cancer

The essential precursor for so-called intestinal-type adenocarcinoma had long been known to be gastritis. Pelayo Correa in Colombia, a country with a very high incidence of gastric cancer, in a very detailed cataloguing of the pathology had defined the sequence of steps resulting in appearance of the tumor (Correa *et al.*, 1976). An initial chronic gastritis developed into severe atrophy with loss of the normal architecture of the stomach, in particular the loss of parietal cells. In cancer-prone individuals, intestinal metaplasia was observed, i.e. intestinal-type epithelium within the gastric mucosa. Patches of this tissue then progressed to dysplasia and ultimately to adenocarcinoma. Carcinogenesis was thought to be a consequence of bacterial overgrowth of the gastric lumen subsequent to development of hypochlorhydria associated with atrophy. Normal oral and intestinal microflora established in the non-acid stomach and broke down nitrates to produce carcinogenic n-nitroso compounds. What was unknown was the cause of the gastritis that progressed to atrophy. Correa, on epidemiological grounds, felt that there had to be some environmental factor. The discovery of *H. pylori* provided the missing link in the chain (Correa and Ruiz, 1989). *H. pylori* causes the gastritis that progresses to the atrophy, the essential precursor lesion for malignancy. Thus it was reasonable to consider the bacterial infection as the cause of the multifactorial disease that is gastric adenocarcinoma (Correa, 1991). Although logically the bacterial infection had to be causal, direct evidence was harder to come by. A meeting of an expert panel in June 1994 was charged by the International Agency for Research on Cancer (IARC) of the World Health Organization to rigorously appraise the evidence. The panel concluded that the evidence that *H. pylori* had a causal role in gastric cancer was sufficient and classed the bacterium as a Class 1 carcinogen, the same category as Hepatitis B virus (IARC, 1994). This decision was based on many epidemiological studies. There was shown to be a clear correlation between the incidence of *H. pylori* infection in a country and the incidence of gastric cancer. Three prospective studies examined the incidence of *H. pylori* infection by serology in persons dying of gastric cancer who had been part of large prospective screening studies and whose serum was thus available for decades before the development of the tumor (Forman *et al.*, 1993; Nomura *et al.*, 1991; Parsonnet *et al.*, 1991). This was compared with the incidence in matched control sera taken from persons bled in the same screen who had not developed cancer. Those people infected with *H. pylori* many years previously were shown to have a sixfold greater chance of developing cancer than uninfected persons. A positive association was found with intestinal-type gastric adenocarcinoma as well as diffuse-type gastric cancer, the other major tumor observed. At the time of the publication of the opinion of the IARC panel in Lyon, there was feeling that the conclusion was premature but all studies subsequently have born out the panel's opinion. The most impressive study was done in Japan, a country where gastric cancer is a major health problem: odds ratios (OR) as high as 20 were found in young Japanese (Kikuchi *et al.*, 1995). Thus it can be concluded that *H. pylori* does cause gastric cancer. This has major implications in the development of preventative measures against infection with this bacterium as gastric cancer remains the No. 2 malignancy in the world, with more than a million deaths annually.

(e) Gastric lymphoma

A small percentage of gastric tumors present as low-grade B-cell lymphomas, i.e. large masses of monoclonal B cells arising out of tissue resembling the lymphoid structures of the mucosal associated lymphoid system (MALT) that are seen in the gastric mucosa of all *H.-pylori*-infected persons (Stolte and Eidt, 1989; Wyatt and Rathbone, 1988). These tumors, because they were known to have the potential to progress to invasive high-grade B-cell lymphomas, were often removed surgically following their diagnosis. Not surprisingly, *H. pylori* was considered as a likely etiology for these tumors. This suspicion was confirmed by the demonstration of regression of the low-grade lymphomas following cure of *H. pylori* infection (Wotherspoon *et al.*, 1993). Anti-*H.-pylori* treatment is now a standard part of management. The low-grade lymphomas can be considered premalignant lesions as they regress; once transition to high grade occurs, antimicrobials have no effect.

(f) Any more?

The role of *H. pylori* in the syndrome of non-ulcer dyspepsia is controversial as treatment studies show conflicting results (Talley, 1994). The reality is that probably about 10% of patients with recurrent severe dyspepsia owe their symptoms to the bacterium and therefore would benefit from therapy. A range of other conditions have been claimed to be associated with infection including rosacea and coronary heart disease (Glynn, 1994; Mendall *et al.*, 1994; Patel *et al.*, 1994, 1995; Rebora, Drago and Picciotto, 1994). Some have claimed that early infection in some cultures produces a particularly erosive gastritis (Pelayo Correa, personal communication). If this is so, then possibly this early infection could impact on later development. The period of hypochlorhydria that is known to accompany first infection could also make a young child more susceptible to intestinal infection, with devastating consequences. In reality, with so many confounding variables in those underprivileged populations who would be affected in early childhood, these early influences of *H. pylori* are going to be impossible to prove. However, the possibility should be a motivation for intervention studies with the vaccine when it is developed.

6.2.4 Gastric diseases as communicable infections: the epidemiology of *H. pylori*

H. pylori infection is clearly a communicable disease and so it is valid to conclude that the associated diseases, e.g. peptic ulcer, are communicable. However, given the multifactorial nature of symptomatic disease and the fact that most infected persons live through life with no major problems, this is not a useful concept. We know much about the epidemiology of *H. pylori* infection but there remains controversy about the actual mode of transmission (Graham *et al.*, 1991; Hazell *et al.*, 1994). The only reservoir appears to be the infected human, with most infection occurring in child-hood (Megraud *et al.*, 1989; Mitchell *et al.*, 1992a). Indeed, the bacterium seems quite hard to catch in adulthood. This has important implications as reinfection is thus very rare following successful treatment, even in those populations where the prevalence is high. Likelihood of infection is greater if the child lives in crowded conditions and has both parents infected (Mitchell *et al.*, 1993; Mitchell, Mitchell and Tobias, 1992; Now-ottny and Heilmann, 1990; Perez Perez *et al.*, 1990; Webb *et al.*, 1994). The balance of the evidence would suggest the mode of transmission is mainly oral–oral via refluxed gastric contents. The fact that the bacterium is bile-sensitive is consistent with this mode of transmission (Mitchell *et al.*, 1992b). All other bacterial pathogens known to be transmitted by the feces are resistant to bile, as they have to be able to survive transit down the intestinal tract. *H. pylori* cannot be cultured from feces except in unusual situations, such as the very young in Gambia who have such rapid transit times that the bacteria can survive (Thomas *et al.*, 1992).

6.3 THE ECOLOGY OF *H. PYLORI* INFECTION: THE KEY TO UNDERSTANDING PATHOGENESIS

Properties of bacteria of special interest to those studying the normal microflora are those adaptations that allow the organism to survive in its chosen specialized ecological niche. By definition, these are the proper-ties that put the bacteria at a selective advantage compared to potential pathogens and these are the properties needed by opportunistic patho-gens should the normal microflora be dislodged from the niche. The commensal bacteria are the 'professional' colonizers of their niche. If they are not dislodged by any predisposing factor then a pathogen simply has to do it better or inhabit an unoccupied niche. *H. pylori* inhabits the most hostile environment in the body. It succeeds where no others can and so a study of the bacterium and how it copes with its niche provides fascinating ecological insights. Indeed, it is changes in the local environ-ment and the ecology of gastric colonization that probably provide an explanation for one of the enigmas of *H. pylori* infection: the different patterns of gastric disease (Lee *et al.*, 1995). In the developing world, the most common *H.-pylori*-associated disease is the gastric ulcer; it is these populations which, if others factors such as high salt diet and low vitamin C are present, progress to gastric cancer. In contrast, in devel-oped societies the normal disease is a duodenal ulcer. Non-NSAID gastric ulcers and gastric cancer are rare. Consistent with this, investigation of previous ulcer disease in a large number of gastric cancers cases in Sweden showed that, while gastric ulcer was associated with an increased risk for cancer, duodenal ulcers were actually protective (Hans-son *et al.*, 1996).

6.3.1 Colonization factors

The natural habitat of *H. pylori* is the gastric mucus and thus, exactly as with the intestinal mucus colonizers, the helical morphology is an important colonizing factor as it allows the organism to swim readily in the mucus layer and down to the surface of the stomach (Hazell *et al.*, 1986). While most of the bacteria swim free in the mucus there are clearly adherent bacteria on the gastric mucosa. This specificity for gastric tissue is best illustrated in the duodenal mucosa of duodenal ulcer patients. *H. pylori* will not colonize normal small-bowel-type epithelium in the duodenum. The bacterium is only seen attached to areas of gastric metaplasia (Carrick *et al.*, 1989), i.e. islets of epithelium with the morphology of gastric cells rather than small bowel cells, which are formed probably in response to acid from the stomach impinging on the mucosa of the duodenal bulb. These are the cells that *H. pylori* colonizes and which become the focus of the acid damage and thus the ulcer. Figure 6.3 shows the bacterium attached to gastric metaplasia.

The first small-bowel-type cell next to the metaplasia has no adherent bacteria, i.e. the adhesion is totally cell-specific.

Much work has gone into the identification of the adhesins involved (Wadstrom, Hirmo and Boren, 1996): multiple adhesins are thought to be involved. Of particular interest are some molecules that have specificity for the Lewis blood group antigens. As these are the molecules found on

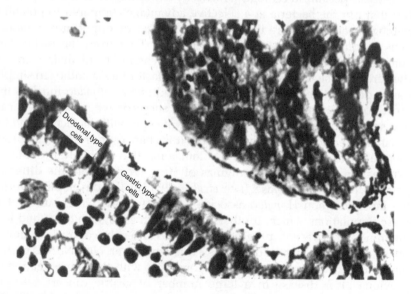

Figure 6.3 An area of gastric metaplasia in the duodenum in which the *H. pylori* organisms can only be seen in the areas of gastric metaplasia. Source: reproduced with permission from Carrick *et al.*, 1989.

tissue surfaces and in mucus, it is not surprising that the bacterium has learnt to recognize them (Boren and Falk, 1994; Boren *et al.*, 1993). Differences in tissue specificity has been suggested as an explanation for differences in ulcer disease among populations with different blood groups.

The environmental condition that must be relevant to colonization yet has been almost completely ignored, not only as concerns *H. pylori*, but also other gut colonizers, either normal microflora or pathogen, is the local oxygen tension. Most of the mucus colonizers in the bowel mucosa are microaerophilic. This would be the predicted oxygen tension at the intestinal surface between the reduced lumen and the aerated tissue. Better understanding of microaerophilism would not only help our ecological understanding of both *Helicobacter* and *Campylobacter* species but might provide novel approaches to therapy.

6.3.2 Hp and pH

The defining environmental factor in the stomach is gastric acidity. The contents of the human stomach have a pH of 1–2. *H. pylori* and the other gastric helicobacters are the only bacteria that have evolved strategies of acid resistance. In some animal species, such as rodents, in which the stomach is less acidic because of the continued presence of food, acid-resistant lactobacilli are normal microflora and inhabit the gastric surface. However, they cannot tolerate pHs of 1–2. A different strategy is needed. The factor solely responsible for the acid resistance of *H. pylori* is the enzyme urease (Mobley, 1996). The bacterium has extraordinary urease activity with up to 20% of its total protein being this enzyme, the action of which is to break down urea to carbon dioxide and ammonia, thus producing a local environment that can neutralize acid. It is easy to conceptualize the entry of the bacterium into the stomach, where it rapidly breaks down the endogenous urea, known to be in the gastric contents, producing its own protective cloud of ammonia, which protects it until it comes in contact with the gastric mucus where it quickly swims into the more alkaline environment and survives. The feasibility of this mechanism can be demonstrated by simple *in vitro* experiments (Clyne, Labigne and Drumm, 1995; Ferrero, Hazell and Lee, 1988; Sjostrom and Larsson, 1996). *H. pylori* is placed into a test tube of medium at pH 2, the bacterium is readily killed, so *H. pylori* is not intrinsically an acidophile. However, if urea at the physiological concentration encountered *in vivo* is added, i.e. 2 mmol urea, the organism will survive in this very low acid environment for up to 1 hour. The same experiment repeated with a urease-negative mutant resulted in the bacterium being killed by the acid despite the presence of physiological urea concentration. If *H. pylori* only required an acid-protective mechanism for initial colonization, urease would be an inducible enzyme only switched on in times of need.

Yet it is constitutive and manufactured in large amounts, implying that it is required for continued survival. Recent animal experiments suggest that urease activity plays an important role in dictating the actual region of the stomach in which the organism grows and this would in turn influence the pathology induced at that site (Danon *et al.*, 1995). *Helicobacter felis*, a close relative of *H. pylori*, preferentially colonizes the non-acid-secreting mucosa of the rodent stomach, the antrum. If the animals are put on acid-suppressive therapy, the numbers of bacteria in the antrum decline and bacteria are seen in the body mucosa. Exactly the same phenomenon is seen in humans infected with *H. pylori* who have their acid output decreased by selective vagotomy or acid-suppressive drugs. The number of bacteria in the antrum declines, as does the antral gastritis, and gastritis increases in the body of the stomach. This has prompted us to suggest that local acid output has a major impact on determining the patterns of disease due to the acid resistance mechanism of *H. pylori*, i.e. the action of urease. This is the explanation for the enigma described above concerning the different patterns of disease caused by *H. pylori* (Lee *et al.*, 1995).

Given the likely interest of the reader in ecological issues, it is appropriate to summarize these ideas here. *H. pylori* evolved to survival in an acid environment by the acquisition of a urease enzyme, which is surface-expressed by a mechanism that is as yet not understood (Phadnis *et al.*, 1996). The protective effect of this enzyme is not enough to allow the bacteria to flourish or grow maximally in the body of the stomach, i.e. in the region where the acid is produced by parietal cells and pumped up from the gastric glands through the mucus into the lumen of the stomach. However, *H. pylori* does survive in the outer mucus layer, inducing no inflammation of consequence. In contrast, in the antrum there is no acid produced in the mucosa and so the ammonia generated by the effect of the urease enzyme is enough to counter the acid diffusing into the mucosa from the lumen; thus the bacteria flourish, inducing inflammation, i.e. an antral gastritis. If acid output is lowered for any reason, this pattern of colonization and associated gastritis may change dramatically. Now, the acid in the body of the stomach may fall below a threshold where the ammonia generated is now protective and the *H. pylori* can multiply, possibly inducing gastritis. In the antrum there is less acid to neutralize and so ammonia accumulates and the local milieu becomes more alkaline because of generation of excess ammonia compared to hydrogen ions at that site. This alkaline ammonia-rich environment is just as toxic to the organism as is too much acid. Thus, numbers of bacteria in the antrum and inflammation go down. In the *in vitro* studies described above, when *H. pylori* was incubated at pH 4–7 in the presence of physiological urea the bacteria were killed whereas urease-negative mutants survived. These *in vitro* effects of acid have just been explained in an important paper that supports our hypothesis and describes urease protective activity in terms of the proton motive force generated across the cell membrane (Meyer Rosberg *et al.*, 1996). Differences of *H. pylori* diseases in different populations can thus be ascribed to differences in acid output. In the developing world, acid output is decreased compared to that in populations in the developed world. This is probably due to poorer nutrition or increased parasite or infection overload. In these populations, gastritis is more widespread in the stomach, occurring in both antrum and body. This is the pattern of inflammation that leads to gastric ulcer and the atrophic gastritis associated with gastric cancer. In the developed world, the higher acid restricts the inflammation to the antrum and does not allow atrophy to develop or ulceration across the antral/body border. This is the typical pattern of gastritis seen in duodenal ulcer patients who are mainly seen in the more affluent populations in the developed world.

6.3.3 Evasion of host defense mechanisms

H. pylori survives for the whole life of its host despite the induction of an immune response so vigorous that serology remains an excellent diagnostic tool (see below). The gastric mucosa is flooded with specific antihelicobacter immunoglobulins and the tissue is heavily infiltrated with neutrophils and committed T cells but the organism thrives (Wyatt, Rathbone and Heatley, 1986). So sophisticated mechanisms of immune evasion must have been acquired. What are they? Some have suggested a downregulation; for example, *H. pylori* extracts have been shown to inhibit mitogen-induced proliferation of peripheral blood mononuclear cells *in vivo* (Knipp *et al.*, 1993), yet the immune responses seem very healthy. A more satisfying, yet unproven, hypothesis is that *H. pylori* infection upsets the Th1/Th2 balance in the gastric mucosa in favor of the organism (Ernst *et al.*, 1996). These T-lymphocyte populations are formed under the stimulus of different cytokines: IL-12 shifts differentiation towards Th1 while IL-10 leads to Th2 production. Th1 cells are responsible for IgG-mediated systemic immunity to both bacteria and viruses while Th2 responses lead to production of mucosal IgA. The Th2 response is the response that is most likely to eliminate *H. pylori* from a mucosal surface. All studies to date suggest that lymphoid cells associated with gastritis are of the Th1 phenotype.

6.3.4 Pathogenic mechanisms

There has been much work trying to identify those factors of *H. pylori* that are responsible for the damage that leads to symptomatic disease. Early work looked for cytotoxins that damaged tissues and these were identified, e.g. the vacuolating cytotoxin coded for by the *vacA* gene

(Covacci *et al.*, 1993; Tummuru, Cover and Blaser, 1993); others postulated a weakening of the mucus layer thus removing a protective barrier (Slomiany *et al.*, 1987). However none of these factors explain why it is that the majority of infected persons never develop symptomatic diseases despite the presence of cytotoxin, protease, phospholipases, etc. What follows is the author's attempt to put *H. pylori* pathogenesis in perspective. It is the most reasonable hypothesis based on the evidence to date. One critical factor is the severity of the inflammation. The more severe the gastritis then the more likely that damage to the gastric mucosa will be enough to allow acid-induced damage to occur. A key is the infiltration of neutrophils, which are attracted by the production of inflammatory cytokines from *H.-pylori*-infected gastric epithelial cells. There are strains of the bacterium that can induce more cytokines, such as IL-8, *in vivo* or *in vitro* (Crabtree *et al.*, 1994a, b). These strains possess a so called 'pathogenicity island': an extra piece of DNA in their chromosome that has about 11 genes coding for products involved in secretion of proteins across the bacterial membrane that are needed for induction of IL-8 (Akopyants, Kersulyte and Berg, 1995; Censini *et al.*, 1996). A marker for this island is the *cagA* gene, which codes for a 120 kDa protein that is very immunogenic in *H.-pylori*-infected persons. In some populations, there is a correlation between infection with the *cagA*-positive strain and ulcer disease or gastric cancer (Blaser *et al.*, 1995). Those infected with this strain have a greater likelihood of having these diseases. In other countries this distinction is not there because most persons are infected with *cagA*-positive *H. pylori* (Mitchell *et al.*, 1996). Nonetheless, these observations are consistent with the hypothesis that symptomatic infection is more likely if a more inflammatory strain is involved. Other factors must be involved in symptomatic disease. Ulcers only occur at transitional zones, e.g. gastric ulcers at the junction between the antral and body mucosa and duodenal ulcers on gastric metaplasia. The mucosa must be weakened at this site, e.g. local blood flow is reduced such that hydrogen ions diffusing into the *H.-pylori*-damaged tissue are not mopped up as readily into the gastric capillary network. This would be compounded by other factors that might reduce blood flow, such as smoking, which is known to be more common in ulcer patients. The final factor is differences in the host (Lee *et al.*, 1993a). Gastric atrophy caused by long-term infection of mice with *Helicobacter* species is entirely dependent on the host factors, as some mouse strains show no inflammation while in others nearly all parietal and chief cells are replaced (Mohammadi *et al.*, 1996; Sakagami *et al.*, 1996). This presumably is a consequence of host differences in immunoreactivity to *Helicobacter* antigens. Other factors likely to be different in the human host are differences in prior antigen exposure and differences in acid output as discussed in section 6.3.2. Peptic ulcer and gastric cancer are thus truly multifactorial diseases: they will only develop if a number of the very diverse factors described above are present. Mere infection alone is not enough however severe it might be. The likely scenarios of disease development are as described below.

(a) Duodenal ulcer

Susceptible patients have an acid output that is higher than average or at the top end of normal (Ball, 1961; Marks and Shay, 1959). The high acid has two consequences. Firstly, it restricts the inflammatory damage to the antrum away from the antral/body border; thus the susceptible site for gastric ulcer is protected. Little inflammation is seen in the body of the stomach and so gastric cancer will not result. Secondly, high acid impinges on the duodenal mucosa causing small islets of gastric meta-plasia to form. The bacterium infects this tissue and induces a duodenitis. If the infection is with *cagA*-positive strains ulceration may result as the threshold of inflammatory damage that results in acid-induced damage is passed. With *cagA*-negative strains an ulcer is unlikely.

(b) Gastric ulcer

These patients have acid that is on the lower side of normal such that the gastritis occurs across the transitional zone (Magnus, 1946; Oi, Oshida and Sugimura, 1959). There is still enough acid present in the stomach to induce ulcer formation at the weak spots in the transition zone, if the tissue is disorganized enough or the local blood flow is slow enough.

(c) Gastric cancer

Gastric cancer is mainly seen in patients infected with *cagA*-positive strains (Blaser *et al.*, 1995). As the damaging effects that lead to atrophic gastritis need to be major, a severe inflammatory response over decades is required. Host-dependent immunoreactivity is probably more important in the development of atrophy as it clearly does not occur in all persons with long-term gastritis. The other essential requirements for tumor development are the dietary factors. These include high levels of nitrates to supply precursors for carcinogenic n-nitroso compounds, a high salt diet to increase cell proliferation and a lack of vitamin C to scavenge tissue-damaging oxygen radicals (Correa, 1988).

6.4 DIAGNOSIS OF *H. PYLORI* INFECTION

Diagnosis of *H. pylori* infection is very straightforward and is easily done. The problem is the interpretation of a positive result with respect to patient management. A brief description of these tests is given below.

Details of test composition and suppliers are provided in detail elsewhere (Lee and Megraud, 1996).

6.4.1 Invasive tests

Introduction of the fiberoptic endoscope has replaced barium contrast X-rays in the diagnosis of gastroduodenal disease. Endoscopy is well tolerated and ulcers or tumors are readily visible. Once the endoscope is in place the collection of gastric biopsies is a simple procedure and results in specimens that can easily be used to diagnose infection.

(a) Biopsy urease tests

H. pylori is present in such large numbers in the tissue and mucus that the presence of the bacterium correlates with urease activity. The biopsy is placed in a solution or agar gel of a medium containing urea and a pH indicator. If H. pylori is present, the ammonia released through the urea breakdown creates an alkaline environment and the indicator changes color. The original commercial test was devised by Barry Marshall himself and many variants of the test are now available (Marshall et al., 1987). A very cheap 'do-it-yourself' test using liquid reagent and plastic microtiter trays, the Hazel biopsy urease test, has been used successfully in Sydney hospitals for more than 10 years (Hazell et al., 1987). Because of the low urease activity of any likely oral contaminants compared to H. pylori and the bacteriostat used in most tests to prevent overgrowth, these tests are surprisingly specific and sensitive, with values for both parameters of greater than 90%. As a result of the effect of acid on colonization described above, antral biopsies can be negative in patients on acid suppression, therefore body biopsies should always be taken from these patients.

(b) Histopathology

Assuming one has access to an experienced histopathologist, histopathology can be a very reliable method of diagnosing H. pylori infection as the bacteria are present in large numbers and other organisms are not always seen. However interobserver differences have been reported (Elzimaity et al., 1996). Given that the distribution of the bacterium can vary, it is essential that biopsies from both antrum and body are examined. The appearance of H. pylori in a gastric biopsy is seen in Fig. 6.4.

An advantage of microscopic examination of biopsies is that much can be learnt about the stage of the disease. This is especially important in those cancer-prone populations where pangastritis, evidence of gastric atrophy and intestinal metaplasia can be observed. Obviously malignancy can only be diagnosed on histopathology.

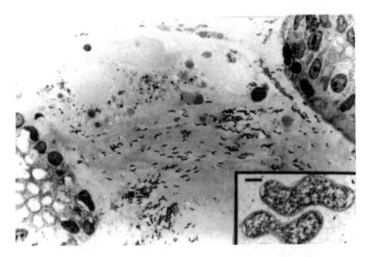

Figure 6.4 Light micrograph of human gastric mucosa showing large numbers of spiral organisms (*H. pylori*) present in the mucus overlying the epithelial surface and in some instances attaching to the epithelial surface (× 500). Inset shows *H. pylori* as viewed by TEM (× 20 000).

(c) Microbiology

Many claim culture to be the 'gold standard' for diagnosis of infection. Certainly if fresh moist culture plates, selective antibiotics and a suitable microaerophilic atmosphere are used, the bacterium will always be cultured. Gastroenterologists have a right to expect that their clinical laboratory can easily grow these bacteria. However, given the reliability of the simple urease test and histopathology, culture is rarely done and is not necessary. The organism is slow-growing and may require 4–5 days on primary culture. The only requirement for culture is for a reference laboratory in each region to monitor for the appearance of resistance to current therapies and from those patients in which therapy has failed. The appearance of typical bacteria in culture is shown in Fig. 6.5.

6.4.2 Non-invasive tests

H. pylori infection may also be reliably diagnosed by the following two classes of test, both of which are becoming increasingly available.

(a) Breath tests

These tests once again rely on the remarkable urease activity of *H. pylori* (Logan, 1993). If the bacteria are in the stomach then there is enough

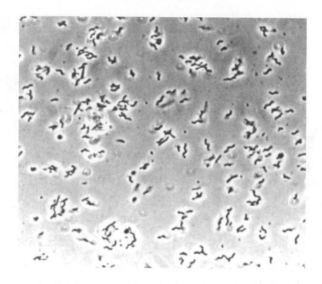

Figure 6.5 Pure culture of *H. pylori* as viewed by phase contrast microscopy (× 400).

enzyme present to break down isotopically labeled urea, resulting in the release of labeled carbon dioxide in the exhaled breath. The labeled urea is ingested by the patient and about 30 minutes later the patient breathes into a capture solution or into a balloon. Both [^{13}C]- and [^{14}C]-labeled urea have been used very successfully, once again with specificity and sensitivity greater than 95% (Logan, 1996; Marshall, 1996). Carbon-13 is non-radioactive and is thus preferred by some, but the equipment required for detecting the natural isotope is very expensive although cheaper machines are becoming available. Carbon-14 is radioactive, but this potential problem has been overcome by the use of microdoses of the labeled urea that have less radioactive exposure than one experiences from natural radiation in one day. The major role for the breath test is for the assessment of success of therapy. It is likely that this will be used more in the future thus eliminating the need for follow-up endoscopy

(b) Serology

Every infected person has high levels of IgG antibody against *H. pylori* that can be simply detected by an ELISA assay. Early tests used whole sonicates of the bacterium as antigen while later tests used purified antigens (Bolton and Hutchinson, 1989; Dent *et al.*, 1988; Hirschl *et al.*, 1988; Mitchell *et al.*, 1988). Recently a multitude of commercial tests have been introduced as the potential market for non-invasive diagnosis is perceived to be vast. All these tests can be reliable, with high sensitivities and specificities in the right hands. However, there have been disturbing reports of poor test performance (Feldman, Deeks and Evans, 1995; Loffeld and Stobberingh, 1991). This is often because a test has been validated in a different population from that in which it is being used (Mitchell *et al.*, 1988). Different cut-off values are needed in different populations, presumably because of different levels of cross-reactive antibody or different patterns of infection. Users of a commercial test should demand validation data for the population in which the test is to be used. Given the ready availability of the tests, the fact that some of them can be done 'in house' and that they are accurate measures of active infection, not merely evidence of past infection, a controversial question becomes 'When should serology be performed?'. If a patient is endoscoped serology is superfluous. Yet endoscopy is the only way an ulcer can be diagnosed. Serology cannot discriminate the ulcer patient. Serology cannot detect the cancer. All it can do is prove that the patient is infected. Yet the majority of *H.-pylori*-infected persons are asymptomatic. Herein lies the problem: what is appropriate action should the test be positive? There have been many strategies proposed for the use of serology. It is suggested that any patient presenting with persistent upper gastrointestinal tests should be offered serology. In countries where gastric cancer is very rare, only serology-negative patients should be investigated by endoscopy; this strategy could save many endoscopies. So what to do with the patient that has symptoms of non-ulcer dyspepsia and is positive by serology? Official guidelines state that the evidence for *H. pylori* as a cause of this syndrome is weak and that the patient should not be treated (Yamada *et al.*, 1994). But the patient has read much in the press about the bacterium and has noted that it is claimed to cause cancer. Despite assurances that this is very unlikely in a developed country anxiety remains. I would suggest that the only golden rule with respect to serology is 'If you request serology on a patient, then you should be prepared to treat if the test is positive.' The question is not what to do with a positive serological result, rather it is when it is useful to do serology. Tests are unreliable with respect to assessment of therapy because of a slow and variable drop of antibody levels following eradication of the organism. Thus the answer to the question when to do serology remains confused. The tests proliferate, the profits are made, yet the benefit to patients remains unproven. The reality is that the tests are already being used and patients are being treated on the basis of a positive result, with little proven benefit and greater risk of the appearance of antibiotic resistance. The question posed later becomes: should everyone with this infection be treated? In a book devoted to understanding the consequence of disturbances in the normal microflora the answer has to remain: no! Therefore, don't test.

6.5 TREATMENT AND PREVENTION OF *H. PYLORI* INFECTION

Given that *H. pylori* is a major global pathogen there is an urgent need for effective therapeutic and prevention strategies. The discovery of this bacterium caused a complete paradigm shift in the management of peptic ulcer such that antimicrobial therapy is now an essential component.

6.5.1 Antimicrobial therapy

(a) Effective regimens

Since the first paper on the effects of antimicrobial treatment on *H. pylori* infection, the medical literature has been overloaded with thousands of full manuscripts or abstracts on almost every conceivable combination of antimicrobials known. Many trials were poor; good trials reported ineffective regimens, but all showed that if you could get rid of the infection ulcers were cured. Given that anti-ulcer therapies have been the biggest grossing pharmaceutical drugs of all time, the rewards for the Holy Grail of an effective therapy have driven this frenetic activity. An increasing shortage of candidates for ulcer trials would indicate that the therapies are near. The first successful regimen was the so-called classical 'triple therapy' devised by Tom Borody in Sydney and based on Barry Marshall's early treatment studies (George *et al.*, 1990). This therapy is a 2-week course of:

- 500 mg tetracycline hydrochloride four times per day;
- 400 mg metronidazole three times per day;
- 120 mg bismuth subcitrate (one De-Nol tablet) four times a day.

If the patient can be persuaded to take all these pills and to tolerate the mild discomfort of an unpleasant taste and black tongue and stools, this regimen can still achieve cure rates of up to 90% and it remains the cheapest therapy. However, many have not been happy with the regimen, mainly because of compliance problems, and alternatives have been sought. There was a desire to reduce the number of drugs being used. Studies showed that the acid-suppressing therapies using omeprazole, a proton-pump inhibitor that was becoming standard for the healing of ulcers, did indeed potentiate the anti-*H.-pylori* effectiveness of monotherapy with a single antimicrobial. For a brief period, dual therapies became the standard, with a proton pump inhibitor and a single antimicrobial. However, in practice these regimens gave extremely variable results and had to be discarded. From current evidence dual therapies should not be used. The most common regimens used today, in those countries where it is approved, are different triple therapies, including the new macrolide antibiotic clarithromycin, which, if given for only 1 week, is consistently being reported as curing more than 90% of *H. pylori* infections. These therapies are 1 week of:

- 500 mg clarithromycin twice per day;
- 1 g amoxycillin twice a day;
- 20 mg of proton-pump inhibitor twice per day;

or

- 250 mg clarithromycin twice per day;
- 400 mg metronidazole twice per day;
- 20 mg of proton-pump inhibitor twice per day.

In those countries where metronidazole resistance is low this nitroimidazole can replace the amoxycillin (Lind *et al.*, 1996).

The classical triple therapy has been extended into a quadruple regimen with the addition of a once-daily dose of a proton-pump inhibitor. With this therapy the highest success rates have been reported and it has been suggested for use in patients where therapy has failed (Tytgat, 1996).

Antimicrobial resistance is the major concern with respect to the continuing effectiveness of current regimens. It is significant that resistance to both metronidazole and clarithromycin, the two mainstays of the successful treatments, have been reported. Given experience with other bacterial pathogens the overuse of anti-*H.-pylori* treatments is likely to make antimicrobial resistance a significant problem in the future. Many companies continue to develop new approaches to therapy.

(b) Who to treat?

Who to treat with the therapies described above remains a controversial issue. Over the past decade, a number of review panels have met in order to produce consensus guidelines based on the available evidence of the time. It is interesting to compare these guidelines and to note the increasingly more liberal approach to therapy as the evidence with respect to the pathogen accumulated and strengthened.

- 1990 World Congress of Gastroenterology, Sydney (Axon, 1991)
  - Where there is no history of NSAID therapy and the duodenal ulcer is *H.-pylori*-related, eradication therapy should be **considered**.
  - In patients with mild disease that can be managed easily by intermittent therapy with acid suppressive or cytoprotective drugs, the working party believes that at present *H. pylori* eradication should **not be employed generally**, etc.
- 1992 Working party of the First United European Gastroenterology Week (Tytgat *et al.*, 1993)

- *H. pylori* eradication therapy is **recommended in all patients** with peptic ulcer disease with proven infection who are not taking NSAIDs, etc.
- 1994 NIH Consensus meeting, Washington (Yamada *et al.*, 1994)
  - Ulcer patients with *H. pylori* infection **require treatment** with anti-microbial agents in addition to antisecretory drugs whether on first presentation with the illness or on recurrence.
  - The value of treating of non-ulcerative dyspepsia patients with *H. pylori* infection **remains to be determined**, etc.
- 1996 Consensus report of the European *Helicobacter pylori* study group, Maastricht. (F. Megraud, personal communication)
  - Eradication therapy is **strongly recommended** for the following patients with proven *H. pylori* infection: duodenal or gastric ulcer (active or not); MALToma; gastritis with severe abnormalities; post early gastric cancer resection.
  - Eradication therapy is **advisable** for the following patients with proven *H. pylori* infection: functional dyspepsia; family history of gastric cancer; gastroesophageal reflux disease treated by long-term proton-pump inhibitor; NSAID therapy; post surgery for peptic ulcer disease.

These latest guidelines may be seen as radical by some. However, they do reflect the reality of what is happening in medical practice. In presentations at international meetings, some have even championed the cause of screening populations for *H. pylori* and treatment of all positives. This is based on an analysis of the causes of death reported for the UK and the conclusion that *H. pylori* is responsible for more deaths than all the other major communicable diseases put together (Axon, personal communication)! Microbiologists may flinch at the widespread use of so many anti-microbials but at present this is the trend of the future.

6.5.2 Immunization

As the importance of *H. pylori* became realized, it became clear that a vaccine would have major uses particularly in those countries where the prevalence of gastric cancer was high. However, there was doubt as to whether immunization would work given that the bacterium survived in the host for life despite the vigorous immune response mounted against it.

Czinn and Nedrud did a simple experiment in which they gave mice an oral vaccine of a sonicate of *H. pylori* and the mucosal adjuvant cholera toxin (CT) and showed the development of increased levels of mucosal IgA (Czinn and Nedrud, 1991). They reasoned that this study suggested that immunization might be possible. At this time we had developed a mouse model of *H. pylori* infection with the bacterium *H. felis*, a close relative of the human helicobacter (Lee *et al.*, 1990). Using a similar immunizing regimen, except the sonicate used was from *H. felis*, mice were shown to be protected from challenge with large numbers of living bacteria (Chen, Lee and Hazell, 1992). Since those early experiments much progress has been made, at least four major pharmaceutical or vaccine companies now have active vaccine development programs and the first human safety trials have been completed (Kreiss *et al.*, 1996). The antigen used in most of the experimental vaccines is recombinant urease although many other proteins have been shown to be protective, including catalase, heat shock proteins, and the *cagA* and *vacA* gene products. The final vaccine will probably be composed of two or three antigens (Lee and Buck, 1996). The limiting factor at present in the quest for a successful commercial product is a suitable adjuvant. Cholera toxin and the related molecule LT (the heat-labile enterotoxin of *Escherichia coli*) are too toxic in humans and so alternatives are required. Various mutant non-toxic forms of these mucosal adjuvants are now being trialed (Ghiara and Michetti, 1995).

An observation, also made in our *H. felis* mouse model, has changed the whole potential and importance of immunization and has been a major impetus to vaccine development. Mice already infected with the bacterium were immunized with an oral regimen known to be protective (Doidge *et al.*, 1994). Many of the animals were actually cured of their infection. That is, a therapeutic effect was observed. This result has been repeated by many other groups and therapeutic immunization has also been demonstrated in ferrets, where a significant number of animals were cured of infection with the natural ferret helicobacter, *Helicobacter mustelae* (Cuenca *et al.*, 1996). A safe vaccine that cures ulcers and also could be used in some populations to prevent gastric cancer without the problem of development of antibiotic resistance would allow the elimination of this major gastroduodenal pathogen from whole populations. There is an excellent chance that this vaccine will come, but from experience with other vaccines it will probably be 6–8 years before it is commercially available.

6.6 *H. PYLORI* AS MEDIEVAL NORMAL MICROFLORA

H. pylori shares many characteristics of bacteria described elsewhere as members of the normal microflora of the human body. This bacterium has evolved to inhabit a restricted ecological niche and remains there for the life of its host. In many cases, the bacterium is passed on from generation to generation from parent to child. It is present in nearly all individuals in some populations, e.g. Estonia, and it causes histopathology in all it infects although this inflammation in most cases appears to be of no consequence. But, as is detailed above, in some individuals serious symptomatic diseases develop. For these reasons a concept of 'almost normal flora' has been coined to distinguish it from close relatives like *C. jejuni*, which may be normal microflora in chickens but in humans are clearly a conventional pathogen (Lee and Hazell, 1988). A hypothesis the author has been developing, which will be described in detail elsewhere, is that *H. pylori* did indeed evolve to be a component of the normal microflora of the human gastrointestinal tract, just as *Bacteroides* spp. did in the large bowel. Host and bacterium lived in well-adjusted balance for millions of years. In recent times, as a result of changes in the host resulting from improved health and nutrition, the gastric environment has altered such that overt disease has appeared. A brief synopsis of these concepts is given below as the hypothesis may be relevant to our understanding of other components of the normal microflora and their contribution to human disease.

In the beginning, the intestinal surfaces were the first to be colonized with the ancestors of what became the lower bowel microflora. There was no normal microflora in the stomach as the animal host had acquired the ability to secret acid in order not only to aid in the breakdown of foodstuffs but also to prevent potentially pathogenic bacteria from entering the lower intestinal tract, where they could cause disease. Over time urease-positive spiral bacteria, probably originating from the lower bowel, adapted to the gastric mucosa. Their motility allowed them to cope with the gastric mucus and the urease enzyme acquired new characteristics to allow it to be acid-protective – increased enzymic activity and a surface location. A non-destructive inflammatory response, gastritis, was induced in the gastric epithelium, which did not harm the host but probably benefited the bacterium by increasing available nutrients from the tissue. Acid output was not at a level that caused either gastric or duodenal ulcer. In other words, acid output in medieval times was less than it was today because of diet and the burden of parasitic and bacterial pathogens. Infection is known to reduce acid output, e.g. by the effect of the cytokine IL-1. Gastric cancer was not a problem, despite the probable presence of gastritis in both antrum and body, as people just did not live long enough for tumors to develop. As societies increased their standard of living, acid output increased. Now gastric ulcers started to appear and gastric cancer became common in those countries where the dietary factors predisposed to malignancy. Prior to the mid 1800s, duodenal ulcers were rarely observed. In the 1930s–1960s duodenal ulcers began to occur in professionals in the higher income bracket and were termed an affliction of the rich (Anon, 1959). This is the group most likely to be well nourished and free from infection, and so acid output was maximum. Now in those developed societies, gastric cancer is rare and ulcer disease is starting to disappear, apart for NSAID-induced lesions. *H. pylori* is being lost from these communities as factors that allowed ready transmission, such as poor hygiene and high density of living, are not present. In all the countries of the world where an improvement in living standards has occurred, *H. pylori* is very rare in childhood and will thus disappear from these populations.

There is a precedent for loss of what was once normal microflora in the human gut. As mentioned above, all animal lower bowel surfaces are heavily colonized with spiral/helical bacteria. These bacteria are conspicuous by their absence in humans in developed societies. However, they have been reported to be present in patients in India (Mathan and Mathan, 1985). Intestinal spirochetosis was reported in Africa early this century (Takeuchi *et al.*, 1974). There is a parallel between the loss of *H. pylori* and the loss of the lower bowel spirals.

6.7 THE NON-GASTRIC HELICOBACTERS: NORMAL MICROFLORA OR UNDISCOVERED PATHOGENS?

The discovery of *H. pylori* is just the first step in the understanding of the importance of mucus associated bacteria in our intestinal tracts. There are many more bacteria waiting in the wings. Most of these bacteria comprise the normal microflora of the animal intestine and probably have no medical significance (Lee, 1991). However, recent findings in animals suggest that, just as *H. pylori* surprised the medical world by being responsible for diseases not previously associated with a microbial etiology, other helicobacters may play a negative role in human health (Fox *et al.*, 1996a).

6.7.1 The exploding genus *Helicobacter*

H. pylori is the type strain of the genus *Helicobacter*. Since the first culture in 1982, there has been a veritable explosion of the genus, with new organisms being cultured and others being identified as *Helicobacter* species. It is now clear that, just as *H. pylori* has evolved to the human stomach, other species have evolved to inhabit the stomachs of other animals. Chance and differences in the gastric environment of the animals means that, although they must share a common primordial ancestor, hence the genetic clustering into the one genus, they have acquired many differences and morphologically are very different (Fig. 6.6).

Thus the cat and dog helicobacter, *Helicobacter felis*, is a much tighter spiral (Lee *et al.*, 1988). *Helicobacter acinonyx*, the cheetah helicobacter, looks much like *H. pylori* (Eaton *et al.*, 1991), while *Helicobacter mustelae*, the ferret representative of the genus, is not even spiral but is a short, curved rod with lateral flagellae (Fox Edrise and Cabot, 1986). Significantly, all these gastric helicobacters are urease-positive, highlighting the importance of the enzyme for gastric survival.

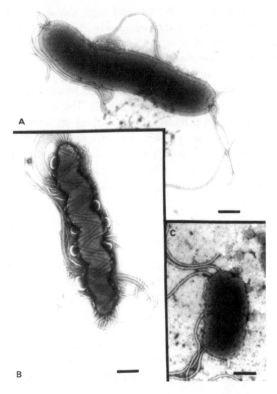

Figure 6.6 Negative stains of gastric helicobacters showing their varied morphology. **(a)** *H. pylori* – S-shaped with bipolar flagella (× 25 000). **(b)** *H. felis* – tight, helical shape with tufts of bipolar flagella (× 12 000). **(c)** *H. mustelae* – short, rod-shaped with polar and lateral flagella (× 25 000).

6.7.2 The lower bowel helicobacters

As gene probes became available and DNA sequencing becomes simpler many of the lower-bowel spiral-shaped bacteria have been shown to belong to the genus *Helicobacter*, which now comprises 16 species (Mendes *et al.*, 1996). More species will doubtless have been named by the time this chapter is published. The taxonomic tree as it exists today is shown in Fig. 6.7.

Most of these bacteria are of no consequence to the clinician interested in disease. However, for those interested in the normal microflora and microbial ecology, there is much to be learnt from the way these bacteria differently inhabit their normal niche. For example, *Helicobacter muridarum*, arguably the most beautiful of the helicobacters under the electron microscope (Fig. 6.8), normally inhabits the lower bowel of rodents, mainly in the ileum and cecum (Lee *et al.*, 1992).

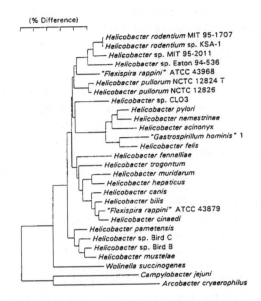

Figure 6.7 Phylogenetic tree for *Helicobacter* species on the basis of 16S rRNA sequence similarity. The scale represents a 1% difference in nucleotide sequence as determined by measuring the length of horizontal lines connecting any two species.

Source: reproduced by courtesy of J. G. Fox.

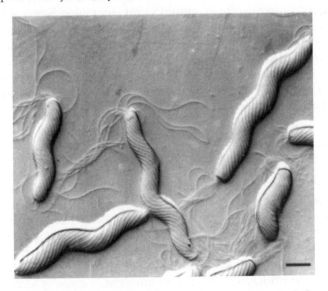

Figure 6.8 Freeze dried preparation of *H. muridarum* showing its S shape, bipolar tufts of flagella and periplasmic fibrils (× 10 000).

Source: reproduced with permission from Lee *et al.*, 1992.

These bacteria are never normally seen in the gastric pits of infected mice. The bacterium has a urease but it is very different to that of *H. pylori* or *H. felis* and it is not protective at acid pH, even if urea is present (Ferrero and Lee, 1991). However, if the mice are coinfected with *H. felis* then *H. muridarum* appears in the gastric pits in high numbers (Lee *et al.*, 1993b). Clearly, infection with *H. felis* has altered the gastric milieu such that *H. muridarum* can now survive. Understanding such subtle differences in colonization will give great insights into the balance between host and its microflora. Basic studies in microbial ecology are no longer fashionable. Hopefully the ability to genetically manipulate *Helicobacter* spp. will be the stimulus for important ecological studies. A strain of *H. pylori* well adapted to mice, the 'Sydney strain,' is now available to all those interested (Lee *et al.*, 1997).

There have recently been some exciting findings in mice in which some of the lower-bowel helicobacters have been shown to be pathogenic. These studies may have parallels in human disease and may herald another phase in the *Helicobacter* story. *Helicobacter hepaticus* is part of the normal microflora in the lower bowel of mice and is present in most conventional colonies throughout the world (Fox *et al.*, 1994). However in aged mice it translocates into the liver where it can induce a very destructive hepatitis that apparently progresses to hepatocellular carcinomas in some mouse strains, e.g. A/JCr male mice (Fox *et al.*, 1996a). *H. hepaticus* is also associated with inflammatory bowel disease in immunodeficient mice and only infrequently in immunocompetent mice. Selected germfree outbred mice infected with this bacterium develop enterocolitis (Fox *et al.*, 1996b). Both these examples of members of the normal bowel microflora causing severe disease if the host changes – in old age or after the establishment of germfree status – have parallels in human disease. Could *Helicobacter* species be involved? Is there a human *H. hepaticus* equivalent in our lower bowel just as *H. pylori* is the human equivalent to *H. felis*?

6.8 CONCLUSION

The human gastrointestinal microflora is one of great complexity and has a major beneficial role on human health. However, disturbance of this complex ecosystem may be the trigger for serious disease. The stomach provides a less complex part of this ecosystem and bacterial diversity is low. While there are parallels to its distant cousins lower down in the bowel, on the basis of the information presented above we have to conclude that *Helicobacter pylori* is certainly not an opportunistic member of the normal microflora, but rather a communicable agent of some diseases of major global significance. Great benefit will flow to humankind if we can accelerate the disappearance of this unwelcome visitor from the stomachs of the world.

6.9 REFERENCES

Akopyants, N. S., Kersulyte, D. and Berg, D. E. (1995) cagII, a new multigene locus associated with virulence in *Helicobacter pylori*. *Gut*, **37**(Suppl. 1), A1.

Anon (1959) Gastric ulcer and the ulcer equation. *Lancet*, **i**, 1131–1133.

Axon, A. T. R. (1991) *Helicobacter pylori* therapy: effect on peptic ulcer disease. *Journal of Gastroenterology and Hepatology*, **6**, 131–137.

Ball, P. A. J. (1961) The secretory background to gastric ulcer. *Lancet*, **i**, 1363–1365.

Bell, G. D. and Powell, K. U. (1996) *Helicobacter pylori* reinfection after apparent eradication – the Ipswich experience. *Scandinavian Journal of Gastroenterology*, **31**(Suppl. 215), 96–104.

Bizzozero, G. (1892) Über die schlauchförmigen Drusen des Magendarmkanals und die Beziehungen ihres Epithels zu dem Oberfachenepithel der Schleim-haut. *Archiv für Mikrobiologie und Anatomie*, **42**, 82–152.

Blaser, M. J., Perez Perez, G. I., Kleanthous, H. *et al.* (1995) Infection with *Helico-bacter pylori* strains possessing *cagA* is associated with an increased risk of developing adenocarcinoma of the stomach. *Cancer Research*, **55**, 2111–2115.

Bolton, F. J. and Hutchinson, D. N. (1989) Evaluation of three *Campylobacter pylori* antigen preparations for screening sera from patients undergoing endoscopy. *Journal of Clinical Pathology*, **42**, 723–726.

Boren, T. and Falk, P. (1994) *Helicobacter pylori* binds to blood group antigens. *Scientific American–Science Medicine*, **1**, 28–37.

Boren, T., Falk, P., Roth, K. A. *et al.* (1993) Attachment of *Helicobacter pylori* to human gastric epithelium mediated by blood group antigens. *Science*, **262**, 1892–1895.

Borody, T. J., Carrick, J. and Hazell, S. L. (1987) Symptoms improve after the eradication of gastric *Campylobacter pyloridis*. *Medical Journal of Australia*, **146**, 450–451.

Carrick, J., Lee, A., Hazell, S. *et al.* (1989) *Campylobacter pylori*, duodenal ulcer, and gastric metaplasia: possible role of functional heterotopic tissue in ulcerogen-esis. *Gut*, **30**, 790–797.

Censini, S., Lange, C., Xiang, Z. Y. *et al.* (1996) Cag, a pathogenicity island of *Helicobacter pylori*, encodes type I-specific and disease-associated virulence fac-tors. *Proceedings of the National Academy of Sciences of the USA*, **93**, 14648–14653.

Chen, M., Lee, A. and Hazell, S. L. (1992) Immunisation against *Helicobacter* infection in a mouse/*Helicobacter felis* model. *Lancet*, **339**, 1120–1121.

Clyne, M., Labigne, A. and Drumm, B. (1995) *Helicobacter pylori* requires an acidic environment to survive in the presence of urea. *Infection and Immunity*, **63**, 1669–1673.

Coghlan, J. G., Gilligan, D., Humphries, H. *et al.* (1987) *Campylobacter pylori* and recurrence of duodenal ulcers – a 12 month follow-up study. *Lancet*, **ii**, 1109–1111.

Correa, P. (1988) A human model of gastric carcinogenesis. *Cancer Research*, **48**, 3554–3560.

Correa, P. (1991) Is gastric carcinoma an infectious disease? *New England Journal of Medicine*, **325**, 1170–1171.

Correa, P. and Ruiz, B. (1989) *Campylobacter pylori* and gastric cancer in *Campylo-bacter pylori and gastroduodenal disease*, (eds Rathbone, B. J. and Heatley, R. V.), Blackwell Scientific Publications, London, pp. 139–145.

Correa, P., Cuello, C., Duque, E. *et al.* (1976) Gastric cancer in Colombia III: natural history of precursor lesions. *Journal of the National Cancer Institute*, **57**, 1027–1035.

Covacci, A., Censini, S., Bugnoli, M. *et al.* (1993) Molecular characterization of the 128-kDa immunodominant antigen of *Helicobacter pylori* associated with cytotoxicity and duodenal ulcer. *Proceedings of the National Academy of Sciences of the USA,* **90,** 5791–5795.

Crabtree, J. E., Farmery, S. M., Covacci, A. *et al.* (1994a) *Helicobacter pylori* induced gastric epithelial IL-8 gene expression is associated with the CagA-positive phenotype. *American Journal of Gastroenterology,* **89,** 1337.

Crabtree, J. E., Wyatt, J. I., Trejdosiewicz, L. K. *et al.* (1994b) Interleukin-8 expression in *Helicobacter pylori* infected, normal, and neoplastic gastroduodenal mucosa. *Journal of Clinical Pathology,* **47,** 61–66.

Cuenca, R., Blanchard, T. G., Czinn, S. J. *et al.* (1996) Therapeutic immunization against *Helicobacter mustelae* in naturally infected ferrets. *Gastroenterology,* **110,** 1770–1775.

Czinn, S. J. and Nedrud, J. G. (1991) Oral immunization against *Helicobacter pylori. Infection and Immunity,* **59,** 2359–2363.

Danon, S. J., O'Rourke, J. L., Moss, N. D. and Lee, A. (1995) The importance of local acid production in the distribution of *Helicobacter felis* in the mouse stomach. *Gastroenterology,* **108,** 1386–1395.

Dent, J. C., McNulty, C. A., Uff, J. S. *et al.* (1988) *Campylobacter pylori* urease: a new serological test. *Lancet,* **i,** 1002.

Dixon, M. F., Genta, R. M., Yardley, J. H. *et al.* (1996) Classification and grading of gastritis – the updated Sydney System. *American Journal of Surgical Pathology,* **20,** 1161–1181.

Doenges, J. L. (1939) Spirochaetes in the gastric glands of *Macacus rhesus* and humans without definite history of related disease. *Archives of Pathology,* **27,** 469–477.

Doidge, C., Gust, I., Lee, A. *et al.* (1994) Therapeutic immunisation against *Helicobacter* infection. *Lancet,* **343,** 914–915.

Dooley, C. P. and Cohen, H. (1988) The clinical significance of *Campylobacter pylori. Annals of Internal Medicine,* **108,** 70–79.

Eaton, K. A., Radin, M. J., Fox, J. G. *et al.* (1991) *Helicobacter acinonyx,* a new species of *Helicobacter* isolated from cheetahs with gastritis. *Microbial Ecology in Health and Disease,* **4**(S), S104.

Elzimaity, H., Graham, D. Y., Alassi, M. T. *et al.* (1996) Interobserver variation in the histopathological assessment of *Helicobacter pylori* gastritis. *Human Pathology,* **27,** 35–41.

Ernst, P. B., Reves, V. E., Gourley, W. H. *et al.* (1996) Is the Th1/Th2 lymphocyte balance upset by *Helicobacter pylori* infection?, in *Helicobacter pylori: Basic Mechanisms to Clinical Cure 1996,* (eds R. H. Hunt and G. N. J. Tytgat), Kluwer Academic Publishers, Dordrecht, pp. 150–157.

Feldman, R. A., Deeks, J. J. and Evans, S. (1995) Multi-laboratory comparison of eight commercially available *Helicobacter pylori* serology kits. *European Journal of Clinical Microbiology and Infectious Disease,* **14,** 428–433.

Ferrero, R. L. and Lee, A. (1988) Motility of *Campylobacter jejuni* in a viscous environment: comparison with conventional rod-shaped bacteria. *Journal of General Microbiology,* **134,** 53–59.

Ferrero, R. L. and Lee, A. (1991) The importance of urease in acid protection for the gastric-colonising bacteria *Helicobacter pylori* and *Helicobacter felis* sp. nov. *Microbial Ecology in Health and Disease,* **4,** 121–134.

Ferrero, R. L., Hazell, S. L. and Lee, A. (1988) The urease enzymes of *Campylobacter pylori* and a related bacterium. *Journal of Medical Microbiology,* **27,** 33–40.

Forman, D., Coleman, M., Debacker, G. *et al.* (1993) An international association between *Helicobacter pylori* infection and gastric cancer. *Lancet,* **341,** 1359–1362.

Fox, J. G., Edrise, B. M., Cabot, E. B. (1986) *Campylobacter*-like organisms isolated from gastric mucosa of ferrets. *American Journal of Veterinary Research*, **47**, 236–239.

Fox, J. G., Dewhirst, F. E., Tully, J. G. *et al.* (1994) *Helicobacter hepaticus* sp nov, a microaerophilic bacterium isolated from livers and intestinal mucosal scrapings from mice. *Journal of Clinical Microbiology*, **32**, 1238–1245.

Fox, J. G., Li, X., Yan, L. *et al.* (1996a) Chronic proliferative hepatitis in A/JCR mice associated with persistent *Helicobacter hepaticus* infection – a model of *Helicobacter*-induced carcinogenesis. *Infection and Immunity*, **64**, 1548–1558.

Fox, J. G., Yan, L., Shames, B. *et al.* (1996b) Persistent hepatitis and enterocolitis in germfree mice infected with *Helicobacter hepaticus*. *Infection and Immunity*, **64**, 3673–3681.

George, L. L., Borody, T. J., Andrews, P. *et al.* (1990) Cure of duodenal ulcer after eradication of *Helicobacter pylori*. *Medical Journal of Australia*, **153**, 145–149.

Ghiara, P. and Michetti, P. (1995) Development of a vaccine. *Current Opinion in Gastroenterology*, **11**(Suppl. 1), 52–56.

Glynn, J. R. (1994) *Helicobacter pylori* and the heart. *Lancet*, **344**, 146.

Goodwin, C. S., Armstrong, J. A., Chilvers, T. *et al.* (1989) Transfer of *Campylobacter pylori* and *Campylobacter mustelae* to *Helicobacter pylori* gen. nov. and *Helicobacter mustelae* comb. nov. respectively. *International Journal of Systematic Bacteriology*, **39**, 397–405.

Graham, J. R. (1995) *Helicobacter pylori*: human pathogen or simply an opportunist? *Lancet*, **345**, 1095–1097.

Graham, D. Y., Malaty, H. M., Evans, D. G. *et al.* (1991) Epidemiology of *Helicobacter pylori* in an asymptomatic population in the United States. Effect of age, race, and socioeconomic status. *Gastroenterology*, **100**, 1495–1501.

Hansson, L. E., Nyren, O., Hsing, A. W. *et al.* (1996) The risk of stomach cancer in patients with gastric or duodenal ulcer disease. *New England Journal of Medicine*, **335**, 242–249.

Hazell, S. L., Lee, A., Brady, L. and Hennessy, W. (1986) *Campylobacter pyloridis* and gastritis: association with intercellular spaces and adaptation to an environment of mucus as important factors in colonization of the gastric epithelium. *Journal of Infectious Diseases*, **153**, 658–663.

Hazell, S. L., Borody, T. J., Gal, A. and Lee, A. (1987) *Campylobacter pyloridis* gastritis I: Detection of urease as a marker of bacterial colonization and gastritis. *American Journal of Gastroenterology*, **82**, 292–296.

Hazell, S. L., Mitchell, H. M., Hedges, M. *et al.* (1994) Hepatitis A and evidence against the community dissemination of *Helicobacter pylori* via feces. *Journal of Infectious Diseases*, **170**, 686–689.

Hentschel, E., Brandstatter, G., Dragosics, B. *et al.* (1993) Effect of ranitidine and amoxycillin plus metronidazole on the eradication of *Helicobacter pylori* and the recurrence of duodenal ulcer. *New England Journal of Medicine*, **328**, 308–312.

Hirschl, A. M., Pletschette, M., Hirschl, M. H. *et al.* (1988) Comparison of different antigen preparations in an evaluation of the immune response to *Campylobacter pylori*. *European Journal of Clinical Microbiology and Infectious Disease*, **7**, 570–575.

IARC (1994) *Schistosomes, Liver Flukes and Helicobacter pylori, IARC Monographs on the Evaluation of Carcinogenic Risks to Humans 61*, World Health Organization, Lyon, pp. 177–240.

Kikuchi, S., Wada, O., Nakajima, T. *et al.* (1995) Serum anti-*Helicobacter pylori* antibody and gastric carcinoma among young adults. *Cancer*, **75**, 2789–2793.

Knipp, U., Birkholz, S., Kaup, W. and Opferkuch, W. (1993) Immune suppressive effects of *Helicobacter pylori* on human peripheral blood mononuclear cells. *Medical Microbiology and Immunology*, **182**, 63–76.

Kreiss, C., Buclin, T., Cosma, M. *et al.* (1996) Oral immunization with recombinant urease without adjuvant in *H. pylori*-infected humans. *Gut*, **39**(Suppl. 2), A39.

Lee, A. (1985) Neglected niches: the microbial ecology of the gastrointestinal tract, in *Advances in Microbial Ecology 8*, (eds K. C. Marshall), Plenum Press, New York, pp. 115–162.

Lee, A. (1991) Spiral organisms: what are they? A microbiological introduction to *Helicobacter pylori*. *Scandinavian Journal of Gastroenterology*, **187**, 9–22.

Lee, A. and Buck, F. (1996) Vaccination and mucosal responses to *Helicobacter pylori* infection. *Alimentary Pharmacology and Therapeutics*, **10**(Suppl. 1), 129–138.

Lee, A. and Hazell, S. L. (1988) *Campylobacter pylori* in health and disease: an ecological perspective. *Microbial Ecology in Health and Disease*, **1**, 1–16.

Lee, A. and Megraud, F. (eds) (1996) Helicobacter pylori: *Techniques for Clinical Diagnosis and Basic Research*, W. B. Saunders, London.

Lee, A., O'Rourke, J. L., Barrington, P. J. and Trust, T. J. (1986) Mucus colonization as a determinant of pathogenicity in intestinal infection by *Campylobacter jejuni*; a mouse cecal model. *Infection and Immunity*, **51**, 536–546.

Lee, A., Hazell, S. L., O'Rourke, J. and Kouprach, S. (1988) Isolation of a spiral-shaped bacterium from the cat stomach. *Infection and Immunity*, **56**, 2843–2850.

Lee, A., Fox, J. G., Otto, G. and Murphy, J. (1990) A small animal model of human *Helicobacter pylori* active chronic gastritis. *Gastroenterology*, **99**, 1315–1323.

Lee, A., Phillips, M. W., O'Rourke, J. L. *et al.* (1992) *Helicobacter muridarum* sp. nov., a microaerophilic helical bacterium with a novel ultrastructure isolated from the intestinal mucosa of rodents. *International Journal of Systematic Bacteriology*, **42**, 27–36.

Lee, A., O'Rourke, J., Dixon, M. and Fox, J. (1993a) *Helicobacter* induced gastritis: look to the host. *Acta Gastro-Enterologica Belgica*, **56**(S), 61.

Lee, A., Chen, M. H., Coltro, N. *et al.* (1993b) Long term infection of the gastric mucosa with *Helicobacter* species does induce atrophic gastritis in an animal model of *Helicobacter pylori* infection. *Zentralblatt für Bakteriologie – International Journal of Medical Microbiology Virology Parasitology and Infectious Diseases*, **280**, 38–50.

Lee, A., Dixon, M. F., Danon, S. J. *et al.* (1995) Local acid production and *Helicobacter pylori*: a unifying hypothesis of gastroduodenal disease. *European Journal of Gastroenterology and Hepatology*, **7**, 461–465.

Lee, A., O'Rourke, J., De Ungria, M. C. *et al.* (1997) A standardised mouse model of *Helicobacter pylori* infection. Introducing the Sydney strain. *Gastroenterology*, **112**, 1386–1397.

Lind, T., Veldhuyzen van Zanten, S., Unge, P. *et al.* (1996) Eradication of *Helicobacter pylori* using one-week triple therapies combining omeprazole with two antimicrobials: the MACH I study. *Helicobacter*, **1**, 138–144.

Loffeld, R. J. and Stobberingh, E. (1991) Comparison of seven commercial assays for detection of IgG antibodies against *Helicobacter pylori*. *Italian Journal of Gastroenterology*, **23**(Suppl. 2), 21.

Logan, R. P. H. (1993) Breath tests to detect *Helicobacter pylori*, in Helicobacter pylori: *Biology and Clinical Practice*, (eds C. S. Goodwin and B. Wormsley), CRC Press, Boca Raton, FL, pp. 307–327.

Logan, R. P. H. (1996) The ^{13}C urea breath test, in *Helicobacter pylori: Techniques for Clinical and Basic Research*, (eds A. Lee and F. Megraud) W. B. Saunders, London, pp. 74–82.

Maaroos, H. I. (1995) *Helicobacter pylori* infection in Estonian population – is it a health problem? *Annals of Medicine*, **27**, 613–616.

McNulty, C. A. M., Gearty, J. C., Crump, B. *et al.* (1986) *Campylobacter pyloridis* and associated gastritis: investigator blind, placebo controlled trial of bismuth salicylate and erythromycin ethylsuccinate. *British Medical Journal*, **293**, 645–649.

Magnus, H. A. (1946) The pathology of simple gastritis. *Journal of Pathology and Bacteriology*, **58**, 431–439.

Marks, I. N. and Shay, H. (1959) Observations on the pathogenesis of gastric ulcer. *Lancet*, **ii**, 1107–1111.

Marshall, B. J. (1983) Unidentified curved bacillus on gastric epithelium in active chronic gastritis. *Lancet*, **i**, 1273–1275.

Marshall, B. J. (1996) The ^{14}C urea breath test, in *Helicobacter pylori: Techniques for Clinical and Basic Research*, (eds A. Lee and F. Megraud) W. B. Saunders, London, pp. 83–93.

Marshall, B. J., Armstrong, J. A., McGechie, D. B. and Glancy, R. J. (1985a) Attempt to fulfil Koch's postulates for pyloric *Campylobacter*. *Medical Journal of Australia*, **142**, 436–439.

Marshall, B. J., McGechie, D. B., Rogers, P. A. and Glancy, R. J. (1985b) Pyloric *Campylobacter* infection and gastroduodenal disease. *Medical Journal of Australia*, **142**, 439–444.

Marshall, B. J., Warren, J. R., Francis, G. J. *et al.* (1987) Rapid urease test in the management of *Campylobacter pyloridis*-associated gastritis. *American Journal of Gastroenterology*, **82**, 200–210.

Mathan, M. M. and Mathan, V. I. (1985) Rectal mucosal morphologic abnormalities in normal subjects in southern India: a tropical colonopathy? *Gut*, **26**, 710–717.

Megraud, F., Brassens Rabbe, M. P., Denis, F. *et al.* (1989) Seroepidemiology of *Campylobacter pylori* infection in various populations. *Journal of Clinical Microbiology*, **27**, 1870–1873.

Mendall, M. A., Goggin, P. M., Molineaux, N. *et al.* (1994) Relation of *Helicobacter pylori* infection and coronary heart disease – abstracts. *British Heart Journal*, **71**, 437–439.

Mendes, E. N., Queiroz, D. M. M., Dewhirst, F. E. *et al.* (1996) *Helicobacter trogontum* sp nov, isolated from the rat intestine. *International Journal of Systematic Bacteriology*, **46**, 916–921.

Meyer Rosberg, K., Scott, D. R., Rex, D. *et al.* (1996) The effect of environmental pH on the proton motive force of *Helicobacter pylori*. *Gastroenterology*, **111**, 886–900.

Mitchell, J. D., Mitchell, H. M. and Tobias, V. (1992) Acute *Helicobacter pylori* infection in an infant, associated with gastric ulceration and serological evidence of intra-familial transmission. *American Journal of Gastroenterology*, **87**, 382–386.

Mitchell, H. M., Lee, A., Berkowicz, J. and Borody, T. (1988) The use of serology to diagnose active *Campylobacter pylori* infection. *Medical Journal of Australia*, **149**, 604–609.

Mitchell, H. M., Li, Y. Y., Hu, P. J. *et al.* (1992a) Epidemiology of *Helicobacter pylori* in Southern China – identification of early childhood as the critical period for acquisition. *Journal of Infectious Diseases*, **166**, 149–153.

Mitchell, H. M., Li, Y., Hu, P. *et al.* (1992b) The susceptibility of *Helicobacter pylori* to bile may be an obstacle to faecal transmission. *European Journal of Gastroenterology and Hepatology*, **4**(Suppl. 1), S79-S83.

Mitchell, H. M., Bohane, T., Hawkes, R. A. and Lee, A. (1993) *Helicobacter pylori* infection within families. *Zentralblatt für Bakteriologie – International*

Journal of Medical Microbiology Virology Parasitology and Infectious Diseases, **280**, 128–136.

Mitchell, H. M., Hazell, S. L., Li, Y. Y. and Hu, P. J. (1996) Serological response to specific *Helicobacter pylori* antigens – antibody against *cagA* antigen is not predictive of gastric cancer in a developing country. *American Journal of Gastroenterology*, **91**, 1785–1788.

Mobley, H. L. T. (1996) The role of *Helicobacter pylori* urease in the pathogenesis of gastritis and peptic ulceration. *Alimentary Pharmacology and Therapeutics*, **10**(Suppl. 1), 57–67.

Mohammadi, M., Redline, R., Nedrud, J. and Czinn, S. (1996) Role of the host in pathogenesis of *Helicobacter*-associated gastritis – *H. felis* infection of inbred and congenic mouse strains. *Infection and Immunity*, **64**, 238–245.

Morris, A. and Nicholson, G. (1987) Ingestion of *Campylobacter pyloridis* causes gastritis and raised fasting gastric pH. *American Journal of Gastroenterology*, **82**, 192–199.

Nomura, A., Stemmermann, G. N., Chyou, P. H. *et al.* (1991) *Helicobacter pylori* infection and gastric carcinoma among Japanese Americans in Hawaii. *New England Journal of Medicine*, **325**, 1132–1136.

Nowottny, U. and Heilmann, K. L. (1990) Epidemiology of *Helicobacter pylori* infection. *Leber Magen Darm*, **20**, 183–186.

Oi, M., Oshida, K. and Sugimura, S. (1959) The location of gastric ulcer. *Gastroenterology*, **36**, 45–56.

Ormand, J. E. and Talley, N. J. (1990) *Helicobacter pylori*: controversies and an approach to management. *Mayo Clinic Proceedings*, **65**, 414–426.

Parsonnet, J., Friedman, G. D., Vandersteen, D. P. *et al.* (1991) *Helicobacter pylori* infection and the risk of gastric carcinoma. *New England Journal of Medicine*, **325**, 1127–1131.

Patel, P., Mendall, M. A., Khulusi, S. *et al.* (1994) *Helicobacter pylori* infection in childhood: Risk factors and effect on growth. *British Medical Journal*, **309**, 1119–1123.

Patel, P., Mendall, M. A., Carrington, D. *et al.* (1995) Association of *Helicobacter pylori* and *Chlamydia pneumoniae* infections with coronary heart disease and cardiovascular risk factors. *British Medical Journal*, **311**, 711–714.

Perez Perez, G. I., Taylor, D. N., Bodhidatta, L. *et al.* (1990) Seroprevalence of *Helicobacter pylori* infections in Thailand. *Journal of Infectious Disease*, **161**, 1237–1241.

Phadnis, S. H., Parlow, M. H., Levy, M. *et al.* (1996) Surface localization of *Helicobacter pylori* urease and a heat shock protein homolog requires bacterial autolysis. *Infection and Immunity*, **64**, 905–912.

Phillips, M. W. and Lee, A. (1983) Isolation and characterization of a spiral bacterium from the crypts of rodent gastrointestinal tracts. *Applied and Environmental Microbiology*, **45**, 675–683.

Price, A. B. (1991) The Sydney System: histological division. *Journal of Gastroenterology and Hepatology*, **6**, 209–222.

Rauws, E. A., Langenberg, W., Houthoff, H. J. *et al.* (1988) *Campylobacter pyloridis*-associated chronic active antral gastritis. A prospective study of its prevalence and the effects of antibacterial and antiulcer treatment. *Gastroenterology*, **94**, 33–40.

Rebora, A., Drago, F. and Picciotto, A. (1994) *Helicobacter pylori* in patients with rosacea. *American Journal of Gastroenterology*, **89**, 1603–1604.

Sakagami, T., Dixon, M., O'Rourke, J. *et al.* (1996) Atrophic gastric changes in both *Helicobacter felis* and *Helicobacter pylori* infected mice are host dependent and separate from antral gastritis. *Gut*, **39**, 639–648.

Salomon, H. (1896) Spirillum of the mammalian stomach and its behaviour with respect to parietal cells. *Zentralblatt für Bakteriologie, Parasitenkunde und Infektioskrankheiten*, **19**, 433–441.

Sjostrom, J. E. and Larsson, H. (1996) Factors affecting growth and antibiotic susceptibility of *Helicobacter pylori* – effect of pH and urea on the survival of a wild-type strain and a urease-deficient mutant. *Journal of Medical Microbiology*, **44**, 425–433.

Slomiany, B. L., Bilski, J., Sarosiek, J. *et al.* (1987) *Campylobacter pyloridis* degrades mucin and undermines gastric mucosal integrity. *Biochemical and Biophysical Research Communications*, **144**, 307–314.

Stolte, M. and Eidt, S. (1989) Lymphoid follicles in antral mucosa: immune response to *Campylobacter pylori*? *Journal of Clinical Pathology*, **42**, 1269–1271.

Takeuchi, A., Jervis, H. R., Nakazawa, H. and Robinson, D. M. (1974) Spiral-shaped organisms on the surface colonic epithelium of the monkey and man. *American Journal of Clinical Nutrition*, **27**, 1287–1296.

Talley, N. J. (1994) A critique of therapeutic trials in *Helicobacter pylori* – positive functional dyspepsia. *Gastroenterology*, **106**, 1174–1183.

Thomas, J. E., Gibson, G. R., Darboe, M. K. *et al.* (1992) Isolation of *Helicobacter pylori* from human faeces. *Lancet*, **340**, 1194–1195.

Tummuru, M. K. R., Cover, T. L. and Blaser, M. J. (1993) Cloning and expression of a high-molecular-mass major antigen of *Helicobacter pylori* – evidence of linkage to cytotoxin production. *Infection and Immunity*, **61**, 1799–1809.

Tytgat, G. N. J. (1996) Aspects of anti-*Helicobacter pylori* eradication therapy, in *Helicobacter pylori: Basic Mechanisms to Clinical Cure 1996*, (eds R. H. Hunt and G. N. J. Tytgat), Kluwer Academic Publishers, Dordrecht, pp. 340–347.

Tytgat, G. N. J., Lee, A., Graham, D. Y. *et al.* (1993) The role of infectious agents in peptic ulcer disease. *Gastroenterology International*, **6**, 76–89.

Wadstrom, T., Hirmo, S. and Boren, T. (1996) Biochemical aspects of *Helicobacter pylori* colonization of the human gastric mucosa. *Alimentary Pharmacology and Therapeutic*, **10**(Suppl. 1), 17–27.

Warren, J. R. (1983) Unidentified curved bacilli on gastric epithelium in active chronic gastritis. *Lancet*, **i**, 1273.

Webb, P. M., Knight, T., Greaves, S. *et al.* (1994) Relation between infection with *Helicobacter pylori* and living conditions in childhood – evidence for person to person transmission in early life. *British Medical Journal*, **308**, 750–753.

Wotherspoon, A. C., Doglioni, C., Diss, T. C. *et al.* (1993) Regression of primary low-grade B-cell gastric lymphoma of mucosa-associated lymphoid tissue type after eradication of *Helicobacter pylori*. *Lancet*, **342**, 575–577.

Wyatt, J. I. and Rathbone, B. J. (1988) Immune response of the gastric mucosa to *Campylobacter pylori*. *Scandinavian Journal of Gastroenterology*, **23**(Suppl. 142), 44–49.

Wyatt, J. I., Rathbone, B. J. and Heatley, R. V. (1986) Local immune response to gastric *Campylobacter* in non-ulcer disease. *Journal of Clinical Pathology*, **39**, 863–870.

Yamada, T., Ahnen, D., Alpers, D. H. *et al.* (1994) *Helicobacter pylori* in peptic ulcer disease. *Journal of the American Medical Association*, **272**, 65–69.

7

The intestinal microflora and intra-abdominal sepsis

Andrew B. Onderdonk

Infections involving the abdominal cavity occur for a variety of reasons, but almost universally are caused by bacteria that normally reside within the gastrointestinal tract. Trauma, GI surgery, perforated large intestine or appendix, and carcinoma of the colon are all common underlying reasons for contamination of the peritoneal cavity with the bacteria within the lumen of the large intestine (Bartlett *et al.*, 1976). Infections related to the abdominal cavity include infections of the abdominal wall, subphrenic and intra-abdominal abscesses, peritonitis, and abscesses of any visceral organ including the spleen, pancreas and liver (Finegold, 1977). Virtually all such infections are polymicrobic and involve a mixture of both obligate anaerobes and facultatively anaerobic bacterial species (Gorbach and Bartlett, 1974; Gorbach, Thadepalli and Norsen, 1974; Hentges, 1983). Morbidity and mortality due to such infections is substantial, as is the cost of such infections to the health-care delivery system. In order to appreciate the complexity of such infections, it is first important to understand something of the complexity of the human intestinal microflora.

7.1 HUMAN INTESTINAL MICROFLORA

The human gastrointestinal tract, from mouth to anus, is a remarkably complex organ with multiple functions and many different microbiological environments. The microorganisms that colonize this organ are both longitudinally and cross-sectionally distributed. An excellent discussion

G. W. Tannock (ed.), *Medical Importance of the Normal Microflora*, 164–176.
© 1999 *Kluwer Academic Publishers. Printed in Great Britain.*

of human intestinal microflora, the factors that control population densities, the development of the microflora in the newborn and the effect of various environmental factors on microbial growth can be found elsewhere (Hentges, 1983; Tannock, 1995). Although bacteria can be found along the entire length of the GI tract, the largest numbers of bacteria reside within the large intestine and cecum. It is generally acknowledged that between 10^{11} and 10^{12} viable bacteria per gram dry weight of intestinal contents are present within the lumen contents of the large bowel. This does not include those organisms that may be intimately attached to the intestinal epithelial tissue, or that reside within the crypts formed by the intestinal mucosa. Irrespective of these exclusions, the number of viable bacteria per gram of intestinal contents represents one of the most dense naturally occurring bacterial populations ever described. Not surprisingly, a diverse array of phenotypes comprise the dominant microflora of the large intestine. As is the case for all monogastric animals, over 99% of the bacteria present as part of human intestinal microflora are obligate anaerobes, most being Gram-negative rods of the genus *Bacteroides*. It is this obligately anaerobic component of the microflora that is most commonly associated with the development of polymicrobic intra-abdominal infections, despite the presence of hundreds of different phenotypes within the lumen contents of the healthy adult.

Studies of the human intestinal microflora indicate that the dominant bacterial species present within the large bowel are members of the genus *Bacteroides*. Among this genus, *B. vulgatus*, *B. thetaiotaomicron* and *B. distasonis* are the numerically dominant species, while *B. fragilis* is present in lower numbers. In addition to the genus *Bacteroides*, members of the genera *Peptostreptococcus*, *Eubacterium*, *Bifidobacterium*, *Clostridium* and *Fusobacterium* are among the numerically dominant organisms present in the large intestine (Hentges, 1983; Tannock, 1995; Evaldson *et al.*, 1982; Hill and Drasar, 1974; Salyers, 1984). Facultative species commonly described as 'enterics', such as *Enterococcus faecalis* and *Escherichia coli*, are present at less than 0.1% of the total population, a ratio of about 1:1000 with the obligate anaerobes. The diversity of the microflora has been well documented through several studies of the intestinal microflora which suggest that up to 400 different phenotypes can be isolated from human intestinal contents. Most of the species present as part of the normal intestinal microflora are not considered to be pathogens for humans, even when accidentally inoculated into the peritoneal cavity. It is of interest to note that, of the hundreds of potential phenotypes present as part of the normal large intestinal microflora, only a few are legitimate pathogens. This suggests that those species capable of causing disease once outside their normal environment of the bowel have special attributes that allow them to survive and proliferate within the peritoneal cavity. These species and their unique attributes, particularly *B. fragilis*, will serve as the basis for much of the discussion in this chapter.

7.2 RECENT HISTORY OF INTRA-ABDOMINAL SEPSIS

With few exceptions, clinicians regarded obligate anaerobes as commensal organisms during intra-abdominal infections until the early 1970s (Bartlett *et al.*, 1976; Gorbach and Bartlett, 1974; Gorbach, Thadepalli and Norsen, 1974). The clinical literature was replete with descriptions of '*E. coli* pus', a common infectious complication of traumatic bowel injury such as knife and gunshot wounds. Although surgeons such as Altenmeier had long held the belief that obligate anaerobes were important agents of infection within the peritoneal cavity, little attention was paid to these early reports. It wasn't until it was shown that patients receiving therapeutic agents with little activity *versus* obligate anaerobes had more complications, including intra-abdominal abscesses, that serious study of obligate anaerobes as agents of infection began.

Attempts to develop animal models simulating human intra-abdominal infections had only modest success, largely based on the lack of good microbiological information for human infections. Based on the clinical studies of Gorbach and colleagues (Thadepalli *et al.*, 1972, 1973), a concerted effort to develop an animal model for intra-abdominal sepsis was made using more complete microbiological and clinical information. A reproducible model system using Wistar rats, developed in 1974, has been used extensively to characterize the role of obligate anaerobes during intra-abdominal infections and to evaluate new therapeutic agents for the treatment of polymicrobic infections (Onderdonk *et al.*, 1974a, b). Indeed, the rat model for intra-abdominal sepsis has proved to be one of the most predictive models of an infectious disease ever developed. This model has been widely used both to explore the basic biology of the bacterial species isolated from human intra-abdominal infections and to evaluate novel therapeutic interventions.

7.3 THE RAT MODEL FOR INTRA-ABDOMINAL SEPSIS

The rat model for intra-abdominal sepsis is a surgical model that employs an inoculum of bacteria obtained from the intestinal contents of other meat-fed rats. This inoculum, in contrast to that obtained from grain-fed animals, mimics the microbiological make-up of human intestinal contents and contains many of the genera commonly found in human intestinal contents. When implanted into other male Wistar rats via an anterior midline incision, a reproducible biphasic disease process occurs. During the first 72 hours after implantation, peritonitis occurs characterized by free-flowing peritoneal exudates, positive blood cultures and a mortality rate of approximately 50% of implanted animals. Surviving recipients of the intestinal content inoculum appear to recover, however, and examination of the peritoneal cavity beginning 5 days after implantation reveals intra-abdominal abscesses in all animals. These two clinical endpoints, mortality and abscess development, are exactly the same as those used to assess human disease. The use of the same inoculum from experiment to experiment resulted in a highly reproducible animal model system. This basic model system has served as the basis for numerous microbiological and immunological observations, as described below.

Microbiological characterization of the two phases of the experimental disease process yielded quite different results. Cultures of peritoneal exudates during the early, often lethal, phase yielded facultative Gram-negative rods, such as *E. coli,* and enterococci as the numerically dominant organisms. Blood cultures obtained during the first 72 hours were often polymicrobic, but almost always included *E. coli.* These findings were consistent with the reported bacteriological findings from humans with untreated peritonitis. In contrast, abscess contents contained high concentrations of obligate anaerobes, particularly *Bacteroides* and *Fusobacterium* sp. in addition to *E. coli* and enterococci. Subsequent studies used selective antimicrobial probes to show that treatment of experimentally infected animals with agents active against *E. coli* prevented mortality, while agents active against obligate anaerobes prevented abscess development (Weinstein *et al.,* 1975). Additional research was directed at evaluating the potential role of the most commonly isolated organisms from peritonitis and abscesses. Four organisms, *E. coli,* enterococci, the *Bacteroides fragilis* group (the taxonomy used at the time classified all members of this species as subspecies, the actual species designation by current taxonomy was *B. thetaiotaomicron*) and *Fusobacterium varium,* were implanted singly and in combinations of two or more organisms into the peritoneal cavity of Wistar rats. It was shown that mortality was exclusively associated with *E. coli*; however, abscesses required the presence of both a facultative and an obligately anaerobic species (Onderdonk *et al.,* 1976). These findings clearly established a role for obligate anaerobes during intra-abdominal infections. More importantly, the therapeutic trials performed in this animal model prompted the clinical use of agents such as clindamycin and gentamicin to treat these serious infections.

7.4 ROLE OF *B. FRAGILIS* IN INTRA-ABDOMINAL ABSCESSES

From a variety of clinical studies defining the organisms present during intra-abdominal infections, it was clear that *B. fragilis* was the most common clinical isolate from such infections and was the most common obligately anaerobic Gram-negative rod isolated from blood cultures, despite the fact that it was not the numerically dominant Gram-negative anaerobe present within the large intestine. It had also been observed that one member of the phenotypically similar *B. fragilis* group organisms, *B. fragilis* ssp. *fragilis*, appeared to have a polysaccharide capsule not present for other phenotypically similar organisms (Kasper and Seiler, 1975; Kasper *et al.*, 1977). The animal model for intra-abdominal sepsis was modified to test both the encapsulated and unencapsulated species of this group. It was shown that encapsulated *B. fragilis* was capable of provoking abscesses alone and that viable organisms were not required in order for abscesses to be induced (Onderdonk *et al.*, 1977). More importantly, it was shown that the capsular polysaccharide component of the outer membrane complex of *B. fragilis* was able to induce abscesses in the experimental model in the absence of viable bacteria. These studies clearly defined the capsular polysaccharide of *B. fragilis* as the major virulence factor for species and helped explain their dominant role during intra-abdominal infections.

On the basis of these observations, this species has been used as a prototype for understanding the role of anaerobes in abscess induction and protection. The unique properties of the capsular polysaccharide of *B. fragilis* and its role in both the induction and prevention of abscesses has been well documented in the literature (Kasper *et al.*, 1977, 1979a, 1980; Kasper, Onderdonk and Bartlett, 1977; Bartlett *et al.*, 1978; Onderdonk *et al.*, 1978, 1979, 1982; Joiner *et al.*, 1980; Kasper and Onderdonk, 1982; Shapiro *et al.*, 1982). Perhaps the most important studies defining our knowledge about a specific bacterial polysaccharide during intra-abdominal infections are those that describe the immunity to abscess development induced by *B. fragilis*. An understanding of the immune response to the capsular polysaccharide of this species has led to the description of an important structure/function relationship between a bacterial polysaccharide and a specific host response. These studies have helped us to understand the very nature of abscess induction; at the same time they have defined the host response to bacterial challenge within the peritoneal cavity.

7.5 IMMUNITY TO *B. FRAGILIS* INDUCED ABSCESSES

Following the observation that *B. fragilis* capsular polysaccharide could induce abscesses, it was shown that parenteral immunization with the capsular polysaccharide prior to challenge with live organisms, or the capsular polysaccharide, protected against the development of abscesses (Kasper *et al.*, 1979a). It was also shown that an antibody response to the capsular polysaccharide developed within a few weeks of immunization. Protection against abscess development in this animal model was assumed to be antibody-mediated, because bacterial polysaccharides were known to induce B-cell-specific immunity. However, passive transfer of hyperimmune antibody from actively immunized animals to naive recipients did not protect against abscesses induced by *B. fragilis* in the animal model system. The possible role of T-cell-dependent immunity was demonstrated by showing that nylon-wool-passaged spleen cells from actively immunized rats were capable of protecting against challenge with *B. fragilis* in naive recipient animals (Kasper *et al.*, 1979b). Additional research revealed that T-lymphocyte populations were capable of transferring the protection against the development of abscesses in this model (Onderdonk *et al.*, 1982).

Using inbred mice, it was shown that T cells were a necessary component for the protection against *B. fragilis* conferred by immunization, and that the subset of cells involved appeared to be a T suppressor cell (Shapiro *et al.*, 1982). In these early experiments, it appeared that the protection against abscess induction was species-specific, because challenge of immunized animals with other combinations of bacteria known to induce abscesses were still capable of inducing abscesses (Onderdonk *et al.*, 1982). On the basis of these observations of a polysaccharide-induced, T-cell-dependent immunity to abscess formation, additional definition of this unusual immune response was sought.

7.5.1 Protection induced by a soluble factor from suppressor T cells

Using the mouse adaptation of the model for intra-abdominal abscesses, attempts were made to better characterize the cells involved in protection against abscesses. This was accomplished by using inbred mice that were actively immunized with the capsular polysaccharide of *B. fragilis*. Spleen cells from these animals were purified into T- and B-cell-enriched populations using standard nylon wool columns. Viable cells and freeze thawed lysates of these cells were passively transferred to naive mice. The passively immunized mice were then challenged with viable *B. fragilis*. It was noted that both the viable T cells and lysates (ITF) of these cells were capable of providing protection against subsequent challenge (Zaleznik *et al.*, 1985). More importantly, selection of the suppressor-cell subset of T cells was shown to be the source of the factor responsible for protection. Further characterization of the active fraction of the T-cell lysates revealed that the molecular size of the active moiety was relatively small since it was dialysable through a 12 000 Da membrane. The active suppressor cell factor could also be neutralized by preincubation with the capsular polysaccharide of *B. fragilis*. It was shown that the ITF was heat-labile and lost activity when treated with proteases.

7.5.2 Host peritoneal cellular response

The knowledge that immunity to *B.-fragilis*-induced abscesses was governed by a cell-mediated mechanism led to an interest in understanding the cellular events occurring within the peritoneal cavity of the host. Small chambers with 0.22 μm filters at each end were used to contain bacterial inoculum within the peritoneal cavity without restricting a free flow of nutrients and bacterial metabolites across the membrane (Onderdonk *et al.*, 1989). Containment chambers were necessary to allow the various types of cells found within the peritoneal cavity to be counted and identified free of damage from bacteria. Total cell counts following challenge with chambers containing bacteria increased to peak levels within 8 hours of challenge. The dominant cell types were neutrophils and macrophages; however, lymphocytes were the dominant cell type identified during the first few hours after challenge. The rapidity with which lymphocytes were found within the peritoneal cavity following challenge suggested that the cells were probably migrating into the peritoneal cavity from the mesothelial barrier rather than from the parenteral circulation. Interestingly, actively immune animals showed a dramatic increase in the numbers of lymphocytes present at 5–7 days following challenge. This corresponded to a decrease in the viable cell density for *B. fragilis* within the filter chamber. In non-immune animals, no increase in lymphocyte populations occurred after 5–7 days and the number of bacteria within the filter chambers did not decrease (Onderdonk *et al.*, 1989). In order to rule out the possibility that the bacterial killing noted was due to antibody, dialysis sacs with a 50 kDa exclusion size were used to contain bacteria. It was shown that killing still occurred within the dialysis sacs, a fact that indicated that antibody was not involved.

7.5.3 Immunochemical characterization of the capsular polysaccharide of *B. fragilis*

Immunochemical characterization of the capsular polysaccharide of *B. fragilis* (strain NCTC 9343) revealed a novel structure composed of two separate polysaccharides (Pantosti *et al.*, 1991). Immunoelectrophoresis of the purified capsular polysaccharide yielded a complex precipitation profile, suggesting that more than one moiety was present. Using column chromatographic methods, two polysaccharides were isolated, one with a neutral charge and the second with a negative charge. These compounds were termed polysaccharide A (PSA) and polysaccharide B (PSB), respectively. Chemical analysis revealed that both polysaccharides contained positively charged amino sugars. The data supported the concept that *B. fragilis* produced two surface expressed polysaccharides with different properties that were distinctly different from the lipopolysaccharide of this organism.

Subsequent work by Tzianabos revealed that PSA and PSB were expressed simultaneously on the surface of the bacterial cell. Structural analysis of the two polysaccharides by NMR showed that PSA was a tetramer, with a three sugar repeating unit, containing one positively charged amino group and a single negatively charged carboxyl group. PSB was shown to be a hexamer, with a three sugar repeating unit, containing a positively charged amino group and two negative charges. Based on the dual charge motif, PSA and PSB were ionically bound to each other under physiological conditions (Tzianabos *et al.*, 1992, 1993). The importance of the dual charge motif on the biological activity of PSA and PSB in both provoking abscesses and preventing their formation when given parentally was demonstrated by chemically modifying the purified polysaccharides. Tzianabos was able to show that loss of either the positive or negative charge on either PSA or PSB resulted in a significant loss in activity. He also showed that the positive charge must be in the form of an amino group, while the negative charge could be contributed by carboxyl or phosphonate groups. Additional studies also revealed that other bacterial polysaccharides with similar charge motifs, such as type I *S. pneumoniae* polysaccharide, were also able to provoke abscesses in the animal model. Chemical alteration of bacterial polysaccharides, such as the Vi antigen for *Salmonella typhi*, allowed them to provoke abscesses as well (Tzianabos *et al.*, 1993, 1994, 1995; Tzianabos, Onderdonk and Kasper, 1994). It was also shown that protection against abscess development was dependent on the presence of a dual charge motif on the polysaccharide used for immunization (Tzianabos, Kasper and Onderdonk, 1995).

7.5.4 Induction of cytokines by *B. fragilis* polysaccharide

The unusual nature of the capsular polysaccharides of *B. fragilis* in both provoking and preventing abscesses has prompted careful study of the host response within the peritoneal cavity. Studies in which the immunization schedule has been altered for PSA indicate that a protective effect can be induced by exposure to the polysaccharide as little as 24 hours prior to challenge. In addition, it was shown that immunization with PSA, the more biologically active of the two polysaccharides, protected against abscess induction by other microorganisms (Tzianabos, Kasper and Onderdonk, 1995). These observations suggested that PSA was acting as an immunomodulator rather than an antigen in the classic sense.

To simulate the early events following *B. fragilis* challenge within the peritoneal cavity, the cytokine response for several isolated cell types was determined. Cells were elicited from the mesothelium of mice using cell culture medium containing mercaptoethanol. Resident peritoneal cells were then concentrated, washed, grown in cell culture media on plates and exposed to either the capsular polysaccharide of *B. fragilis*, the lipopolysaccharide of this organism, the LPS of *E. coli*, the polysaccharide of type III *S. pneumoniae*, or beta-glucan (another immunomodulatory

polysaccharide). It was shown that both the capsular polysaccharide of
B. fragilis and the lipopolysaccharide of this organism were capable of
inducing TNF-α and IL-1α in a dose dependent manner. Cell cultures
of mouse resident peritoneal cells in the absence of a stimulus did not
produce appreciable amounts of these cytokines. Known stimulation of
TNF-α and IL-1α by *E. coli* LPS and the polysaccharide of type III
S. pneumoniae occurred predictably in these cell cultures. A negative
control for these experiments, beta glucan, did not induce either TNF-α
or IL-1α (Gibson *et al.*, 1996).

Additional studies using mouse resident peritoneal cells included measurement of IL-10 levels following incubation with capsular polysaccharide of *B. fragilis*, the polysaccharide of type III *S. pneumoniae* or beta glucan
were performed. IL-10 is a potent immunoregulating cytokine that stimulates a Th$_2$ response, also known as a humoral immune response. In
addition, IL-10 is known to suppress Th$_1$, or cell-mediated, immune
responses. The results of these studies indicated that the capsular polysaccharide of *B. fragilis* stimulates an IL-10 response from resident peritoneal cells. Other studies used human peripheral mononuclear (MNL)
and polymorphonuclear cells (PMNL), grown in culture and exposed to
the capsular polysaccharide of *B. fragilis*. In this series of experiments it
was shown that IL-8 was produced in high levels by PMNL within 6
hours after exposure to capsular polysaccharide, with a maximum effect
noted at 24 hours. MNL also showed a robust response after 6 hours of
exposure, but a non-specific stimulation of IL-8 appeared to occur for
MNL at 24 hours when compared to the media control (Gibson, Tzianabos and Onderdonk, 1996).

Polysaccharides from bacteria have been shown by others to provoke a
variety of cytokine responses when incubated with mammalian cells. The
study described above demonstrates that cytokines are released from
both resident peritoneal cells, PMNL and MNL, following exposure to
the capsular polysaccharide of *B. fragilis*. TNF-α and IL-1α both result in a
variety of changes to host cells, including the up regulation of cell adhesion molecules. Although a similar response can be elicited by LPS from
other Gram-negative organisms, the capsular polysaccharide appears to
be unique in its ability to provoke an IL-8 and IL-10 response. IL-10 is
produced by T and B lymphocytes, as well as by macrophages. It has
several apparent functions, not the least of which is to serve as a potent
down regulator of the immune response. IL-10 has been shown to also
decrease the sensitivity of cells to TNF-α, a function that may be a key
mechanism for *B. fragilis* to suppress the normal immune response during
initial development of an infectious process. IL-8, on the other hand, is a
potent chemokine that may elicit a local signal that attracts polymorphonuclear cells to the site of abscess formation. In any event, it is clear that
much of the host defense system within the peritoneal cavity is due to

events that occur at the mesothelial boundary. The cells present as part of this layer are fully capable of producing cytokines in response to bacterial challenge with the capsular polysaccharide of *B. fragilis*. Additional study is needed to more fully understand the complex series of signals from cell to cell and from bacteria to host that result in abscess development.

7.6 ROLE OF OTHER BACTERIAL SPECIES DURING INTRA-ABDOMINAL SEPSIS

While this discussion has centered on the virulence of *B. fragilis* and its capsular polysaccharide as a prototype for anaerobes and their participation during intra-abdominal infections, other organisms can and do play an important role in intra-abdominal infections. Among those that have been isolated with regularity, *Prevotella* species are most common. Although the detailed studies performed with *B. fragilis* have not been carried out with other bacterial species, *Prevotella* (formerly *Bacteroides* sp.) appear to have many of the same characteristics as *B. fragilis* (Mansheim, Onderdonk and Kasper, 1978; Mansheim and Kasper, 1979). Species such as *Fusobacterium* are more commonly associated with liver abscesses than with mesenteric abscesses, yet these organisms have the added problem of containing a lipopolysaccharide that acts in much the same way as endotoxin from facultative species such as *E. coli* (Finegold, 1977). *Peptostreptococcus* species are also regularly found as members of the infecting microflora present in abscesses. It has also been reported that the synergistic effects noted between obligate anaerobes and facultative species for abscess development may also contribute to the early mortality attributed to facultative Gram-negative rods.

While it is clear from the clinical literature that a broad array of organisms may be isolated from cases of intra-abdominal sepsis, it is also clear that only a few of the hundreds of species present as part of the intestinal microflora are regularly isolated from such infections. With the exception of *B. fragilis*, little is known about the virulence factors or synergistic effects that may occur between organisms present as part of the polymicrobic infectious process. There is no question that organisms such as enterococci and *Prevotella* have polysaccharide constituents external to the cell wall that may be capsules. Whether such materials are essential components for the survival and virulence of these species within the peritoneal cavity is unknown.

7.7 SUMMARY

Members of the microflora of the human large intestine are capable of provoking serious infectious processes within the peritoneal cavity of humans. Among the many species present within the lumen of the large intestine, only a few are considered to be legitimate pathogens when they gain access to the peritoneal cavity. *B. fragilis* is the most common obligately anaerobic member of the intestinal microflora associated with such infections. It has been shown, using an animal model system, that the polysaccharide capsule of this species accounts for much of its virulence. Other bacterial species have also been shown to participate in the polymicrobic infections that often occur within the peritoneal cavity. Studies on the virulence of such species and their respective roles in the infectious process remain to be performed.

7.8 REFERENCES

Bartlett, J. G., Onderdonk, A. B., Louie, T. and Gorbach, S. L. (1976) Experimental intra-abdominal sepsis, in *Chemotherapy*, (eds J. D. Williams and A. M. Geddes), Plenum Press, New York, vol. 2, pp. 249–358.

Bartlett, J. G., Onderdonk, A. B., Louie, T. *et al.* (1978) A review. Lessons from an animal model of intra-abdominal sepsis. *Archives of Surgery*, **113**, 853–857.

Evaldson, G., Heimdahl, A., Kager, L. and Nord, C. E. (1982) The normal human anaerobic microflora. *Scandinavian Journal of Infectious Diseases (Supplement)*, **35**, 9–15.

Finegold, S. M. (1977) *Anaerobic Bacteria in Human Disease*, Academic Press, New York.

Gibson, F. C., Tzianabos, A. O. and Onderdonk, A. B. (1996) The capsular polysaccharide complex of *Bacteroides fragilis* induces cytokine production from human and murine phagocytic cells. *Infection and Immunity*, **64**, 1065–1069.

Gorbach, S. L. and Bartlett, J. G. (1974) Anaerobic infections. *New England Journal of Medicine*, **290**, 1177–1184.

Gorbach, S. L., Thadepalli, H. and Norsen, J. (1974) *Anaerobic Microorganisms in Intraabdominal Infections*, Charles C. Thomas, Springfield, IL.

Hentges, D. J. (1983) *Human Intestinal Microflora in Health and Disease*, Academic Press, New York.

Hill, M. J. and Drasar, B. S. (1974) *Human Intestinal Microflora*, Academic Press, London.

Joiner, K. A., Gelfand, J. A., Onderdonk, A. B. *et al.* (1980) Host factors in the formation of abscesses. *Journal of Infectious Diseases*, **142**, 40–49.

Kasper, D. L. and Onderdonk, A. B. (1982) Infection with *Bacteroides fragilis*: pathogenesis and immunoprophylaxis in an animal model. *Scandinavian Journal of Infectious Diseases (Supplement)*, **31**, 28–33.

Kasper, D. L. and Seiler, M. W. (1975) Immunochemical characterization of the outer membrane complex of *Bacteroides fragilis* subspecies *fragilis*. *Journal of Infectious Diseases*, **132**, 440–450.

Kasper, D. L., Onderdonk, A. B. and Bartlett, J. G. (1977) Quantitative determination of the antibody response to the capsular polysaccharide of *Bacteroides fragilis* in an animal model of intraabdominal abscess formation. *Journal of Infectious Diseases*, **136**, 789–795.

Kasper, D. L., Hayes, M. E., Reinap, B. G. *et al.* (1977) Isolation and identification of encapsulated strains of *Bacteroides fragilis*. *Journal of Infectious Diseases*, **136**, 75–81.

Kasper, D. L., Onderdonk, A. B., Crabb, J. and Bartlett, J. G. (1979a) Protective efficacy of immunization with capsular antigen against experimental infection with *Bacteroides fragilis*. *Journal of Infectious Diseases*, **140**, 724–731.

Kasper, D. L., Onderdonk, A. B., Polk, B. F. and Bartlett, J. G. (1979b) Surface antigens as virulence factors in infection with *Bacteroides fragilis*. *Reviews of Infectious Diseases*, **1**, 278–290.

Kasper, D. L., Onderdonk, A. B., Reinap, B. G. and Linberg, A. A. (1980) Variations of *Bacteroides fragilis* with *in vitro* passage: presence of an outer membrane-associated glycan and loss of capsular antigen. *Journal of Infectious Diseases*, **142**, 750–756.

Mansheim, B. J. and Kasper, D. L. (1979) Detection of anticapsular antibodies to *Bacteroides asaccharolyticus* in serum from rabbits and humans by use of an enzyme-linked immunosorbent assay. *Journal of Infectious Diseases*, **140**, 945–951.

Mansheim, B. J., Onderdonk, A. B. and Kasper, D. L. (1978) Immunochemical and biologic studies of the lipopolysaccharide of *Bacteroides melaninogenicus* subspecies *asaccharolyticus*. *Journal of Immunology*, **120**, 72–78.

Mansheim, B. J., Onderdonk, A. B. and Kasper, D. L. (1979) Immunochemical characterization of surface antigens of *Bacteroides melaninogenicus*. *Reviews of Infectious Diseases*, **1**, 263–276.

Onderdonk, A. B., Weinstein, W. M., Sullivan, N. M. *et al.* (1974a) Experimental intra-abdominal abscesses in rats. I. Development of Animal Model. *Infection and Immunity*, **10**.

Onderdonk, A. B., Weinstein, W. M., Sullivan, N. M. *et al.* (1974b) Experimental intra-abdominal abscesses in rats. II. Quantitative bacteriology of infected animals. *Infection and Immunity*, **10**, 1.

Onderdonk, A. B., Bartlett, J. G., Louie, T. *et al.* (1976) Microbial synergy in experimental intra-abdominal abscess. *Infection and Immunity*, **13**, 22–26.

Onderdonk, A. B., Kasper, D. L., Cisneros, R. L. and Bartlett, J. G. (1977) The capsular polysaccharide of *Bacteroides fragilis* as a virulence factor: comparison of the pathogenic potential of encapsulated and unencapsulated strains. *Journal of Infectious Diseases*, **136**, 82–89.

Onderdonk, A. B., Moon, N. E., Kasper, D. L. and Bartlett, J. G. (1978) Adherence of *Bacteroides fragilis in vivo*. *Infection and Immunity*, **19**, 1083–1087.

Onderdonk, A. B., Kasper, D. L., Mansheim, B. J. *et al.* (1979) Experimental animal models for anaerobic infections. *Reviews of Infectious Diseases*, **1**, 291–301.

Onderdonk, A. B., Markham, R. B., Zaleznik, D. F. *et al.* (1982) Evidence for T cell-dependent immunity to *Bacteroides fragilis* in an intraabdominal abscess model. *Journal of Clinical Investigation*, **69**, 9–16.

Onderdonk, A. B., Cisneros, R. L., Crabb, J. H. *et al.* (1989) Intraperitoneal host cellular responses and *in vivo* killing of *Bacteroides fragilis* in a bacterial containment chamber. *Infection and Immunity*, **57**, 3030–3037.

Pantosti, A., Tzianabos, A. O., Onderdonk, A. B. and Kasper, D. L. (1991) Immunochemical characterization of two surface polysaccharides of *Bacteroides fragilis*. *Infection and Immunity*, **59**, 2075–2082.

Salyers, A. A. (1984) *Bacteroides* of the human lower intestinal tract. *Annual Review of Microbiology*, **38**, 293–313.

Shapiro, M. E., Onderdonk, A. B., Kasper, D. L. and Finberg, R. W. (1982) Cellular immunity to *Bacteroides fragilis* capsular polysaccharide. *Journal of Experimental Medicine*, **155**, 1188–1197.

Tannock, G. W. (1995) *Normal Microflora*, Chapman & Hall, London.

Thadepalli, H., Gorbach, S. L., Broido, P. and Norsen, J. (1972) A prospective study of infections in penetrating abdominal trauma. *American Journal of Clinical Nutrition*, **25**, 709–721.

Thadepalli, H., Gorbach, S., Broido, P. *et al.* (1973) Abdominal trauma, anaerobes, and antibiotics. *Surgery Gynecology and Obstetrics*, **137**, 270–276.

Tzianabos, A. O., Kasper, D. L. and Onderdonk, A. B. (1995) Structure and function of *Bacteroides fragilis* capsular polysaccharides: relationship to induction and prevention of abscesses. *Clinical Infectious Diseases*, **20**(Suppl. 2), S132–S140.

Tzianabos, A. O., Onderdonk, A. B. and Kasper, D. L. (1994) Bacterial structure and functional relationship to abscess formation. *Infectious Agents and Disease*, **3**, 256–265.

Tzianabos, A. O., Pantosti, A., Baumann, H. *et al.* (1992) The capsular polysaccharide of *Bacteroides fragilis* comprises two ionically linked polysaccharides. *Journal of Biological Chemistry*, **267**, 18230–18235.

Tzianabos, A. O., Onderdonk, A. B., Rosner, B. *et al.* (1993) Structural features of polysaccharides that induce intra-abdominal abscesses. *Science*, **262**, 416–419.

Tzianabos, A. O., Onderdonk, A. B., Smith, R. S. and Kasper, D. L. (1994) Structure–function relationships for polysaccharide-induced intraabdominal abscesses. *Infection and Immunity*, **62**, 3590–3593.

Tzianabos, A. O., Kasper, D. L., Cisneros, R. L. *et al.* (1995) Polysaccharide-mediated protection against abscess formation in experimental intraabdominal sepsis. *Journal of Clinical Microbiology*, **96**, 2727–2731.

Weinstein, W. M., Onderdonk, A. B., Bartlett, J. G. *et al.* (1975) Antimicrobial therapy of experimental intraabdominal sepsis. *Journal of Infectious Diseases*, **132**, 282–286.

Zaleznik, D. F., Finberg, R. W. and Shapiro, M. E. (1985) A soluble suppressor T cell factor protects against experimental intraabdominal abscesses. *Journal of Clinical Investigation*, **75**, 1023–1027.

8

The intestinal microflora and inflammatory bowel disease

Vinton S. Chadwick and Wangxue Chen

8.1 CLINICAL ASPECTS OF INFLAMMATORY BOWEL DISEASE

Inflammatory bowel disease (IBD) is an umbrella term used to describe three conditions, Crohn's disease (CD), ulcerative colitis (UC) and non-specific colitis. The latter condition may be an intermediate form of the other two and is a common form of IBD in children. Both CD and UC are chronic relapsing inflammatory disorders of the intestine and in this respect differ from self-limited inflammation caused by infections or toxins. They are also distinct from more recently recognized milder forms of chronic inflammation called microscopic colitis, which comes in at least two varieties known as lymphocytic and collagenous colitis (Kingham, 1991). Microscopic colitis and inflammation caused by ischaemia or therapeutic radiation will not be discussed, even though the intestinal bacterial microflora may play an important role in these conditions.

Ulcerative colitis only affects the large intestine, while CD can affect any site from mouth to anus, although most commonly it affects distal small and proximal large intestine. Differentiating CD from UC can be problematic in some patients with disease confined to the colon. The pathological and clinical features of these diseases are summarized in Table 8.1.

Although CD and UC are very often confined to the intestine and in that sense are not multisystem diseases, several extraintestinal conditions are associated with IBD and these complications (Table 8.2) may

G. W. Tannock (ed.), *Medical Importance of the Normal Microflora*, 177–221.
© 1999 *Kluwer Academic Publishers. Printed in Great Britain.*

Table 8.1 Characteristic clinical and pathological features of UC and CD (note: anaemia, hypoalbuminaemia, raised sedimentation rate (ESR) and C-reactive protein (CRP) are common features in both diseases)

	Ulcerative colitis	*Crohn's disease*
Clinical features		
Gradual onset	Usual	Usual
Diarrhoea	Usually blood and mucus	Usually watery
Abdominal pain	Infrequent	Frequent
Fever	Only if severe	Frequent
Endoscopic features		
Rectal involvement	Invariable	Frequently absent
Distribution	Continuous and symmetrical	Segmental and asymmetrical
Ulceration	Diffuse with friability	Focal aphthoid or linear
Cobblestoning	Absent	frequent submucosal nodules
Fistulas	Absent	Quite common
Pathological features		
Inflammation	Mucosal	Mucosal and submucosal
Goblet cell mucus	Depleted	Retained
Crypt abscesses	Frequent	Less frequent
Crypt architecture	Distorted	Often normal
Granulomas	Absent	Frequent (> 60%)

Table 8.2 Extraintestinal complications of inflammatory bowel disease

Eyes	Uveitis
	Episcleritis
Mouth	Aphthous ulceration
Skin	Erythema nodosum
	Pyoderma gangrenosum
Joints	Pauciarticular peripheral arthritis
	Ankylosing spondylitis
Liver/biliary	Primary sclerosing cholangitis
	Peri-cholangitisgallstones*
Renal tract	Hyperoxaluria and nephrolithiasis*
	Hydronephrosis*

* Associated with ileal resection or ileal Crohn's disease

occasionally antedate the bowel disease and may implicate the intestinal microflora in disease pathogenesis.

Inflammatory bowel disease occurs worldwide but is more common in North America and Northern Europe and among Caucasians in comparable latitudes in the southern hemisphere. Up to 2 million people may

Table 8.3 Established advances in medical and surgical therapy of IBD (note: selective anti-inflammatory agents such as leukotriene B5 antagonists and selective immunosuppressives (monoclonal antibodies, etc.) are currently under clinical trial)

Anti-inflammatory agents	5-aminosalicylates (5ASA derivatives)
	• delayed release oral preparations
	• stable enema preparations
	Corticosteroids
	• conventional systemic or local
	• non-absorbable or non-systemic preparations
Antibiotics	Metronidazole for colonic CD
Immunosuppressants	Azathioprine/6-mercaptoporine
	Methotrexate
	Cyclosporin A
Operations for UC	Panproctocolectomy/ileostomy
	Ileoanal pouch
Operations for CD	Segmental resections
	Strictureplasty
Dietary therapy	Elemental and defined formula diets in CD

be affected and most of those afflicted are between the ages of 16 and 40. These diseases are a major cause of morbidity in this young population. It has been estimated in Scotland (population 5 million) that by the end of the century, CD alone will account for 35 300 hospital bed-days per annum (Sedgwick *et al.*, 1990). In any 10-year period, 15% of patients require surgery. Hospital outpatient workloads are enormous. While some patients enjoy good health and long periods of remission, others suffer chronic ill health and major disruption of lifestyle, and a small number lose so much intestinal function that they require intravenous nutrition sometimes for long periods. Advances in medical and surgical therapy have been considerable in recent years and are summarized in Table 8.3. While therapeutic advances are encouraging, the major need is for further understanding of the genetic and environmental causes of these diseases.

8.2 IBD, GENETICS AND THE INTESTINAL BACTERIAL MICROFLORA

Evidence for genetic factors in IBD includes observations of familial aggregation, with high relative risks of disease in first-degree relatives, high concordance rates in twins, association with certain genetic syndromes, linkage with genetic markers and an ethnic predisposition in Jews (Yang, Shohat and Rotter, 1992).

The best model to explain genetic influences in aetiology is 'genetic heterogeneity' (Yang, Shohat and Rotter, 1992). According to this model, several different genes could result in predisposition to IBD. Genes coding for MHC genes or for immunoregulatory molecules, genes controlling intestinal permeability and genes coding for intestinal mucus production are candidates (Pavli, Cavenaugh and Grimm, 1996). For example, MHC class II gene sharing between affected siblings with UC was 15:13:1 for 2, 1, and 0 alleles instead of that expected by chance of 1:2:1. This is strong evidence for linkage between MHC genes and UC. Observed ratios for CD were close to 1:2:1, suggesting no such linkage for this disease (Hugot *et al.*, 1996; Pavli, Cavenaugh and Grimm, 1996; Satsangi *et al.*, 1996). A linkage with HLA genes does not necessarily implicate these directly in pathogenesis since other genes such as tumour necrosis factor or complement components are closely related loci.

Using microsatellite probes rather than candidate gene probes a complete sib-pair genome analysis is possible. Preliminary results suggest several different chromosomal regions of interest and sequencing of these regions is underway (Hugot *et al.*, 1996; Pavli, Cavenaugh and Grimm, 1996; Satsangi *et al.*, 1996).

Scandinavian twin studies (Tysk *et al.*, 1988) show that while concordance for CD reaches 50% in monozygotic twins it may be only 6% in UC. This has been interpreted as indicating that genetic factors are more important in CD or that much IBD is not attributable to genetic susceptibility. Both of these assertions could be wrong. The chances of encountering the environmental trigger at the appropriate age could explain discordance in identical twins. Interestingly, a British study (Thompson *et al.*, 1996) showed that concordance in monozygotic twins was similar for both UC and CD at around 17%.

It may well be that if an individual has no inherited genes conferring susceptibility, it is impossible for him/her to develop IBD, even if exposed to the relevant environmental triggers. Thus, according to how one interprets the epidemiological data, the importance of genetic susceptibility in IBD is either paramount or a factor in only a proportion of patients.

Unlikely as it may seem at first, the composition of the intestinal microflora may be determined by the genetic constitution of the host and genes involved may thus be relevant to disease pathogenesis (van de Merwe, Stegeman and Hazenberg, 1983). The nasal (Hoeksma and Winkler, 1963) and faecal microflora (van de Merwe, Stegeman and Hazenberg, 1983) are much more similar in monozygotic than in dizygotic twins. Patients with CD have an obligate anaerobic faecal microflora composed of greater numbers of Gram-positive coccoid rods (*Eubacterium*, *Peptostreptococcus* and *Coprococcus*) and Gram-negative rods (*Bacteroides* and *Fusobacterium*) than the microflora of healthy subjects (van de

Merwe *et al.*, 1988). Approximately a third of relatives of CD patients also have this 'characteristic' flora, some of whom develop CD subsequently.

8.3 IBD, ENVIRONMENTAL FACTORS AND THE INTESTINAL MICROFLORA

The geographical variation, rising and falling incidence over short time-scales, evidence for birth cohort effects, different prevalence of IBD in identical racial groups in different locations and the association with well established epidemiological risk factors confirm the importance of environmental influence in pathogenesis of IBD (Ekbom and Adami, 1992).

In the northern hemisphere, there is an unexplained north–south gradient in incidence of IBD. A common theme in recent epidemiological studies is the importance of geographical location in early life, probably including the neonatal period, and the probable role of adverse perinatal events as initiators of disordered immune responses manifest later in life as IBD. Adverse perinatal events were 4.4 times more frequent in Swedish patients with CD (Ekbom *et al.*, 1990) and perinatal diarrhoeal illness had an odds ratio of 3.2 times in Canadian patients with UC (Koletzko *et al.*, 1991). A birth cohort effect has been shown in Sweden, where the incidence of extensive UC and CD has been falling in the age group 20–29 and rising in the group aged 30–39 years (Ekbom *et al.*, 1991). Infections *in utero* or in early neonatal life may act as a trigger for abnormal enteric immune responses that persist and predispose to later development of IBD.

8.3.1 Primary infections in aetiology of IBD

While this review is focussed on intestinal microflora, it is clear from the epidemiological evidence above that an infectious pathogenic agent could be involved in aetiology either as a trigger of altered immunity or alternatively as a persistent infection.

Wakefield *et al.* (1993) have suggested that persistent measles virus infection, within granuloma in the endothelium of mesenteric blood vessels, results in a granulomatous vasculitis leading to the lesions of Crohn's disease on the basis of multifocal ischaemia. Detection of measles antigen was by electron microscopy, immunohistochemistry and *in situ* hybridization. Others (Liu *et al.*, 1995; Iizuka *et al.*, 1995) have failed to confirm this finding using immunodetection, or PCR detection of viral RNA, respectively.

Ekbom *et al.* (1994) reported that the incidence of CD in those born within 3 months of a measles epidemic was increased (incidence ratio

1.46, 95% CI 1.11–1.89). Furthermore, measles vaccination has been reported to increase the risk of development of CD, although the methodology of this study has been challenged (Thompson *et al.*, 1995). It may be that perinatal measles infection is just one of a number of infections in early life that increases vulnerability to CD. For example, infection with *Mycobacterium paratuberculosis* (McFadden and Seechurn, 1992) or *Listeria monocytogenes* (Liu *et al.*, 1995) have been implicated in CD and are considered later.

While perinatal infection may disturb acquisition of normal immunity, an alternative suggestion has been mooted, that protection from early exposure to environmental agents could predispose to CD. Improved domestic hygiene in infancy (e.g. access to hot tap water) has been linked to susceptibility to CD but not UC (Ekbom *et al.*, 1990) implying that the absence of some environmental exposure in neonatal life in CD patients may lead to later exposure, with adverse outcomes.

Recent work shows that intestinal and peripheral blood lymphocytes proliferate in response to heterologous but not autologous bacterial sonicates. This immune tolerance to autologous bowel microflora is apparently lost locally at sites of inflammation in the gut of patients with IBD (Duchmann *et al.*, 1995). Thus, IBD can be viewed as a problem either of initial development or of maintenance of immunological tolerance whenever the intestinal mucosal barrier is transiently damaged. Other environmental factors such as the apparent protective effect of prior appendicectomy in UC may affect these mechanisms (Rutgeerts *et al.*, 1994).

8.4 EVIDENCE IMPLICATING INTESTINAL LUMINAL BACTERIA IN IBD

The ubiquitous antigens in the intestinal lumen play a major role in the inflammation of IBD. However, the evidence linking the inflammatory process in IBD with these antigens and in particular with the intestinal microflora, while compelling, is largely circumstantial.

Lines of evidence come firstly from clinical and pathological observations on disease distribution, recurrence after surgery and response to medical therapy including antibiotics. Secondly, there is more formal microbiological evidence based on both microscopy and culture techniques. Thirdly, there is evidence of permeation of inflammatory bacterial products across the intestinal mucosal barrier in IBD. Fourthly, there is evidence of host immunity against the bacteria and their products. Fifthly, there is evidence from the role of bacteria in experimental animal models of IBD utilizing specific pathogen or germfree lines, or specially susceptible strains with spontaneous disease or models created by transgenic or gene knockout manipulations.

8.4.1 Luminal factors, disease distribution and site of recurrence after surgery

Since in UC, inflammation is confined to the large intestine, one would suspect that the normal intestinal microflora represented the source of the putative luminal factors or ubiquitous antigens which might be the target of host inflammatory responses. However, while UC is confined to the colon and rectum, it is the more distal regions of the large intestine which are most likely to be involved in those with less than total colitis. These are not sites of highest bacterial concentrations but more likely sites of increasing bacterial death due to decreasing availability of nutrients and, at least in the rectum, reduced anaerobicity.

After total colectomy intestinal inflammation does not recur in those with free draining ileostomies, but pouchitis occurs in a substantial proportion of patients with continent ileostomies (Phillips, 1987) or ileo-anal pouch (Madden, Farthing and Nicholls, 1990) operations. These pouches are distal intestinal reservoirs colonized by a bacterial microflora intermediate between the normal ileal and colonic microflora (see later). Pouchitis occurs only in patients with previous UC and only very rarely in those with previous polyposis coli who have colectomies for colon cancer prevention.

These important clinical observations suggest that intestinal inflammation in patients with UC is not strictly dependent on the presence of colonic epithelium (since the pouches are lined by ileal mucosa) but is directed against luminal factors present in reservoirs containing a large bowel type of microflora and only occurs in those genetically predisposed or previously sensitized. The precise targets of the host response have not been determined but are presumably factors at highest concentrations in the most distal region of the bowel whether that be rectum, colon or ileal reservoir, the most likely factors being bacteria or their metabolic products.

The importance of luminal factors in the pathogenesis of CD has received rather more attention by clinical researchers than is the case for UC. In contrast to the situation described above for UC, CD can affect the mouth, oesophagus, stomach, small or large intestine and this distribution is consistent with either dietary or bacterial factors being implicated. It is important to recognize that patients with CD may have a much more extensive intestinal abnormality than is evident from macroscopically visible lesions and extends into relatively sterile parts of the intestinal tract. If intestinal microflora or their products are involved, then their role may be in amplifying the inflammatory response thus determining the site of the major inflammatory lesions and the predilection for terminal ileum and colon.

Recent work on endoscopically defined anastomotic recurrence after right hemicolectomy for ileocolonic CD (Rutgeerts *et al.*, 1990, 1991) showed that disease recurrence was essentially inevitable with rates as high as 80% at 6 months. The only patients protected from recurrence were those with temporary proximal ileostomy, none of whom developed recurrence (Rutgeerts *et al.*, 1991). However, following ileostomy closure, these patients were as susceptible to recurrence as those with immediate closure suggesting that the normal faecal stream over the anastomotic site was critical in determining recurrence.

Previous work had shown that diversion of the faecal stream might lead to improvement in CD and that reinfusion of ileostomy effluent into excluded colon might trigger events associated with disease relapse (Harper *et al.*, 1985). Factors in ileostomy effluent provoking symptomatic responses were apparently greater than $0.22\,\mu m$ in size and there was little response to ultrafiltrate challenge. These studies and the apparent benefits of antibiotics, especially metronidazole, in colonic CD (Blichfeldt *et al.*, 1978) may implicate intestinal bacteria in CD but do not exclude a major role for dietary factors. The combination of an elemental diet and broad-spectrum non-absorbable antibiotics in CD was as effective as corticosteroids in one short-term trial (Saverymuttu, Hodgson and Chadwick, 1985), reinforcing the importance of luminal factors in the inflammatory process.

It is of course not surprising that patients with inflamed and sometimes ulcerated intestines should show untoward reactions to intestinal luminal contents, either via direct responses or via priming or stimulation of immune responses normally down-regulated in health. The evidence presented above could all be consistent with a secondary or permissive role for luminal factors, including intestinal microflora, in UC or CD.

8.4.2 Quantitative and qualitative changes in intestinal luminal bacteria in IBD

Comparisons among studies of faecal bacteria in IBD are difficult because of differences in methodology, including elapsed time between collection and culture, number and variety of selective media, the nature of the sample (whether solid, liquid, blood or mucus) and whether counts are expressed per gram wet or dry weight. The appropriateness of the statistical analysis and the precision of the methods may be criticized in some studies. It is likely that methods have progressively improved, so that more recent studies may more closely reflect the true situation.

The earliest report of faecal bacterial counts in UC was in 1950, when Seneca and Henderson (1950) applied aerobic culture techniques to samples from 17 patients. They found an increase in total bacterial concentration and coliform count in UC. In contrast, Cooke (1967) found no significant differences in number or type of organisms in 20 patients with UC, although species reported on were confined to *Escherichia coli, E. faecalis, Lactobacillus, Bacteroides, Clostridium*, anaerobic streptococci and *Proteus*. The patients were described as being in acute relapse and were hospitalized, suggesting moderate to severe activity of disease.

Gorbach and others (1968) took a more sophisticated approach by studying the ecology of the faecal microflora in untreated and treated patients. They set out to test the hypothesis that an imbalance or 'dysbiosis' of the intestinal microflora might occur in IBD. They chose typical cases of UC (17) and CD (8), with a range of extent and severity. None of the patients had been treated with steroids or antimicrobials within 6 weeks. The post-treatment study group comprised 12 patients with UC on sulphasalazine and 8 UC and two CD patients treated with prednisolone. Controls were 30 normal individuals (120 stool samples), five of whom took sulphasalazine and five placebo to simulate a control-treated group. Stools were refrigerated for up to 24 hours and 1 g of stool was homogenized in sterile saline. Nine different media were used and concentrations of total aerobes, total anaerobes, coliforms, streptococci, staphylococci and micrococci were determined. Their first conclusion was that the microflora of patients with mild to moderately active UC, whether distal or extensive, was not significantly different from that of normal controls. In severe extensive UC, concentrations of coliforms increased to 8.6 ($\log_{10}$ per ml), compared with 6.4 in normals and 7.1 in distal colitis ($p < 0.01$). Significant increase in coliforms was also observed in both moderate and severe CD, 7.7 compared with 6.4 in normals ($p < 0.05$) and there was also a significant reduction in lactobacilli in CD (6.1 *versus* 7.2, $p = 0.05$). There were no statistically significant changes with therapy in patients and only a trend for increase in Gram-positive organisms in normal subjects on sulphasalazine. The authors comment that the finding of increased faecal coliform concentrations with active disease and decrease with improvement could be a result rather than a cause of the diarrhoea. They emphasized that the methods used yielded only gross estimates.

There have been several studies of the faecal microflora of patients with CD and their first-degree relatives (Wensinck, van de Merwe and Mayberry, 1983; Ruseler-van Embden and Both-Patoir, 1983). A characteristic anaerobic flora with more Gram-positive coccoid rods and Gram-negative rods had been reported. A major study (van de Merwe *et al.*, 1988) involved 80 subjects in all, with 15 CD patients, 45 first-degree relatives and 20 unrelated healthy subjects. Distribution of disease and previous operations were documented and an elapsed time of 2 hours only between collection and processing of samples was permitted. Total anaerobes, Gram-negative anaerobes, Gram-positive anaerobes (including *Eubacterium, Bifidobacterium*, cocci, coccoid rods and others) and total aerobes were quantified. Concentrations of Gram-positive coccoid rods were 15-fold and of Gram-negative rods 2.5-fold greater in CD than in healthy subjects and this was not a consequence of operative resection. In CD, some Gram-negative rods (*Bacteroides* and *Fusobacterium*) increased at the expense of others (*Eubacterium* and *Bifidobacterium*). No effect of treatment with sulphasalazine on faecal bacteria was noted. In a 5–7-year follow up, three of the asymptomatic children who were relatives of patients with CD and who had an abnormal flora developed symptoms suggestive of CD, suggesting that an abnormal microflora antedates clinically apparent disease.

The above study and previous ones by the same group, which show genetic influences over the colonic microflora by comparison of mono-zygotic and dizygotic twins (van de Merwe, Stegeman and Hazenberg, 1983) suggest that the 'abnormal' microflora in CD patients may be determined by genetic factors. Thus, the 'characteristic flora' may be indigenous to subjects predisposed to CD. Others have shown that the composition of the colonic microflora is stable over time (Holdleman, Good and Moore, 1976), so that it is possible that the 'abnormal' flora is present from a very early age. The suggestion of genetic control of gut microflora is consistent with observations made previously on the nasal microflora (Hoeksma and Winkler, 1963). The genes predisposing to CD may simply be in linkage disequilibrium with those controlling the gut microflora or the 'abnormal' microflora may be a causative factor in the pathogenic process of CD. One suggestion has been that *Coprococcus comes*, strain Me46, one of the coccoid rods may be proinflammatory under certain conditions by virtue of its ability to activate complement and its resistance to opsonization by specific immunoglobulin (Ig) G antibodies (van de Merwe and Stegeman, 1985; van de Merwe *et al.*, 1988). Other examples of possible involvement of specific organisms in CD and UC are considered later. In summary, then, a genetic predisposition to CD may be accompanied by a genetic predisposition to permit intestinal colonization by an abnormal microflora.

The anaerobic faecal microflora in CD showing the increase in Gram-positive coccoid rods and in total Gram-negative rods has been confirmed by others (Ruseler-van Embden and Both-Patoir, 1983). Among the Gram-negatives, *Bacteroides vulgatus* accounted for 40% of the total in patients with CD but only 6% of the microflora in healthy subjects. In a subsequent study (Ruseler-van Embden and van Lieshout, 1987) higher percentages of Gram-negative rods were again found in patients with CD and coccoid rods were found in all patients with CD but in none of the healthy subjects. Bacterial concentrations were expressed per gram dry weight of faeces in these studies.

More recently, a semi-quantitative method for determination of faecal bacterial counts has been described (Giaffer, Holdsworth and Duerden, 1991) and applied to patients with both UC (37 patients) and CD (42 patients), and to 21 healthy controls. Among the CD patients half had clinically active disease, and distribution, operative history and activity were recorded. Among the UC patients 18 had active disease and, overall, 11 patients had disease confined to the rectum. The results of the study showed a significant increase in total numbers of aerobes in patients with active CD compared to quiescent CD, UC or controls, and in *E. coli* scores between active and inactive CD. Total anaerobic and *Bacteroides* scores were similar in CD, UC and controls. *Lactobacillus* and *Bifidobacterium* scores were lower in CD. No differences in scores for *B. vulgatus* or *Bacteroides fragilis* were observed. There were no differences between UC and controls for any parameter. These results for aerobes and coliforms in CD are consistent with those of Gorbach *et al.* (1968); however they are markedly discrepant from those of van de Merwe and Mol (1980), van de Merwe *et al.* (1988), Ruseler-van Embden and Both-Patoir (1983) and Ruseler-van Embden and van Lieshout (1987), all of whom found increased numbers of coccoid rods and Gram-negative rods in patients with CD. The earlier study of West *et al.* (1974) found no changes in faecal microflora in UC or CD, but only five patients with CD were studied. Keighley *et al.* (1978) found increased coliforms, *B. fragilis* and *Lactobacillus* in operative ileocolonic aspirates in patients with CD, but patients with UC had microflora indistinguishable from controls irrespective of disease activity; however, there was no increase in Gram-positive anaerobic coccoid rods.

It is impossible to come to a definitive conclusion concerning the faecal microflora in IBD, since there is no consensus. There is general agreement that diarrhoea per se can lead to an increase in faecal aerobes and coliforms, so when these have been reported they may be secondary and non-specific. In UC, most authors have found no other distinctive changes in number or composition of the microflora irrespective of disease activity. The real question, therefore, is how we evaluate the so called characteristic faecal microflora of CD. Do patients with CD have a tenfold increase in anaerobic Gram-positive coccoid rods such as *Peptostreptococcus*, *Coprococcus* and *Eubacterium* or not? Furthermore, do they have an increase in the proportion or absolute number of Gram-negative rods and is the proportion of *B. vulgatus* increased as high as 40%? Is the positive finding of such changes more likely to be correct than a negative finding? These questions definitely need resolving.

8.4.3 Changes in intestinal microflora with surgical diversion or ileostomy reservoir procedures

Surgical diversion of the faecal stream has been reported to be beneficial to the inflammatory process in patients with CD (Roediger, 1980; Harper *et al.*, 1983). This is to some extent surprising, since faecal diversion may have deleterious effects when performed for other disorders. In particular diversion colitis is a well recognized entity. The pathogenesis of this is incompletely understood, but the beneficial effects of the normal microflora in colonization resistance and in formation of short-chain fatty acids especially butyrate (Roediger, 1980), may be reduced when nutrients are diverted from the microbial environment. The condition responds to short-chain fatty-acid enemas (Guillemot *et al.*, 1991). It is therefore particularly relevant to consider the microbial changes in the excluded colorectum after diversion, since, in spite of the potentially deleterious effects, overall, they appear to be beneficial to those with CD.

In a study of 16 patients (Neut *et al.*, 1989), ten with IBD, eight of these with CD, the microflora from the excluded colorectum was compared with faeces of 16 healthy controls. There was a tenfold reduction in total anaerobes in the patients and these reductions were mainly in *Eubacterium* or *Bifidobacterium*. Interestingly, higher isolation rates for anaerobes, particularly *Peptostreptococcus* and *Bacteroides*, were found in patients with IBD compared to those with other disorders such as diverticular disease. Again, we see a possible proclivity for patients with IBD, presumably those with CD, who predominated in this study, to harbour an increased anaerobic microflora including Gram-positive coccoid rods and Gram-negative rods. Among aerobes, enterobacteria were increased and, probably reflecting loss of colonization resistance, species such as *Proteus*, *Providencia* and *Morganella* were found in patients but not in controls. If these changes are beneficial to patients with CD, then the reduction in anaerobic microflora is possibly the most likely change. It has to be remembered that diversion also reduces exposure to food mitogens and endogenous secretions including bile salts. Changes in bacterial populations are not the only factors to consider.

In UC, there is no clinical improvement following diversion of the faecal stream (Harper *et al.*, 1983). The role of luminal bacteria in the inflammatory response, however, may be inferred from the complication of pouchitis, which occurs after continent ileostomy or ileo-anal pouch operations for UC but not after similar operations for polyposis coli. Conventional (Brooke) ileostomy is regarded as a curative operation in UC (Brooke, 1956; Rhodes and Kirsner, 1965). Inflammation in the terminal ileum is very rare and even then very mild. The ileostomy microflora is not the same as normal ileal microflora, but major changes are observed when the microflora of conventional ileostomies are compared with those from an ileal or ileo-anal reservoir (Philipson *et al.*, 1975; Metcalf and Phillips, 1986; Phillips, 1987). Several studies have been reported. In one (Luukkonen *et al.*, 1988), there were 15 patients in each of three groups with conventional Brooke ileostomies, continent Kock pouch ileostomies or pelvic ileal pouches. Two patients in the Kock group and four of the pelvic pouch group had acute or subacute pouchitis, which in every case subsequently responded to a course of metronidazole. Anaerobic bacterial counts were usually greater than 7 logs in patients with pouches and less than 3 logs in patients with free-draining ileostomies. Concentrations of aerobes were similar among groups. The commonest anaerobe was *B. fragilis* (10^7–10^{11}) in both pouch groups. Among aerobes, *E. coli*, *Proteus* and *Klebsiella* predominated. Mucosal histology was either normal or showed mild chronic inflammation except in those with pouchitis, five of whom had acute inflammatory changes. Failure to demonstrate specific changes in faecal microflora in patients with pouchitis compared to those with similar pouches but without symptoms or acute mucosal inflammation was also found in other studies (Kelly *et al.*, 1983; O'Connell *et al.*, 1986). Nevertheless, the response to metronidazole and the association of inflammation with pouches containing high counts of anaerobes and not in free-draining ileostomies with much lower anaerobe counts, implicates the anaerobic microflora. Whether or not inflammation occurs clearly depends on the aggressiveness of the host response and does not require quantitative or qualitative changes in microflora above those already present as a consequence of the presence of the pouch. In this respect, findings are consistent with those in UC, where there are no consistent changes in quantity or quality of the faecal microflora when remission is compared with relapse.

Failure to find microbiological correlates of acute pouchitis has not been universal. Santavirta *et al.* (1991) compared mucosal morphology and faecal bacteriology in 30 patients with ileo-anal J-pouches and ten patients with conventional ileostomies. They confirmed the increase in total bacterial counts and anaerobic counts in patients with pouches. The ratio of anaerobes to aerobes was on average 5000 with a pouch and only 1 with conventional ileostomy. Some degree of mucosal inflammation was present in up to 70% of patients with pouches whereas mucosal inflammation was graded as 0 in nine out of ten, and 1 in only one out of ten patients with conventional ileostomies. The authors claim a correlation between numbers of aerobes and the grade of acute inflammation, but this is markedly influenced by results in only two patients. Similarly, the correlation between grade of chronic inflammation and number of anaerobes was significant, but weak. Five of the nine patients with chronic inflammation had anaerobic counts within the range of those without any inflammation. This study therefore confirms that pouchitis can occur without significant changes in faecal flora when compared to those with pouches and no inflammation, while suggesting that the total load of anaerobes and possibly aerobes may be on the high side in some of those with inflammation.

In contrast to the situation in respect of the pouch itself, bacterial overgrowth proximally has been related to symptoms of diarrhoea and evidence of malabsorption of nutrients in some patients after Kock ileostomies (Kelly *et al.*, 1983). Jejunal aspirates showed an increase in anaerobic bacterial counts (range 10^3–10^8 per gram of aspirate). The symptoms, the malabsorption and the number of jejunal bacteria decreased after treatment with metronidazole. Presumably similar problems complicate ileo-anal anastomosis and are best viewed as metabolic and pathophysiological consequences of small bowel bacterial overgrowth distinct from the inflammatory processes of pouchitis. Jejuno-ileal motility is not much altered following creation of an ileo-anal pouch (Stryker *et al.*, 1985). Transit is, however, slower than after conventional ileostomy and considerably slower than normal orocaecal transit (Soper *et al.*, 1989). It is possible that this adaptation to colectomy in combination with loss of the ileocaecal valve, predisposes to bacterial overgrowth in more proximal regions of the intestine.

8.5 INTESTINAL MUCOSAL BACTERIA AND IBD

Examination of the faecal microflora is an appropriate way of assessing the composition of the colonic luminal microflora, but almost certainly doesn't provide insights into the specific mucosally associated microflora (Savage, 1970), which has been demonstrated in experimental animals and probably exists in humans.

8.5.1 Conventional mucosal bacteria

Since intramural bacteria were visualized in tissues from patients with CD and a response to metronidazole and other antibiotics was reported, the possibility that an abnormal mucosal microflora might exist in CD has been investigated (Peach *et al.*, 1978), using full-thickness operative samples from histologically abnormal and normal tissue from CD specimens and control tissue from resections for colon cancer. Tissue for bacteriological study was snap-frozen in glycerol transport broth and stored at −20°C for up to a month before thawing and processing in an anaerobic chamber. A mucosal microflora or at least a positive culture was found with all the large bowel and three-quarters of the small bowel samples. Greater numbers of bacteria were associated with colonic tissue than jejunal tissue (10^7–10^8 compared with 10^3–10^4) and ileal tissue was quantitatively intermediate. However, no increase in numbers of bacteria were found in CD tissue either affected or unaffected compared with controls. No bacteria were isolated from five mesenteric lymph nodes from five patients with CD. In controls, roughly half the isolates were aerobic and half anaerobic whereas in CD three-quarters were aerobic isolates. Thus. in contrast to the luminal microflora, where anaerobes outnumber facultative aerobes by 100:1, the mucosal microflora is equally or predominantly facultative, at least in terms of what can be cultured. In contrast with animal studies where certain strains of bacteria are isolated from different sites in the gut, in this study no highly selective distribution was evident. Many of the aerobic isolates were Gram-negative rods, *Bacteroides* being the commonest anaerobe isolated.

The advent of fibreoptic colonoscopy provided the opportunity to obtain multiple mucosal biopsies under direct vision. After standardized preparation, biopsies were taken using a sterile biopsy instrument introduced via a protective tube down the biopsy channel of the instrument (Edmiston, Avant and Wilson, 1982). The forceps were pushed through a paraffin seal at the end of the tube prior to taking the biopsy from the mucosa. Tissue from each part of the colon was examined and diseased and adjacent normal-appearing mucosa was examined. Strict anaerobic culture techniques were used. Effects of sampling artefacts were examined by studying paired biopsies from closely adjacent regions and no differences were observed. There were no differences in numbers of organisms isolated from ascending, transverse or descending colon (6.3, 5.6 and 5.3 logs respectively) while mean counts in the sigmoid although significantly lower at 4.7 logs per milligram were attributed to the enema preparation prior to colonoscopy. Mean values for numbers of anaerobes isolated were higher in polyps, cancers and IBD-affected mucosa compared to their respective control adjacent tissues, but only the difference for polyps reached significance. No differences in total numbers of organisms and recovery of genera were seen between disease states. Recovery of genera in all cases was *Bacteroides* > *Fusobacterium* > *Clostridium* > *Eubacterium* > *Peptostreptococcus*. This study has limited relevance to IBD since only five patients unspecified in respect of disease type or activity were included. From disease distribution, however, it is likely that at least three of the five patients had CD. While total numbers of anaerobes and number of genera showed no differences in diseased *versus* unaffected adjacent tissue in IBD, a greater number of species were isolated from inflamed tissue.

In a study of the frequency of isolation of selected obligate or facultative enteropathogens from mucosal biopsies in patients with 'chronic colitis', there were 29 patients with UC and 14 with CD (Horing *et al.*, 1991). In UC, microorganisms were isolated from 14 (48%) patients' biopsies and in CD from seven (50%). In the 20 healthy controls who had a rectal rather than a colonic biopsy, none of the specifically searched-for organisms were grown. In patients with 'non-specific colitis or proctitis', 21% and 35% had positive isolates. Obligate enteropathogens (*Salmonella*, *Shigella*, *Yersinia*, *Campylobacter* and *Cryptosporidium*) were not found in any case of UC or CD. Of the facultative pathogens, *Klebsiella* and *Pseudomonas* were isolated in 31% and 24% respectively of patients with UC and 21% and 7% of patients with CD. These organisms were also found in 'non-specific colitis'. In all colitis groups, *Chlamydia* was grown from approximately 20% of biopsies. Finally, *Aeromonas hydrophilia* was found in 3.4% of UC biopsies and one or other obligate pathogen in 8% of patients with 'non-specific colitis'.

Interpretation of these findings is difficult. In the case of IBD, these facultative pathogens may be acting as pathogens via superinfection of mucosal sites or as microflora occupying a mucosal niche. The failure to find *Klebsiella* or *Pseudomonas* in normal rectal mucosa may be related to the fact that although these organisms can both be found among the normal gut microflora, they occur in small numbers only. Which changes in IBD may favour colonization of the mucosa by these genera is worthy of study. It still seems likely that healthy mucosa is colonized by other Gram-negative rods, since they were found in a high percentage of control 'normal' colonic tissue in the studies described previously. The role of *Chlamydia*, which has been implicated in venereal proctitis but not previously in colitis, is equally obscure, although interesting.

In CD, one of the more interesting microbiological studies was an in-depth study of two French families with a remarkably high incidence of the disease, six cases in one family and seven cases in the other. Direct microscopy, immunocytochemistry and ELISA failed to reveal evidence for mycobacterial infection or infection with an extensive list of pathogenic bacteria or viruses (van Kruiningen *et al.*, 1992), confirming previous negative studies. However immunocytochemical evidence for *Listeria*, *E. coli* and streptococcal antigens in CD tissue from these patients was subsequently reported by the same group (Liu *et al.*, 1995).

Superinfection with pathogenic bacteria such as *Salmonella*, *Campylobacter* and *Clostridium difficile* can be associated with IBD relapses or even at first presentation of IBD, but is the exception rather than the rule. If in addition we must consider superinfection with facultative pathogens and consider this in terms of juxtamucosal ecology, then a substantial proportion of presentations and relapses might be explained by 'superinfection events' of one or other kind.

To shed some light on the rather controversial role of superinfection in IBD relapse, a prospective study involving 64 patients with IBD, 49 with CD and 15 with UC was undertaken (Weber *et al.*, 1992). Patients were entered into the study at the time of relapse and colonoscopy was performed to obtain two or three biopsies from each region of the colon and the terminal ileum. Samples were transported in Ringer's solution and plated on selective media. Stool specimens were also examined on the same media. Among the obligate enteropathogens, *C. difficile* was found in five patients with CD and one with UC. *Campylobacter jejuni* was found in one patient (CD) and *S. typhimurium* in one patient (UC). Enteropathogenic *E. coli* was isolated from three patients (CD). The total number positive was 11, or 17%. The commonest pathogen was *C. difficile*, as found in a similar study previously (Greenfield *et al.*, 1983), but its role in relapse is uncertain. A carrier state of 15% (Gilligan, McCarthy and Genta, 1981; Nakamura *et al.*, 1981) has been reported in asymptomatic adults and in patients with IBD (Weber *et al.*, 1992), and *C. difficile* disappeared during treatment with steroids or sulphasalazine. The rare occurrence of *C. jejuni* (Goodman *et al.*, 1980; Blaser *et al.*, 1984) and *Salmonella typhimurium* (Thaylor-Robison *et al.*, 1989) confirms previous studies. *Yersinia enterocolitica* was not found in any patient, although positive serology was observed in four patients and was attributed to non-specific immune stimulation. *Aeromonas, Shigella* and *Chlamydia* were not detected in this study and no evidence of superinfection with protozoa, including *Giardia lamblia*, was found, in contrast to other reports (Burke and Axon, 1988). The authors conclude that enteropathogenic organisms play only a minor role in aetiopathogenesis of IBD and if found require careful observation but not specific intervention with drugs. The same paper provides interesting data on culture of other organisms from tissue specimens. They isolated 15 bacterial and six fungal species that were described as not generally related with enteropathogenicity. *Klebsiella pneumoniae* and *P. aeruginosa* were found in 11% and 8% respectively. *E. coli* was the commonest organism, being found in 84%; other genera included *Proteus, Enterobacter, Citrobacter* and *Hafnia*. Among fungi, *Candida albicans* was found in 39%, with a particularly high frequency in CD (45%).

Studies of colonic mucosal microflora in active and inactive UC were carried out by Fabia *et al.* (1993) using colonoscopic biopsies from 30 patients and 30 controls. In active UC significant decreases in anaerobic bacteria (brain heart infusion medium), anaerobic Gram-negatives and *Lactobacillus* were found approaching 1.5 log reductions in each case compared with controls. No such reductions were found in UC in remission. The reduction in *Lactobacillus*, an organism with putative protective effects against obligate or facultative pathogens, was related to the presence of *P. mirabilis* in up to 10% of UC samples but in none of the biopsies from control patients. Whether these changes in mucosal microflora antedate or postdate relapse is not known. The investigators report similar changes in bacterial microflora in acetic-acid-induced colitis in rats, which appear in parallel with the full-blown inflammatory response 4 days after exposure to acetic acid. The changes noticed in the mucosal microflora have not been reported for the faecal microflora, although studies of mucosal microflora to date suggest that it is composed of organisms also culturable from faeces. The composition of the mucosal microflora, however, is clearly different from that of the faecal microflora in health: in particular, the ratio of aerobes to anaerobes may differ markedly.

As well as the quantitative differences in active UC noted above, there may also be qualitative differences in mucosal bacteria in this disease.

One such difference is in adhesivity of *E. coli* isolated from the stools of patients with IBD, both UC and CD, compared with isolates from healthy controls or those with infectious diarrhoea due to *C. jejuni* (Burke and Axon, 1988). In these studies an adhesion index is calculated after incubation of subcultured *E. coli* with buccal epithelial cells from a single donor in the presence of D-mannose and determination of the proportion of cells with more than 50 adherent Gram-negative rods. Adherent and non-adherent control organisms are included in the assay. Taking an adhesion index of 25% as the upper limit with control 'non-adhesive' strains, 86% of isolates from patients with IBD were adhesive, compared with none in normal controls and 27% in infectious diarrhoea. Adhesive strains were found with equal frequency in UC in remission or relapse and there was no difference between UC and CD. The adhesive property appears not to be a simple consequence of inflammation, persists in culture, is mannose-resistant and therefore not due to type 1 fimbria and may by analogy with other virulence factors be plasmid-encoded. The authors suggest that transfer of plasmids from enteropathogenic intestinal bacteria to commensal *E. coli* may account for this phenomenon. Others have suggested that *E. coli* obtained from patients with ulcerative colitis can degrade mucins (Cooke *et al.*, 1974) and produce necrotoxins and haemolysins to a greater extent than those from control subjects. Immunoglobulin A1 protease activity of the colonic flora has also been implicated as a possible pathogenic factor, but there is little evidence to associate it specifically with IBD (Barr *et al.*, 1987).

The significance of 'sticky' *E. coli* in pathogenesis of UC was called into question by studies (Hartley *et al.*, 1993) of mucosal rather than faecal *E. coli*, isolated and tested for mannose-resistant adhesion and surface hydrophobicity. In active colitis mucosal *E. coli* isolation rates were reduced and they were paradoxically less adhesive. The authors argued that these data did not suggest a pathogenic role for *E. coli*. However, we await studies of simultaneously collected faecal and mucosal *E. coli* to resolve the issue.

In the hope that light would be shed on the pathogenesis of pouchitis after colectomy and ileo-anal pouch construction, faecal samples were tested for adhesive *E. coli* but, in contrast with observations by the same researchers in UC, no increased carriage of adhesive bacteria could be demonstrated; in fact an inverse correlation was found. The authors argue that this suggests a different pathogenesis for UC and pouchitis.

This whole area is now in need of clarification. Adhesive *E. coli* are apparently present in the faeces but not the mucosa in active UC and are apparently absent from the faeces in pouchitis. Tobramycin was trialled on the basis of the 'sticky' *E. coli* theory and has been reported to be beneficial in active UC but its effects on the faecal microflora were not reported (Burke *et al.*, 1990).

8.5.2 L-forms of mucosal bacteria in IBD

Another qualitative change in mucosal microflora in IBD is the frequency of isolation of bacterial L-forms, otherwise known as cell-wall-deficient (CWD) organisms or spheroplasts. Two studies (Belsheim *et al.*, 1983; Ibbotson, Pease and Allan, 1987a) have demonstrated a high recovery of bacterial L-forms from gut tissue in both CD and UC in up to 40–50% of samples, compared with tissue from control subjects, where recovery was 1–6%. L-forms were recovered from both involved and uninvolved tissue and in 13% of uninvolved lymph nodes in CD. Isolation was independent of *in vivo* therapy with antibiotics, which are known to induce L-formation *in vitro*. Attempts to produce reversion to parental forms met with little success and revealed commensal organisms such as *Pseudomonas* species, *E. coli* and *E. faecalis* but not *Mycobacterium*. However, other groups specifically aiming to culture tissue for mycobacteria and also recovering CWD (spheroplasts; Chiodini *et al.*, 1986; Markesich, Graham and Yoshimura, 1988) that were variably acid-fast (see below) have succeeded in demonstrating *Mycobacterium paratuberculosis* in one or more of the many CWD isolates using ribosomal RNA or DNA probes.

Theoretically, where the putative microbial aetiological agent in a disease may be a cell-wall-defective variant and therefore difficult to isolate, serological evidence of infection may be helpful. Shafii *et al.* (1981) showed that 22 of 25 sera from patients with CD reacted in an immunofluorescence assay against CWD revertent strains of *Pseudomonas*-like organisms isolated from patients with CD. Weak positive fluorescence with UC sera and control sera was readily preabsorbed with a mixture of *P. aeruginosa*, *E. coli* and *Bacteroides thetaiotomicron* organisms. This procedure did not diminish responses with CD sera. Fluorescence intensity with CD sera correlated positively with disease activity. *Pseudomonas maltophilia* (Parent and Mitchell, 1978) has been suggested as a possible CWD organism in CD, but serological evidence does not support this (Ibbotson, Pease and Allan, 1987b). Serological evidence for and against involvement of chlamydia in CD has been published. Most recent studies failed to find either an increase frequency of chlamydial antibodies (*Chlamydia trachomatis*) in CD or chlamydial DNA sequences in CD tissue using PCR (McGarity *et al.*, 1991). As noted above, *Listeria* antigens have been reported in CD tissues (Liu *et al.*, 1995).

Non-cultivatable CWD organisms have been shown to be the cause of human chronic ocular inflammatory disease. These organisms, which parasitize vitreous leukocytes, have been found in the eyes of patients with UC (Wirostko, Johnson and Wirostko, 1990) and CD (Johnson and Wirostko, 1989) with idiopathic uveitis, leading to the suggestion that similar infections of the gut may underlie the intestinal diseases.

If L-forms do play a pathogenic role in IBD, they could explain past failures to visualize organisms in, or culture them from tissues. They are resistant to host defences, probably because they lack cell wall antigens and are thus less antigenic, and they survive intracellularly. It is also possible that the high recovery of L-forms from both UC and CD tissue reflects secondary invasion or translocation of bacteria that then undergo L-transformation under the influence of hostile host factors, such as leukocyte lysozyme, which can damage bacterial cell walls and promote spheroplast formation.

8.5.3 Mycobacteria and IBD

Many mycobacteria are ubiquitous in the environment so that isolation of mycobacteria from patients with CD does not necessarily imply involvement in pathogenesis. As with other facultative pathogens, *Mycobacterium avium*, *M. paratuberculosis* and *Mycobacterium kansasii* can only be implicated in pathogenesis of inflammation if they are repeatedly isolated from normally sterile body sites or are found repeatedly in association with a particular disease and not in appropriate disease controls. Finally, evidence of humoral or cellular immunity would be expected, since it is hard to envisage an inflammatory response to an organism in the absence of an immune response.

Mycobacterium paratuberculosis remains after much study a possible aetiological agent in CD (McFadden and Seechurn, 1992). It is a member of the *M. avium* group but can be distinguished from the other members of the group by use of DNA probes, which recognize the multiple copies of a DNA insertion element known as IS900 in the *M. paratuberculosis* genome (Green *et al.*, 1989). This organism has been isolated by culture from only a small proportion of CD tissues examined to date (3–4%; Chiodini *et al.*, 1984; Graham, Markesich and Yoshimura, 1987; McFadden *et al.*, 1987; Gitnick *et al.*, 1989), but never apparently from control or UC tissue. By implication, then, if it causes CD, bacilli must be extremely sparsely distributed in tissue (paucibacillary) or present as spheroplasts. *Mycobacterium paratuberculosis* causes Johne's disease in ruminants (Chiodini, van Kruiningen and Merkal, 1984), when it is multibacillary and usually easy to isolate, even though the organism has fastidious growth requirements. Johne's disease is characterized by an ileitis not dissimilar to that of CD. The difference in bacillary load in the two disorders is said to reflect the nature of the immune response, with depressed cell-mediated immunity (CMI) being found in Johne's and well-developed CMI being found in patients with CD, accounting for the tuberculoid type of histopathology with granulomas and the low density of bacilli. Thus, it is argued that isolation rates of 3–4% in CD are still consistent with an aetiological role for *M. paratuberculosis* (McFadden and Seechurn, 1992).

It might be hoped that DNA techniques would shed light on the situation. Using DNA–DNA hybridization (Yoshimura *et al.*, 1987), mycobacterial DNA was reported in CD tissues, but other studies using Southern blotting of extracted DNA were negative. These had a reported sensitivity of 1 mycobacterial genome per 100 human cells (Butcher, McFadden and Hermon-Taylor, 1988).

With the advent of polymerase chain reaction (PCR) and subsequently 'nested PCR' techniques, sensitivity could be increased 1000-fold. Primers based on the IS900 insertion sequence and a probe hybridizing to this sequence were used to detect *M. paratuberculosis* genomes with a sensitivity of 1 bacterial genome per 10^4 human cells. It is argued that, even with this sensitivity, a negative PCR may not exclude significant infection (McFadden and Seechurn, 1992). A positive PCR does not prove that the organism is causing the disease, nor even if it is viable: it may be a contaminant *in vivo* and cross-contamination *in vitro* is always a risk. Nevertheless, the pattern of results in patient groups should be subject to analysis and interpretation.

Using DNA extracts of full-thickness samples of intestine removed at surgery from 40 patients with CD, 23 with UC and 40 controls, *M. paratuberculosis* genomes were detected in 65% of CD (small intestine and colon), 4.3% of UC and 12.5% of control tissues (Sanderson *et al.*, 1992). The authors concluded that positive results in controls were consistent with previously unsuspected alimentary prevalence in humans, but the overall results were consistent with an aetiological role for *M. paratuberculosis* in CD. In another study (Moss *et al.*, 1992), six out of 18 cultures of CD tissues yielded positive results in PCR assays for IS900 DNA (only one out of six non-IBD controls was positive), but were described as of low signal intensity. *Mycobacterium avium* (IS902 PCR-positive) was found in two of the CD, two of UC and two of the non-IBD controls. This study was blinded and PCR results bore no relation to visible bacillary forms of mycobacteria or spheroplasts in culture, suggesting that *M. paratuberculosis*, if present in the original tissue did not replicate in culture.

The above studies are consistent, but not overwhelmingly so, with a role for *M. paratuberculosis* in CD, although this role may be one of passive association rather than active pathogenicity. This has been suggested by PCR studies of serial colonoscopic biopsies (Murray *et al.*, 1995). Others, using PCR with a reported sensitivity of 1 bacterial genome per 10^6 human cells, have failed to find even an association (Bell and Jewell, 1991). One such study used a fluorescence-based PCR capable of detection of 1–2 mycobacterial genomes (Rowbotham *et al.*, 1995). In tissue samples from 68 patients with CD, 49 patients with UC and 26 non-IBD controls, no samples were found to be positive for *M. paratuberculosis* IS900 insertion element. The authors considered factors such as disease activity, surgical *versus* biopsy samples, presence of granulomata in tissue, age of patients, hospital source, etc., and concluded that none of these factors could explain the discrepancies among studies. They assert that variation in techniques and the possibility of artefactual positives may explain differences. Since artefacts would be randomly distributed between test and controls, diseased and control samples would have to be analysed in different PCR runs to account for putative artefacts (Rowbotham *et al.*, 1995). At this point in time, there are more negative PCR studies than positive ones.

Hermon-Taylor and colleagues argue strongly that the balance of evidence still strongly suggests that *M. paratuberculosis* is an aetiological agent in CD (Hermon-Taylor, 1993). The source of infection may be milk from cows harbouring subclinical infection. Milk samples (7% of cartons from retail outlets) may be PCR-positive even after pasteurization (Millar *et al.*, 1996). Failure by others to detect *M. paratuberculosis* genomes in CD tissues is attributed to defective DNA isolation techniques. Possible contamination artefacts are said to be unlikely with newer hybridization capture techniques, which further increase positivity rates in CD.

The issue has now been intensively researched for over 10 years. If there has been an expansion in *M. paratuberculosis* infection in cattle due to intensive dairy farming and milk is contaminated, then we need to know if viable organisms are present in milk that is PCR-positive. An exchange of tissue samples and DNA extracts between laboratories may help to clarify why some excellent labs cannot reliably identify the organism in affected tissues while others have always been able to do so.

8.6 TRANSLOCATION OF BACTERIA ACROSS THE INTESTINE IN IBD

It is now well recognized that normal intestinal bacteria can pass from the intestinal lumen to extra-intestinal sites under a variety of clinically important circumstances, including immunosuppression, surgical stress and trauma (Wells, Maddaus and Simmons, 1988). There is extensive experimental evidence indicating that anaerobic bacteria in the gut lumen, which do not translocate readily, may play a pivotal role in limiting translocation of facultative aerobes such as *E. coli*, other enterobacteria, *Pseudomonas* species and *Enterococcus* species (Wells *et al.*, 1988). High-risk patients have been treated with selective antimicrobials to reduce translocation and preserve the anaerobic microflora of the gut, thus reducing systemic infections (Guiot *et al.*, 1983; Wade *et al.*, 1983). Mechanisms controlling bacterial translocation are obscure but both host and microbial factors are clearly important. Bacteria that can survive within macrophages, such as *Salmonella* or *Listeria*, can translocate readily and commensal bacteria may reach mesenteric lymph nodes within host phagocytes (Wells, Maddaus and Simmons, 1987). Although anaerobes do not easily translocate they may do so in conjunction with aerobes when there is gross intestinal damage, as, for example, caused by irradiation or mesenteric ischaemia. There is also compelling evidence that some aerobes (*E. faecalis*, for example) can translocate across an intact intestinal mucosa, possibly by being taken up by enterocytes and then by macrophages (Wells, 1990). Particles such as yeast and starch translocate too and this process may be part of the normal antigen sampling mechanism of the gut (Wells, Maddaus and Simmons, 1988). As far as bacteria are concerned, translocation is increased as the concentration of a particular species increases in the lumen and there is experimental evidence to show that elimination of the mucus layer results in an increase in the populations of bacteria directly adherent to the enterocyte surface and increased translocation to extraintestinal sites. It is likely that IgA prevents adhesion, since it is not an opsonin and would not facilitate phagocytosis, which is a key step in translocation. Conversely, IgM and IgG could promote translocation by this mechanism. The ability or otherwise of phagocytes to kill intracellular bacteria will ultimately determine the outcome of translocation (Wells, Maddaus and Simmons, 1988).

The possible relevance of the translocation phenomena discussed above to IBD relates firstly to direct microscopic observations of bacteria in the intestinal tissue of patients with CD (Aluwihare, 1971) and UC (Ohkusa *et al.*, 1993). A second line of evidence comes from bacteriological studies performed at the time of surgery, showing increased recovery of bacteria from serosal surfaces and mesenteric lymph nodes in IBD (Ambrose *et al.*, 1984; Laffineur *et al.*, 1992). These studies support the idea that translocation is increased in IBD but we are uncertain of the extent of this phenomenon in milder degrees of inflammatory activity and do not know how frequently bacteria can be visualized or cultured in intestinal samples of healthy patients. When intestinal mucosa is ulcerated one might well expect translocation, but investigators also report detection of bacteria in non-ulcerated mucosa (Ohkusa *et al.*, 1993). Without further clarification it is impossible to judge the significance of these findings in relation to the inflammatory response in IBD. Inflammatory products of intestinal bacteria certainly cross the epithelial barrier in active IBD, so the significance of small numbers of bacteria in the mucosal tissues as opposed to the adjacent lumen is arguable. In terms of the systemic consequences of translocation in IBD, there is a 44-fold increased representation of IBD among patients with native valve endocarditis (Kreuzpaintner *et al.*, 1992), which may be related to the disease itself, to diagnostic manipulations or to immunosuppressive therapy.

8.7 HUMORAL IMMUNITY AGAINST BACTERIA IN IBD

A major difficulty in defining a role for a particular bacterium in IBD is that isolation of the organism from the faeces or the tissue does not necessarily implicate it in the disease process. Demonstration of a specific serological response would at least indicate host recognition of the agent. This could then be used as an argument for involvement of an organism, if this response was greater in degree than responses to other bacteria and was found in a large proportion of the patients with the disease but not in appropriate disease controls or healthy subjects. Unfortunately, this logic has substantial limitations in practice.

Notwithstanding the paucity of direct evidence implicating obligate enteropathogens in IBD, serological responses to the major pathogens have been studied. Blaser *et al.* (1984) examined responses to *Campylobacter, Yersinia, Listeria* and *Brucella* in 40 patients with active CD, who were well characterized and compared with age- and sex-matched healthy controls. In a complement fixation assay, reciprocal geometric mean titres of the sera from patients with CD were higher to each antigen than were titres in controls, but only statistically significant in the case of *Yersinia pseudotuberculosis*. The authors concluded that polyclonal B cell stimulation during periods of active bowel inflammation might result in increased serum levels of cross-reactive antibodies directed at numerous species. Antibody titres did not correlate with any measure of disease activity, distribution or duration.

It is not surprising that serum antibody titres to commensal organisms, including facultative enteropathogens, may be raised in IBD. Brown and Lee (1974) found elevated titres to *B. fragilis* and *E. faecalis* in patients with CD and UC which correlated with disease severity rather than faecal concentrations. Tabaqchali, O'Donoghue and Bettelheim (1978) found increased anti-*E.-coli* antibodies in sera from patients with CD and UC and Mathews *et al.* (1980) found elevated antibody responses to *Peptostreptococcus* and *Eubacterium*, two of the major faecal anaerobes. The Rotterdam group (Wensinck *et al.*, 1981), who described the characteristic faecal microflora of CD described previously, also found increased agglutinating antibody titres against strains of *Eubacterium* and *Peptostreptococcus* and *Coprococcus* species in patients with CD compared with appropriate disease controls and healthy subjects. Subsequently, the world-wide occurrence of agglutinating antibodies against four particularly selected coccoid anaerobes, *Eubacterium contortum* (Me44 and Me47), *Peptostreptococcus productus* (C18) and *C. comes* (Me46) was assessed in 937 sera from patients with CD, UC and control groups, from 19 centres in 17 countries (Wensinck, van de Merwe and Mayberry, 1983). Results were expressed as positive or negative and tests were read blind. Positive agglutination tests were found in 59% of CD patients, 29% of UC patients and 8% of healthy or diseased controls. Relative risk for CD patients of a positive test was 16.5 compared with healthy controls and was 5.0 for patients with CD confined to the colon compared with UC patients. These data are consistent with the proposition that the immune system is more frequently exposed to antigenic challenge with anaerobic coccoid rods and other intestinal anaerobes and aerobes in patients with IBD than in healthy controls.

To test the hypothesis that in both UC and CD there is an abnormally increased immune response against non-pathogenic intestinal bacteria, immunoglobulins were isolated directly from wash fluid obtained during colonoscopy (Macpherson *et al.*, 1996). Bacterial proteins, supernatants and pellets from lysed preparations of organisms were tested against Ig preparations after SDS-PAGE separation. Immunoglobulin G concentrations, as expected, were higher in IBD samples than in irritable bowel syndrome (IBS) controls. In ELISA assays, IgG from patients with active IBD and from inflamed segments bound proteins from non-pathogenic strains of *E. coli*, *B. fragilis*, *Klebsiella aerogenes* and *Clostridium perfringens*, confirming a prominent local IgG humoral immune response against intestinal bacterial proteins in active UC and CD. The antibodies bound to bacterial cytoplasmic rather then membrane proteins, which differentiates them from serum IgG, which binds well to both. The authors argue this may reflect inhibition of responses against membrane proteins by B-cell tolerance mechanisms active in the gut.

Studies of humoral immunity to mycobacterial antigens have given both positive and negative results in CD. Using a panel of three mycobacterial antigens, which were heterogeneous and not species-specific, positivity to all three antigens was found in 18% of patients with CD and to at least one antigen in 84% (Elsaghier *et al.*, 1992). Using an ELISA and eight mycobacterial sonicates (Stainsby, 1993), a large proportion of patients with CD had antibodies that bound most antigens but there were no statistical differences from responses with control sera or sera from patients with UC, including responses to *M. paratuberculosis*. The findings were interpreted as indicating widespread contact with environmental mycobacteria and not supporting a role for *M. paratuberculosis* in CD. Other studies of humoral immunity did not support a role for *M. paratuberculosis* either (Brunello *et al.*, 1991; Tanaka *et al.*, 1991).

8.8 CELL-MEDIATED IMMUNITY AGAINST BACTERIA IN IBD

8.8.1 Lymphocyte proliferative responses to bacterial antigens

Studies by Fiocchi, Battisto and Farmer (1981) demonstrated that mucosal T cells from patients with IBD showed enhanced proliferation to *Bacteroides* and staphylococcal antigens. Ibbotson and colleagues (1992) studied the proliferative responses of both mesenteric and peripheral lymphocytes to mycobacterial and non-mycobacterial antigens in 32 patients with CD, five with UC and 21 controls. They showed that many controls failed to show responses to any test antigens with either mesenteric lymph node mononuclear cells (MLNMC) or peripheral blood mononuclear cells (PBMNC). In contrast, cells from patients with IBD showed increased interleukin-2 (IL-2) responsiveness (suggesting activation with upregulation of IL-2 receptors) and enhanced responses to some non-mycobacterial antigens such as *Y. enterocolitica*, with lesser responses to other enterobacteria such as *E. coli* and *Salmonella agona*, but no responses to mycobacterial antigens, including *M. paratuberculosis*, were found. Cell-mediated immune responses to bacterial antigens were generally higher in MLNMC than in PBMNC. The authors address the possible role of *Yersinia*, pointing out that a chronic form of this infection is now recognized that is culture-negative and with negative serology using standard *Yersinia* antigens.

Ebert *et al.* (1991) showed that while PBMNC from patients with CD demonstrated little proliferative response to *M. paratuberculosis* antigen, they developed marked suppressor activity. The possibility that suppression of CMI or tolerance (anergy) to this organism is present in those with CD may still need to be considered, although the association of tolerance with simultaneous inflammatory responses may be mechanistically impossible to explain. In contrast, reduced suppressor activity in response to mycobacterial antigens was observed by others (Dalton, Hoang and Jewell, 1992) in PBMNCs and lamina propria mononuclear cells, similar to the non-specific antigen suppression of PBMNC previously reported (Hodgson, Wands and Isselbacher, 1978).

8.8.2 Intestinal B and T cell hybridomas in IBD

Hybridomas have been prepared from active B cells in lymphoid tissue draining lesions of CD and UC, by fusion of mesenteric lymph node (MLN) suspensions with murine myeloma cells (Chao *et al.*, 1988). Supernatants were screened against pooled mycobacterial antigens, *B. fragilis* and *B. vulgatus*, two long-term mycobacterial isolates from patients with CD and a pool of other organisms. Some 30% of IgG-secreting hybridomas from CD lymphocytes reacted with pooled mycobacteria and a similar proportion with pooled antigens from other intestinal organisms compared with about 15% in each case from UC-derived hybridomas. Using this technique, mesenteric lymphocytes do appear to be sensitized to some mycobacterial or common antigen, but there was no obvious clonality to suggest a role for indicating that one particular genus in aetiology.

Similar conclusions were reached in a study that performed a molecular genetic analysis of the arrangement of immunoglobulin and antigen-specific T-cell receptor genes of isolated lamina propria lympho-cytes from resected gut from patients with CD, UC and other gastroin-testinal disorders (Kaulfersch, Fiocchi and Waldmann, 1988). The sensitivity of the technique was sufficient to detect a monoclonal popula-tion of as little as 1% clonal expansion in a mixed cell population. In essence, the B and T cells were polyclonal, consistent with a broad immune response to a multiplicity of antigens, thus providing no insight into antigens that might be important in the aetiology of inflammation.

8.8.3 Superantigens in IBD

Superantigens are produced by a number of bacteria and viruses and are capable of activating large numbers of T cells via direct binding to the T-cell receptor (Vβ-specific), with release of inflammatory cytokines. While diarrhoea is a symptom of superantigen exposure, the possibility of a role for superantigens in a chronic disease like CD is based on very indirect evidence.

Posnett *et al.* (1990) reported that a subset of CD patients have elevated levels of Vβ8$^+$ T cells in mesenteric lymph nodes. Others have shown that Vβ8$^+$ T-cell function may also be abnormal in CD (Baca-Estrada, Wong and Croitoni, 1995). Since superantigens may induce anergy in some T cells and activation in others, the mechanism of putative action in CD is obscure at present. However, mucosal T-cell activation is a dominant feature in CD and, since these cells are polyclonal, the possibility of superantigen activation must be considered.

Clinical and histological similarities between yersiniosis and Crohn's disease are considerable. Both CD and *Yersinia* infection are associated with extraintestinal manifestations of arthritis and erythema nodosum. Since *Yersinia* produce a superantigen, a role for superantigens in CD is possible. It has been suggested that the finding of increased CMI against *Yersinia* antigens may be a non-specific marker of superantigen produc-tion (Ibbotson and Lowes, 1995).

8.9 METABOLIC ACTIVITY OF INTESTINAL BACTERIA IN IBD

Changes in the composition, morphology and properties of the microflora in IBD outlined earlier are difficult to evaluate and depend largely on culture techniques that convey little information about bacterial turnover and metabolism. However, several studies have addressed the metabolic activity of commensal bacteria.

Determination of concentrations of short-chain fatty acids as markers for anaerobic bacterial metabolism has shown increased succinic acid concentrations in stool samples from patients with UC (Onderdonk and Bartlett, 1979), presumably reflecting increased numbers or metabolic activity of anaerobes. Others have found increased concentrations of short-chain fatty acids, especially N-butyrate, but related this to deficient absorption/utilization by the colonic mucosa, rather than increased production by luminal bacteria (Roediger *et al.*, 1982).

Benno *et al.* (1993) demonstrated altered function of intestinal bacteria in patients with UC. Microflora-associated characteristics were studied in faecal samples. Samples from patients with UC showed reduced coprostanol and urobilinogen formation and increased faecal tryptic activity. There were no significant differences between proctitis, left-sided colitis and total colitis. All patients were on sulphasalazine so an antibacterial effect or an effect on the microflora through free-radical scavenging by this drug is a possible mechanism for the observed effects. In any event these changes in microflora-associated characteristics suggest quite marked changes in microbial metabolism. Studies in patients off treatment are needed for interpretation of these findings.

Bacterial conversion of dietary and other substrates to hydrogen sulphide has been implicated as a possible factor in the mucosal inflammation of UC (Pitcher and Cummings, 1996). Patients with UC apparently harbour more viable sulphate-reducing bacteria (SRB) than controls, a higher proportion of whom harbour predominantly methanogenic microflora. Hydrogen sulphide and other sulphur compounds inhibit butyrate oxidation, reducing energy supply to the colonic epithelial cells. Patients with UC already have a defect in butyrate oxidation, which is present in uninflamed mucosa and may reflect a genetic defect (Roediger, 1980, 1993; Chapman *et al.*, 1994). 5-ASA, which is an effective agent in UC, reduces sulphate reduction from dietary sulphate and could protect against hydrogen sulphide toxicity. Experimental animal models of colitis based on administration of inhibitors of butyrate oxidation such as 2-bromo-octanoate, or feeding of sulphated polysaccharides and the abrogation of their effects with antianaerobic antibiotics, underscore the attractiveness of the hydrogen sulphide hypothesis, elegantly reviewed by Pitcher and Cummings (1996).

8.10 ANIMAL MODELS OF IBD AND INTESTINAL BACTERIA

Studies on animal models of IBD have provided compelling evidence to show that enteric bacteria play important roles in the pathogenesis of intestinal inflammation. Experimental carrageenan colitis in guinea pigs, for example, will not develop in a germfree colony (Onderdonk, Franklin and Cisneros, 1981). Further compelling evidence exists for a role of the microflora in non-steroidal-anti-inflammatory-induced ileitis, radiation-induced colitis, graft-*versus*-host disease of the intestine, indomethacin-induced colitis and ischaemic colitis in experimental animals, as the germfree state markedly attenuates or abolishes the inflammatory response in each case (Robert and Asano, 1977; Elson *et al.*, 1995; Sartor, Rath and Sellon, 1996). Antibiotics, especially those active against anaerobic bacteria, may prevent development of experimentally induced enterocolitis (Videla *et al.*, 1994; Elson *et al.*, 1995).

The importance of resident enteric bacteria in the initiation and perpetuation of chronic intestinal inflammation is confirmed further by studies of the development of spontaneous colitis in genetically manipulated animals. Gene-targeted disruption of one of two cytokine genes (*IL-2* and *IL-10*) or several T-cell receptor genes (*TCRα, TCRβ* and *TCRα x δ*) in mice or generation of HLA-B27- and human β_2-microglobulin-transgenic rats have produced animals that develop IBD-like diseases (Hammer *et al.*, 1990; Kuhn *et al.*, 1993; Mombaerts *et al.*, 1993; Sadlack *et al.*, 1993). Disease onset in these animals coincides with bacterial colonization of the gut, colitis is most severe in the caecum and right colon where the largest numbers of bacteria are found, and conventional animals develop colitis whereas germfree animals do not.

Gene-knockout mice represent novel models for studying the role of commensal bacteria in the development of chronic intestinal inflammation. For example, IL-2$^{-/-}$ mice spontaneously develop colitis when raised under conventional conditions but exhibit no colitis in germfree conditions (Sadlack *et al.*, 1993). Similarly, conventional IL-10$^{-/-}$ mice develop severe and widespread enteritis whereas specific-pathogen-free (SPF) IL-10$^{-/-}$ mice have lesions restricted to the proximal colon and not elsewhere (Kuhn *et al.*, 1993). Extensive search for pathogenic organisms in these mice has been unsuccessful, suggesting a possible role of non-pathogenic enteric microflora. Furthermore, IBD-like colitis was also observed in TCR, MHC II and Gα_{12} knockout mice kept in SPF conditions (Mombaerts *et al.*, 1993; Rudolph *et al.*, 1995), although the effect of a germfree environment on the development of the disease has not yet been determined.

Experimental colonization of germfree animals with defined organisms further confirms the importance of intestinal microflora in the pathogenesis of chronic gut inflammation. Monoassociation of germfree rats with *E. coli* accentuates small intestinal ulcers induced by indomethacin (Robert and Asano, 1977). Moreover, germfree HLA-B27 transgenic rats develop colitis within 1 month of exposure to defined bacterial cocktails that contained *Bacteroides* spp. (Rath *et al.*, 1996). In contrast, those colonized with a combination of *E. contortum, P. productus, E. coli, E. faecum,* and *E. faecum* isolated from CD patients unexpectedly had no colitis. These results indicate that perhaps not all enteric bacteria have equal ability to induce and perpetuate chronic gut inflammation.

The presence of anaerobic bacteria, especially *Bacteroides* spp., seems to be necessary for the development of spontaneous or induced gut inflammation and ulceration in animal models. Luminal concentrations of *Bacteroides* species, especially *Bacteroides distasonis*, are increased in acute and chronic phases of inflammation induced by dextran sulphate sodium (DSS; Okayasu *et al.*, 1990; Yamada, Ohkusa and Okayasu, 1992). Metronidazole attenuates or even abolishes intestinal inflammation and ulceration induced by indomethacin (Yamada *et al.*, 1993), DSS (Yamada, Ohkusa and Okayasu, 1992), or carrageenan (Elson *et al.*, 1995), and spontaneous enteritis in HLA-B27 transgenic rats (Rath *et al.*, 1996). In carrageenan-induced colitis, obligate anaerobes (metronidazole-sensitive), in particular *B. vulgatus*, appear to be important only in the initiation of colitis, because established colitis was not responsive to metronidazole, which implies that other organisms may maintain the inflammatory response once established (Onderdonk, 1985).

Although studies of animal models of IBD have shown that normal luminal bacteria or their products play critical roles in the initiation and perpetuation of chronic enterocolitis, their mechanisms remain unknown at this time. Enhanced humoral and cellular immune responses to normal luminal bacteria have been noted in several mouse models of colitis (Brandwein *et al.*, 1994, 1995; Cong *et al.*, 1995; Duchmann *et al.*, 1996). Thus, it is quite likely that the normal mucosal immune response to enteric microorganisms and luminal antigens is altered (loss of tolerance) in these models. This altered immune response possibly leads either to induction of mucosal injury due to imbalance of cytokines present at the mucosal sites or to the generation of autoimmune-like responses. While the pathogenesis of chronic intestinal inflammation in mice and IBD in humans may not turn out to be the same, studies of experimentally induced and spontaneously developed colitis in animal models, including gene-targeting animals, will not only provide clues to the pathogenesis of human IBD but also allow the role of factors such as bacterial antigens considered to be important in the pathogenesis of IBD to be elucidated.

8.11 INFLAMMATORY PRODUCTS OF INTESTINAL BACTERIA AND IBD

It is now well established that soluble products of bacteria can produce all the histopathological and immunopathological responses attributable to intact viable organisms. The intestinal microflora secretes or contains within them a number of potent inflammatory products such as lipopolysaccharide (LPS or endotoxin), peptidoglycan–polysaccharide complexes (PG–PS), and N-formylmethionyl oligopeptides such as N-formylmethionyl-leucyl-phenylalanine (FMLP). Why these particular products stimulate such a wide range of inflammatory responses while other products do not is unknown. There are substantial interactions among these products and the priming, immunoadjuvant and direct effects of these products could account for much of the intestinal inflammatory and immunological activity observed in IBD.

8.11.1 Endotoxin

Endotoxin (LPS) is a major component of the outer cell membrane of Gram-negative bacteria. These bacteria do not secrete endotoxin but rather shed this cell component during normal growth and cell division and at the time of cell lysis or death. As these processes are occurring continuously in the heavily colonized gut, free endotoxin is present in the gut contents. Probably 1–3 mg of LPS is present per gram of faeces (Rogers, Moore and Cohen, 1985).

The healthy intestine is an efficient barrier to all except trace amounts of endotoxin taken up by pinocytosis (Nolan *et al.*, 1977). A striking increase in endotoxin permeation is produced experimentally by altering intestinal permeability, probably by opening up paracellular pathways (Gans and Matsumoto, 1974). Systemic endotoxinaemia occurs following saturation of hepatic clearance mechanisms and with lymphatic transport (Fink, Suin de Boutemard and Haeckel, 1988). Indeed, in experimental peritonitis about half the absorbed endotoxin is via lymph, the remainder by portal venous blood (Olafsson, Nylander and Olsson, 1986). In humans, transient endotoxinaemia may follow colonoscopy, as a result of either bowel instrumentation alone or mucosal biopsy. The incidence quoted varied from 9% to 65% (Kelley *et al.*, 1985). Although this may have no clinical significance, it supports the view that injury to the mucosal barrier of the gut results in an impressive transmucosal leak of luminal endotoxin.

In IBD, endotoxinaemia is present in all those with very active disease but in only about 10–12% of regular outpatients in the IBD clinic. In one study, plasma endotoxin levels correlated with the Van Hees index of disease activity and endotoxin levels fell progressively during therapy with total parenteral nutrition and steroids (Wellemann *et al.*, 1986). The addition of total gut irrigation to the aforementioned therapy resulted in a substantially more rapid decline in endotoxin levels and more rapid onset of remission.

8.11.2 Peptidoglycan–polysaccharide complexes

Peptidoglycan–polysaccharide polymers are structural components of the cell walls of both Gram-positive and Gram-negative bacteria (Schleifer and Krause, 1971; Kotani *et al.*, 1986). Peptidoglycan–polysaccharide polymers induce arthritis, uveitis, pancarditis and hepatic granulomata after enteral injection. Intramural injection of streptococcal PG–PS into the caecum in mice induces chronic granulomatous inflammation (Sartor *et al.*, 1986). Peptidoglycan can be demonstrated in the granulomata by immunofluorescence. The inflammatory response to PG-PS resembles CD. A phenomenon analogous to flare-ups in IBD could be induced in these colitic animals by intravenous injection of endotoxin.

Peptidoglycan–polysaccharide produces different responses in different strains of rat, suggesting genetic variation, and PG– PS from different species of bacteria vary considerably in inflammatory potential. Some, such as group A streptococci, enterobacteria and mycobacteria, produce protracted granulomatous inflammation and others such as *Peptostreptococcus* produce only transient inflammation. Clearly, the pattern and chronicity of inflammation to PG–PS is critically dependent on the intrinsic properties of both microorganisms and host responses.

8.11.3 Formyl peptides

The observation that bacterial secretions were chemotactic for mammalian neutrophils was made over 20 years ago (Keller and Sorkin, 1967; Ward, Lepow and Newman, 1968). Schiffman *et al.* and others later demonstrated a directed rabbit neutrophil response (chemotaxis) to low-molecular-weight (150–1500 Mr) relatively heat-stable factors from filtrates of *E. coli* culture medium (Schiffman, Corcoran and Wahl, 1975; Scheurlen *et al.*, 1988).

Formylmethionine is unique to prokaryotes (Inouye and Halegova, 1980) and mitochondria (Miura, Amaya and Mori, 1986) as the first amino acid assembled at the ribosome during synthesis of proteins. During post-transcriptional processing formylmethionine and the subsequent 20 or so amino acids direct the newly synthesized protein to its site of action, e.g. cell membrane or extracellular medium if secreted (Briggs *et al.*, 1986). This amino-acid sequence is called the 'signal sequence' and is usually clipped off once the nascent polypeptide arrives at its site of action (Novak, Ray and Dev, 1986). Bacteria are a ready source of formylmethionyl peptides fitting the general requirements for chemotactic activity defined by earlier studies (Schiffman *et al.*, 1975).

Improved chromatography and the development of immunoassays has allowed the purification of authentic FMLP from *E. coli* culture supernatants (Marasco *et al.*, 1984; Broom *et al.*, 1989). Several other N-blocked methionyl peptides were sequenced from bioactive fractions of HPLC-fractionated *E. coli* supernatant, but FMLP accounted for most of the bioactivity present (Broom *et al.*, 1989). FMLP-like immunoreactivity was also demonstrated co-eluting with FMLP on HPLC from the culture supernatants of various other gut commensals: *P. vulgaris*, *K. pneumoniae*, *E. faecalis* and *B. fragilis* (Hobson *et al.*, 1990). Authentic FMLP has also been isolated and sequenced from *Helicobacter pylori* culture supernatants (Mooney *et al.*, 1991; Broom *et al.*, 1992). Others have isolated a potent pro-inflammatory formyl tetrapeptide (F-Met-Leu-Phe-Ile) from *Staphylococcus aureus* culture medium (Rot, Henderson and Leonard, 1986; Rot *et al.*, 1989).

It appears the production of formyl peptide receptor agonists is relatively ubiquitous among both aerobic and anaerobic bacteria. FMLP is one of a group of naturally occurring formyl peptides activating acute inflammatory cells. It may be generated during activation of the SOS operon in bacteria exposed to oxidative stress (Broom *et al.*, 1993)

Mammalian inflammatory cells have thus evolved a receptor recognizing a defined group of secreted bacterial peptides. Immunologically naive inflammatory and immune-competent cells do not recognize bacterial cells per se. Presumably this detection system has evolved to allow an immediate, directed inflammatory response to invasive bacteria independent of delayed immune mechanisms.

Formyl peptide receptors have been described on human, non-human primate, rat, rabbit, murine, porcine and equine granulocytes (Styrt, 1989). In humans, the FMLP receptor is inducible and the specific granule represents a mobilizable reservoir of FMLP receptors that are shifted to the plasma membrane with degranulation (Fletcher and Gallin, 1983; Gallin and Seligmann, 1984; Gardner, Melnick and Malech, 1986). It is noteworthy that circulating neutrophils from patients with active CD and UC have increased numbers of FMLP receptors and enhanced responsiveness to FMLP (Anton, Targan and Shanahan, 1989).

In view of the potent biological effects of FMLP, it is not surprising that a pronounced inflammatory response follows the introduction of FMLP across epithelial barriers such as lung and skin (Desai *et al.*, 1979; Mellor, Myers and Chadwick, 1986). In the intestine, colonic infusions in mice and rats (Chester *et al.*, 1985) and rectal administration in rabbits resulted in experimental colitis (LeDuc and Nast, 1990), although the concentrations used in these studies were at least 1000 times greater than those estimated by bioassay of intestinal contents (Chadwick *et al.*, 1988). Changes in vascular permeability, blood flow and mucosal permeability were produced with FMLP in the rat small intestine (von Ritter *et al.*, 1988). These effects were not found in animals rendered neutropenic. Infusion of FMLP into mesenteric arteries resulted in the appearance of leukotrienes (LTB$_4$) in mesenteric veins, suggesting the presence of resident cells in the gut which respond to formyl peptides by eicosanoid synthesis (Granger *et al.*, 1988). Furthermore, FMLP is spasmogenic for intestinal smooth muscle, an effect mediated by enteric nerves and M1 cholinergic pathways and dependent on the integrity of the mucosal layer of the gut (Hobson *et al.*, 1988).

Studies in the rat with both tritiated FMLP and radioiodinated formyl-methionyl-leucyl-tyrosine demonstrated that the intestinal mucosa is relatively impermeable to these peptides, whereas substantial enzymic degradation of peptides occurs intraluminally (Woodhouse *et al.*, 1987). Nevertheless, both in the ileum and colon the mucosal barrier is not 100% efficient and intact peptide (1%) was absorbed. Absorbed intact peptide was excreted in bile and underwent enterohepatic circulation (Anderson *et al.*, 1987). In rats with experimental colitis, the colonic absorption and enterohepatic circulation of intact peptides increased eightfold (Hobson *et al.*, 1988) and, when colonic loops were pretreated with dithiothreitol to increase the mucosal permeability, absorption and biliary recovery increased 50-fold (Ferry *et al.*, 1989). There was a good correlation between changes in colonic mucosal permeability measured by [^{51}Cr]-EDTA absorption and biliary recovery of formylmethionyl peptide (Ferry *et al.*, 1989). FMLP has potent effects on biliary lipid excretion (Mizuno *et al.*, 1996).

The formyl peptides, especially FMLP, may be important mediators of intestinal inflammation in IBD and could contribute to some of the extra-intestinal effects, particularly in the liver and biliary system. Direct experimental evidence for involvement of formyl peptides in pathophysiology of IBD is, however, still awaited.

8.12 CONCLUSIONS

Dysregulation in the intestinal immune system of individuals with IBD results from an interaction between genetic factors and adverse neonatal events (Fig. 8.1).

PATHOGENESIS OF IBD

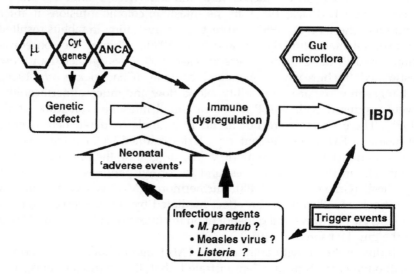

Figure 8.1 Current hypothesis for the pathogenesis of IBD. μ = structural genes controlling intestinal mucosal permeability; Cyt genes = cytokine genes and other immunoregulatory genes; ANCA = genes associated with production of anti-neutrophil cytoplasmic autoantibodies; *M. paratub* = *M. paratuberculosis*.

These events may include viral or bacterial infections which may or may not persist. There appears to be a lack of negative regulatory control affecting T helper cells so that inflammatory responses when triggered by intercurrent events are not down-regulated. Different patterns of immune response or different initiating events may explain differences between UC and CD. The intestinal microflora or their metabolic products play an integral role either as targets for immune cells or facilitators or amplifiers of the inflammatory response. Increased understanding of safe ways to manipulate the gastrointestinal microflora may have as much potential in therapy of IBD in the future as developments in immunotherapy to control host responses.

8.13 ACKNOWLEDGEMENTS

The authors' research is supported by grants from the Health Research Council of New Zealand and the Lottery Grants Board (Health Research) of New Zealand.

8.14 REFERENCES

Aluwihare, A. P. R. (1971) Electron microscopy in Crohn's disease. *Gut*, **12**, 509.

Ambrose, N. S., Johnson, M., Burdon, D. W. *et al.* (1984) Incidence of pathogenic bacteria from mesenteric lymph nodes and ileal serosa during Crohn's disease surgery. *British Journal of Surgery*, **71**, 623.

Anderson, R. P., Woodhouse, A. F., Myers, D. B. *et al.* (1987) Hepatobiliary excretion and enterohepatic circulation of bacterial chemotactic peptide (FMLP) in the rat. *Journal of Gastroenterology and Hepatology*, **2**, 45.

Anton, P. A., Targan, S. R. and Shanahan, F. (1989) Increased neutrophil receptors for and response to proinflammatory bacterial peptide formyl-methionyl-leucyl-phenylalanine in Crohn's disease. *Gastroenterology*, **97**, 20.

Baca-Estrada, M., Wong, D. and Croitoni, K. (1995) Cytotoxic activity of Vβ8⁺ T cells in Crohn's disease: the role of bacterial superantigens. *Clinical and Experimental Immunology*, **99**, 398.

Barr, G., Hudson, M., Priddle, J. *et al.* (1987) Colonic bacterial proteases to IgA1 and sIgA in patients with ulcerative colitis. *Gut*, **28**, 186.

Bell, J. I. and Jewell, D. P. (1991) *Mycobacterium paratuberculosis* DNA cannot be detected in Crohn's disease tissues. *Gastroenterology*, **100**, A611.

Belsheim, M. R., Darwish, R. Z., Watson, W. C. *et al.* (1983) Bacterial L-form isolation from inflammatory bowel disease patients. *Gastroenterology*, **85**, 364.

Benno, P., Leijonmarck, C. E., Monsen, U. *et al.* (1993) Functional alterations of the microflora in patients with ulcerative colitis. *Scandinavian Journal of Gastroenterology*, **28**, 839.

Blaser, M. J., Hoverson, D., Ely, I. G. *et al.* (1984) Studies of *Campylobacter jejuni* in patients with inflammatory bowel disease. *Gastroenterology*, **86**, 33.

Blichfeldt, P., Blomhoff, J. P., Myhre, E. *et al.* (1978) Metronidazole in Crohn's disease. A double blind cross-over clinical trial. *Scandinavian Journal of Gastroenterology*, **13**, 123.

Brandwein, S., McCabe, R., Ridwan, B. *et al.* (1994) Immunologic reactivity of colitic C3H/HeJBir mice to enteric bacteria. *Gastroenterology*, **106**, A656.

Brandwein, S., McCabe, R., Ridwan, B. *et al.* (1995) Spontaneously colitic C3H/HeJBir mice demonstrate antibody reactivity to isolated colonies of enteric bacteria. *Gastroenterology*, **108**, 787.

Briggs, M. S., Cornell, D. G., Dulhy, R. A. *et al.* (1986) Conformations of signal peptides induced by lipids suggests initial steps in protein export. *Science*, **233**, 206.

Brooke, B. N. (1956) Outcome of surgery for ulcerative colitis. *Lancet*, **ii**, 532.

Broom, M. F., Sherriff, R. M., Tate, W. P. *et al.* (1989) Partial purification and characterisation of a formyl-methionine deformylase from rat small intestine. *Biochemical Journal*, **257**, 51.

Broom, M. F., Sherriff, R. M., Munster, D. *et al.* (1992) Identification of formyl Met-Leu-Phe in culture filtrates of *Helicobacter pylori*. *Microbios*, **72**, 239.

Broom, M., Sherriff, R., Ferry, D. *et al.* (1993) Formylmethionyl-leucylphenylalanine and the SOS operon in *Escherichia coli*: a model of host–bacterial interaction. *Biochemical Journal*, **291**, 895.

Brown, W. R. and Lee, E. (1974) Radioimmunological measurements of bacterial antibodies. Human serum antibodies reactive with *Bacteroides fragilis* and enterococcus in gastrointestinal and immunological disorders. *Gastroenterology*, **66**, 1145.

Brunello, F., Pera, A., Martini, S. *et al.* (1991) Antibodies to *Mycobacterium paratuberculosis* in patients with Crohn's disease. *Digestive Diseases and Sciences*, **36**, 1741.

Burke, D. and Axon, A. (1988) Adhesive *Escherichia coli* in inflammatory bowel disease and infective diarrhoea. *British Medical Journal*, **297**, 102.

Burke, D., Axon, A., Clayden, S. *et al.* (1990) The efficacy of tobramycin in the treatment of ulcerative colitis. *Alimentary Pharmacology and Therapeutics*, **4**, 123.

Butcher, P. D., McFadden, J. J. and Hermon-Taylor, J. (1988) Investigation of mycobacteria in Crohn's disease tissue by Southern blotting and DNA hybridisation with cloned mycobacterial genomic DNA probes from a Crohn's disease isolated mycobacteria. *Gut*, **29**, 1222.

Chadwick, V. S., Mellor, D. M., Myers, D. B. *et al.* (1988) Production of peptides inducing chemotaxis and lysosomal enzyme release in human neutrophils by intestinal bacteria *in vitro* and *in vivo*. *Scandinavian Journal of Gastroenterology*, **23**, 121.

Chao, L., Stelle, J., Rodrigues, C. *et al.* (1988) Specificity of antibodies secreted by hybridomas generated from activated B cells in the mesenteric lymphnodes of patients with inflammatory bowel disease. *Gut*, **29**, 35.

Chapman, M., Grahn, M., Boyle, M. *et al.* (1994) Butyrate oxidation is impaired in the colonic mucosa of sufferers of quiescent ulcerative colitis. *Gut*, **35**, 73.

Chester, J. F., Ross, J. S., Malt, R. A. *et al.* (1985) Acute colitis produced by chemotactic peptides in rats and mice. *American Journal of Pathology*, **121**, 284.

Chiodini, R. J., van Kruiningen, H. J. and Merkal, R. S. (1984) Ruminant paratuberculosis (Johne's disease). The current status and future prospects. *Cornell Veterinarium*, **74**, 218.

Chiodini, R. J., van Kruiningen, H. J., Thayer, W. R. *et al.* (1984) Possible role of mycobacteria in inflammatory bowel disease. I. An unclassified *Mycobacterium* species isolated from patients with Crohn's disease. *Digestive Diseases and Sciences*, **29**, 1073.

Chiodini, R. J., van Kruiningen, H. J., Thayer, W. R. *et al.* (1986) Spheroplastic phase of mycobacteria isolated from patients with Crohn's disease. *Journal of Clinical Microbiology*, **24**, 357.

Cong, Y., Brandwein, S., McCabe, R. *et al.* (1995) Th1 response to enteric bacteria in colitic C3H/HeJBir mice. *Clinical Immunology and Immunopathology*, **76**(Suppl.), S44.

Cooke, E. (1967) A quantitative comparison of the faecal flora of patients with ulcerative colitis and that of normal persons. *Journal of Pathology and Bacteriology*, **9**, 439.

Cooke, E. M., Ewins, S. P., Gywel-Jones, J. *et al.* (1974) Properties of strains of *Escherichia coli* carried in different phases of ulcerative colitis. *Gut*, **15**, 143.

Dalton, H., Hoang, P. and Jewell, D. (1992) Antigen induced suppression in peripheral blood and lamina propria mononuclear cells in inflammatory bowel disease. *Gut*, **33**, 324.

Desai, U., Kreutzer, D. L., Showell, H. *et al.* (1979) Acute inflammatory pulmonary reactions induced by chemotactic factors. *American Journal of Pathology*, **96**, 71.

Duchmann, R., Kaiser, I., Hermann, E. *et al.* (1995) Tolerance exists towards resident intestinal flora but is broken in active inflammatory bowel disease (IBD). *Clinical and Experimental Immunology*, **102**, 448.

Duchmann, R., Schmitt, E., Knolle, P. *et al.* (1996) Tolerance towards resident intestinal flora in mice is abrogated in experimental colitis and restored by treatment with interleukin-10 or antibodies to interleukin-12. *European Journal of Immunology*, **26**, 934.

Ebert, E. C., Bhat, B. D., Liu, S. *et al.* (1991) Induction of suppressor cells by *Mycobacterium paratuberculosis* antigen in inflammatory bowel disease. *Clinical and Experimental Immunology*, **83**, 320.

Edmiston, J. C. E., Avant, G. R. and Wilson, F. A. (1982) Anaerobic bacterial populations on normal and diseased human biopsy tissue obtained at colonoscopy. *Applied and Environmental Microbiology*, **43**, 1173.

Ekbom, A. and Adami, H.-O. (1992) The epidemiology of inflammatory bowel disease, in *Current Topics in Gastroenterology. Inflammatory Bowel Disease*, (eds R. P. MacDermott and W. F. Stenson), Elsevier, New York, p. 1.

Ekbom, A., Adami, H.-O., Helmick, C. G. *et al.* (1990) Perinatal risk factors for inflammatory bowel disease: a case-controlled study. *American Journal of Epidemiology*, **132**, 1111.

Ekbom, A., Helmick, C., Zack, M. *et al.* (1991) The epidemiology of inflammatory bowel disease: a large population-based study in Sweden. *Gastroenterology*, **100**, 350.

Ekbom, A., Wakefield, A., Zack, M. *et al.* (1994) Perinatal measles infection and subsequent Crohn's disease. *Lancet*, **344**, 508.

Elsaghier, A., Prantera, C., Moreno, C. *et al.* (1992) Antibodies to *Mycobacterium paratuberculosis* – specific protein antigens in Crohn's disease. *Clinical and Experimental Immunology*, **90**, 503.

Elson, C., Sartor, R., Tennyson, G. *et al.* (1995) Experimental models of inflammatory bowel disease. *Gastroenterology*, **109**, 1344.

Fabia, R., Ar'Rajab, A., Johansson, M. L. *et al.* (1993) Impairment of bacterial flora in human ulcerative colitis and experimental colitis in the rat. *Digestion*, **54**, 248.

Ferry, D. M., Butt, T. J., Broom, M. F. *et al.* (1989) Bacterial chemotactic oligopeptides and the intestinal mucosal barrier. *Gastroenterology*, **97**, 61.

Fink, P. C., Suin de Boutemard, C. and Haeckel, R. (1988) Endotoxaemia in patients with Crohn's disease: a longitudinal study of elastase/alpha1-proteinase inhibitor and limulus-amoebocyte-lysate reactivity. *Journal of Clinical Chemistry and Clinical Biochemistry*, **26**, 117.

Fiocchi, C., Battisto, J. R. and Farmer, R. G. (1981) Studies on isolated gut mucosal lymphocytes in inflammatory bowel disease. Detection of activated T-cells and enhanced proliferation to *Staphylococcus aureus* and lipopolysaccharides. *Digest-ive Diseases and Sciences*, **26**, 728.

Fletcher, M. P. and Gallin, J. I. (1983) Human neutrophils contain an intracellular pool of putative receptors for the chemoattractant N-formyl-methionyl-leucyl-phenylalanine. *Blood*, **62**, 792.

Gallin, J. I. and Seligmann, B. E. (1984) Mobilization and adaptation of human neutrophil chemoattractant fMet-Leu-Phe receptors. *Federation Proceedings*, **43**, 2732.

Gans, H. and Matsumoto, K. (1974) The escape of endotoxin from the intestine. *Surgery, Gynecology and Obstetrics*, **139**, 395.

Gardner, J. P., Melnick, D. A. and Malech, H. L. (1986) Characterization of the formyl-peptide chemotactic receptor appearing at the phagocytic cell surface after exposure to phorbol myristate acetate. *Journal of Immunology*, **136**, 1400.

Giaffer, M. H., Holdsworth, C. D. and Duerden, B. I. (1991) The assessment of faecal flora in patients with inflammatory bowel disease by a simplified bacter-iological technique. *Journal of Medical Microbiology*, **35**, 238.

Gilligan, P. H., McCarthy, L. R. and Genta, V. W. (1981) Relative frequency of *Clostridium difficile* in patients with diarrheal disease. *Journal of Clinical Micro-biology*, **14**, 26.

Gitnick, G., Collins, J., Beaman, B. *et al.* (1989) Preliminary report on isolation of mycobacteria from patients with Crohn's disease. *Digestive Diseases and Sciences*, **34**, 925.

Goodman, M. J., Pearson, K. W., McGhie, D. *et al.* (1980) *Campylobacter* and *Giardia lamblia* causing exacerbation of inflammatory bowel disease. *Lancet*, **ii**, 1247.

Gorbach, S., Nahas, L., Plaut, A. *et al.* (1968) Studies of intestinal microflora–fecal microbial ecology in ulcerative colitis and regional enteritis: relationship to severity of disease and chemotherapy. *Gastroenterology*, **54**, 575.

Graham, D. Y., Markesich, D. C. and Yoshimura, H. H. (1987) Mycobacteria and inflammatory bowel disease. Results of culture. *Gastroenterology*, **92**, 436.

Granger, D. N., Zimmerman, B. J., Sekizuka, E. *et al.* (1988) Intestinal microvas-cular exchange in the rat during luminal perfusion with formyl-methionyl-leucyl-phenylalanine. *Gastroenterology*, **94**, 673.

Green, E. P., Tizard, M. L., Moss, M. T. *et al.* (1989) Sequence and characteristics of IS900, an insertion element identified in a human Crohn's disease isolate of *Mycobacterium paratuberculosis*. *Nucleic Acids Research*, **17**, 9063.

Greenfield, C., Aguilar Ramirez, J. R., Pounder, R. E. *et al.* (1983) *Clostridium difficile* and inflammatory bowel disease. *Gut*, **24**, 713.

Guillemot, F., Colombel, J. F., Neut, C. *et al.* (1991) Treatment of diversion colitis by short chain fatty acids: prospective and double-blind study. *Diseases of the Colon and Rectum*, **34**, 861.

Guiot, H. F. L., van den Broek, P. J., van der Meer, J. W. M. *et al.* (1983) Selective antimicrobial modulation of the intestinal flora of patients with acute nonlym-phocytic leukemia: a double-blind, placebo-controlled study. *Journal of Infectious Diseases*, **147**, 615.

Hammer, R. E., Maika, S. D., Richardson, J. A. *et al.* (1990) Spontaneous inflam-matory disease in transgenic rats expressing HLA-B27-associated human dis-orders. *Cell*, **63**, 1099.

Harper, P., Truelove, S., Lee, E. *et al.* (1983) Split ileostomy and ileocolostomy for Crohn's disease of the colon and ulcerative colitis: a 20 year survey. *Gut*, **24**, 106.

Harper, P. H., Lee, E. C. G., Kettlewell, M. G. W. *et al.* (1985) Role of the faecal stream in the maintenance of Crohn's colitis. *Gut*, **26**, 279.

Hartley, M. G., Hudson, M. J., Swarbrick, E. T. *et al.* (1993) Adhesive and hydrophobic properties of *Escherichia coli* from the rectal mucosa of patients with ulcerative colitis. *Gut*, **34**, 63.

Hermon-Taylor, J. (1993) Causation of Crohn's disease: the impact of clusters (editorial). *Gastroenterology*, **104**, 643.

Hobson, C. H., Butt, T. J., Ferry, D. M. *et al.* (1988) Enterohepatic circulation of bacterial chemotactic peptide in rats with experimental colitis. *Gastroenterology*, **94**, 1006.

Hobson, C. H., Roberts, E. C., Broom, M. F. *et al.* (1990) Radio-immunoassay for formyl methionyl leucyl phenylalanine. I. Development and application to assessment of chemotactic peptide production by enteric bacteria. *Journal of Gastroenterology and Hepatology*, **5**, 32.

Hodgson, H. J. F., Wands, J. R. and Isselbacher, K. J. (1978) Decreased suppressor cell activity in inflammatory bowel disease. *Clinical and Experimental Immunology*, **32**, 451.

Hoeksma, A. and Winkler, K. C. (1963) The normal flora of the nose in twins. *Acta Leidensia*, **32**, 123.

Holdleman, L. V., Good, I. J. and Moore, W. E. C. (1976) Human fecal flora: variation in bacterial composition within individuals and a possible effect of emotional stress. *Applied and Environmental Microbiology*, **31**, 359.

Horing, E., Gopfert, D., Schroter, G. *et al.* (1991) Frequency and spectrum of microorganisms isolated from biopsy specimens in chronic colitis. *Endoscopy*, **23**, 325.

Hugot, J. P., Laurent-Puig, P., Gower-Rousseau, C. *et al.* (1996) Mapping of a susceptibility locus for Crohn's disease on chromosome 16. *Nature*, **379**, 821.

Ibbotson, J. P. and Lowes, J. R. (1995) Potential role of superantigen induced activation of cell mediated immune mechanisms in the pathogenesis of Crohn's disease (review). *Gut*, **36**, 1.

Ibbotson, J. P., Pease, P. E. and Allan, R. N. (1987a) Cell-wall deficient bacteria in inflammatory bowel disease. *European Journal of Clinical Microbiology*, **6**, 429.

Ibbotson, J. P., Pease, P. E. and Allan, R. N. (1987b) Serological studies in Crohn's disease. *European Journal of Clinical Microbiology*, **6**, 286.

Ibbotson, J., Lowes, J., Chahal, H. *et al.* (1992) Mucosal cell-mediated immunity to mycobacterial, enterobacterial and other microbial antigens in inflammatory bowel disease. *Clinical and Experimental Immunology*, **87**, 224.

Iizuka, M., Nakagomi, O., Chiba, M. *et al.* (1995) Absence of measles virus in Crohn's disease. *Lancet*, **345**, 660.

Inouye, M. and Halegova, S. (1980) Secretion and membrane localization of proteins in *Escherichia coli. CRC Critical Reviews in Biology*, **7**, 339.

Johnson, L. E. W. and Wirostko, W. (1989) Crohn's disease uveitis – parasitization of vitreous leukocytes by mollicute-like organisms. *American Journal of Clinical Pathology*, **91**, 259.

Kaulfersch, W., Fiocchi, C. and Waldmann, T. A. (1988) Polyclonal nature of the intestinal mucosal lymphocyte populations in inflammatory bowel disease. A molecular genetic evaluation of the immunoglobulin and T-cell antigen receptors. *Gastroenterology*, **95**, 364.

Keighley, M., Arabi, Y., Dimock, F. *et al.* (1978) Influence of inflammatory bowel disease on intestinal microflora. *Gut*, **19**, 1099.

Keller, H. U. and Sorkin, E. (1967) Studies on chemotaxis. V: On the chemotactic effect of bacteria. *International Archives of Allergy*, **31**, 505.

Kelley, C. J., Ingoldby, C. J. H., Blenkharn, J. I. *et al.* (1985) Colonoscopy related endotoxemia. *Surgery Gynecology and Obstetrics*, **161**, 332.

Kelly, D. G., Phillips, S. F., Kelly, K. A. *et al.* (1983) Dysfunction of the continent ileostomy: clinical features and bacteriology. *Gut*, **24**, 193.

Kingham, J. G. (1991) Microscopic colitis (review). *Gut*, **32**, 234.

Koletzko, S., Griffiths, A., Corey, M. *et al.* (1991) Infant feeding practices and ulcerative colitis in childhood. *British Medical Journal*, **302**, 1580.

Kotani, S., Tsujimoto, M., Koga, T. *et al.* (1986) Chemical structure and biological activity relationship of bacterial cell walls and muramyl peptides. *Federation Proceedings*, **45**, 2534.

Kreuzpaintner, G., Horstkotte, D., Heyll, A. *et al.* (1992) Increased risk of bacterial endocarditis in inflammatory bowel disease. *American Journal of Medicine*, **92**, 391.

Kuhn, R., Lohler, J., Rennick, D. *et al.* (1993) Interleukin-10-deficient mice develop chronic enterocolitis. *Cell*, **75**, 263.

Laffineur, G., Lescut, D., Vincent, P. *et al.* (1992) Translocation bacterienne dans la maladie de Crohn. *Gastroenterology and Clinical Biology*, **16**, 777.

LeDuc, L. E. and Nast, C. C. (1990) Chemotactic peptide-induced acute colitis in rabbits. *Gastroenterology*, **98**, 989.

Liu, Y., van Kruiningen, H. J., West, A. B. *et al.* (1995) Immunocytochemical evidence of *Listeria*, *Escherichia coli*, and *Streptococcus* antigens in Crohn's disease. *Gastroenterology*, **108**, 1396.

Luukkonen, P., Valtonen, V., Sivonen, A. *et al.* (1988) Fecal bacteriology and reservoir ileitis in patients operated on for ulcerative colitis. *Diseases of the Colon and Rectum*, **31**, 864.

McFadden, J. J. and Seechurn, P. (1992) Mycobacteria and Crohn's disease. Molecular approaches, in *Current Topics in Gastroenterology. Inflammatory Bowel Disease*, (eds R. P. MacDermott and W. F. Stenson), Elsevier, New York, p. 259.

McFadden, J. J., Butcher, P. D., Chiodini, R. *et al.* (1987) Crohn's disease-isolated mycobacteria are identical to *Mycobacterium paratuberculosis*, as determined by DNA probes that distinguish between mycobacterial species. *Journal of Clinical Microbiology*, **25**, 796.

McGarity, B. H., Robertson, D. A. F., Clarke, I. N. *et al.* (1991) Deoxyribonucleic acid amplification and hybridisation in Crohn's disease using a chlamydial plasmid probe. *Gut*, **32**, 1011.

Macpherson, A., Khoo, U. Y., Forgacs, I. *et al.* (1996) Mucosal antibodies in inflammatory bowel disease are directed against intestinal bacteria. *Gut*, **38**, 365.

Madden, M. V., Farthing, M. J. G. and Nicholls, R. J. (1990) Inflammation in ileal reservoirs: 'pouchitis'. *Gut*, **31**, 247.

Marasco, W. A., Phan, S. H., Krutzsch, H. *et al.* (1984) Purification and identification of formyl-methionyl-leucyl-phenylalanine as the major peptide neutrophil chemotactic factor produced by *Escherichia coli*. *Journal of Biological Chemistry*, **259**, 5430.

Markesich, D. C., Graham, D. Y. and Yoshimura, H. H. (1988) Progress in culture and subculture of spheroplasts and fastidious acid-fast bacilli isolated from intestinal tissues. *Journal of Clinical Microbiology*, **26**, 1600.

Matthews, N., Mayberry, J. F., Rhodes, J. *et al.* (1980) Agglutinins to bacteria in Crohn's disease. *Gut*, **21**, 376.

Mellor, D. M., Myers, D. B. and Chadwick, V. S. (1986) The cored sponge model of *in vivo* leucocyte chemotaxis. *Agents and Actions*, **18**, 550.

Metcalf, A. M. and Phillips, S. F. (1986) Ileostomy diarrhoea. *Clinics in Gastroenterology*, **15**, 705.

Millar, D., Ford, J., Sanderson, J. *et al.* (1996) IS900 PCR to detect *Mycobacterium paratuberculosis* in retail supplies of whole pasteurized cows' milk in England and Wales. *Applied and Environmental Microbiology*, **62**, 3446–52.

Miura, A., Amaya, Y. and Mori, M. (1986) A metalloprotease involved in the processing of mitochondrial precursor proteins. *Biochemical and Biophysical Research Communications*, **134**, 1151.

Mizuno, K., Hoshino, M., Hayakawa, T. *et al.* (1996) Uncoupling of biliary lipid from bile acid secretion by formyl-methionyl-leucyl-phenylalanine in the rats. *Hepatology*, **24**, 1224.

Mombaerts, P., Mizoguchi, E., Grusby, M. *et al.* (1993) Spontaneous development of inflammatory bowel disease in T cell receptor mutant mice. *Cell*, **75**, 275.

Mooney, C., Keenan, J., Munster, D. *et al.* (1991) Neutrophil activation by *Helicobacter pylori*. *Gut*, **32**, 853.

Moss, M., Sanderson, J., Tizard, M. *et al.* (1992) Polymerase chain reaction detection of *Mycobacterium paratuberculosis* and *Mycobacterium avium* subsp *silvaticum* in long term cultures from Crohn's disease and control tissues. *Gut*, **33**, 1209.

Murray, A., Oliaro, J., Schlup, M. *et al.* (1995) *Mycobacterium paratuberculosis* and inflammatory bowel disease: frequency distribution in serial colonoscopic biopsies. *Microbios*, **83**, 217.

Nakamura, S., Mikawa, M., Nakashio, S. *et al.* (1981) Isolation of *Clostridium difficile* from feces and the antibody sera of young and elderly adults. *Microbiology and Immunology*, **25**, 345.

Neut, C., Colombel, J. F., Guillemot, F. *et al.* (1989) Impaired bacterial flora in human excluded colon. *Gut*, **30**, 1094.

Nolan, J. P., Hare, D. K., McDevitt, J. J. *et al.* (1977) In vitro studies of intestinal endotoxin absorption. I. Kinetics of absorption in the isolated everted gut sac. *Gastroenterology*, **72**, 434.

Novak, P., Ray, P. H. and Dev, I. K. (1986) Localization and purification of two enzymes from *Escherichia coli* capable of hydrolyzing a signal peptide. *Journal of Biological Chemistry*, **261**, 420.

O'Connell, P. R., Rankin, D. R., Weiland, L. H. *et al.* (1986) Enteric bacteriology, absorption, morphology and emptying after ileal pouch-anal anastomosis. *British Journal of Surgery*, **73**, 909.

Ohkusa, T., Okayasu, I., Tokoi, S. *et al.* (1993) Bacterial invasion into the colonic mucosa in ulcerative colitis. *Journal of Gastroenterology and Hepatology*, **8**, 116.

Okayasu, I., Hatakeyama, S., Yamada, M. *et al.* (1990) A novel method in the induction of reliable experimental acute and chronic ulcerative colitis in mice. *Gastroenterology*, **98**, 694.

Olafsson, P., Nylander, G. and Olsson, P. (1986) Endotoxin: route of transport in experimental peritonitis. *American Journal of Surgery*, **151**, 443.

Onderdonk, A. (1985) Experimental models for ulcerative colitis. *Digestive Diseases and Sciences*, **30**, 40.

Onderdonk, A. B. and Bartlett, J. G. (1979) Bacteriological studies of experimental ulcerative colitis. *American Journal of Nutrition*, **32**, 258.

Onderdonk, A. B., Franklin, M. L. and Cisneros, R. L. (1981) Production of experimental ulcerative colitis in gnotobiotic guinea pigs with simplified microflora. *Infection and Immunity*, **32**, 225.

Parent, K. and Mitchell, P. (1978) Cell wall-defective variants of pseudomonas-like (group Va) bacteria in Crohn's disease. *Gastroenterology*, **75**, 368.

Pavli, P., Cavanaugh, J. and Grimm, M. (1996) Inflammatory bowel disease: germs or genes (comment)? *Lancet*, **347**, 1198.

Peach, S., Lock, M. R., Katz, D. *et al.* (1978) Mucosal-associated bacterial flora of the intestine in patients with Crohn's disease and in a control group. *Gut*, **19**, 1034.

Philipson, B., Brandberg, A., Jagenburg, R. *et al.* (1975) Mucosal morphology, bacteriology, and absorption in intra-abdominal ileostomy reservoir. *Scandinavian Journal of Gastroenterology*, **10**, 145.

Phillips, S. F. (1987) Biological effects of a reservoir at the end of the small bowel. *World Journal of Surgery*, **11**, 763.

Pitcher, M. and Cummings, J. H. (1996) Hydrogen sulphide: a bacterial toxin in ulcerative colitis? *Gut*, **39**, 1.

Posnett, D. N., Schmelkin, I., Burton, D. A. *et al.* (1990) T cell antigen receptor V gene usage: increases in Vβ8⁺ T cells in Crohn's disease. *Journal of Clinical Investigation*, **85**, 1770.

Rath, H. C., Herfarth, H. H., Ikeda, J. S. *et al.* (1996) Normal luminal bacteria, especially bacteroides species, mediate chronic colitis, gastritis, and arthritis in HLA-B27/human beta2 microglobulin transgenic rats. *Journal of Clinical Investigation*, **98**, 945.

Rhodes, J. B. and Kirsner, J. B. (1965) The early and late course of patients with ulcerative colitis after ileostomy and colectomy. *Surgery, Gynecology and Obstetrics*, **121**, 1303.

Robert, A. and Asano, T. (1977) Resistance of germ free rats to indomethacin-induced intestinal lesions. *Prostaglandins*, **14**, 331.

Roediger, W. E. W. (1980) Role of anaerobic bacteria in the metabolic welfare of the colonic mucosa in man. *Gut*, **21**, 793.

Roediger, W. E. (1993) Reducing sulfur compounds of the colon impair colonocyte nutrition: implications for ulcerative colitis. *Gastroenterology*, **104**, 802.

Roediger, W. E. W., Heyworth, M., Willoughby, P. *et al.* (1982) Luminal ions and short chain fatty acids as markers of functional activity of the mucosa in ulcerative colitis. *Journal of Clinical Pathology*, **35**, 323.

Rogers, M. J., Moore, R. and Cohen, J. (1985) The relationship between faecal endotoxin and faecal microflora of the C57BL. *Journal of Hygiene*, **95**, 397.

Rot, A., Henderson, L. E. and Leonard, E. J. (1986) *Staphylococcus aureus*-derived chemoattractant activity for human monocytes. *Journal of Leukocyte Biology*, **40**, 43.

Rot, A., Henderson, L. E., Sowder, R. *et al.* (1989) *Staphylococcus aureus* tetrapeptide with high chemotactic potency and efficacy for human leukocytes. *Journal of Leukocyte Biology*, **45**, 114.

Rowbotham, D. S., Mapstone, N. P., Trejdosiewicz, L. K. *et al.* (1995) *Mycobacterium paratuberculosis* DNA not detected in Crohn's disease tissue by fluorescent polymerase chain reaction. *Gut*, **37**, 660.

Rudolph, U., Finegold, M. J., Rich, S. S. *et al.* (1995) Gi2 alpha protein deficiency: a model of inflammatory bowel disease. *Journal of Clinical Immunology*, **15**, 101.

Ruseler-van Embden, J. G. and Both-Patoir, H. C. (1983) Anaerobic gram-negative faecal flora in patients with Crohn's disease and healthy subjects. *Antonie van Leeuwenhoek*, **49**, 125.

Ruseler-van Embden, J. and van Lieshout, L. (1987) Increased faecal glycosidases in patients with Crohn's disease. *Digestion*, **37**, 43.

Rutgeerts, P., Geboes, K., Vantrappen, G. *et al.* (1990) Predictability of the post-operative course of Crohn's disease. *Gastroenterology*, **99**, 956.

Rutgeerts, P., Geboes, K., Peeters, M. *et al.* (1991) Effect of faecal stream diversion on recurrence of Crohn's disease in the neoterminal ileum. *Lancet*, **338**, 771.

Rutgeerts, P., D'Haens, G., Hiele, M. *et al.* (1994) Appendectomy protects against ulcerative colitis. *Gastroenterology*, **106**, 1251.

Sadlack, B., Merz, H., Schorle, H. *et al.* (1993) Ulcerative colitis-like disease in mice with a disrupted interleukin-2 gene. *Cell*, **75**, 253.

Sanderson, J. D., Moss, M. T., Tizard, M. L. V. *et al.* (1992) *Mycobacterium paratuberculosis* DNA in Crohn's disease tissue. *Gut*, **33**, 890.

Santavirta, J., Mattila, J., Kokki, M. *et al.* (1991) Mucosal morphology and faecal bacteriology after ileoanal anastomosis. *International Journal of Colorectal Diseases*, **6**, 38.

Sartor, R., Rath, H. and Sellon, R. (1996) Microbial factors in chronic intestinal inflammation. *Current Opinion in Gastroenterology*, **12**, 327.

Sartor, R. B., Bond, T. M., Compton, K. Y. *et al.* (1986) Intestinal absorption of bacterial cell wall polymers in rats. *Advances in Experimental Medicine and Biology*, **216A**, 835.

Satsangi, J., Grootscholten, C., Holt, H. *et al.* (1996) Clinical patterns of familial inflammatory bowel disease. *Gut*, **38**, 738.

Savage, D. C. (1970) Associations of indigenous microorganisms with gastrointestinal mucosal epithelia. *American Journal of Clinical Nutrition*, **23**, 1495.

Saverymuttu, S., Hodgson, H. J. F. and Chadwick, V. S. (1985) Controlled trial comparing prednisolone with an elemental diet plus non-absorbable antibiotics in active Crohn's disease. *Gut*, **26**, 994.

Scheurlen, C., Kruis, W., Spengler, U. *et al.* (1988) Crohn's disease is frequently complicated by Giardiasis. *Scandinavian Journal of Gastroenterology*, **23**, 833.

Schiffman, E., Corcoran, B. A. and Wahl, S. M. (1975) N-formylmethionyl peptides as chemoattractants for leucocytes. *Proceedings of the National Academy of Sciences of the USA*, **72**, 1059.

Schiffman, E., Corcoran, B. A., Ward, P. A. *et al.* (1975) The isolation and partial purification of neutrophil chemotactic factors from *Escherichia coli*. *Journal of Immunology*, **114**, 1831.

Schleifer, K. H. and Krause, R. M. (1971) The immunochemistry of peptidoglycan. *Journal of Biological Chemistry*, **264**, 986.

Sedgwick, D., Drummond, J., Clarke, J. *et al.* (1990) Workload implications of the relentless increase in incidence of Crohn's disease. *Gut*, **31**, A1205.

Seneca, H. and Henderson, E. (1950) Normal intestinal bacteria in ulcerative colitis. *Gastroenterology*, **15**, 34.

Shafii, A., Sopher, S., Lev, M. *et al.* (1981) An antibody against revertant forms of cell-wall deficient bacterial variant in sera from patients with Crohn's disease. *Lancet*, **ii**, 332.

Soper, N. J., Orkin, B. A., Kelly, K. A. *et al.* (1989) Gastrointestinal transit after proctocolectomy with ileal pouch-anal anastomosis or ileostomy. *Journal of Surgical Research*, **46**, 300.

Stainsby, K. J. (1993) Antibodies to *Mycobacterium paratuberculosis* and nine species of environmental mycobacteria in Crohn's disease and control subjects. *Gut*, **34**, 371.

Stryker, S. J., Borody, T. J., Phillips, S. F. *et al.* (1985) Motility of the small intestine after proctocolectomy and ileal pouch-anal anastomosis. *Annals of Surgery*, **201**, 351.

Styrt, B. (1989) Species variation in neutrophil biochemistry and function. *Journal of Leukocyte Biology*, **46**, 63.

Tabaqchali, S., O'Donoghue, D. P. and Bettelheim, K. A. (1978) *Escherichia coli* antibodies in patients with inflammatory bowel disease. *Gut*, **19**, 108.

Tanaka, K., Wilks, M., Coates, P. J. *et al.* (1991) *Mycobacterium paratuberculosis* and Crohn's disease. *Gut*, **32**, 43.

Thaylor-Robison, S., Miles, R., Whitehead, A. *et al.* (1989) *Salmonella* infection and ulcerative colitis. *Lancet*, **i**, 1145.

Thompson, N., Montgomery, S., Pounder, R. *et al.* (1995) Is measles vaccination a risk factor for inflammatory bowel disease? *Lancet,* **345**, 1071.

Thompson, N. P., Driscoll, R., Pounder, R. E. *et al.* (1996) Genetics versus environment in inflammatory bowel disease: results of a British twin study. *British Medical Journal,* **312**, 95.

Tysk, C., Lindberg, E., Jarnerot, G. *et al.* (1988) Ulcerative colitis and Crohn's disease in an unselected population of monozygotic and dizygotic twins. A study of heritability and the influence of smoking. *Gut,* **29**, 990.

van de Merwe, J. P. and Mol, G. J. (1980) A possible role of *Eubacterium* and *Peptostreptococcus* species in the aetiology of Crohn's disease. *Antonie van Leeuwenhoek,* **46**, 587.

van de Merwe, J. P. and Stegeman, J. H. (1985) Binding *Coprococcus comes* to the Fc portion of IgG. A possible role in the pathogenesis of Crohn's disease? *European Journal of Immunology,* **15**, 860.

van de Merwe, J. P., Stegeman, J. H. and Hazenberg, M. P. (1983) The resident faecal flora is determined by genetic characteristics of the host. Implications for Crohn's disease? *Antonie van Leeuwenhoek,* **49**, 119.

van de Merwe, J. P., Schroder, A. M., Wensinck, F. *et al.* (1988) The obligate anaerobic faecal flora of patients with Crohn's disease and their first-degree relatives. *Scandinavian Journal of Gastroenterology,* **23**, 1125.

van Kruiningen, H., Colombel, J., Cartun, R. *et al.* (1992) An in-depth study of Crohn's disease in two French families. *Gastroenterology,* **103**, 351.

Videla, S., Vilaseca, J., Guarner, F. *et al.* (1994) Role of intestinal microflora in chronic inflammation and ulceration of the rat colon. *Gut,* **35**, 1090.

von Ritter, C., Sekizuka, E., Grisham, M. B. *et al.* (1988) The chemotactic peptide N-formyl methionyl-leucyl-phenylalanine increases mucosal permeability in the distal ileum of the rat. *Gastroenterology,* **95**, 651.

Wade, J. C., de Jongh, C. A., Newman, K. A. *et al.* (1983) Selective antimicrobial modulation as prophylaxis against infection during granulocytopenia: trimethoprim–sulfamethoxazole vs nalidixic acid. *Journal of Infectious Diseases,* **147**, 624.

Wakefield, A. J., Pittilo, R. M., Sim, R. *et al.* (1993) Evidence of persistent measles virus infection in Crohn's disease. *Journal of Medical Virology,* **39**, 345.

Ward, P. A., Lepow, I. H. and Newman, L. J. (1968) Bacterial factors chemotactic for polymorphonuclear leukocytes. *American Journal of Pathology,* **52**, 725.

Weber, P., Koch, M., Heizmann, W. *et al.* (1992) Microbic superinfection in relapse of inflammatory bowel disease. *Journal of Clinical Gastroenterology,* **14**, 302.

Wellemann, W., Flink, P., Benner, F. *et al.* (1986) Endotoxinaemia in active Crohn's disease. Treatment with whole gut irrigation and 5 aminosalicylic acid. *Gut,* **27**, 177.

Wells, C. (1990) Relationship between intestinal microecology and the translocation of intestinal bacteria. *Antonie van Leeuwenhoek,* **58**, 87.

Wells, C., Maddaus, M. and Simmons, R. (1987) Role of the macrophage in the translocation of intestinal bacteria. *Archives of Surgery,* **122**, 48.

Wells, C., Maddaus, M., Jechorek, R. *et al.* (1988a) Role of intestinal anaerobic bacteria in colonization resistance. *European Journal of Clinical Microbiology and Infectious Diseases,* **7**, 107.

Wells, C., Maddaus, M. A. and Simmons, R. L. (1988b) Proposed mechanisms for the translocation of intestinal bacteria. *Reviews of Infectious Diseases,* **10**, 958.

Wensinck, F. and van de Merwe, J. P. (1981) Serum agglutinins to *Eubacterium* and *Peptostreptococcus* species in Crohn's and other diseases. *Journal of Hygiene,* **87**, 13.

Wensinck, F., Custers-Van, L., Poppelaars-Kustermans, P. A. *et al.* (1981) The faecal flora of patients with Crohn's disease. *Journal of Hygiene*, **87**, 1.

Wensinck, F., van de Merwe, J. and Mayberry, J. (1983) An international study of agglutinins to *Eubacterium*, *Peptostreptococcus* and *Coprococcus* species in Crohn's disease, ulcerative colitis and control subjects. *Digestion*, **27**, 63.

West, B., Lendrum, R., Hill, M. *et al.* (1974) Effects of sulphasalazine (Salazopyrin) on faecal flora in patients with inflammatory bowel disease. *Gut*, **15**, 960.

Wirostko, E., Johnson, L. and Wirostko, B. (1990) Ulcerative colitis associated chronic uveitis. Parasitization of intraocular leucocytes by mollicute-like organisms. *Journal of Submicroscopic Cytology and Pathology*, **22**, 231.

Woodhouse, A. F., Anderson, R. P., Myers, D. B. *et al.* (1987) Intestinal absorption, metabolism and effects of bacterial chemotactic peptides in rat intestine. *Journal of Gastroenterology and Hepatology*, **2**, 35.

Yamada, M., Ohkusa, T. and Okayasu, I. (1992) Occurrence of dysplasia and adenocarcinoma after experimental chronic ulcerative colitis in hamsters induced by dextran sulphate sodium. *Gut*, **33**, 1521.

Yamada, T., Deitch, E., Specian, R. *et al.* (1993) Mechanisms of acute and chronic intestinal inflammation induced by indomethacin. *Inflammation*, **17**, 641.

Yang, H., Shohat, T. and Rotter, J. I. (1992) The genetics of inflammatory bowel disease, in *Current Topics in Gastroenterology. Inflammatory Bowel Disease*, (eds R. P. MacDermott and W. F. Stenson), Elsevier, New York, p. 17.

Yoshimura, H. H., Graham, D. Y., Estes, M. K. *et al.* (1987) Investigation of association of mycobacteria with inflammatory bowel disease by nucleic acid hybridization. *Journal of Clinical Microbiology*, **25**, 45.

9

Mucin degradation and its significance in inflammatory conditions of the gastrointestinal tract

Anthony M. Roberton and Anthony P. Corfield

9.1 MUCUS AND MUCINS

A protective mucus gel layer covers the surface of the gastrointestinal tract. The main structural component of mucus is the mucins, large, heavily glycosylated glycoproteins that form gels when sufficiently concentrated (Allen, 1981a). Native mucus gels contain 2–10% mucin (mucus glycoprotein) dry weight, most of the balance being water. Other constituents present in mucus from the surface of the gastrointestinal tract are proteins, nucleic acids, lipids, sloughed epithelial cells and bacteria (Allen and Hoskins, 1988).

9.1.1 Roles of the mucus barrier

The mucus gel is at the interface between the delicate epithelial cells lining the gut and the gut lumen, which contains many aggressive or damaging agents. One major role of the mucus is to protect the epithelium from damage by these luminal factors, the nature of which changes with each region of the gut.

In the stomach, the damaging factors include acid, pepsin and hypo- or hyperosmotic conditions. The mucus layer maintains the pH gradient

G. W. Tannock (ed.), *Medical Importance of the Normal Microflora*, 222–261.
© 1999 *Kluwer Academic Publishers. Printed in Great Britain.*

between the cell surface and the lumen acid and acts as a primary barrier. In the small intestine the mucus layer protects against damage by the pancreatic proteases and high-molecular-weight toxins if present. Pancreatic proteases are still present in the large intestine, and in addition there are large numbers of bacteria, some of which produce potentially harmful enzymes and antigens. The mucus layer is believed to protect against bacterial invasion of the host tissue, but much remains to be understood about the relationship between colonic bacteria and the colon mucus layer. The mucin in the mucus is potentially an excellent source of energy, but since excessive destruction of the mucus layer would damage the host and might destroy the habitat for the bacteria, a 'mutualistic symbiosis' must have evolved, resulting in 'peaceful co-existence' between the bacteria and the host (Luckey, 1972).

Other roles for the mucus barrier have been suggested. These include lubrication of food particles, enhanced uptake of fat digestion products by acidic mucin-mediated dissociation of mixed micelles (Shiau, Kelemen and Reed, 1990) and protection against sudden dehydration due to osmotic changes. The surface of the mucus layer is hydrophobic (Goddard, Kao and Lichtenberger, 1990) because of the association of lipids with mucins (Witas *et al.*, 1983) and the configuration of the carbohydrate chains (Sundari, Raman and Balasubramanian, 1991). All molecules being secreted or absorbed must pass through the mucus layer and, because of the latter's viscosity, hydrophobic properties and acidic groups, there is potential for differential rates of passage and adsorption.

Attention has recently focussed on the trefoil peptides, a family of small proteins that are secreted by the gastrointestinal mucosae (Sands and Podolsky, 1996). They are involved in the process of wound healing and mucosal restitution, and are associated with mucins in the adherent mucus gel. They may have a protective function at the intestinal mucosal surface, preserving the integrity of the mucus barrier (Sands and Podolsky, 1996). This is likely to be an area of continuing interest for understanding mucin stability. In addition, a mitogenic activity of trefoil peptides has been identified, which plays an important role in the recovery of the mucosa after damage (Dignass *et al.*, 1994).

9.1.2 Structure and thickness of the mucus barrier

Mucus forms a continuous insoluble gel, adherent to the mucosal surface. The thickness of the mucus varies with location in the human: 50–450 μm in stomach, 10–250 μm overlying the villi in the ileum (Allen and Hoskins, 1988) and 107 ± 48 μm, 134 ± 68 μm and 155 ± 54 μm in proximal colon, distal colon and rectum respectively (Pullan *et al.*, 1994). Isolated and purified mucins form reconstituted viscoelastic gels at similar concentrations of mucin to those present in native mucus gels (gastric, duodenal, small intestinal and colonic mucus contain approximately 50, 40, 30 and 20 mg mucin/ml; Allen and Hoskins, 1988). The covalent polymeric structure of the mucin is an important feature of mucus gels, since either exposure to protease or reduction by mercaptoethanol causes the gel to collapse. The gel is sufficiently dense that breakdown by protease results in progressive removal of the mucus only from the surface (Bell *et al.*, 1985). The carbohydrate side chains of the mucin help protect the polypeptide core from proteases (Variyam and Hoskins, 1983). Non-covalent interactions between carbohydrate side chains by relatively stable interdigitation contributes to the viscoelastic gelling properties (Sellers *et al.*, 1988). The secreted mucus layer adheres to the mucosal membrane, probably by interaction with the glycolipids and the membrane-bound glycoproteins and mucins present in the glycocalyx (van Klinken *et al.*, 1995).

Mucins are large glycoproteins. Two categories of mucins exist at the mucosal surface: those that are anchored to the membrane and those that are secreted. The relative amounts of carbohydrate and protein present varies for mucins from different locations but, partly because of glycosylation differences and partly due to the presence of different apomucins within the mucus gel, the carbohydrate can constitute 70–85% of the mass (Roberton *et al.*, 1989; Levine *et al.*, 1987). Additionally, genetic heterogeneity exists within mucin types, and the length of carbohydrate chains may vary in disease. Thus, values for carbohydrate and protein compositions should be taken as guides rather than fixed values.

Secreted mucins are very high-molecular-weight molecules (with the exception of MUC7), usually present as linear polymers. Their polymerized molecular size is difficult to determine because degradation by proteases and shear forces is hard to prevent during isolation. Values up to 30×10^6 Da have been proposed. The size of mucin monomers is usually 0.25–2×10^6 Da, and these are polymerized in the native state by end to end disulphide bonds.

9.1.3 Mucin genes, their expression and mucin chain peptide domains

A useful analogy for mucin structure is the bottle brush model (Allen, 1981b), where the protein core is represented by the central wire and the oligosaccharides by the bristles. Amino acid sequences of the protein cores, deduced from cDNA sequences, have identified regions of tandem repeat sequences. The number and sequence of amino acids in each tandem repeat are used to distinguish between different apomucins (Gum, 1995). The tandem repeat regions of mucins have a high content of threonine, proline and often serine, are heavily glycosylated and thus

Table 9.1 Amino acid sequences of the tandem repeat regions of human mucins MUC1 to MUC7; the number of amino acids in each tandem repeat sequence is given in brackets

Mucin	Amino acid sequence of the tandem repeat	Reference
MUC1	PGSTAPPAHGVTSAPDTRPA (20)	Gendler *et al.*, 1990
MUC2	PTTTPITTTTTVTPTPTPTGTQT (23) PPTTTPSPPTTTTTTP (imperfectly conserved repeats) (length 7–40, average 16)	Gum *et al.*, 1989
MUC3	HSTPSFTSSITTTETTS (17)	Gum *et al.*, 1990
MUC4	TSSA(V)STGHATP(S)LPVTD (16)	Porchet *et al.*, 1991
MUC5 AC	TTSTTSAP (8)	Aubert *et al.*, 1991
MUC6	SPFSSTGPMTATSFQTTTTYPTPSHPQTTL PTHVPPFSTSLVTPSTGTVITPTHAQMAT SASIHSTPTGTIPPPTTLKATGSTHTAPPM TPTTSGTSQAHSSFSTAKTSTSLHSHTSST HHPEVTPTSTTTITPNPTSTGTSTPVA HTTSATSSRLPTPFTTHSPPTGS (169)	Toribara *et al.*, 1993
MUC7	TTAAPPTPSATTPAPPSSSAPPE (23)	Bobeck *et al.*, 1993

are resistant to proteases. More than one type of tandem repeat region can occur within a mucin core and the number of tandem repeat units often shows variance between individuals. The tandem repeat sequences of the first seven mucins are shown in Table 9.1.

The structure of MUC2 is discussed as an example of a secreted mucin because it is the only high-molecular-weight human mucin for which the protein core has been completely sequenced and because it is the major mucin expressed in the intestines (Fig. 9.1).

The most common allele of MUC2 contains about 100 tandem repeat sequences in the middle of the molecule, and this will yield a synthesized mucin protein chain of 5179 amino acids. The different domains within the mucin protein have been summarized by Gum *et al.* (1994). At the N-terminus is a cysteine-rich domain of about 1380 amino acids, which includes three subdomains with sequence homologies to each other and to the D-domains found in prepro–von-Willebrand protein. This region contains few hydroxylated amino acids and will have low glycosylation. Adjacent to this is a small (347 amino acid) region that is likely to be highly glycosylated, containing imperfectly conserved repeat sequences of variable length, rich in hydroxy amino acids (59%) and proline (36%). A short (148 amino acid) subdomain rich in cysteine separates this variable length repeat region from the central part of the molecule comprising the regular 23 amino acid tandem repeats. This contains approximately 2300 amino acids, although the number depends on the

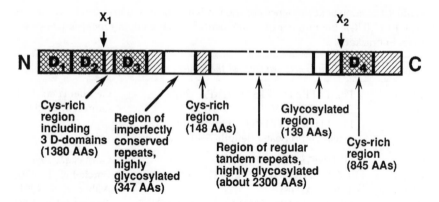

Figure 9.1 Structure of MUC2 mucin. Regions rich in hydroxy amino acids that are highly glycosylated (unfilled areas), cysteine-rich sequences (striped areas) and the cysteine-rich sequences showing homology to the D-domains of prepro–von-Willebrand protein (cross-hatched areas showing the letter D) are shown. N and C are the N-terminus and C-terminus. X1 and X2 are possible post-translational modification cleavage sites. The latter would release the so-called 'link-peptide' if disulphide intra- and interchain bonds become reduced. Subunit multimerization is dependent on disulphide interchain bonds. Note that only about 54% of the protein chain is classified as 'highly glycosylated', the remainder containing a much lower hydroxy amino acid content.

allele being expressed, with proline (22%) and threonine (55%) being the predominant amino acids. It is a region of heavy O-glycosylation. On the C-terminal side of the tandem repeats is a small (139 amino acids) threonine-, serine- and proline-rich sequence, followed by a cysteine-rich domain of 845 amino acids at the C-terminus. This includes a region with sequence homology to the D-domains of prepro–von-Willebrand protein.

It is speculated that during the multimerization and processing of the mucin subunits to form covalently linked polymers by disulphide interchain formation between regions of high cysteine content, the protein chain may be cleaved near the C-terminus, creating a glyco-peptide previously designated the mucin 'link peptide'. Also, by analogy with von Willebrand protein processing, a cleavage may occur near the N-terminus, before secretion as a multimer of mucin subunits (Fig. 9.1).

MUC1 is a membrane-bound mucin, which is anchored to the membrane by a trans-membrane domain. It is found in the glycocalyx at the apical surface of epithelial cells, but is not present as multimers linked by disulphide bridges. Like other mucins it contains a tandem repeat region that is highly glycosylated. The number of repeats is hypervariable between individuals (Swallow *et al.*, 1987). The function of MUC1 is to act at the cell surface during cell recognition and adhesion.

9.1.4 Mucin carbohydrate chains

The *MUC*-gene-encoded tandem repeat peptide sequences, rich in serine and threonine, are the sites of attachment for the majority of the O-linked oligosaccharide chains typical of mucins. Each serine and threonine residue is a potential linkage site for N-acetylgalactosamine, the first sugar in each chain. Each tandem repeat sequence has the potential for attachment of large numbers of oligosaccharide chains and total carbohydrate content is amplified by the repetition of these sequences. The O-linked oligosaccharides contain a restricted number of monosaccharides, including galactose (Gal), fucose (Fuc), N-acetylgalactosamine (GalNAc), N-acetylglucosamine (GlcNAc) and sialic acids, but with no uronic acids. Sulphate groups may be attached to the oligosaccharides. Small though significant amounts of mannose are also measurable in mucins (Carlstedt *et al.*, 1985; Forstner and Forstner, 1994). The oligosaccharides vary in length and branching, and their charges range from neutral to highly acidic, depending on the presence of sialic acid or sulphate. Three regions – the core, the backbone and the periphery – can be identified in oligosaccharides, as shown in Fig. 9.2.

The identification and structural properties of each of these regions are important as their complete degradation is required if mucin carbohydrate is to be reused by the host or bacterial flora. Enzymes specific for cleavage of these structures are part of the total degrading potential needed for normal mucin turnover.

The core sequences link the polypeptide to the extension of the oligosaccharide chain and are shown in Table 9.2. Only cores 1–3 are commonly found.

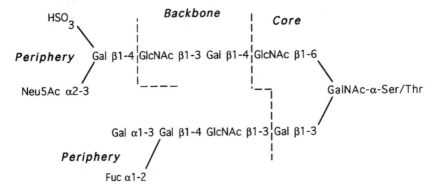

Figure 9.2 Regions of O-linked oligosaccharide structure. The three regions of O-linked oligosaccharide structure, the core, backbone, and periphery, are shown in a hypothetical example. The branched structure shows a class 2 core, a single Gal-GlcNAc backbone unit with peripheral units of acidic sialic acid and sulphate residues linked to galactose, and a peripheral blood group B antigen.

Table 9.2 Structures of mucin O-linked oligosaccharide cores. The linkage of the oligosaccharide chains on to MUC peptide serine and threonine residues occurs as characteristic core structures of which there are eight or more. Note that classes 2, 4 and 7 give rise to branched oligosaccharides. Class 1 is by far the most common core found in nature

Core class	Structure	Abundance
1	GalNAc-α-O-Ser/Thr-R / Gal β1-3	Widespread
2	GlcNAc β1-6 \ GalNAc-α-O-Ser/Thr-R / Gal β1-3	Common
3	GalNAc-α-O-Ser/Thr-R / G1NAcβ1-3	Limited
4	GlcNAcβ1-6 \ GalNAc-α-O-Ser/Thr-R / GlcNAc β1-3	Limited
5	GalNAc-α-O-Ser/Thr-R / GalNAc α1-3	Uncommon
6	GlcNAc β1-6 \ GalNAc-α-O-Ser/Thr-R	Rare
7	Gal β1-6 \ GalNAc-α-O-Ser/Thr-R / Gal β1-3	Rare
8	GalNAc α1-6 \ GalNAc-α-O-Ser/Thr-R	Rare

The presence of backbone structures has been identified in many large O-linked oligosaccharides. These are based on repeated Gal–GlcNAc structures of backbone type 1 (Gal β1-3 GlcNAc) and type 2 (Gal β1-4 GlcNAc). Poly N-acetyllactosamino-glycans (Gal β1-4 GlcNAc β1-3-)$_n$ are recognized by anti-blood-group-i antibodies, while branched chains of the type

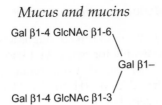

are also found and react with anti-blood-group-I antibodies (Hounsell, Davies and Renouf, 1996; Schachter and Brockhausen, 1992).

Table 9.3 Peripheral groups found in O-linked mucin type oligosaccharides. The terminal non-reducing structures found in oligosaccharides are varied. They exhibit neutral and acidic characters and contain informational structures such as the ABH and Lewis blood group antigens. A small selection of such structures is shown, to illustrate both the variety and similarities commonly found in mucin oligosaccharides, which bacteria have to recognize

Blood group H (type 1 chain)	Gal β1-3 GlcNAc- / Fuc α1-2
Blood group H (type 2 chain)	Gal β1-4 GlcNAc- / Fuc α1-2
Blood group A (type 1 chain)	GalNAc α1-3 Gal β1-3 GlcNAc- / Fuc α1-2
Lewis^a blood group	Gal β1-3 GlcNAc- / Fuc α1-4
Lewis^x blood group	Gal β1-4 GlcNAc- / Fuc α1-3
Sialyl-Lewis^x blood group	Neu5Ac α2-3 Gal β1-4 GlcNAc- / Fuc α1-3
6-sulpho-Lewis^x blood group	HSO$_3$ \ 6Gal β1-4 GlcNAc- / Fuc α1-3
6-sulpho, 3-sialyl type 2 chain	HSO$_3$ \ ^{6}Gal β1-4 GlcNAc- / Neu5Ac α2-3
6-sialyl-N-acetylgalactosamine (sialyl Tn)	Neu5Ac α2-6 GalNAc-α-O-Ser/Thr-R

Peripheral glycosylation of O-linked oligosaccharides is varied and includes structures responsible for many well known antigenic determinants (Table 9.3) such as ABH-blood-group-related sequences and a range of sialylated and sulphated units (Hounsell, Davies and Renouf, 1996; Schachter and Brockhausen, 1992).

The peripheral glycosylation of oligosaccharide chains is important as these residues must be removed by bacterial enzymes before the inner part of the chain can be degraded (Hoskins *et al.*, 1985). This implies a potential regulatory role for these terminal residues in the overall rate of oligosaccharide breakdown. Apart from properties conferred on these molecules by the large proportion of carbohydrate, precise structural information is contained in many oligosaccharides, as is demonstrated by the blood group antigens being recognized by receptors.

It is important to note that the glycosylation of each MUC polypeptide is different from tissue to tissue, and this constitutes an additional feature of mucin structural complexity (Roberton *et al.*, 1989). The tissue-specific nature of mucin glycosylation is clearly related to the functional requirements at each mucosal site (Corfield and Warren, 1996).

The high saccharide content of mucins confers the typical physico-chemical properties of high-molecular-weight carbohydrate polymers. In addition there is a considerable expression of specific and unique carbohydrate structural information. This is implicated in many individual biological functions such as the ligands identified for many cell adhesion molecules present in the glycocalyx (van Klinken *et al.*, 1995), specific structures acting as ligands for bacterial adhesins (Lasky, 1995), and receptors for viruses (Yolken *et al.*, 1994).

As each mucin contains oligosaccharides typical for its site of expression it is important to identify these structures in order to link them with function at their specific tissue location. Many of the oligosaccharide structures identified occur in different mucins and in other glycoconjugates (Roussel and Lamblin, 1996). The sharing of carbohydrate epitopes with other molecules has made the identification of individual mucin glycosylation patterns more difficult.

During biosynthesis, the addition of N-linked oligosaccharide chains to the protein chain targets the mucin precursors to the correct subcellular compartments for subsequent O-glycosylation (Dekker, Van Beurden-Lammers and Strous, 1989). In addition it is assumed that there is a separation of different mucins into vesicles destined for constitutive or stimulated secretion pathways. This has been demonstrated in cultured colonic cells (McCool, Forstner and Forstner, 1995).

Individual oligosaccharide sidechains on mucins vary from two to 12 or more monosaccharides (Podolsky, 1985a,b). Knowledge of the final oligosaccharide structures is necessary to understand the mechanisms needed for the degradative processes orchestrated by bacteria.

9.1.5 Sulphomucins

Two major regions of the human digestive tract that secrete sulpho-mucins are the mouth and colon. These regions also contain large populations of microorganisms. Historically mucins have been categorized as sulphomucins on the basis of their stainability in tissue sections by high iron diamine pH 1.5, Alcian blue pH 1, or high iron diamine and Alcian blue pH 2.5. However the term sulphomucin does not imply that sulpho-mucins may contain only sulphate and no sialic acid. It is more accurate to picture mucins as containing mixtures of sialic acid and sulphate groups, or a paucity of acidic groups, and the terms sulphomucin, sialo-mucin and neutral mucin as reflecting which of these three features predominate. Indeed individual oligosaccharide chains can contain a mixture of acidic groups (Mawhinney *et al.*, 1992a; Lo-Guidice *et al.*, 1994).

Difficulties occur when one attempts to chemically define a sulpho-mucin in terms of a minimum sulphate content. Mucins with a higher sulphate content come from regions where significant 'sulphomucin' staining can be demonstrated; however, highly purified human small intestinal mucins, which are classified histochemically as sialomucins, give sulphate analyses between 0.45 and 1.6 g sulphate per 100 g of mucin (Wesley, Forstner and Forstner, 1983; Mantle and Allen, 1989). When purified pig stomach mucin from the fundus region was incubated with a purified mucin-desulphating glycosulphatase from *Prevotella* strain RS2 (section 9.2.4), the loss of HID stainability was greater than the loss of sulphate, suggesting that the HID stain is qualitative rather than quantitative for sulphate content (Roberton and Wright, 1997).

The designation 'sulphomucin' may become better definable when more is known about the effects of different positional substitution of sulphate on sugars, the effects of sulphating different sugars, and the effects of where the sulphated sugars are attached to the oligosaccharide chain.

A variety of sulphated sugar structures are found in mucins. These structures are likely to be important in understanding the degradation of sulphated oligosaccharides by bacterial enzymes, as well as conferring different properties such as binding specificity. The main sulphated sugars reported in human (bronchial) mucin include N-acetyl glucosa-mine-6-sulphate, galactose-6-sulphate, galactose-3-sulphate and perhaps galactose-4-sulphate (Mawhinney *et al.*, 1987, 1992a, b; Lo-Guidice *et al.*, 1994; Lamblin *et al.*, 1991). N-acetyl glucosamine-6-sulphate has been reported in pig and rat stomach (Slomiany and Meyer, 1972; Liau and Horowitz, 1982), and galactose-6-sulphate in rat stomach (Liau and Horowitz, 1982). There is presently little information on the nature of sulphated sugars in colon mucin, although a human colonic cancer cell line synthesized mucin containing galactose-3-sulphate (Capon *et al.*, 1992), which was also present in meconium (Capon *et al.*, 1989).

9.1.6 Sialomucins

Acidic mucins from saliva, small intestine and colon contain mucin fractions that are rich in sialic acids. A family of over 30 different sialic acids has been identified (Schauer *et al.*, 1995). Many of these occur in gastrointestinal mucins. As noted above, the terms sialo-, neutral and sulphomucins are not strictly accurate, and sialomucin preparations will also contain neutral and sulphated oligosaccharides. Preparations of total mucins can be separated into mucin subtypes that correspond to the sialo-, sulpho- and neutral mucin definitions from histological work. Whole mucus samples have usually been fractionated by ion-exchange chromatography (e.g. Podolsky and Isselbacher, 1984; Thornton *et al.*, 1995) enabling separation of the less acidic sialomucins from the more strongly bound sulphomucins.

In saliva high- and low-molecular-weight mucins have been detected. The high-molecular-weight mucin set (MG1) is comprised of a mixture of sialo- and sulphomucins. Sialomucins are secreted by the submandibular and sublingual glands and contain 20% and 12% sialic acid and 1% and 2% sulphate of mucin dry weight respectively. Oligosaccharide chains vary in length from two to 20 sugar residues and contain both acidic and neutral structures. The lower-molecular-weight mucin (MG2, specified by the *MUC7* gene) is a sialomucin with only short (di- and trisaccharide) chains and no sulphate.

Small intestinal mucins from both human and rat have been isolated showing the patterns of oligosaccharides typical for sialomucins. Mucins from human autopsy specimens and ileal conduits usually contain a high proportion of sialic acid (range 1.4–34 mol sialic acid per 100 mol carbohydrate) and in addition lower amounts of sulphate (4.1–7.6 mol sulphate per 100 mol carbohydrate; Wesley *et al.*, 1983; Roberton *et al.*, 1991). In rat small intestinal insoluble mucin two glycopeptides derived from muc2 were found to contain roughly equivalent sialylated and neutral oligo-saccharides (40% each) and a smaller fraction of sulphated chains (Carl-stedt *et al.*, 1993; Karlsson *et al.*, 1996).

Colonic mucins have been fractionated by discontinuous ion exchange into five or six different species, showing varying proportions of sialic acid and sulphate and demonstrating that sialomucins comprise only a part of the total mucin subfractions (Podolsky and Isselbacher, 1984). Examination of the oligosaccharide chains from these mucins confirmed the presence of a population of sialylated structures (Podolsky, 1985a, b).

9.1.7 Neutral mucins

The mucins isolated from human, pig (antrum) and rat stomachs have confirmed the predominantly neutral character of these mucins

(Slomiany and Meyer, 1972; Allen, 1981a). The presence of small but significant proportions of sialic acid and sulphate has been identified (Liau and Horowitz, 1982). The oligosaccharide chains are characteristically large, branched and neutral (Slomiany and Meyer, 1973). In colon mucins as noted above a small population of mucins with low sialic acid and sulphate can be identified on ion-exchange chromatography that correspond to the definition of neutral mucin subtype. As separation of mucins on the basis of their charge has not been carried out for all regions of the gastrointestinal tract, classification has relied on the histological staining techniques. It must be assumed that neutral mucins exist, albeit as minor components in mucus from these sites.

9.1.8 Mucin histology

Mucus is produced at mucosal surfaces throughout the gastrointestinal tract. The histology of the colorectum will be emphasized in this account because bacterial degradation of mucin mainly occurs in this region.

In the oral cavity mucus is synthesized by mucus acinar cells in the paired submandibular, sublingual, palatal and other minor glands. These cells secrete mucins into ducts that drain into the oral cavity.

Mucus is secreted in the stomach by two types of cell: the mucous epithelial cells at the epithelial surface, and the mucous neck cells of the glands. These cell types synthesize different apomucins (Ho *et al.*, 1995).

The main site of mucus production in the colon and rectum is the goblet cells, although small quantities of mucus are produced by columnar epithelial cells. The histology of this region and the limits of the specificity of mucin-, lectin- and immunohistochemical techniques has been reviewed by Jass and Roberton (1994).

The histochemical classification of mucins into neutral, sialo- and sulphomucins can be made using combinations and variations of three classical staining methods – the periodic-acid–Schiff (PAS) stain, Alcian blue (AB) stain and high iron diamine (HID) stain. PAS predominantly reacts with the more terminal sugars and side branches of the oligosaccharide chains, giving a magenta colour. Variations in the periodate concentration determine which sugars can react. AB gives a deep blue stain at pH 2.5 with sialic acid, and HID a brown/purple stain with sulphate groups. Use of the HID/AB pH 2.5 stain mixture identifies mucins that are predominantly sulphated or predominantly sialated. As noted previously (section 9.1.5) mucins often contain both these acidic groups and the final colour is presumably one of competition between the two stains. The characteristics of mucus produced depend on the site within a region, the position of the goblet cell in an intestinal crypt, age and genetic constitution of the individual. This means that the mucins present in the surface mucus gel layer may be a mixture of several different types in any one area, and bacterial enzymes may be able to degrade some but not others of the mucin molecules and oligosaccharide chains. It also implies that when colon mucus is collected from the whole colon of one individual, it will contain a mixture of the mucin types found in this organ.

(a) Acidic mucins

In the distal colon and rectum, the goblet cells at and near the surface stain blue for sialomucin, while the goblet cells lower in the crypts stain for sulphomucin (Jass and Roberton, 1994). In the proximal colon sialomucins are present in the lower part of the crypt, while sulphomucins occur in the goblet cells of the upper crypt (Filipe, 1979). Sialic acids can be classified into two subgroups. O-acetylated sialic acids can be strongly stained with PAS if the O-acetyl groups are first removed by KOH pretreatment. In most people the goblet cells stain only weakly with diastase (to remove glycogen)/PAS (dPAS), indicating that the sialic acid present is O-acetylated. In the Caucasian population about 9% of people express non-O-acetylated sialic acid on the mucin chains, because they are homozygous for lack of O-acetyl transferase. Non-O-acetylated sialic acid can be stained with mild PAS stain, containing low concentrations of periodate. Monoclonal antibodies (mAbs) have been prepared that react with O-acetylated sialic acids, giving similar distribution patterns to those found by histochemical techniques (Jass and Roberton, 1994; Corfield and Warren, 1996).

(b) Carbohydrate groups of mucins

The structure and termini of the carbohydrate chains may be determined by the pattern of binding of various lectins and mAbs. Great care must be taken that the reagents have sufficient specificity and are not interfered with by other groups such as sialic acids. Histology based on the detectable groups has been used to demonstrate site differences in the mucins produced, precancerous and cancerous changes, and age and developmental alterations to the mucins. For example, the expression of blood group antigens A, B, and H_2 and Lewis[b] in secretors, reflecting the terminal region of the mucin oligosaccharide chain, are limited to the proximal colon and absent in the distal colon and rectum in adults. In contrast Lewis[a] antigen is expressed throughout the colon.

(c) Mucin protein expression in stomach and intestines

The distribution of mucin proteins can be elucidated by use of polyclonal and mAbs directed against the repetitive region of the nascent protein backbone, and by mRNA hybridization by Northern blot or *in situ* hybridization of tissues. MUC1 is a membrane-associated mucin, and is sparsely expressed in stomach, small intestine and colon epithelium forming part of the surface glycocalyx. MUC2 is the predominant mucin expressed by the goblet cells of the small intestine and colon. MUC3 is produced mainly by the columnar cells and in lesser amounts by the goblet cells of the intestines. In the small intestine it is present in the villous compartment while in the colon it is located at the mucosal surface and in the upper part of the crypts (Ho *et al.*, 1996). MUC4 is expressed in the small intestine and colon (Porchet *et al.*, 1991; Audie *et al.*, 1993). Stomach mucin expression is quite different. MUC 5AC is localized in the gastric surface mucus cells, while MUC6 is expressed in the mucous neck cells of the cardia and antrum (Toribara *et al.*, 1993; Ho *et al.*, 1995; Carlstedt *et al.*, 1995). In the mouth, low-molecular-weight salivary mucin (MUC7) is found in mucous acinar cells.

Great efforts have been made to detect mucin changes that precede or accompany development of colorectal carcinoma. It is important to recognize that the changes may not be due to changes in the pathways of goblet cell mucin synthesis. Cells of goblet lineage may not be represented in the carcinoma, and mucin-like material present may include up-regulated mucin-like material that is normally expressed in small quantities along the apical membrane of crypt base columnar cells (Ajioka, Allison and Jass, 1996).

9.2 MUCIN DEGRADATION

9.2.1 The concept of bacterial 'mucinase'

The concept that bacterial enzymes can degrade mucin in the colon has been accepted for several decades. Vercellotti *et al.* (1977) demonstrated the loss of sugars characteristic of mucins from the high-molecular-weight fraction in the colon as compared with that from the ileum, indicating that mucins were degraded by bacteria. Other early studies aimed at measuring mucin breakdown were carried out using pure or mixed cultures and commercially available mucin. An overall property classified as 'mucinase' was determined using various criteria, including volatile fatty acid production (Salyers *et al.*, 1977), degradation of the mucin judged by loss of precipitation properties (van der Wiel-Korstanje and Winkler, 1975) or loss of mucin protein and mucin carbohydrates (Variyam and Hoskins, 1981).

Hoskins and coworkers were the first group to introduce a number of more sophisticated criteria for assessing whether isolated strains of bacteria could degrade mucins. These included loss of blood group antigenicity, loss of mucin sugars that could be specifically measured and measurement of glycosidase activities using model carbohydrate substrates (Hoskins *et al.*, 1985).

It has now become clear that mucin breakdown by bacteria is a very complex process requiring many diverse enzymes for full degradation. Some parts of the molecule have to be removed before others become accessible, enzymes may need to be induced, not all the enzymes are necessarily present in a single strain of bacteria, and mucin breakdown is most easily catalysed by cooperation between several types of bacteria. Mucin gels may behave differently than soluble mucins, and mucins from different regions or from diseased tissue show different resistance to breakdown. The bacterial enzymes involved in degradation may be secreted, bound to the bacterial outer surface, present in the periplasm but still able to access external macromolecules, cytoplasmic, or released by cell lysis. Enzyme activity is thus dependent on the fractions being assayed. Furthermore, it is likely that the mucus layer is colonized by a specialized microflora, not necessarily reflective of the average faecal population. This degree of complexity means that the concept of a bacterium having 'mucinase' activity is too general.

Salyers, Valentine and Hwa (1993) have written an excellent review on polysaccharide utilization pathways in the Bacteroides, discussing in detail their common features, such as polysaccharide binding proteins, location of polysaccharide degrading enzymes in the periplasm or cytoplasm but not extracellularly in the medium, more than one enzyme possessing the same activity, and induction of degradative enzymes and necessary membrane proteins only when the bacteria are growing on the polysaccharide.

It is important to define whether the end-point of mucus degradation is the destruction of the protective properties of the gel, the weakening of the mucus gel, the destruction of binding epitopes, the loss of defence against pathogenic bacteria seeking to invade the host tissue or the availability of mucin sugar and protein as food for the bacteria.

The ability to dissect out the component parts of this puzzle will be aided by the sophisticated techniques that are becoming available for analysis of glycoprotein components (Quigley and Kelly, 1995).

9.2.2 Assay difficulties

The most appropriate substrate for studying human colonic mucin degradation is human colonic mucin. This is available from tissue culture of biopsy material in small quantities that can be made radioactive if required, or in larger quantities from autopsy material that has been obtained within a few hours of death. These sources, after isolation and purification of mucin, will give sufficient material for analytical experiments but insufficient material for growth of bacteria.

Pig gastric mucin has become the mucin of choice, where bacterial growth or induction of mucin degrading enzymes in bacteria is to be studied, usually because it is available commercially. However, pig gastric mucin and human colonic mucin have a number of important structural differences. The backbone proteins are different in sequence and size, and type 1 core is present in stomach mucin compared to types 3 and 4 core in colon (Schachter and Brockhausen, 1992). The sulphate and sialic acid levels are higher in colon mucin (Allen, 1981a; Podolsky and Isselbacher, 1983). Little O-acetylation of sialic acid occurs in the stomach mucin whereas colon mucin from most individuals contains O-acetylated sialic acids. It is not surprising that pig stomach mucin is more easily and more extensively degraded than colon mucin.

The quality of some commercial preparations of pig gastric mucins is variable between batches, with fractionation often needed to improve the content of mucin. Preparation of one's own material can be a better option. Three people working in a slaughterhouse can easily collect a litre of good quality pig gastric mucus in 3 hours. If the fundal mucus is kept separate, a highly sulphated preparation of mucin can be obtained. Pig colonic mucin is probably a better model mucin for bacterial growth experiments, but is not commercially available and the collection is more difficult, unpleasant and harder to obtain in large quantities.

Many assays for mucin-degrading enzymes have been performed with model substrates. Care should be exercised that the choice is appropriate. For example, because many sulphated compounds are present in the digestive tract, various desulphating enzymes that will not desulphate mucins can be detected using commercially available substrates for aryl sulphatases. The same may be true if non-mucin proteins are used to measure proteases in the context of mucin proteolytic activity, or if non-O-acetylated sialylated saccharides are used in the context of colon mucin desialylation.

The problems inherent in extrapolating enzyme measurements made on soluble preparations of mucins to predict their actions on mucin gels, the problem of where mucin-degrading enzymes are located with respect to the bacterial cell that synthesized them, and the problems of enzyme induction have been alluded to in section 9.2.1. In addition, a single subculture using mucin as substrate may not be informative, and many subcultures may be needed before bacteria grow optimally.

A final problem lies in the comparison of enzyme activities from different assays. When radioactive substrate is used the substrate concentration will be minute compared to that used in a spectrophotometric assay. The result is that enzyme activities from the former will exhibit much lower values, and comparisons are not possible.

9.2.3 Regional aspects of mucin degradation in the gastrointestinal tract

The degradation of the mucus barrier by the bacterial populations has focused on those regions where large numbers of bacteria are found.

Bacterial mucin degradation has mainly been studied in the oral cavity, stomach and large intestine.

(a) Oral cavity

Salivary mucins are degraded by enzymes secreted by the oral bacterial microflora (DeJong and Van der Hoeven, 1987). This is part of the overall pattern of salivary mucin function, which includes alimentation, tissue coating and modulation of oral microflora. The partial degradation of mucins by glycosidases creates or destroys oligosaccharide structures acting as receptors between the mucins, the dental pellicle and the bacteria themselves (Nieuw Amerongen, Bolscher and Veerman, 1995; Tabak, 1995). Secretion of proteolytic (Rafay, Homer and Beighton, 1996; DeJong and Van der Hoeven, 1987; Iwase *et al.*, 1992) and glycosidic enzymes (Iwase *et al.*, 1992; Rafay, Homer and Beighton, 1996) by oral bacterial strains suggests that mucins may also serve as energy substrates for oral bacteria. Synergistic action of different strains of bacteria in achieving mucin and glycoprotein degradation suggests that individual strains of oral bacteria do not necessarily produce all enzymes necessary for complete mucin degradation (Bradshaw *et al.*, 1994; Homer and Beighton, 1992).

Removal of sulphate from salivary sulphomucins (Nieuw Amerongen, Bolscher and Veerman, 1995; Tabak, 1995) is achieved by the action of carbohydrate-specific sulphatases, which have been detected in several oral bacterial strains (Smalley *et al.*, 1994). Sialidases are found in many oral bacterial strains (Corfield, 1992) and their induction by mucin together with enzymes on the catabolic pathway implies that sialic acids are used for energy metabolism in the oral cavity (Rafay, Homer and Beighton, 1996).

(b) Stomach

The normal stomach receives bacteria from the mouth and food but in humans is thought not to support a normal microflora. However studies on the stomach-colonizing pathogen *Helicobacter pylori*, and its association with gastric and duodenal ulceration has been a major area of research (McGowan, Cover and Blaser, 1996). Evidence for 'mucinase' activity with an accompanying reduction in mucus viscosity has been presented (Hashiguchi, 1993; Sarosiek, Slomiany and Slomiany, 1987) along with actions of *H. pylori* proteases (Slomiany *et al.*, 1987) and sulphatase (Slomiany *et al.*, 1992) on gastric mucin. Others have been unable to identify either a loss in viscosity (Markesich *et al.*, 1994) or protease activity (Oliver *et al.*, 1997). The effect of ammonia formed by secreted *H. pylori* urease has been proposed as an alternative mechanism for destabilization of the mucus layer and disruption of mucin biosynthesis (Sidebotham and Baron, 1990). Very recent work (Oliver *et al.*, 1997) revealed no significant reduction in the mucus barrier thickness in human gastric antrum of patients infected with *H. pylori*; however, a 20% reduction in gel-forming polymeric mucin was found. No mucolytic proteinase activity was detected, and only tiny reductions in viscosity of mucin isolated from infected individuals could be measured. *In vitro* studies on the effect of ammonia at concentrations that might be generated locally by *H. pylori*, and *in vitro* incubations of *H. pylori* in the presence of urea with isolated mucin revealed significant loss of viscosity and depolymerization of mucins. Thus a local rise in pH *in vivo*, due to *H. pylori* urease, may be partly responsible for a localized loss of gel structure in the immediate areas of colonization.

(c) Large intestine

The large intestine is the main habitat for bacterial colonization in the gastrointestinal tract, and the complex factors affecting the microbiology of the mucus biofilm are now being integrated into an overall picture. Mucin degradation by bacteria is one of five parameters characteristic for the switch from germfree to microflora-associated characteristics (Midtvedt, Carlstedt-Duke and Midtvedt, 1994).

The appearance of a mucin-degrading intestinal microflora between 3 and 12 months of age has been followed by staining patterns of electrophoresed faecal samples (Midtvedt, Carlstedt-Duke and Midtvedt, 1994). Establishment of a complete adult-type flora may take up to 20 months (Midtvedt, Carlstedt-Duke and Midtvedt, 1994). The overall degradation of mucin occurs by the action of sulphatases, glycosidases and peptidases secreted by synergistic mixtures of bacterial or by mucin oligosaccharide degrader (MOD) strains (Hoskins *et al.*, 1985, 1992).

Study of individual bacterial strains has shown a variation of mucin-degrading activity. This ranges from a *Peptostreptococcus* strain from rat intestine that produced peptidases but no glycosidases, but showed mucin-degrading activity (Carlstedt-Duke *et al.*, 1986), to the MOD strains exhibiting largely glycosidase activities (Corfield *et al.*, 1992a; Hoskins *et al.*, 1985, 1992). The secreted mucins synthesized by the colorectal mucosa show a low sulphate content at birth, rising after 3 months, and increasing to approach adult levels over 3–16 years (Aslam, Spicer and Corfield, 1997).

The initial colonization of the neonatal gut by bacteria metabolizing the milk and milk products in the infant diet may not be linked with a requirement to degrade mucins. Part of the protection necessary at the mucosal surface may be afforded by the oligosaccharides present in high concentration in breast milk (Zopf and Roth, 1996). As the gut adapts to the changes in diet following milk feeding new bacterial strains begin to colonize and new enzymes appear. The inability of *Lactobacillus* and *Bifidobacterium* strains to degrade intestinal mucins reflects their link with the lactation phase after birth and indicates a role for other bacteria in mucus degradation (Ruseler van Embden *et al.*, 1995). The pattern of milk oligosaccharides is known to change during the lactation period and up to the initiation of weaning. The duration of breast feeding in infants results in a significant delay in the appearance of a mucin-degrading microflora (Midtvedt, Carlstedt-Duke and Midtvedt, 1994).

9.2.4 Mucin desulphating glycosulphatases

There is suggestive evidence that the sulphation of mucins is an import-ant factor that rate-limits degradation of mucins. Better criteria for eval-uating the characteristics of sulphomucins and the possibility of specifically desulphating mucins with purified enzymes should provide more direct methods for evaluating this in the future.

The evidence for sulphation of mucins being protective is consistent with the observation that those regions of the digestive tract that are the habitat of large communities of bacteria secrete high levels of sulpho-mucins. *In vitro* experiments studying the effect of sulphated glycopep-tides on sialidase (Mian, Anderson and Kent, 1979a) or sulphated glycoproteins on pepsin (Mikuni-Takagaki and Hotta, 1979) showed that they inhibited the degradative enzymes.

More direct studies on degradation of three preparations of pig mucins, from the gastric antrum and fundus and from colon, by three strains of mucin-degrading colon anaerobes (Stanley *et al.*, 1986) demonstrated that the less sulphated mucin preparations were more susceptible to degrada-tion, and that sulphate removal by the bacteria allowed more carbo-hydrate removal. Much the same conclusion can be drawn from the work of Houdret *et al.* (1989) on mucin composition of sputum from patients with *Pseudomonas aeruginosa* infection of the lung. They showed that the lung neutral mucin fraction was selectively degraded compared to the sulphomucin fraction.

Experiments with crude enzymes from faecal homogenates were also consistent with sulphomucin being protective. Tsai *et al.* (1992) showed that addition of a purified bacterial mucin glycosulphatase to a faecal fraction rich in glycosidase activity facilitated more extensive mucin oligosaccharide chain degradation. Corfield *et al.* (1993) have shown that faecal extracts from ulcerative colitis patients, which contain higher activ-ities of sulphatases, degrade mucin more extensively and rapidly than faecal extracts from normal subjects. These results illustrate the import-ance of mucin sulphation in protection of mucin against degradation.

The study of bacterial sulphatases associated with mucin desulphation is complicated by the many other sulphated compounds present in the large bowel, including sulphated glycosides, choline sulphate, glycosaminoglycans, sulphated glycoproteins other than mucins, sulpholipids, bilirubin sulphate and sulphated metabolites synthesized as part of the detoxification process and secreted in bile. A variety of arylsulphatases and non-arylsulphatases is likely to be present in the colon. Thus sulphated model compounds to be used in assays for mucin-desulphating glycosulphatases need to be chosen with care, with loss of sulphate from mucin being the criteria for a role in mucin desulphation. If [^{35}S]mucin desulphation is being measured using crude extracts, it is important to exclude putative endoglycosidases as a reason for separation of the label from the mucin.

The existence of enzymes that will partially desulphate mucins has been demonstrated in a number of bacteria. These include the colonic bacterium *Bacteroides thetaiotaomicron* and *Bacteroides fragilis* (Tsai, Hart and Rhodes, 1991; Tsai *et al.*, 1992), *Prevotella* strain RS2 isolated from pig colon, which contains at least two different mucin desulphating enzymes (Stanley *et al.*, 1986; Wilkinson and Roberton, 1988; Roberton *et al.*, 1993), and two mucin oligosaccharide degrading (MOD) strains of faecal anaerobe *Ruminococcus torques* IX-70 and *Bifidobacterium* VIII-210 (Corfield *et al.*, 1992a). A mucin desulphating glycosulphatase was reported in the extracellular material elaborated by the stomach pathogen *Helicobacter pylori* (Murty *et al.*, 1992), and Smalley *et al.* (1994) have found mucin desulphating activity in some strains of oral streptococci.

In experiments where mucin desulphation has been demonstrated in whole bacterial cells or crude extracts by partial desulphation of [^{35}S]mucin, the sugar-sulphate specificity of the desulphating enzyme(s) is unknown. Use of model substrates with identified monosaccharide-sulphate substitution is required to confirm the type of sulphatase involved.

To date only two mucin desulphating enzymes have been purified to homogeneity. *Prevotella* strain RS2 glycosulphatase is a cell-associated inducible periplasmic enzyme, which is thought to be specific for N-acetylglucosamine-6-sulphate groups in mucins and has a neutral pH optimum (Roberton *et al.*, 1993). A mucin-desulphating glycosulphatase has also been purified from the supernatant fluid of human faeces (Tsai *et al.*, 1992). It has a number of different properties to the *Prevotella* enzyme above, with a pH optimum of 4.5 and a much smaller molecular mass. Its bacterial origin is unknown.

9.2.5 Mucin degrading sialidases and deacetylases

Sialic acids occupy terminal positions in sialylated oligosaccharides from mucins. They are present as a variety of derivatives, and are linked to different sugars (Fig. 9.3).

SIALIC ACID MOLECULE

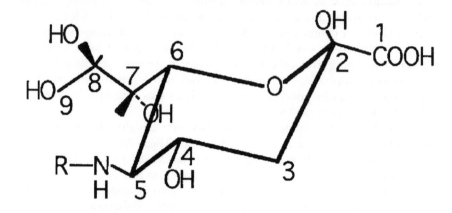

Sialic acids
C5, N-acetyl or N-glycolyl
C4 and C7-9 mono or poly-O-acetylated
Other less common substitutions, eg. methyl, lactyl

Linkage to neighboring monosaccharides
α2-3   Galactose
α2-6   Galactose, N-acetylglucosamine,
          N-acetylgalactosamine
α2-8   Sialic acids

Figure 9.3 Sialic acid linkages in oligosaccharides. The many possible combinations of sialic acid substitution in oligosaccharides found in nature are outlined in the figure. The sialic acids exist as a family of derivatives substituted at C5 amino group and at hydroxyl groups at C4 and C7–9. These individual sialic acids are linked glycosidically through different linkages to adjacent monosaccharides.

In most humans, colonic mucins contain sialic acids that are O-acetylated at positions C7–9. In order to degrade sialylated oligosaccharides, the O-acetyl esters must first be removed, followed by the sialic acids. This is achieved by the action of sialate O-acetyl esterases and sialidases (neuraminidases; Corfield, 1992).

Although studies of sialidases have focused historically on the involvement of sialidases in pathogenic processes, sialidases are also present in the normal enteric microflora, including *Bacteroides fragilis* (Russo *et al.*, 1990). The structural gene encoding bacterial sialidase, *nanH*, has been cloned from a number of pathogenic and non-pathogenic bacteria. Enzymes in the *nan* gene cluster in *Escherichia coli* include: sialic acid permease, required for the uptake of sialic acids into the bacterial cells; acylneuraminate pyruvate lyase (aldolase), necessary for the cleavage of sialic acids into N-acetylmannosamine and pyruvate; and other enzymes in the metabolic pathways for the conversion of N-acetylhexosamines into glucose (Vimr and Troy, 1985). Thus, mucin-bound sialic acids can be cleaved, recovered and converted to glucose (Corfield, 1992; Vimr and Troy, 1985).

The occurrence of sialidase activity in enteric bacterial strains is widespread among numerically dominant anaerobes (e.g. *Bacteroides* spp.), facultative anaerobes (e.g. *Escherichia coli* and *Enterococcus faecalis*) and MOD bacteria (Corfield, 1992; Hoskins *et al.*, 1985). The last group generally show higher sialidase activities than other enteric bacteria. The sialidase activities in faecal extracts and enteric bacterial strains show high mucin-desialylating activity, present as cell bound and secreted proteins in variable proportions (Corfield *et al.*, 1992a, 1993; Hoskins *et al.*, 1985; Tanaka, Ito and Iwasaki, 1992).

Most bacterial sialidases show decreased sialic acids removal if O-acetyl esters are present. The presence of two or three O-acetyl esters at positions C7–9 is more inhibitory. O-acetyl esters in colonic mucins thus have a significant inhibitory effect on sialic acid removal by enteric bacterial sialidases (Corfield *et al.*, 1992a). The discovery of a bacterial sialate O-acetylesterase in faecal extracts (Corfield *et al.*, 1988; Poon *et al.*, 1983) showed the mechanism by which sialidase inhibition could be released. The relatively low levels of O-acetylesterase activity compared with sialidase suggested a possible regulatory role of this enzyme on the rate of O-acetyl sialic acid removal and hence overall mucin degradation. The bacterial sialate O-acetyl esterase has not yet been purified. It should be noted that people lacking O-acetylation of sialic acid suffer no known consequences.

The methods used to assay sialidases require careful consideration. Many workers have used the 4-methyl-umbelliferyl- and 4-nitrophenyl-alpha-glycosides of sialic acid because of their commercial availability and the ease and sensitivity of the assays. However, these substrates do not discriminate between different natural sialoglycoconjugate substrate activities and in some cases do not appear to represent any natural substrate activity (Scharfman *et al.*, 1991). This is significant in the gut where serum-, membrane- and mucin-type glycoproteins are potential substrates. Specific sialoglycoproteins must be selected to allow discrimination to be made. Accordingly a range of natural sialoglycoproteins have been used to characterize sialidase assays (Corfield, 1992).

9.2.6 Mucin degrading glycosidases

The degradation of mucin oligosaccharide chains is completed by a variety of hydrolytic enzymes that have specificity for individual monosaccharides in glycosidic linkage to adjacent sugars in oligosaccharide chains. The glycosidases that carry out these functions have been studied largely from pathogenic bacteria (Roussel and Lamblin, 1996). It has not yet been ascertained to what extent the glycosidases produced by normal gut microflora have the same properties. Certainly the sialidases show important differences.

Two main types of glycosidase have been identified: exo- and endo-enzymes. Exoglycosidases cleave at the non-reducing terminus of an oligosaccharide chain, removing one sugar at a time, while endoglycosidases cleave at sites within the oligosaccharide chain. Glycosidases acting on O-linked oligosaccharides can be divided into groups relevant to the region of the oligosaccharide chain (Fig. 9.2, Table 9.2, Table 9.3).

(a) Removal of oligosaccharides from peptide

Enzymes that remove complete O-linked oligosaccharides from mucin glycoproteins or glycopeptides are termed O-glycanases (peptide N-acetyl-α-galactosaminidases). They are specific for the GalNAc-α-O-serine/threonine linkage, but have not been widely detected. Enzymic activity has been detected removing the terminal GalNAc alone or the disaccharide Gal β1–3 GalNAc (oligosaccharide core 1, Table 9.2) in *Streptococcus pneumoniae* (Glasgow, Paulson and Hill, 1977) and *Ruminococcus torques* (Hoskins and Corfield, unpublished data). These enzymes are inactive on substrates that contain the Gal β1–3 GalNAc-α-O-ser/thr core unit substituted with additional sugars, and there are no reports of enzymes from normal enteric bacterial sources that will release extended, sialylated, fucosylated or sulphated oligosaccharides from mucins.

(b) Removal of peripheral residues

Apart from the sialidases and sulphatases discussed above, other peripheral monosaccharide residues must be removed before the remaining oligosaccharide chains can be degraded (Table 9.3). These enzymes include N-acetyl-α-galactosaminidases and α-galactosidases acting on blood group A and B respectively. The presence of these enzymes in many enteric bacteria has been demonstrated, either by direct assay of the enzyme (Hoskins *et al.*, 1985) or by a haemagglutination inhibition assay measuring loss of blood group activity (Ruseler van Embden *et al.*, 1992).

Destruction of the H- and Lewis-blood-group antigens requires the action of α-fucosidases. Although hydrolysis of synthetic 4-nitrophenyl and 4-methyl umbelliferyl glycosides of α-Fuc have been widely detected in enteric bacterial strains, there are few reports showing specific removal of Fuc in α1–2, α1–3 and α1–4 linkage to Gal or GlcNAc residues in natural substrates.

Terminal β-galactose residues appear to be readily removed by abundant β1–4-galactosidase activity in MOD and numerically dominant strains (Hoskins *et al.*, 1985), as measured with both synthetic and natural substrates. The relative abundance of β1–4-galactosidase together with N-acetylhexosaminidase (N-acetyl-β-1-3-glucosaminidase) in MOD strains (Hoskins *et al.*, 1985) and faecal samples (Corfield *et al.*, 1993) may account for the ability to degrade poly N-acetyl lactosamine units present in some extended mucin oligosaccharide chains. N-acetyl-β-hexosaminidase activities have been detected using synthetic substrates containing β-GlcNAc or β-GalNAc, but the specificities for neither of these hexosamines have been investigated in detail.

(c) Cleavage of poly-N-acetyllactosamine backbone structures

The cleavage of extended oligosaccharide chains containing repeats of (Gal β1–4 GlcNAc β1–3)$_n$ is catalysed by endo-β-galactosidases acting on 'internal' glycosidic linkages in poly-N-acetyllactosamine backbone structures. Typically, the enzymes from *Citrobacter freundii* (Fukuda and Matsumura, 1975) and *Bacteroides fragilis* (Scudder *et al.*, 1984) cleave the internal Gal β1–4 linkage in the oligosaccharide sequence: -Gal β1–4 GlcNAc β1–3 Gal β1–4 GlcNAc β1–3-, provided that the Gal residue has no substitution at C6. Fucosylation of the GlcNAc at the linkage site reduces the rate of cleavage of the Gal β1–4 linkage. The role of endogalactosidases in mucin degradation has not been evaluated. The concerted action of galactosidase and glucosaminidase, as noted above, may complete the degradation of these backbones.

(d) Cleavage of oligosaccharide core structures

The most common core structures requiring degradation in gastrointestinal mucins are based on Gal β1–3 GalNAc α-O (core 1) and GlcNAc β1–3 GalNAc α-O (core 3; Table 9.2). The degradation products of pig gastric mucin by MOD bacterial strains showed the disaccharide Gal β1–3 GalNAc among the final products, indicating the absence of a β1–3 galactosidase activity (Hoskins *et al.*, 1992). Gastric mucin does not contain GlcNAc β1–3 GalNAc α-O, and core 3 specific N-acetyl-β1-3-glucosaminidase has not been identified yet (because no suitable substrate is available).

(e) Linkages which are not readily cleaved

The substrate specificities of purified glycosidases have shown that certain substituted structures are cleaved slowly or not at all. Such specificity implies regulation and preferred pathways of degradation. Although such properties have been examined in only a few cases, and many enteric bacterial glycosidases remain to be examined, it is useful to consider an example of the type of specificity shown by these enzymes.

The β-galactosidase from *Escherichia coli* cleaves β1–4 linked galactose in lactose in preference to Gal β1–4 GlcNAc or glycoproteins containing Gal β1–4 GlcNAc in N-linked oligosaccharides. The enzyme also shows good activity against Gal β1–3 GlcNAc present in milk oligosaccharides (Miller *et al.*, 1994). The substitution of an adjacent branch sugar prevents the cleavage. For example: Gal β1–3 GlcNAc β1–3 Gal β1–4 Glc is cleaved; and Gal β1–3 (Fuc α1–4)GlcNAc β1–3 Gal β1–4 Glc is not cleaved.

Another example is the influence of sulphate substitution on glycosidase action. The action of both β-galactosidases and N-acetyl-β-hexosaminidases is reduced by the presence of sulphate esters at positions 3 and 6 on galactose, N-acetylglucosamine or N-acetylgalactosamine residues in oligosaccharides and in sulphated mucins (Kent *et al.*, 1978). The influence of mucin oligosaccharide sulphation has been shown for β-galactosidase (Mian, Anderson and Kent, 1979b), but remains to be correlated with the action of enteric bacterial degradation of gastrointestinal mucins. The interaction of sulphates with these enzymes deserves further study.

9.2.7 Proteinases

Faecal bacterial extracts contain a variety of proteases, both cell-associated and secreted. These proteases are believed to be important in the solubilization and weakening of the mucus gel, by analogy with the solubilization by pepsin and acid of the stomach mucin gel, although no comparable experiments have been published using colon mucus gels and faecal extracts.

Macfarlane *et al.* (1988) have shown that *Bacteroides-fragilis*-type organisms contain largely cell-bound proteolytic activity, while a smaller proportion of proteolytic activity in faeces, derived from *Enterococcus faecalis*, *Propionibacterium* and *Clostridium*, was secreted and extracellular. Mucin addition caused up to 40% decrease in faecal bacterial cell-associated proteases, and a much greater increase in extracellular proteases, as measured by azocasein hydrolysis (Macfarlane, Hay and Gibson, 1989).

Hutton *et al.* (1990) have made faecal protease measurements physiologically relevant to mucin, by using colonic mucins as substrate and measuring mucin protein breakdown by three criteria – increased N-terminal groups, decreased mucin viscosity and decreased molecular weight distribution of mucin fragments. However, they centrifuged the faecal extracts to obtain the source of enzyme, removing the cell-associated proteases from these experiments. Even so, they found mucolytic activity by each criterion, indicating that the potential exists for solubilization, weakening and degradation of mucus gels.

Recently Macfarlane, McBain and Macfarlane (1996) have examined the proteolytic activity of mucus-colonizing faecal bacteria, made by suspending mucus gels in a chemostat and inoculating with faecal suspensions. These mucus-colonizing bacteria produced proteolytic activities, as assayed with native protein substrates and chromogenic substrates.

Critical experiments remain to be done in order to clarify our understanding of the breakdown of mucins in the mucus gel layer. Firstly, mucolytic experiments must be done with mucus gels (of colonic origins), in addition to soluble mucins, and the chronology of glycosidase and proteolytic events determined. No simultaneous assessment of proteolytic and glycolytic cleavages has been made to date. It needs to be shown whether predominantly cell-bound or secreted proteases (and glycosidases) are involved. It would be preferable that such experiments, involving live bacteria, are conducted anaerobically under a partial CO_2 atmosphere, as many saccharolytic anaerobes, including the Bacteroides, have an absolute requirement for 10–30% CO_2 for growth and metabolism (Caldwell, Keeney and van Soest, 1969). The ability of certain gut anaerobes to transport large peptide products to the inside of the cell prior to metabolism (Pittman, Lakshmanan and Bryant, 1967) also needs to be further examined, since such products will not necessarily give rise to increased N-terminal groups in the medium. The known division of colonic mucins into sulphomucins and sialomucins will also need to be considered in case they have different susceptibilities to degradation (Mikuni-Takagaki and Hotta, 1979).

The above experimental work on mucolytic proteolysis is of considerable importance when inflammatory bowel diseases are studied in relation to the effectiveness of the mucus barrier. Preliminary work indicates that faecal protease activities are increased in ulcerative colitis and Crohn's patients (Samson *et al.*, 1991; Corfield *et al.*, 1988), and the hypothesis that a weakened and less effective mucus gel contributes to a lower barrier effect requires determination of mucin gel quality and the relative levels of bound or secreted mucolytic, proteolytic and glycolytic enzymes interacting with the secreted gel layer.

9.2.8 Volatile short-chain fatty acids

Consistent with the mutual symbiosis concept of the bacterial relationship with the host, the mucosal fermentation products from many bacteria are beneficial to colon health. Acetate increases colonic blood flow and ileal motility (Scheppach, 1994) and butyrate is a major energy source for colonocytes, has a strong differentiating action and slows the growth rate of colon cancer cells (Hague *et al.*, 1993). Butyrate irrigation has been successfully used as a treatment to reduce inflammation in distal ulcerative colitis (Scheppach *et al.*, 1992) and it can also increase mucin biosynthesis *in vitro* in normal and ulcerative colitis tissue (Finnie *et al.*, 1995).

9.3 INFLAMMATORY CONDITIONS OF THE GASTROINTESTINAL TRACT

Correlation of the sites of inflammatory bowel diseases with the location of enteric bacterial colonization in the intestinal tract has led to speculation that their aetiology is linked. However, neither ulcerative colitis nor Crohn's disease has been associated with a particular bacterial agent (Chadwick and Anderson, 1995). The case for bacteriological involvement either as a primary or secondary result of disease is presented in detail elsewhere in this volume and is discussed here relative to the mucin degrading potential of bacterial populations.

The consideration of 'mucinase' assays and the structure of mucin oligosaccharide chains referred to earlier in this chapter are of special relevance to these diseases. The correct physiological and clinical interpretation of results is vital for real progress in the understanding of these diseases. Evaluation of limitations in the methods used has led to the identification of several important areas for future research.

9.3.1 Crohn's disease

The nature of the adherent mucus barrier in Crohn's disease differs from that seen in normal individuals and ulcerative colitis patients (Hodgson and Bhatti, 1995). A thicker gel layer has been identified by direct measurement, compared with normal controls (Pullan *et al.*, 1994) and an increase in the rate of mucin synthesis has been described (Ryder *et al.*, 1995). Examination of the nature of the mucins secreted in Crohn's disease by metabolic labelling methods has shown that modification of mucin carbohydrate occurs (Raouf *et al.*, 1992; Ryder *et al.*, 1995; Smith and Podolsky, 1987). Qualitative details of the nature of this change are not available; however, some indications for the loss of mucin sulphation have been presented (Raouf *et al.*, 1992; Ryder *et al.*, 1995).

The alterations in the nature of the secreted mucins in Crohn's disease may lead to a change in their susceptibility to degradation by enteric bacterial enzymes. However, there is no direct evidence for this yet.

Examination of faecal 'mucinase' activity (measured as a size reduction in colonic mucin after incubation) in patients with Crohn's disease showed no significant difference compared to normal controls, in contrast to increased levels found in ulcerative colitis (Dwarakanath *et al.*, 1995). Examination of mucin-degrading enzymes showed no significant differences for α- and β-galactosidase, sialidase, N-acetyl-β-glucosaminidase and α-fucosidase with synthetic and natural substrates (Corfield *et al.*, 1988; Rhodes *et al.*, 1985b; Ruseler van Embden and van Lieshout, 1987). Increased acylneuraminate pyruvate lyase in Crohn's patients suggests a possible role for sialic acid metabolism in the disease (Corfield *et al.*, 1988).

Degradation of mucins by faecal proteinases in Crohn's disease may be increased. A significantly higher level of faecal proteolytic activity was found compared with normal controls using an azocasein substrate (Corfield *et al.*, 1988). Examination of additional substrates confirmed this result and identified increased leucine aminopeptidase in stool samples from patients with Crohn's disease (Ruseler van Embden and van Lieshout, 1987).

9.3.2 Ulcerative colitis

Ulcerative colitis (UC) is associated with a reduction in the thickness of the adherent mucus gel layer at the surface of the colorectal mucosa (Pullan *et al.*, 1994). The reasons for this loss may be due to altered synthesis and secretion of the mucins and/or increased bacterial degradation. Investigations of the nature of mucins synthesized and secreted by the colorectal mucosa have shown that a major change is associated with the sulphation of the oligosaccharide chains. This has been demonstrated using metabolic labelling (Corfield *et al.*, 1996; Corfield *et al.*, 1992b; Raouf, *et al.*, 1992) and chemical analysis of sulphate in isolated mucins (Corfield *et al.*, 1996). The loss of sulphation may be related to the severity of the disease or a genetic factor, as a group of South Asian colitics with mild disease showed no loss of sulphation (Probert *et al.*, 1995).

Other results have shown that a number of changes in glycosylation of colorectal mucin may be associated with UC (Boland *et al.*, 1982; Podolsky and Isselbacher, 1984). Metabolic labelling experiments have shown that increases in total sialic acid levels occur in UC (Corfield *et al.*, 1996; Parker *et al.*, 1995; Raouf *et al.*, 1992) and reduced sialic acid O-acetylation has been reported based on histological (Reid *et al.*, 1984) and biochemical studies (Corfield *et al.*, 1993). It is not certain whether these results may be due to the naturally occurring absence of sialic acid O-acetylation found in approximately 9% of Europeans (Jass and Roberton, 1994; Corfield and Warren, 1996), as this level of natural occurrence appears in UC patients.

In addition to the qualitative changes, evidence for a reduction in the rate of mucin synthesis in UC has emerged (Corfield *et al.*, 1996; Probert *et al.*, 1995; Ryder *et al.*, 1995).

These results suggest that the nature and amount of mucin produced in UC is not adequate to provide normal viscoelastic properties and resistance to bacterial degradative processes present in the colorectum. Such changes in mucin structure would make these molecules more susceptible to degradation by the normal spectrum of secreted bacterial enzymes.

Studies of total 'mucinase' activity have been carried out using radio-labelled mucin substrates and have demonstrated an elevated activity associated with the disease (Corfield *et al.*, 1993; Dwarakanath *et al.*, 1995). In addition, there is evidence supporting higher levels of glycosidase activities, which are thought to be involved in the regulation of the rate of mucin breakdown and would result in increased mucin loss as a result. A number of early studies on UC yielded inconclusive results most likely due to the use of synthetic substrates (Rhodes *et al.*, 1985a, b). Further studies using physiologically relevant substrates for sialidase and the detection of an O-acetylesterase enzyme active with O-acetylated sialic acids led to the identification of a pathway for the cleavage of O-acety-lated sialic acids (Corfield *et al.*, 1988). This entailed initial removal of O-acetyl esters by the esterase followed by sialidase action to release the sialic acids. Examination of the sialidase activity in UC showed that there was no difference between controls and disease using a mucin oligosac-charide substrate, but that the activity was significantly higher than the O-acetyl esterase (Corfield *et al.*, 1988, 1992b, 1993). Furthermore, the level of O-acetyl esterase was higher in UC patients compared with controls. Thus, the increase in O-acetyl esterase activity would allow desialylation to proceed more rapidly and facilitate an increased rate of mucin degradation.

The comparison of sulphatase activities in ulcerative colitis patients using mucin relevant substrates including lactitol 6-O-sulphate (Corfield *et al.*, 1993) and [^{35}S]-radiolabelled mucins (Corfield *et al.*, 1993; Tsai *et al.*, 1992, 1995) showed increased activity. Here the combination of reduced sulphation of the mucin in the colon and the increased sulphatase activity are expected to cause a significant increase in mucin degradation in UC and affect both the physicochemical behaviour and rate of loss of adher-ent mucus gel at the mucosal surface.

Increased activity of bacterial proteinase activity in UC patients has been described (Corfield *et al.*, 1988; Samson *et al.*, 1991). Initial studies using azocasein as a substrate demonstrated an eightfold increase in proteolytic activity in faecal extracts from UC patients compared with normal controls (Corfield *et al.*, 1988). A closer analysis of this observation confirmed the finding in UC using a succinyl albumin substrate (Samson *et al.*, 1991).

Recent interest in restorative proctocolectomy with ileoanal pouch for treatment of long-term or severe UC has focused attention on 'pouchitis', an inflammatory condition of the pouch resembling UC itself. Examination of the pouch contents has shown a distinctive microflora and pattern of glycosidase expression (Ruseler van Embden, Schouten and van Lieshout, 1994; Ruseler van Embden *et al.*, 1992). Ten different glycosidase activities were assayed with synthetic substrates in canine pouches and found to increase on occlusion together with increased total bacterial numbers. However no increase in protease activities was observed under these conditions.

Correlation of the sites of inflammatory bowel diseases with the location of enteric bacterial colonization in the intestinal tract has led to speculation that their aetiology is linked (Chadwick and Anderson, 1995). However, neither UC nor Crohn's disease have been associated with a particular bacterial agent (Chadwick and Anderson, 1995).

Studies on the degradation of mucins by bacterial 'mucinase' activity in inflammatory bowel disease and colorectal cancer have been limited in number. This is partly due to the absence of a clear aetiological basis for these diseases (Chadwick and Anderson, 1995). The case for bacteriological involvement either as a primary or secondary result of disease is presented in detail elsewhere in this volume.

9.4 FUTURE DIRECTIONS OF RESEARCH ON MUCIN DEGRADATION

Research on mucins is in a phase of rapid expansion. Mucin degradation is still a relatively young field, and there are many aspects that need to become focussed and integrated. Knowledge is needed in diverse research areas – microbiology, enzymology, analytical biochemistry, 'mucinology', pathology and molecular biology, with the following areas particularly in need of investigation.

There is a need to understand the metabolism of bacteria in and on the surface of the mucus gel biofilm, as an extension to present studies on faecal bacterial interaction with soluble mucins. Information is required on which bacteria are involved and the properties of the enzymes they produce to metabolize mucins, including their specificity, whether the enzymes are secreted or cell-associated and whether anaerobic conditions play a part and should, therefore, be simulated in studies. The regulation of enzyme production needs to be understood, to assess the levels and significance of constitutive and induced pathways for enzyme expression. At the gene level the identification of operons or gene cassettes of mucin-degrading enzymes and accompanying proteins has not been studied in the context of normal or disease situations. Does the presence of a normal microflora prevent colonization by pathogenic bacteria? Can enzymes modify epitopes on the mucin and glycocalyx, which then act as receptors for bacterial adhesins?

Are there undetected differences in the bacterial mucosal microflora associated with inflammatory bowel diseases, or just differences in bacterial enzyme expression?

The mucins themselves require further study, in particular from a structural viewpoint. The precise nature of mucin proteolysis by bacterial proteinases has not been studied at the mucin peptide level and many questions of substrate specificity remain to be answered. It seems to be particularly important to study colonic mucin, and make use of the new insights provided by the sequences of MUC proteins which are now becoming available. The purity of commercial mucins gives cause for concern and the use of gastric in place of large intestinal mucin as a substrate needs to be reassessed. Mucin from the pig stomach has been widely used as a substrate in bacterial growth, enzyme induction and degradation experiments. The relationship of sugar sequences to ordered degradation and the susceptibility of peptide regions in mature mucins containing low glycosylation to proteolytic attack have not been systematically analysed. We do not understand how the areas of low glycosylation in mucins (in MUC2 just under half of the protein) are protected by the regions of high glycosylation. The effects of mixtures of mucin apoproteins in preventing degradation has not been studied.

The question of what limits the rate of mucin breakdown has not been answered. What are the relative contributions of sulphated sugars, sialic acids and their O-acetyl esters, disulphide bond cross linking, lipid content, proteases, length of oligosaccharide chains, trefoil peptides, putative inhibitors of mucin degradative enzymes in mucus, mucin turnover and maturity, and mucin gel viscoelasticity?

This chapter draws attention to a surface defence system in dynamic equilibrium with a normal bacterial microflora. New progress has been reported in both the mucin and bacterial fields and has heightened awareness of the questions still outstanding. It is certain that during the next few years these areas will attract a great deal of attention.

9.5 REFERENCES

Ajoika, Y., Allison, L. J. and Jass, J. R. (1996) Significance of MUC1 and MUC2 mucin expression in colorectal cancer. *Journal of Clinical Pathology*, **49**, 560–564.

Allen, A. (1981a) The structure and function of gastrointestinal mucus, in *Basic Mechanisms of Gastrointestinal Mucosal Cell Injury and Protection*, (ed. J. W. Harmon), Williams & Wilkins, Baltimore, MD, pp. 351–367.

Allen, A. (1981b) Structure and function of gastrointestinal mucus, in *Physiology of the Gastrointestinal Tract*, (ed. L. R. Johnson), Raven Press, New York, pp. 617–639.

Allen, A. and Hoskins, L. C. (1988) Colonic mucus in health and disease, in *Diseases of the Colon and Rectum*, (eds R. Kirsner and R. G. Shortes), Williams & Wilkins, Baltimore, MD, pp. 65–94.

Aslam, A., Spicer, R. D. and Corfield, A. P. (1997) Biochemical analysis of mucin glycoproteins in paediatric colonic mucus. *Biochemical Society Transactions*, **25**, 7S.

Aubert, J. P., Porchet, N., Crepin, M. *et al.* (1991) Evidence for different human tracheobronchial mucin peptides deduced from nucleotide cDNA sequences. *American Journal of Respiratory Cell and Molecular Biology*, **5**, 178–185.

Audie, J. P., Janin, A., Porchet, N. *et al.* (1993) Expression of human mucin genes in respiratory, digestive, and reproductive tracts ascertained by *in situ* hybridization. *Journal of Histochemistry and Cytochemistry*, **41**, 1479–1485.

Bell, A. E., Sellers, L. A., Allen, A. *et al.* (1985) Properties of gastric and duodenal mucus: effect of proteolysis, disulfide reduction, bile, acid, ethanol, and hypertonicity on mucus gel structure. *Gastroenterology*, **88**, 269–280.

Bobeck, L. A., Tsai, H., Biesbrock, A. R. and Levine, M. J. (1993) Molecular cloning, sequence and specificity of expression of the gene encoding the low molecular weight human salivary mucin (MUC7). *Journal of Biological Chemistry*, **268**, 20563–20569.

Boland, C. R., Lance, P., Levin, B. *et al.* (1982) Lectin binding indicates an abnormality of the goblet cell glycoconjugates in ulcerative colitis. *Gastroenterology*, **82**, 1021–1028.

Bradshaw, D. J., Homer, K. A., Marsh, P. D. and Beighton, D. (1994) Metabolic cooperation in oral communities during growth on mucin. *Microbiology*, **140**, 3407–3412.

Caldwell, D. R., Keeney, M. and van Soest, P. J. (1969) Effects of carbon dioxide on growth and maltose fermentation by *Bacteroides amylophilus*. *Journal of Bacteriology*, **98**, 668–676.

Capon, C., Leroy, Y., Wieruszeski, J. M. *et al.* (1989) Structures of O-glycosidically linked oligosaccharides isolated from human meconium glycoproteins. *European Journal of Biochemistry*, **182**, 139–182.

Capon, C., Laboisse, C. L., Wieruszeski, J.-M. *et al.* (1992) Oligosaccharide structures of mucins secreted by the human colonic cancer cell line CL.16E. *Journal of Biological Chemistry*, **267**, 19248–19257.

Carlstedt, I., Sheehan, J. K., Corfield, A. P. and Gallagher, J. T. (1985) Mucous glycoproteins: a gel of a problem. *Essays in Biochemistry*, **20**, 40–76.

Carlstedt, I., Herrmann, A., Karlsson, H. *et al.* (1993) Characterisation of two different glycosylated domains from the insoluble mucin complex of rat small intestine. *Journal of Biological Chemistry*, **268**, 18771–18781.

Carlstedt, I., Herrman, A., Hovenberg, H. *et al.* (1995) 'Soluble' and 'insoluble' mucins – identification of distinct populations. *Biochemical Society Transactions*, **23**, 845–851.

Carlstedt-Duke, B., Midtvedt, T., Nord, C. E. and Gustafsson, B. E. (1986) Isolation and characterisation of a mucin-degrading strain of *Peptostreptococcus* from rat intestinal tract. *Acta Pathologica Microbiologica et Immunologica Scandinavica (B)*, **94**, 293–300.

Chadwick, V. S. and Anderson, R. P. (1995) The role of intestinal bacteria in etiology and maintenance of inflammatory bowel disease, in *Human Colonic Bacteria. Role in Nutrition, Physiology and Pathology*, (eds G. R. Macfarlane and G. T. Gibson), CRC Press, Boca Raton, FL, pp. 227–256.

Corfield, T. (1992) Bacterial sialidases – roles in pathogenicity and nutrition. *Glycobiology*, **2**, 509– 521.

Corfield, A. P. and Warren, B. F. (1996) The modern investigation of mucus glycoproteins and their role in gastrointestinal disease. *Journal of Pathology*, **180**, 8–17.

Corfield, A. P., Williams, A. J. K., Clamp, J. R. *et al.* (1988) Degradation by bacterial enzymes of colonic mucus from normal subjects and patients with inflammatory bowel disease: the role of sialic acid metabolism and the detection of a novel O-acetylsialic acid esterase. *Clinical Science,* **74,** 71–78.

Corfield, A. P., Wagner, S. A., Clamp, J. R. *et al.* (1992a) Mucin degradation in the human colon: production of sialidase, sialate O-acetylase, N-acetylneuraminate lyase, arylesterase, and glycosulfatase activities by strains of faecal bacteria. *Infection and Immunity,* **60,** 3971–3978.

Corfield, A. P., Warren, B. F., Bartolo, D. C. C. *et al.* (1992b) Mucin changes in ileoanal pouches monitored by metabolic labelling and histochemistry. *British Journal of Surgery,* **79,** 1209–1212.

Corfield, A. P., Wagner, S. A., O'Donnell, L. J. D. *et al.* (1993) The roles of enteric bacterial sialidase, sialate O-acetyl esterase, and glycosulfatase in the degradation of human colonic mucin. *Glycoconjugate Journal,* **10,** 72–81.

Corfield, A. P., Myerscough, N., Bradfield, N. *et al.* (1996) Colonic mucins in ulcerative colitis: evidence for loss of sulphation. *Glycoconjugate Journal,* **13,** 809–822.

DeJong, M. H. and Van der Hoeven, J. S. (1987) The growth of oral bacteria on saliva. *Journal of Dental Research,* **66,** 498–505.

Dekker, J., Van Beurden-Lammers, W. M. O. and Strous, G. J. (1989) Biosynthesis of gastric mucus glycoprotein of the rat. *Journal of Biological Chemistry,* **264,** 10431–10437.

Dignass, A., Lynch-Devaney, K., Thim, L. and Podolsky, D. K. (1994) Trefoil peptides promote epithelial migration through a transforming growth factor beta-dependent pathway. *Journal of Clinical Investigation,* **94,** 376–383.

Dwarakanath, A. D., Campbell, B. J., Tsai, H. H. *et al.* (1995) Faecal mucinase activity assessed in inflammatory bowel disease using ^{14}C threonine labelled mucin substrate. *Gut,* **37,** 58–62.

Filipe, M. I. (1979) Mucins in the human gastrointestinal epithelium: a review. *Investigative Cell Pathology,* **2,** 195–216.

Finnie, I. A., Dwarakanath, A. D., Taylor, B. A. and Rhodes, J. M. (1995) Colonic mucin synthesis is increased by sodium butyrate. *Gut,* **36,** 93–99.

Forstner, J. F. and Forstner, G. G. (1994) Gastrointestinal mucus, in *Physiology of the Digestive Tract,* (ed. L. R. Johnson), Raven Press, New York, pp. 1245–1283.

Fukuda, M. and Matsumura, G. (1975) Endo-β-galactosidase of *Escherichia freundii.* Hydrolysis of pig colonic mucin and milk oligosaccharides by endoglycosidic action. *Biochemical and Biophysical Research Communications,* **64,** 465–471.

Gendler, S. J., Lancaster, C. A., Taylor-Papadimitriou, J. *et al.* (1990) Molecular cloning and expression of human tumour-associated polymorphic epithelial mucin. *Journal of Biological Chemistry,* **265,** 15286–15293.

Glasgow, L. R., Paulson, J. C. and Hill, R. L. (1977) Systematic purification of five glycosidases from *Streptococcus (Diplococcus) pneumoniae. Journal of Biological Chemistry,* **252,** 8615–8623.

Goddard, P. J., Kao, Y. J. and Lichtenberger, L. M. (1990) Luminal surface hydrophobicity of canine gastric mucosa is dependant on a surface mucous gel. *Gastroenterology,* **98,** 361–370.

Gum, J. R. (1995) Human mucin glycoproteins: varied structures predict diverse properties and specific functions. *Biochemical Society Transactions,* **23,** 795–799.

Gum, J. R., Byrd, J. C., Hicks, J. W. *et al.* (1989) Molecular cloning of human intestinal mucin cDNAs. Sequence analysis and evidence for genetic polymorphism. *Journal of Biological Chemistry,* **264,** 6480–6487.

Gum, J. R., Hicks, J. W., Swallow, D. M. *et al.* (1990) Molecular cloning of cDNAs derived from a novel human intestinal mucin gene. *Biochemical and Biophysical Research Communications*, **171**, 407–415.

Gum, J. R., Hicks, J. W., Toribara, N. W. *et al.* (1994) Molecular cloning of human intestinal mucin (*MUC2*) cDNA. Identification of the amino terminus and over-all sequence similarity to prepro–von Willebrand factor. *Journal of Biological Chemistry*, **269**, 2440–2446.

Hague, A., Manning, A. M., Hanlon, K. *et al.* (1993) Sodium butyrate induces apoptosis in human colonic tumour cell lines in a p53 independent pathway: implications for the possible role of dietary fibre in the prevention of large bowel cancer. *International Journal of Cancer*, **55**, 498–505.

Hashiguchi, T. (1993) Mucinase activity of *Helicobacter pylori*: application of simplified mucinase test. *Nippon Rinsho – Japanese Journal of Clinical Medicine*, **51**, 3166–3169.

Ho, S. B., Roberton, A. M., Shekels, L. L. *et al.* (1995) Mucins in human gastric epithelium: isolation of a second gastric mucin gene and localisation of mucin gene expression. *Gastroenterology*, **109**, 735–747.

Ho, S. B., Ewing, S. L., Montgomery, C. K. and Kim, Y. S. (1996) Altered mucin core peptide immunoreactivity in the colon polyp-carcinoma sequence. *Oncology Research*, **8**, 53–61.

Hodgson, H. J. F. and Bhatti, M. (1995) Assessment of disease in ulcerative colitis and Crohn's disease. *Inflammatory Bowel Disease*, **1**, 117–134.

Homer, K. A. and Beighton, D. (1992) Synergistic degradation of transferrin by mutans streptococci in association with other dental plaque bacteria. *Microbial Ecology in Health and Disease*, **5**, 111–116.

Hoskins, L. C., Agustines, M., McKee, W. B. *et al.* (1985) Mucin degradation in human colon ecosystems. Isolation and properties of fecal strains that degrade ABH blood group antigens and oligosaccharides from mucin glycoproteins. *Journal of Clinical Investigation*, **75**, 944–953.

Hoskins, L. C., Boulding, E. T., Gerken, T. A. *et al.* (1992) Mucin glycoprotein degradation by mucin oligosaccharide-degrading strains of human faecal bacteria. Characterisation of saccharide cleavage products and their potential role in nutritional support of larger faecal bacterial populations. *Microbial Ecology in Health and Disease*, **5**, 193–207.

Houdret, N., Ramphal, R., Scharfman, A. *et al.* (1989) Evidence for the *in vivo* degradation of human respiratory mucins during *Pseudomonas aeruginosa* infection. *Biochimica et Biophysica Acta*, **992**, 96–105.

Hounsell, E. F., Davies, M. J. and Renouf, D. V. (1996) O-linked protein glycosylation structure and function. *Glycoconjugate Journal*, **13**, 19–26.

Hutton, D. A., Pearson, J. P., Allen, A. and Foster, S. N. F. (1990) Mucolysis of the colonic mucus barrier by faecal proteinases: inhibition by interacting polyacrylate. *Clinical Science*, **78**, 265–271.

Iwase, H., Ishikarasaka, I., Hotta, K. *et al.* (1992) Analysis of porcine gastric mucus glycoprotein added to a culture medium of *Streptomyces* sp OH-11242 as the only source of carbon. *Comparative Biochemistry and Physiology B*, **101**, 651–655.

Jass, J. R. and Roberton, A. M. (1994) Colorectal mucin histochemistry in health and disease: a critical review. *Pathology International*, **44**, 487–504.

Karlsson, N. G., Johansson, M. E., Asker, N. *et al.* (1996) Molecular characterisation of the large heavily glycosylated domain glycopeptide from the rat small intestinal Muc2 mucin. *Glycoconjugate Journal*, **13**, 823–831.

Kent, P. W., Coles, C. J., Cooper, J. R. and Mian, N. R. (1978) Sulphate ester groups as potential information regulators in glycoproteins, in *Carbohydrate Sulphates*, (ed. R. G. Schweiger), American Chemistry Society Symposium Series 77, American Chemistry Society, Washington, DC, pp. 29–43.

Lamblin, G., Rahmoune, H., Wieruszeski, J.-M. *et al.* (1991) Structure of two sulphated oligosaccharides from respiratory mucins of a patient suffering from cystic fibrosis. A fast atom bombardment m.s. and ^{1}H-n. m.r. spectroscopic study. *Biochemical Journal*, **275**, 199–206.

Lasky, L. A. (1995) Sialomucins in inflammation and hematopoiesis. *Advances in Experimental Medicine and Biology*, **376**, 259–260.

Levine, M. J., Reddy, M. S., Tabak, L. A. *et al.* (1987) Structural aspects of salivary glycoproteins. *Journal of Dental Research*, **66**, 436–441.

Liau, Y. H. and Horowitz, M. I. (1982) Incorporation *in vitro* of [^{3}H]glucosamine or [^{3}H] glucose and [^{35}S]SO_4^{2-} into rat gastric mucosa. *Journal of Biological Chemistry*, **257**, 4709–4718.

Lo-Guidice, J.-M., Wieruszeski, J.-M., Lemoine, J. *et al.* (1994) Sialylation and sulfation of the carbohydrate chains in respiratory mucins from a patient with cystic fibrosis. *Journal of Biological Chemistry*, **269**, 18794–18813.

Luckey, T. D. (1972) Introduction to intestinal microecology. *American Journal of Clinical Nutrition*, **25**, 1292–1294.

McCool, D. J., Forstner, J. F. and Forstner, G. G. (1995) Regulated and unregulated pathways for MUC2 mucin secretion in human colonic LS180 adenocarcinoma cells are distinct. *Biochemical Journal*, **312**, 125–132.

Macfarlane, G. T., Hay, S. and Gibson, G. R. (1989) Influence of mucin on glycosidase, protease and arylamidase activities of human gut bacteria grown in a three stage continuous culture system. *Journal of Applied Bacteriology*, **66**, 407–417.

Macfarlane, G. T., Allison, G., Gibson, S. A. W. and Cummings, J. H. (1988) Contribution of the microflora to proteolysis in the human large intestine. *Journal of Applied Bacteriology*, **64**, 37–46.

Macfarlane, S., McBain, A. J. and Macfarlane, G. T. (1996) Characterisation of proteolytic and peptidolytic activities in human colonic biofilm populations, in *ASM Conference on Microbial Biofilms, Salt Lake City, UT, 29 Sept.–5 Oct. 1996*, abstracts, p. 31.

McGowan, C. C., Cover, T. L. and Blaser, M. J. (1996) *Helicobacter pylori* and gastric acid: biological and therapeutic implications. *Gastroenterology*, **110**, 926–938.

Mantle, M. and Allen, A. (1989) Gastrointestinal mucins, in *Gastrointestinal Secretions*, (ed. J. S. Davison), John Wright, Bristol, pp. 202–229.

Markesich, D. C., Anand, B. S., Lew, G. M. and Graham, D. Y. (1994) *Helicobacter pylori* infection does not reduce the viscosity of human gastric mucus gel. *Gut*, **35**, 327– 329.

Mawhinney, T. P., Adelstein, E., Morris, D. A. *et al.* (1987) Structure determination of five sulfated oligosaccharides derived from tracheobronchial mucus glycoproteins. *Journal of Biological Chemistry*, **262**, 2994–3001.

Mawhinney, T. P., Landrum, D. C., Gayer, D. A. and Barbero, G. J. (1992a) Sulfated sialo-oligosaccharides derived from tracheobronchial mucous glycoproteins of a patient suffering from cystic fibrosis. *Carbohydrate Research*, **235**, 179– 197.

Mawhinney, T. P., Adelstein, E., Gayer, D. A. *et al.* (1992b) Structural analysis of monosulfated side-chain oligosaccharides isolated from human tracheobronchial mucous glycoproteins. *Carbohydrate Research*, **223**, 187–207.

Mian, N., Anderson, C. E. and Kent, P. W. (1979a) Neuraminidase inhibition by chemically sulphated glycopeptides. *Biochemical Journal*, **181**, 377–385.

Mian, N., Anderson, C. E. and Kent, P. W. (1979b) Effect of O-sulphated groups in lactose and N-acetylneuraminyl-lactose on their enzymic hydrolysis. *Biochemical Journal*, **181**, 387–399.

Midtvedt, A. C., Carlstedt-Duke, B. and Midtvedt, T. (1994) Establishment of a mucin-degrading intestinal microflora during the first two years of human life. *Journal of Pediatric Gastroenterology and Nutrition*, **18**, 321–326.

Mikuni-Takagaki, Y. and Hotta, K. (1979) Characterisation of peptic inhibitory activity associated with sulfated glycoprotein isolated from gastric mucosa. *Biochimica et Biophysica Acta*, **584**, 288–297.

Miller, J. B., McVeague, P., McNeil, Y. and Gillard, B. (1994) Human milk oligosaccharides. *Acta Paediatrica*, **83**, 1051.

Murty, V. L. N., Piotrowski, J., Morita, M. *et al.* (1992) Inhibition of *Helicobacter pylori* glycosulfatase activity toward gastric sulfomucin by nitecapone. *Biochemistry International*, **26**, 1091–1099.

Nieuw Amerongen, A. V., Bolscher, J. G. M. and Veerman, E. C. I. (1995) Salivary mucins: protective functions in relation to their diversity. *Glycobiology*, **5**, 733–740.

Oliver, L., Newton, J. L., Goddard, P. *et al.* (1997) Effects of *Helicobacter pylori* on the adherent gastric mucus barrier. *Biochemical Society Transactions*, **25**, 372s.

Parker, N., Tsai, H. H., Ryder, S. D. *et al.* (1995) Increased rate of sialylation of colonic mucin by cultured ulcerative colitis mucosal explants. *Digestion*, **56**, 52–56.

Pittman, K. A., Lakshmanan, S. and Bryant, M. P. (1967) Oligopeptide uptake by *Bacteroides ruminicola. Journal of Bacteriology*, **93**, 1499–1508.

Podolsky, D. K. (1985a) Oligosaccharide structures of human colonic mucin. *Journal of Biological Chemistry*, **260**, 8262–8271.

Podolsky, D. K. (1985b) Oligosaccharide structures of isolated human colonic mucin species. *Journal of Biological Chemistry*, **260**, 15510–15515.

Podolsky, D. K. and Isselbacher, K. J. (1983) Composition of human colonic mucin: selective alteration in inflammatory bowel disease. *Journal of Clinical Investigation*, **72**, 142–153.

Podolsky, D. K. and Isselbacher, K. J. (1984) Glycoprotein composition of colonic mucosa: specific alterations in ulcerative colitis. *Gastroenterology*, **87**, 991–998.

Poon, H., Reid, P. E., Ramey, C. W. *et al.* (1983) Removal of O-acetylated sialic acids from rat colonic epithelial glycoproteins by cell-free extracts of rat faeces. *Canadian Journal of Biochemistry and Cell Biology*, **61**, 868–874.

Porchet, N., Nguyen, V. C., Dufosse, J. *et al.* (1991) Molecular cloning and chromosomal localisation of a novel human tracheo-bronchial mucin cDNA containing tandemly repeated sequences of 48 base pairs. *Biochemical and Biophysical Research Communications*, **175**, 414–422.

Probert, C. S. J., Warren, B. F., Perry, T. *et al.* (1995) South Asian and European colitics show characteristic differences in colonic mucus glycoprotein type and turnover. Potential identification of a lower risk group for severe disease and cancer. *Gut*, **36**, 696–702.

Pullan, R. D., Thomas, G. A. O., Rhodes, M. *et al.* (1994) Thickness of adherent mucus gel on colonic mucosa in humans and its relevance to colitis. *Gut*, **35**, 353–359.

Quigley, M. E. and Kelly, S. M. (1995) Structure, function, and metabolism of host mucus glycoproteins, in *Human Colonic Bacteria. Role in Nutrition, Physiology and Pathology*, (eds G. R. Macfarlane and G. T. Gibson), CRC Press, Boca Raton, FL, pp. 175–199.

Rafay, A. M., Homer, K. A. and Beighton, D. (1996) Effect of mucin and glucose on proteolytic and glycosidic activities of *Streptococcus oralis*. *Journal of Medical Microbiology*, **44**, 409–417.

Raouf, A. H., Tsai, H. H., Parker, N. *et al.* (1992) Sulphation of colonic and rectal mucin in inflammatory bowel disease: reduced sulphation of rectal mucus in ulcerative colitis. *Clinical Science*, **83**, 623–626.

Reid, P. E., Culling, C. F. A., Dunn, W. L. *et al.* (1984) Chemical and histochemical studies of normal and diseased human gastrointestinal tract. I. A comparison between histologically normal colon, colonic tumours, ulcerative colitis and diverticular disease of the colon. *Histochemistry Journal*, **16**, 235–251.

Rhodes, J. M., Gallimore, R., Elias, E. and Kennedy, J. F. (1985a) Faecal sulphatase in health and inflammatory bowel disease. *Gut*, **26**, 466–469.

Rhodes, J. M., Gallimore, R., Elias, E. *et al.* (1985b) Faecal mucus degrading glycosidases in ulcerative colitis and Crohn's disease. *Gut*, **26**, 761–765.

Roberton, A. M. and Wright, D. P. (1997) Bacterial glycosulfatases and sulfo-mucin degradation. *Canadian Journal of Gastroenterology*, **11**, 361–366.

Roberton, A. M., Mantle, M., Fahim, R. E. F. *et al.* (1989) The putative 'link' glycopeptide associated with mucus glycoproteins: composition and properties of preparations from the gastrointestinal tracts of several mammals. *Biochemical Journal*, **261**, 637–647.

Roberton, A. M., Rabel, B., Harding, C. A. *et al.* (1991) Use of the ileal conduit as a model for studying human small intestinal mucus glycoprotein secretion. *American Journal of Physiology*, **261** (*Gastrointestinal and Liver Physiology*, **24**) G728–G734.

Roberton, A. M., McKenzie, C., Scharfe, N. and Stubbs, L. (1993) A glycosulpha-tase that removes sulphate from mucus glycoprotein. *Biochemical Journal*, **293**, 683–689.

Roussel, P. and Lamblin, G. (1996) Human mucosal mucins in diseases, in *Glyco-proteins and Disease*, vol., 1, (eds J. Montreuil, J. F. G. Vliegenthart and H. Schachter), Elsevier, Amsterdam, pp. 349–391.

Ruseler van Embden, J. G. H. and van Lieshout, L. M. C. (1987) Increased faecal glycosidases in patients with Crohn's disease. *Digestion*, **37**, 43–50.

Ruseler van Embden, J. G. H., Schouten, W. R. and van Lieshout, L. M. C. (1994) Pouchitis: result of microbial imbalance? *Gut*, **35**, 658–664.

Ruseler van Embden, J. G. H., Schouten, W. R., van Lieshout, L. M. C. and Auwerda, H. J. A. (1992) Changes in the bacterial composition and enzymic activity in ileostomy and ileal reservoir during intermittent occlusion: a study using dogs. *Applied and Environmental Microbiology*, **58**, 111–118.

Ruseler van Embden, J. G. H., van Lieshout, L. M. C., Gosselink, M. J. and Marteau, P (1995) Inability of *Lactobacillus casei* strain GG, *Lactobacillus acidophi-lus* and *Bifidobacterium bifidum* to degrade intestinal mucus glycoproteins. *Scan-dinavian Journal of Gastroenterology*, **30**, 675–680.

Russo, T. A., Thompson, J. S., Godoy, V. J. and Malamy, M. H. (1990) Cloning and expression of *Bacteroides fragilis* TA2480 neuraminidase gene nanH in *Escherichia coli*. *Journal of Bacteriology*, **172**, 2594–2600.

Ryder, S. D., Raouf, A. H., Parker, N. *et al.* (1995) Abnormal mucosal glycoprotein synthesis in inflammatory bowel diseases is not related to cigarette smoking. *Digestion*, **56**, 370–376.

Salyers, A. A., Valentine, P. and Hwa, V. (1993) Genetics of polysaccharide utilisation pathways of colonic *Bacteroides* species, in *Genetics and Molecular Biology of Anaerobic Bacteria*, (ed. M. Sebald), Brock/Springer series in Contem-porary Bioscience, Springer-Verlag, New York, pp. 505–516.

Salyers, A. A., West, S. E. H., Vercellotti, J. R. and Wilkins, T. D. (1977) Fermentation of mucins and plant polysaccharides by anaerobic bacteria from the human colon. *Applied and Environmental Microbiology*, **34**, 529–533.

Samson, H. J., Allen, A., Pearson, J. P. *et al.* (1991) Faecal proteinase activity: raised values in patients with ulcerative colitis. *Gut*, **32**, A1235.

Sands, B. E. and Podolsky, D. E. (1996) The trefoil peptide family. *Annual Review of Physiology*, **58**, 253–273.

Sarosiek, J., Slomiany, B. L. and Slomiany, A (1987) Evidence for weakening of gastric mucosal integrity by *Campylobacter pylori. Scandinavian Journal of Gastroenterology*, **23**, 585–590.

Schachter, H. and Brockhausen, I. (1992) The biosynthesis of serine (threonine)-N-acetylgalactosamine-linked carbohydrate moities, in *Glycoconjugates*, (eds H. J. Allen and E. C. Kisailus), Marcel Dekker, New York, pp. 263–332.

Scharfman, A., Ramphal, R., Neut, C. *et al.* (1991) Arylneuraminidase activity of *Pseudomonas aeruginosa* does not degrade natural substrates such as human respiratory mucins. *Infection and Immunity*, **59**, 4283–4285.

Schauer, R., Kelm, S., Reuter, G. *et al.* (1995) Biochemistry and role of sialic acids, in *Biology of the Sialic Acids*, (ed. A. Rosenberg), Plenum Press, New York, pp. 7–67.

Scheppach, W. (1994) Effects of short chain fatty acids on gut morphology and function. *Gut (Supplement)*, **1**, S35-S38.

Scheppach, W., Sommer, H., Kirchner, T. *et al.* (1992) Effect of butyrate enemas on the colonic mucosa in distal ulcerative colitis. *Gastroenterology*, **103**, 51–56.

Scudder, P., Hanfland, P., Ventura, K.-J. and Feizi, T. (1984) Endo-β-D-galactosidases of *Bacteroides fragilis* and *Escherichia freundii* hydrolyse linear but not branched oligosaccharide domains of glycolipids of the neolactose series. *Journal of Biological Chemistry*, **259**, 6586–6592.

Sellers, L. A., Allen, A. Morris, E. R. and Ross-Murphy, S. B. (1988) Mucus glycoprotein gels. Role of glycoprotein polymeric structure and carbohydrate side-chains in gel-formation. *Carbohydrate Research*, **178**, 93–110.

Shiau, Y.-F., Kelemen, R. J. and Reed, M. A. (1990) Acidic mucin layer facilitates micelle dissociation and fatty acid diffusion. *American Journal of Physiology*, **259** (*Gastrointestinal and Liver Physiology*, **22**), G671– G675.

Sidebotham, R. L. and Baron, J. H. (1990) Hypothesis: *Helicobacter pylori*, urease, mucus and gastric ulcer. *Lancet*, **335**, 193–195.

Slomiany, B. L. and Meyer, K. (1972) Isolation and structural studies of sulfated glycoproteins of hog gastric mucosa. *Journal of Biological Chemistry*, **247**, 5062–5070.

Slomiany, B. L. and Meyer, K. (1973) Oligosaccharides produced by acetolysis of blood group active (A + H) sulfated glycoproteins from hog gastric mucin. *Journal of Biological Chemistry*, **248**, 2990–2995.

Slomiany, B. L., Bilski, J., Sarosiek, J. *et al.* (1987) *Campylobacter pyloridis* degrades mucin and undermines gastric mucosal integrity. *Biochemical and Biophysical Research Communications*, **144**, 307–314.

Slomiany, B. L., Murty, V. L. N., Piotrowski, J. *et al.* (1992) Glycosulfatase activity of *Helicobacter pylori* towards human gastric mucin: effect of sucralfate. *Biochemical and Biophysical Research Communications*, **183**, 506–513.

Smalley, J. W., Dwarakanath, A. D., Rhodes, J. M. and Hart, C. A. (1994) Mucin-sulphatase activity of some oral Streptococci. *Caries Research*, **28**, 416–420.

Smith, A. C. and Podolsky, D. K. (1987) Biosynthesis and secretion of human colonic mucin glycoproteins. *Journal of Clinical Investigation*, **80**, 300–307.

Stanley, R. A., Ram, S. P., Wilkinson, R. K. and Roberton, A. M. (1986) Degradation of pig gastric and colonic mucins by bacteria isolated from the pig colon. *Applied and Environmental Microbiology*, **51**, 1104–1109.

Sundari, C. S., Raman, B. and Balasubramanian, D. (1991) Hydrophobic surfaces in oligosaccharides: linear dextrins are amphiphilic chains. *Biochimica et Biophysica Acta*, **1065**, 35–41.

Swallow, D. M., Gendler, S., Griffiths, B. *et al.* (1987) The human tumour-associated epithelial mucins are coded by an expressed hypervariable gene locus PUM. *Nature*, **328**, 82–84.

Tabak, L. A. (1995) In defense of the oral cavity: structure, biosynthesis and function of salivary mucins. *Annual Review of Physiology*, **57**, 547–564.

Tanaka, H., Ito, F. and Iwasaki, T. (1992) Purification and charaterisation of a sialidase from *Bacteroides fragilis* SBT31382. *Biochemical and Biophysical Research Communications*, **189**, 524–529.

Thornton, D. J., Howard, M., Devine, P. L. and Sheehan, J. K. (1995) Methods for separation and deglycosylation of mucin subunits. *Analytical Biochemistry*, **227**, 162–167.

Toribara, N. W., Roberton, A. M., Ho, S. B. *et al.* (1993) Human gastric mucin. Identification of a unique species by expression cloning. *Journal of Biological Chemistry*, **268**, 683–689.

Tsai, H. H., Hart, C. A. and Rhodes, J. M. (1991) Production of mucin degrading sulphatase and glycosidases by *Bacteroides thetaiotaomicron*. *Letters in Applied Microbiology*, **13**, 97–101.

Tsai, H. H., Sunderland, D., Gibson, G. R. *et al.* (1992) A novel mucin sulphatase from human faeces: its identification, purification and characterisation. *Clinical Science*, **82**, 447–454.

Tsai, H. H., Dwarakanath, A. D., Hart, C. A. *et al.* (1995) Increased faecal mucin sulphatase activity in ulcerative colitis: a potential target for treatment. *Gut*, **36**, 570–576.

van der Wiel-Korstanje, J. A. A. and Winkler, K. C. (1975) The faecal flora in ulcerative colitis. *Journal of Medical Microbiology*, **8**, 491–501.

van Klinken, B. J. W., Dekker, J., Buller, H. A. and Einerhand, A. W. C. (1995) Mucin gene structure and expression: protection vs. adhesion. *American Journal of Physiology*, **269**, G613–G627.

Variyam, E. P. and Hoskins, L. C. (1981) Mucin degradation in human colon ecosystems. Degradation of hog gastric mucin by fecal extracts and fecal cultures. *Gastroenterology*, **81**, 751–758.

Variyam, E. P. and Hoskins, L. C. (1983) In vitro degradation of gastric mucin. Carbohydrate side chains protect polypeptide core from pancreatic proteases. *Gastroenterology*, **84**, 533–537.

Vercellotti, J. R., Salyers, A. A., Bullard, W. S. and Wilkins, T. D. (1977) Breakdown of mucin and plant polysaccharides in the human colon. *Canadian Journal of Biochemistry*, **55**, 1190–1196.

Vimr, E. R. and Troy, F. A. (1985) Identification of an inducible catabolic system for sialic acids (nan) in *Escherichia coli*. *Journal of Bacteriology*, **164**, 845–853.

Wesley, A. W., Forstner, J. F. and Forstner, G. G. (1983) Structure of intestinal-mucus glycoprotein from human post-mortem or surgical tissue: inferences from correlation analysis of sugar and sulfate composition of individual mucins. *Carbohydrate Research*, **115**, 151–163.

Wilkinson, R. K. and Roberton, A. M. (1988) A novel glycosulphatase isolated from a mucus glycopeptide-degrading *Bacteroides* species. *FEMS Microbiology Letters*, **50**, 195–199.

Witas, H., Sarociek, J., Aono, M. *et al.* (1983) Lipids associated with rat small-intestinal mucus glycoprotein. *Carbohydrate Research*, **120**, 67–76.

Yolken, R. H., Ojeh, C., Khatri, I. A. *et al.* (1994) Intestinal mucins inhibit rotavirus replication in an oligosaccharide-dependant manner. *Journal of Infectious Diseases*, **169**, 1002–1006.

Zopf, D. and Roth, S. (1996) Oligosaccharide anti-infective agents. *Lancet*, **347**, 1017–1021.

10

Colon cancer: the potential involvement of the normal microflora

John Birkbeck

10.1 INTRODUCTION

Adenocarcinoma of the colon is one of the commonest malignancies seen in Western countries, often vying for first place with bronchial carcinoma in the league tables. The human colon is a unique piece of anatomy, in that it contains vast quantities of living and actively metabolizing microorganisms, which play a crucial role in the health of the colonic mucosa from which adenocarcinomas are derived. Hence it is reasonable to enquire whether the nature of the activities of this microflora may influence the risk of malignancy in the colorectal area. It is further relevant to ask to what extent dietary factors, which may influence the risk of colon cancer, may act through, or be modulated by, the microflora.

There is an abundance of material already on this topic. Especially valuable are the 1995 compilation edited by Gibson and Macfarlane, and the recently published, encyclopedic WCRF/AICR text on the relationship of diet to cancer (AICR, 1997).

10.2 CARCINOGENESIS

Malignant change in epithelial cells is considered typically to involve endogenous and exogenous factors. For the colonic epithelium, it is likely that the progression from normal cell to abnormal but benign cell

G. W. Tannock (ed.), *Medical Importance of the Normal Microflora*, 262–294.

(adenoma) to malignant cell (carcinoma) is usual (Hill, 1974), adenoma formation probably being preceded by the development of 'aberrant crypt foci' (ACF) which feature prominently in animal models. The work of Vogelstein's group in characterizing the steps in alterations in colonocyte growth and differentiation have been acknowledged (AICR, 1997).

There are certainly genetic factors prominent in some cases of colon cancer, notably in familial adenomatous polyposis, and hereditary non-polyposis colorectal cancer, but these represent a small minority of cases. International and other epidemiological studies support a view that colon adenocarcinoma is a malignancy where the causative factors are pre-dominantly environmental. While smoking and schistosomiasis may be contributory, it is overwhelmingly dietary factors that appear to promote, or to inhibit colon carcinogenesis.

Modulation, direct or indirect, of food components or the body's response to food components by the colonic microflora could potentially then play a major role in the risk of an individual developing colon cancer. It is worth noting that carcinomas of the intestinal tract are most frequent in areas with a high bacterial exposure (colon and to a lesser extent stomach), and almost non-existent in areas which have a very low bacterial exposure (small intestine; Rowland, 1995).

The human colon may be characterized as a large fermentation chamber, the enclosed microflora serving a variety of functions.

10.3 THE CONTAINER

The colon extends from the ileocaecal valve to the rectum. Conventionally it is divided geographically into a series of sectors, proximodistally, but functionally it can be generally divided into a proximal part, especially the caecum, which contains very liquid material, and the more distal part where the contents are more formed as water reabsorption proceeds to produce faeces. The entire colon is reported to be 1.5 m long, although, as with all the mobile parts of the gut, length depends on smooth muscle activity. The proximal colon especially is characterized by rhythmic antiperistaltic waves, which mix the contents; the middle section by rhythmic peristaltic waves, with intermittent more powerful expulsive waves that move the contents towards the rectum; and the distal part by little peristaltic action but infrequent powerful expulsive waves. These differences probably reflect differences in function. The distoproximal waves in the caecum and ascending colon could serve to mix the contents, ensuring exposure of ileal residue to the microflora. The remainder of the colon is concerned with the relatively gradual movement of contents towards the rectum, to allow time for the remaining absorptive functions (water and electrolytes) to occur.

The functions of the colon include hosting an enormous bacterial population, which appears to be necessary for the health of the colonic mucosa and performs many other functions generally beneficial to the host. The colon abstracts water, electrolytes and other useful substances from the material that passes beyond the small intestine, and it prepares the gut contents for intermittent expulsion from the body. The typical capacity is about 500 ml, and an average content of 220 g is reported (Cummings *et al.*, 1990).

10.4 THE MUCOSA

The colon is lined with columnar epithelium, which is extensively folded into club-like villi separated by fairly deep crypts. As is usual in the gut, epithelial cells originate deep within the crypts and gradually migrate towards the tips of the villi. Some cells are dislodged by the passage of intestinal contents but their lifespan is in any case limited by a process of apoptosis, or programmed cell death. Since the development of any tumour requires a lengthy period, malignant change must involve interference with the apoptotic mechanism.

10.5 THE BACTERIA

The colonic contents typically contain about 10^{11} microorganisms per gram (dry weight), or over 10^{13} *in toto*. This is perhaps an order of magnitude greater than the number of cells in the human body. Bacteria that are usual colonic inhabitants are very diverse. There are major difficulties in studying the microflora in their natural habitat. Most work has been performed on faeces, but this does not necessarily reflect the proximal colonic contents, particularly from a quantitative viewpoint. Many species, for example, are obligate anaerobes, and exposure to even small doses of air may affect their numbers. Sampling colon contents is usually performed in surgical or endoscopic situations, where bowel 'preparation' can have a marked influence on bacterial types and numbers, either due to fasting or even antibiotic administration. However it is likely that several hundred different types may be present at least on some occasions.

Extensive studies of faecal microflora by Finegold, Sutter and Mathesen (1983) show that only a few species are almost always present in substantial quantities. His group studied various populations including strict vegetarians, Japanese on a diet free of ruminant meat, and Americans on a typical Western diet including beef. No class of bacteria was invariably present, but *Bacteroides*, anaerobic cocci, *Eubacterium* and *Clostridium* were usually found, and streptococci and Gram-negative facultatives were isolated from over 85% of all subjects. While the bacterial spectrum may differ between individuals, diet seems to have relatively little effect on the mix of bacterial species present.

Some bacterial species are strict anaerobes, such as *Bacteroides* and *Clostridium*, while others are facultative anaerobes. Although in theory oxygen could diffuse from the mucosal circulation into at least the layers close to the mucosa, in practice this does not seem to inhibit the obligate anaerobes, and oxidative reactions are notably absent.

10.6 THE SUBSTRATE

Contrary to popular belief, the digestion and absorption of macronutrients from ingested food is not complete. Atwater and Benedict (1899) studied three young American men to determine the degree of utilization and found that, on a typical American diet of their day, 99% of carbohydrate (i.e. metabolizable carbohydrate: the concept of dietary fibre had not developed), 95% of fat and 92% of protein was utilized. It was on this basis that the famous 'Atwater factors' for converting analysed food components into energy were based. These allowed for the quoted degrees of incomplete absorption. Subsequent work (Southgate and Durnin, 1970) showed that the original figures were indeed applicable to ordinary Western diets. We still calculate 'energy intake' using those factors, even though in many circumstances, especially during growth, a substantial part of the protein, a small amount of the fat and a smaller still part of the carbohydrate are used for structural purposes and are not converted into energy, and some 'dietary fibre' is used for energy. Clearly, then, substantial amounts of protein and fat in a typical diet of today are not absorbed. There are several reasons for this. Some material is encompassed by plant cell walls and is relatively inaccessible to digestive enzymes. Some may become associated with 'dietary fibre' and again escape digestion. The movement of gut contents may be too rapid to permit complete assimilation. Some dietary protein, and a few lipids, may be resistant to digestive enzymes, e.g. that altered by the Maillard reaction in non-enzymic browning, where amino acids such as lysine are complexed with carbohydrate by heat.

The concept of 'dietary fibre' further complicates the picture. With the exception of lignin, a phenolic substance that is only a minor constituent of most diets, 'fibre' can be thought of as 'carbohydrate substances of plant origin that are not digested by the secretions of the human intestinal tract'. Although it is evident that many of these substances, such as pectin, are by no means fibrous, the term has been retained. This is a physiological definition. However since there is no method as yet devised which will measure all such substances reliably, we tend to fall back on analytical definitions which must give a somewhat distorted picture.

Atwater and Benedict's studies suggested that almost all (metabolizable) carbohydrate is digested and absorbed. However this must be slightly qualified. There are a few uncommon oligosaccharides in edible plants that are not digestible. But, while uncommon, they could by choice form a significant part of an individual's food intake. Secondly there is the anomalous significance of lactose.

Lactose is a disaccharide that is found exclusively in mammalian milks (except, curiously, in the pinnipedia: the seals and sealions). It is digested by a brush border enzyme of the small intestine, β-galactosidase or lactase-phlorizin hydrolase, which hydrolyses it to its constituent monosaccharides, glucose and galactose, which are then readily, and actively, absorbed. In the normal human infant, most lactose is so handled. In most of the world's population, the level of this enzyme declines from about 1–2 years of age and reaches low levels by 3–4 years, although the timing is variable and the final level may not be reached until adolescence in some populations (Johnson *et al.*, 1993). The enzyme does not disappear completely, and residual levels are usually 5–10% of the former level, although sometimes can be as high as 30% (Sterchi *et al.*, 1990). However in most members of certain populations, especially Europeans and some African pastoral tribes, this fall in lactase does not occur and the child and adult retain, at least into advanced age, the ability to digest the sugar. The term 'lactase deficiency' was originally ascribed to this phenomenon (by Europeans), but now the term 'lactase persistence' is most appropriate since this is the minority situation.

As an example of a population mixture, figures for the multicultural New Zealand population are given (Table 10.1).

In the past, populations using mammalian milk either had lactase persistence and hence could consume milk with impunity, or devised procedures to ferment milk to products such as yoghurt or koumiss, in which most of the lactose has been fermented by bacterial action to lactic acid, which also acts as a preservative.

The widespread availability of liquid (e.g. UHT) or dried whole or skim milk powders, and the immigration of non-Europeans into traditionally

Table 10.1 Lactose malabsorption in the New Zealand population

Population group	Prevalence of lactose malabsorption*(%)
European	9
Maori	64
Samoan	54
Chinese	80
Indian†	65

*Cobiac 1994
† Scrimshaw and Murray 1988

milk-drinking cultures such as the United States, Europe or Australasia, has resulted in large numbers of people consuming modest amounts of lactose that they are unable completely to digest. American studies have shown that such people generally can consume up to 150–200 ml of milk (6–8 g of lactose) at any one time without unacceptable consequences. Such milk of course can provide many other valuable nutrients. If this is repeated several times per day, the amount of undigested lactose could be as high as 20 g per day or more.

In fact all such people can digest a little, but a variable amount, of lactose, and a small amount may be absorbed unchanged and excreted in the urine. The evidence is that lactase persistence is a genetic variation and not simply due to adaptation to a milk-drinking environment.

10.7 DIETARY FIBRE

The concept of 'roughage' as being important for human health, especially bowel health, extends to the 19th century. However the idea that this undigested plant material did anything more than provide 'bulk' and perhaps water-retaining properties was slow to develop. It was not until the 1940s that it was really recognized that the human colon resembled the ruminant in that it contained organisms that fermented carbohydrates (Bancroft, McNally and Phillipson, 1944). As far back as 1935, Williams and Olmstedt had shown that the laxative effect of 'unrefined' carbohydrate was due not to the undigested cellulose but to fermentable hemicellulose (Williams and Olmstedt, 1935; Baghurst, Baghurst and Record, 1996). The term 'dietary fibre' was introduced in 1953 by Hipsley and developed by Trowell in 1972 to describe 'remnants of plant cell walls'. FASEB in 1987 introduced the concept of 'endogenous components of plant materials in the diet which are resistant to digestion by enzymes produced by humans', which is getting close to the current concept but tends to exclude some forms of resistant starch produced by heat.

Meanwhile, various workers in southern Africa were drawing attention to the very different health statistics of Africans compared to Europeans and speculating that dietary differences might be responsible. The high fibre intake of the former was noted by Walker and Arvidsson (1954), and in 1960 Higginson and Oettle suggested that the very low incidence of colon cancer in indigenous Africans might be related to high fibre intake. Burkitt (1971, 1973; Burkitt and Trowell, 1975) also related a number of intestinal diseases to a low intake of dietary fibre and noted the very bulky stool passed by Africans compared to the meagre produce of Europeans.

As the functions of intestinal bacteria and 'dietary fibre' have unfolded, it is now accepted that an adequate intake of a variety of types of 'dietary fibre' is necessary for the health of the colon and of the individual overall (Byers, 1995).

The current concept of 'dietary fibre' includes lignin, a phenolic substance which, being unpleasantly fibrous, forms but a small part of the human diet, and is untouched by any part of the intestinal tract. The rest is carbohydrate in nature. It includes some starch that for a combination of reasons is not digested by small intestinal amylases and is called 'resistant starch'. Resistance may be due to the physical location of the starch in the plant cell, making it inaccessible to the enzymes, even after chewing; to the crystalline form of the molecules; and to modification of structure produced by heating and by subsequent cooling (retrograded starch; Englyst, Wiggins and Cummings, 1982; Baghurst, Baghurst and Record, 1996).

The balance are called non-starch polysaccharides (NSPs) and include cellulose and hemicelluloses; pectins (which are by no means 'fibrous'); gums and mucilages; inulin, betaglucans and fructans; and algal polysaccharides.

These compounds are to varying degrees digested and metabolized by colonic bacteria (Tables 10.2, 10.3), the predominant end products being short-chain fatty acids, predominantly acetate, butyrate and propionate. Estimates of production are at least 300 mmol/d (Cummings *et al.*, 1987).

It has been proposed also that dietary fibre may function by binding various sterol substances. These include cholesterol and bile acids, which would otherwise be recycled into an enterohepatic circulation, which will reduce the body content of cholesterol-type compounds and help reduce hypercholesterolaemia; and binding of primary bile acids thus preventing production of secondary bile acids by bacterial dehydroxylation, which have been shown to be co-carcinogenic and mutagenic (Narisawa *et al.*, 1974). Lignin may be an important factor in this binding (Shulz and Howie, 1986).

Table 10.2 Starch digestibility; data obtained in ileostomy subjects (after Muir and O'Dea, 1992); Muir *et al.*, 1994

Food/100 g	Total starch (g)	% not digested in ileum
White bread	62	3
Cornflakes	74	5
Oats	58	2
Bananas (yellow green ends)	19	89
Potato: Fresh cooked	45	3
Cooked cooled	47	12
Cooked/reheated	47	8

Table 10.3 Patterns of short-chain fatty acid production from various substrates; % total acid production (from Baghurst, Baghurst and Record, 1996, based on Kritchevsky, 1995)

Substrate	Butyrate (%)	Propionate (%)	Acetate (%)
Resistant starch	38	21	41
Starch	29	22	50
Oat bran	23	21	57
Wheat bran	19	15	57
Cellulose	19	20	61
Guar gum	11	26	59
Ispaghula	10	26	56
Pectin	9	14	79

10.8 SHORT-CHAIN FATTY ACIDS

The bacterial metabolism of the products of the breakdown of carbohydrates in the colon produces short-chain fatty acids (SCFAs). The predominant species are acetate, butyrate and propionate, with small amounts of others, including branched-chain ones. The proportions of these produced by incubating fresh human faeces with pure dietary fibre are shown in Table 10.4 (Bugaut and Bentéjac, 1993).

The proportions of SCFA in human faeces are reported to average acetate 53%, butyrate 20% and propionate 27%, but these may not reflect intracolonic levels since so much of the SCFA is absorbed (Savage, 1986).

While total SCFA production has been estimated at 300 mmol/d, the faeces contain only about 10 mmol/d; i.e. about 97% is absorbed (Hoverstad, 1986).

SCFAs are the primary fuel of the colonocyte, especially butyrate (Wolever, Spadafora and Eshuis, 1991; Williams *et al.*, 1992). What is not used by the colonocytes is passed to the portal blood to the liver, where it is largely cleared (Table 10.5; Cummings *et al.*, 1987).

Table 10.4 Production of SCFAs from various substrates; SCFA yields from incubating fibres with fresh human faecal flora *in vitro* (mol/100 ml; 24-hour incubation except for cellulose, 48 hours). Source: from Bugaut and Bentéjac, 1993

Substrate	Acetate	Propionate	Butyrate
Pectin	81	11	8
Gum arabic	68	23	9
Oat bran	65	19	16
Wheat bran	63	16	21
Cellulose	53	21	26

Table 10.5 SCFA concentrations in caecal and blood samples from sudden-death humans (μmol/ml); data from Cummings *et al.*, 1987, cited by Bugaut and Bentéjac, 1993.

Location	Acetate	Propionate	Butyrate
Caecal contents	69.1	25.3	26.1
Hepatic portal blood	0.258	0.088	0.029
Hepatic vein blood	0.115	0.021	0.012
Peripheral blood	0.070	0.005	0.004

SCFAs stimulate colonocyte proliferation in the crypts (butyrate > propionate > acetate), although the precise mechanism is controversial. However butyrate seems to have a specific effect on genic factors that are antineoplastic. The apparent paradox between butyrate stimulating cell proliferation and also being antineoplastic may be explained by studies showing that butyrate has effects on malignant cells that promote differentiation, whereas it has little such effect on normal colonocytes (Gibson *et al.*, 1990). It is very evident that, while butyrate has beneficial effects in maintaining normal colonic health and inhibiting malignancy, the mechanisms involved are very complex and as yet incompletely understood.

Recent work has shown that butyrate enhances histone acetylation (Scheppach, Bartram and Richter, 1995), and inhibits the enzyme histone deacetylase (Hassig, Tong and Schreiber, 1997), which may be an important mechanism for its antineoplastic effects. This interrupts histone–DNA interaction allowing gene expression. This effect has been shown to enhance normal apoptotic cell death (McBain *et al.*, 1997; Heerdt, Houston and Augenlicht, 1997). Deoxycholate induces crypt surface proliferation, considered an intermediate step in carcinogenesis, and this effect is inhibited by butyrate (Velasquez *et al.*, 1997).

Butyrate enhances cell differentiation and cell apoptosis in colonocytes (Hague *et al.*, 1993; Heerdt, Houston and Augenlicht, 1997). Butyrate in fact seems to influence the expression of a variety of genes in colonocytes, including various kinases (Schwartz *et al.*, 1995; Dang, Wang and Dee, 1995; Brunton *et al.*, 1997; Nakano *et al.*, 1997), plasminogen activators (Antalis and Reeder, 1995) and prostaglandins (Bartram *et al.*, 1995).

Jass (1985) characterized butyrate as a differentiating agent, by inducing changes in malignant cell lines, such as:

- reduced growth rate;
- suppression of anchorage-independent growth;
- decreased efficiency of cloning;
- expression of new phenotypes, not necessarily adult normal cells.

Studies, however, in rats consuming oat bran or wheat bran found that, although both increased butyrate content of the colon, wheat bran feeding had an inhibitory effect on azomethane-induced colon carcinoma but oat bran did not. Wheat bran feeding, however, produced more usual proportions of the main SCFAs (65:10:20) and produced greater stool volume. It appears from this study that lumen pH and butyrate levels in isolation are not the factors explaining this antineoplastic effect (Zoran *et al.*, 1997).

A diet poor in dietary fibre contributes to inflammatory changes in the colonic mucosa and to ulcerative colitis, a chronic and severe inflammatory condition ultimately associated with neoplastic change. It seems likely that an inadequate fibre intake is an important precursor of colonic malignancy. The magnitude of the effect of one specific type of dietary fibre, resistant starch, has just been exemplified by the studies of Hylla *et al.* (1998). Administration of high intake (average 55 g/d) of resistant starch as amylopectin was compared to low intake (average 8 g/d) in healthy adults. Although faecal SCFA excretion did not change, around 90% of the resistant starch was metabolized in the colon, as evidenced by breath hydrogen levels. This was accompanied by about a 50% increase in stool weight, a 26% reduction in bacterial β-glucosidase activity, a 30% reduction in total and secondary bile acids and a similar reduction in neutral sterols.

10.9 PEPTIDES

Peptides reach the colon from a variety of sources. Some represent incompletely digested food proteins. These include certain meat proteins such as myofibril and sarcoplasma proteins. Thermally modified proteins, especially from milk that has been heated in sterilization (such as UHT milk) or drying (whole or skim milk powder), but also heated egg albumin, are incompletely digested in the small intestine. Maillard products, too, formed by the effects of heat on protein–carbohydrate mixtures, where links between glucose and amino acids such as lysine form, are also present.

However much of the protein reaching the colon is of endogenous origin. This includes digestive enzymes, debris of shed cells, antibodies or antigen–antibody complexes, all of which can be substrates for bacterial proteases. Amino acids so released can be decarboxylated to form amines, or deaminated to form short-chain fatty acids. The amines may be further metabolized to produce hydrocarbons such as methane, plus ammonia.

Many colonic bacteria have proteolytic ability, especially *Bacteroides fragilis*. Not only can they degrade peptides, including particularly pancreatic trypsin and chymotrypsin (Macfarlane, Gibson and Gibson, 1992),

Table 10.6 Some bacterial species generating SCFAs in the colon; numbers of selected bacteria in the faeces of humans consuming different diets. Veg = strict vegetarians; Japn = Japanese diet; West = typical USA diet. Values represent $\log_{10}$ per gram dry faeces, hence a difference of 1.0 represents a tenfold difference. (After Finegold, Sutter and Matheson (1983) and others)

Species	Veg	Japn	West	Products
Bacteroides	11.7	10.8	11.3	Acetate, propionate, succinate
Bifidobacterium	10.9	9.7	10.4	Acetate, lactate, ethanol, formate
Eubacterium	11.0	10.6	10.6	Acetate, butyrate, lactate
Ruminococcus	10.2	10.3	10.0	Acetate
Peptostreptococcus	11.1	10.2	10.2	Acetate, lactate
Lactobacillus	11.1	9.0	9.3	Lactate
Clostridium	9.4	9.7	10.2	Acetate, propionate, butyrate, lactate
Streptococcus	8.6	8.7	9.1	Lactate, acetate
Anaerobic strep.	11.4	9.5	10.5	Lactate, acetate

but they also can utilize ammonia (Allison, 1995). Examples of products produced by different organisms were listed by Finegold, Sutter and Mathesen (1983; Table 10.6).

Urea does not seem to be present in significant amounts in colonic contents, so is not an ammonia source. Faecal ammonia levels average about 14 mmol/l (Wilson, Muhrer and Bloomfield, 1968; Wrong, 1988).

Ammonia is toxic to colonocytes, interfering with their metabolism and shortening their life. It thus increases the replication and turnover of colonic mucosal cells (Warwick, 1971). Ammonia is further apparently less toxic to transformed cells (Visek, 1972, 1978). Administration of lactulose leads to increased uptake of ammonia by some bacteria, reducing luminal ammonia levels, and so is used in management of liver disease where blood ammonia levels are raised.

Normally, however, ammonia is rapidly absorbed into the portal blood, and in the liver is converted to urea for excretion in the urine. Thus while ammonia is toxic to colonocytes, this does not normally pose a problem.

Amines are formed by decarboxylation of amino acids. These are usually rapidly degraded by monoamine oxidase or diamine oxidase, either in the colonic mucosa or the liver. Some products reach the urine in altered form. Putrescine and cadaverine are oxidatively deaminated to pyrollidine and piperidine. While many amines have biological effects, such as elevated blood pressure by tyramine, their main concern is in reactions to form N-nitroso compounds (nitrosamines). These mutagenic compounds are formed by interaction between secondary amines and nitrite. However the evidence is that this is not a major effect in the colon

(Eisenbrand, Spiegelhalder and Pressman, 1981). Phenolic compounds are produced from aromatic amino acids, and are usually converted to glucuronides either in the colonic mucosa or the liver, and ultimately excreted in the urine. Levels of these compounds in the colonic contents are normally very low. Indoles formed from tryptophane are also rapidly excreted, either as indican (indole glucoside) or as indoleacetic acid. Decarboxylation of the latter compound produces skatole (3-methylindole), a major source of the unpleasant odour of faeces.

10.10 NITROSAMINES

Nitrosamines are compounds, carcinogenic in experimental models, formed in the body by the interaction of nitrite and a variety of nitrogenous substances including amines, amides and methylureas. Nitrite arises readily in the body from the reduction of nitrates which are commonly present in foods, either naturally or added as preservatives in meat products (Green *et al.*, 1981). They are also sometimes present in water supplies, specifically those contaminated by sewage or fertilizer run-off. Nitrite production can occur in acidic conditions in the stomach where N-nitroso compounds can form (Wagner and Tannenbaum, 1985); the reaction is powerfully inhibited by ascorbic acid. But the reaction can also occur at neutral pH where bacteria are present. N-nitroso compounds are not formed in germfree rats (Massey *et al.*, 1988), so it is likely that both nitrate reduction and nitrosation are bacterially induced in the colon

Human studies have found faecal excretion rates of 40–590 µg nitrosamines per kilogram of faeces. Removal of dietary nitrates reduced these values to very low levels and addition of dietary nitrate restored the values, suggesting that most nitrosamines are endogenous rather than dietary in origin (Rowland *et al.*, 1991).

Notwithstanding these findings, current evidence does not ascribe a significant role to nitrosamines in carcinogenesis. Previous evidence linked nitrosamines to stomach cancer (NAS/NRC, 1981; Magee, 1981). There is no good evidence to link nitrosamines to colon cancer, and the massive reduction in stomach cancer suggests that either their role was overestimated or the powerful inhibitory effect of ascorbic acid and tocopherols on nitrosamine formation is effective (Walker, 1990).

10.11 DIETARY FATS

The overwhelming majority of dietary fat is in the form of long-chain triacylglycerides. Some foods contribute triacylglycerides (TAGs) containing medium-chain fatty acids, and a few, notably dairy foods, SCFAs. Small amounts of phospholipids, cholesterol and waxes are also present, and today monoacylglycerides and diacylglycerides may be used as food additives. As noted, the majority of these lipids are digested and absorbed in the small intestine. Unabsorbed TAGs or fatty acids do not in general seem to be metabolized by colonic bacteria. Thus if for any reason fat digestion is diminished, lager amounts of TAGs appear in the faeces as steatorrhoea. This is normally a pathological event but today may be induced either by the use of agents that to some degree inhibit the action of pancreatic lipase (such as orlistat) or by the use of fibre that specifically binds TAGs and prevents lipolysis, such as chitosan, an acetylated form of chitin derived from shellfish. Both these agents run the risk of 'anal leakage' due to the steatorrhoea produced. This does not mean, however, that dietary fats are irrelevant in the colon. Epidemiological studies, either case-control or cohort studies, generally do not suggest that high levels of dietary fat intake raise the risk of colon adenocarcinoma (Giovannucci and Goldin, 1997). Information that might relate to specific types of fatty acid is meagre. Broadly, the only fat-containing food that seems to be related to a risk of colon cancer is ruminant meat, principally beef. Fat from dairy foods, poultry, fish or plant sources does not seem to raise the risk (Goldbohm *et al.*, 1994; Bostick *et al.*, 1994; Giovannucci *et al.*, 1994). Since the fatty-acid composition of beef fat is not unique, it is hard to see how some facet of this could explain this relationship, and some other mechanism, such as compounds formed during high-temperature cooking, may be more relevant. Yet the large Nurses' Health Study found a doubled risk of colon cancer in those women with the highest, compared to the lowest, intakes of fat from animal sources, mainly red meat (Willett *et al.*, 1990). Other such studies, while smaller, failed to confirm this association (Stemmermann, Nomura and Heilbrun, 1984; Garland *et al.*, 1985). Overall the evidence linking animal fat to colon cancer risk is unconvincing. The same is true for monounsaturates: countries with high olive oil intakes such as Greece have low rates of colon cancer. It is worth noting that a principal saturated fatty acid of beef, stearic acid, is rapidly converted to oleic acid in the body and so acts in many ways like a monounsaturate. Nor is there any convincing evidence for a relationship with intake of polyunsaturates, although evidence is scanty (Giovannucci and Goldin, 1997). Although in general the evidence does not favour a link between *trans*-fatty acids and colon cancer, one possible factor may be conjugated linoleic acid from dairy fats.

Fish or ω3 fatty acids, however, may protect against the promotional phase of colon cancer (Giovannucci and Goldin, 1997), perhaps by influencing prostaglandin production.

Although a specifically raised requirement for linoleic acid by colonic tumour cells has not been found, in contrast to tumours in some other tissues, polyunsaturated fatty acids can be hydrated into hydroxy acids, and it is known this conversion can be produced by colonic bacteria. Such compounds are known to be mitogenic (Klurfeld and Bull, 1997). Growth of experimental tumours has been found to be inhibited by eicosapentaenoic acid (20:5ω3) and this inhibition is reversed by linoleic acid (18:2ω6). Oxidation products, however, seem the most likely candidate for tumour promotion at present (Klurfeld and Bull, 1997). In animal feeding experiments, beef tallow and vegetable oils rich in linoleic acid have been found to have tumour-promoting properties. The precise mechanisms involved are elusive.

10.12 SECONDARY BILE ACIDS

One explanation advanced for the epidemiological link between diets high in fat or in meat and colon cancer is the conversion of primary to secondary bile acids (deoxycholic acid, lithocholic acid and ursodeoxycholic acid) by colonic bacteria at relatively neutral pH (Nagengast, Grubben and vanMunster, 1995). These compounds have clearly been shown to be tumour promoters (Rowland, Mallett and Wise, 1985). They may act in part by activating protein kinase $C\beta_1$ (Pongracz *et al.*, 1995).

Groups of people on high-fat, high-risk diets certainly have a high faecal content of secondary bile acids, but case control studies have not provided convincing evidence to support such a relationship (Eyssen and Caenepeel, 1988). Yet a recent study (van der Meer *et al.*, 1997) has shown a significant effect of secondary bile acids, especially lithocholic acid, in subjects with colorectal adenoma or carcinoma, which supports previous observations on colon cancer subjects (Tocchi *et al.*, 1996). People with colorectal adenomas have also been reported to have elevated serum deoxycholic acid levels compared to controls (Bayerdorffer *et al.*, 1995).

Epidemiological observations have found that high intakes of starch, non-starch polysaccharide and vegetables seem to protect against colon cancer, while high meat (fat?) intakes may enhance the risk (Bingham, 1997). Laboratory studies have found protective effects of starch (Christi *et al.*, 1995, 1997); lactulose (Hennigan *et al.*, 1995; Owen, 1997), wheat bran (Alberts *et al.*, 1996), fish oil (Bartram *et al.*, 1996) and calcium from milk (Alberts *et al.*, 1996; van der Meer *et al.*, 1997) that reduce this effect.

Cholecystectomy has important significance for colon cancer. It induces a high faecal loss of bile acids, with an increased proportion of secondary bile acids, especially deoxycholate (McMichael and Potter, 1985). Several studies, at Framingham, Honolulu and Evans County, recorded an increased risk with low serum cholesterol levels. It may be that efficient clearance of cholesterol results in a greater level of bile acids in bile, and it is relevant that the excess risk seems to be in lesions of the proximal colon.

Plasma oestrogens also seem to affect bile acid synthesis and composition (Nakagaki and Nakayama, 1982), and this can explain the raised risk in women, especially nulliparous women, and also in men following oestrogen therapy for prostate cancer.

The deleterious effects of deoxycholate on human colonocytes can be counteracted by butyrate (Bartram et al., 1995), so the effects of starch and other fibres can be thus explained.

There is evidence that dietary calcium may be protective by precipitating fatty acids and secondary bile acids (Govers et al., 1996). Fish-oil administration has been found to decrease the excretion of the possible carcinogen 4-cholesten-3-one (Bartram et al., 1996), regardless of dietary fat level. The reported relationship between red and processed meat intake, but not intake of white meat, to colon cancer could be due to fat content, to nitrates, or to the formation of heterocyclic amines in the cooking process (Bingham, 1997). High meat intake certainly increases faecal ammonia and nitrosamine content (Bingham, 1997). The relationship, however, remains controversial.

One difficult point is that both animal and vegetable fats promote primary bile acid production, yet epidemiological evidence suggests that only animal, predominantly red meat, fat raises the risk of colon cancer (Giovannucci and Goldin, 1997).

It has been suggested that perhaps the unique effect of red meat might be that beef tallow could be less well digested than other animal fats. Beef tallow is rich in stearic acid (18:0) (Giovannucci and Goldin, 1997). Alternatively, other animal fats could contain protective factors. Although calcium in dairy foods (but not in butter) could perhaps form insoluble soaps with fatty acids and secondary bile acids (Newmark, Wargovich and Bruce, 1984), the effect on faecal bile acid concentration is not supported by subsequent evidence (Alder, McKeown-Eyssen and Bright-See, 1993; Giovannucci and Goldin, 1997).

The proposed mechanism of carcinogenesis by secondary bile acids was an irritative one (Narisawa et al., 1974). A case-control study showed that individuals with colon cancer excrete twice the level of sterols and bile acids that controls do (Mastromarino, Reddy and Wynder, 1976). However a more recent study found that the hyperproliferative effect of high fat intakes was not mediated by either fatty acid or bile acid levels in the colon (van der Meer et al., 1997). Dietary calcium was found to be protective as noted previously. Overall, however, the true role of secondary bile acids remains uncertain.

10.13 FECAPENTAENES

These are conjugated polyunsaturated ether lipids, probably synthesized by colonic bacteria from ether phospholipids (plasmalogens; van Tassell *et al.*, 1989; van Tassell, Kingston and Wilkins, 1990). The origin of their precursors is not established but production of fecapentaenes is not affected by either dietary fat level (e.g. 20% or 40% energy from fat), nor the P:S ratio, 1.0 or 0.3 (Taylor *et al.*, 1988). Fecapentaene-12 and -14 seem to be the common varieties.

The fecapentaenes are mutagenic (Peters *et al.*, 1988; Kingston, van Tassell and Wilkins, 1990) but not carcinogenic (Venitt and Bosworth, 1988; Povey *et al.*, 1990) *in vitro*. Yet fecapentaene excretion is greater in vegetarians than in omnivores (DeKok *et al.*, 1992), although vegetarians have a lower risk of colon cancer. Dietary factors that are related to levels of fecapentaene excretion include calcium, cereal fibre and carotenes (DeKok *et al.*, 1992).

If fecapentaenes are somehow related to cancer production, binding to dietary fibre may prevent any adverse effects, which could explain the paradoxical findings in vegetarians. Low vitamin C status has been reported to increase fecapentaene excretion (Jacob *et al.*, 1991).

While the levels of potential carcinogens at the time of diagnosis of colon cancer are not necessarily helpful in elucidating causative mechanisms, it has been found that fecapentaene excretion is not higher in individuals with adenocarcinoma (Schiffman *et al.*, 1989).

Fecapentaenes appear in the faeces shortly after birth, although absent from meconium (Block *et al.*, 1990). Levels seem to be constant in those consuming typical high-fat low-fibre diets (Block *et al.*, 1990). Despite evidence from animal studies that fecapentaenes may be colon tumour promoting (Zarkovic *et al.*, 1993), the balance of evidence at present makes an important role of fecapentaenes in colon cancer dubious, but the matter is still open (Rowland, 1995).

10.14 MUCINS

Mucins are glycoproteins that create the special properties of mucus. They are metabolized by colonic bacteria, but only by the cooperative efforts of a variety of microbial species. The first stage appears to be proteolysis of peptide parts that are not protected by carbohydrate. The oligosaccharides are then separated and yet further groups of bacteria can cleave and metabolize them.

Blood group substances have a similar structure and suffer the same fate. There is some evidence that the presence or absence of some such compounds, for example A, H or Lewis, can be reflected in the balance of bacterial species, bifidobacteria being favoured by secretor status because they can metabolize fucose (Hoskins *et al.*, 1985; Hoskins, 1992). Mucopolysaccharides are sulphated and this can stimulate H_2S production.

10.15 PLANT GLYCOSIDES

Colonic bacteria produce β-glycosidases, which hydrolyse a variety of plant glycosides. In general the products are considered to be harmless, or possibly even beneficial to health. A few, such as amygdalin, a cyanide-containing glycoside present in almond kernels and a number of other fruit components, have been looked on as potentially toxic to the colonic mucosa, or as carcinogenic, such as rutin or franguloside, but overall it is not considered that these play any significant role in carcinogenesis. Special mention must be made, however, of the phyto-oestrogens.

10.15.1 Phyto-oestrogens

A well established model for the interaction of diet and the colon bacteria relates to the plant flavonoids. Compounds of great interest have been genistin and diadzin in soybeans (Setchell, 1985), but similar compounds (lignans) occur in whole cereal foods. Genistin and diadzin are predominantly β-glucosides, although they also exist as 6-O-malonylglucosides and 6-O-acetylglucosides (Setchell *et al.*, 1997). These compounds are not absorbed from the human gut. However colonic bacteria metabolize the glycosides to release the aglycones genistein and diadzein. These compounds are readily absorbed (Setchell and Adlercreutz, 1988).

They also are further metabolized by colonic bacteria to equol, des-methylangolensin and p-ethylphenol (Axelson *et al.*, 1982; Joannou *et al.*, 1995; Franke and Custer, 1996). Equol also has oestrogenic properties.

The effects of the phyto-oestrogens are twofold. Although not steroidal in structure, they are capable of binding to oestrogen receptors, and have weak oestrogenic properties. Their dose–response effect is typically 10^{-4}–10^{-5} that of oestradiol.

In subjects with relatively high endogenous oestrogen levels (especially adult women) phyto-oestrogens may actually reduce oestrogenic effects by preventing the binding of endogenous hormones. Phyto-oestrogens tend to become unbound from receptors less readily than natural hormones. These compounds certainly have effects *in vivo*. Consumption of soy foods will lengthen the menstrual cycle and depress the mid-cycle level of gonadotrophins in normal women (Cassidy, Bingham and Setchell, 1994). This effect is also considered to be the basis for the reduced risk of breast and prostate cancers in Oriental populations with regular consumption of soy foods such as tofu and miso (Rose, 1993; Coward *et al.*, 1993; Messina *et al.*, 1994; Adlerkreutz, 1995; Lamartiniere *et al.*, 1995; Kirkman *et al.*, 1995).

Epidemiological studies have claimed that there is a relationship between eating soybean products and reduction in risk of rectal and colon cancer in Japan (Watanabe *et al.*, 1984), China (Hu *et al.*, 1991) and the USA. However these compounds are not generally recognized as being of significance in preventing colon cancer (AICR, 1997).

Plant flavonoids are mutagenic *in vitro*, yet are credited with being anticancer agents *in vivo*. Quercetin reduces colonocyte proliferation (Deschner *et al.*, 1991), and has been reported to reduce colonic tumours, but the evidence is as yet sketchy (Dragsted, Strube and Larsen, 1993). Some flavonoids have been reported to reduce the activity of benz(α)pyrene hydroxylase and hence reduce its carcinogenicity, while others may actually have the opposite effect. It is not clear just what the overall role of these compounds may be.

10.16 VITAMIN PRODUCTION

The colonic bacteria produce substantial amounts of two vitamins. There is much speculation as to the importance of this in the economy of the host.

Biotin is a B vitamin quite widespread in the diet, albeit in small amounts. It functions as a carboxyl carrier in many carboxylation and decarboxylation reactions. Acetyl- and propionyl-CoA carboxylases require biotin as a coenzyme. These are of course required for the metabolism of SCFAs.

The amount of biotin present in colonocytes is unknown (Cherbonnel-Lasserre *et al.*, 1997). These authors examined the biotin content of normal and carcinoma cells of the human colon. They found that, while normal colonocytes had a high biotin content, malignant cells had a low content. Transcription of the two genes for propionyl-CoA carboxylase were related to the biotin levels and not to the chromosomal precursors.

Biotin is strongly bound by a protein of raw egg white called avidin. An avidin–biotin system has been used to enhance experimental chemotherapeutic targeting for colon cancer cells (Nakaki, Takikawa and Yamanaka, 1997). It is doubtful that colonic synthesis makes a significant contribution to biotin status of the human body, but this is uncertain (Bonjour, 1991).

It is also well established that colonic bacteria synthesize vitamin K in its menoquinone form, especially as MK-9 and MK-10. The question is whether this material is absorbed sufficiently to be of significance. Despite the high levels, much may be bound in bacterial membranes and be unavailable (Suttie, 1995). Certainly there is a substantial proportion, perhaps as much as 50%, of liver vitamin K in menaquinone form. Recent studies of colonic instillation of vitamin K in rats suggest that absorption of the plant form (phylloquinone) and of menaquinones of short (MK-4) and long (MK-9) chain length were very poor (Groenen-van

Dooren *et al.*, 1995). The true role of gut vitamin K synthesis in terms of the overall requirement is undetermined.

10.17 OLIGOSACCHARIDES

Small saccharides, typically with only up to five or six monosaccharide units, are present in many plant foods and recently have become of considerable commercial interest as prebiotics, i.e. compounds that modify the spectrum of colonic bacterial flora in what is presumed to be a way beneficial to host health (Gibson *et al.*, 1996; Table 10.7). For this reason they are being produced, often synthetically, for introduction into health food products (Roberfroid, 1997).

Table 10.7 Oligosaccharides currently of commercial interest

Fructo-oligosaccharides
Xylo-oligosaccharides
Isomalto-oligosaccharides
Polydextrose
Galacto-oligosaccharides
Palatinose oligosaccharides
Soy oligosaccharides

Generally, specific health claims have to be avoided, but a laxative effect and sweetness that can be used in diabetic foods are useful attributes.

10.17.1 Raffinose and stachyose

Raffinose (Gal-Glc-Fru) and stachyose (Gal-Gal-Glc-Fru) are oligosaccharides prominent in the bean family. Since they are not digested in the small intestine, they enter the colon where their metabolism tends to engender considerable gas production. This unwelcome propensity has resulted in considerable American research efforts to find ways of minimizing the levels in beans such as haricot and navy (Calloway and Margen, 1971; Rackis, 1981). Soybeans are also rich in these compounds, and they have been shown to favour the growth of bifidobacteria (Rackis *et al.*, 1970a, 1970b; Hata, Yamamoto and Nakajima, 1991). Other colonic species apparently cannot utilize these oligosaccharides (Yazawa, Imai and Tamura, 1978). In turn this enhancement of bifidobacterial growth has been reported to reduce colon cancer in a mouse model (Koo and Rao, 1991).

Inulin, the storage saccharide of the *Compositae*, such as dandelions and dahlias, and the Jerusalem artichoke, consists of a linear polymer of over 20 fructose molecules, linked terminally to one D-glucose. Inulin is not digested by small intestinal amylases but is readily utilized by colonic bacteria with considerable gas production (Wang and Gibson, 1993).

Similar but smaller fructo-oligosaccharides (FOS) are present in small amounts in many food plants, such as the onion family, chicory and wheat.

While FOS (and inulin) support the growth of most faecal organisms, they favour the growth of bifidobacteria but not lactobacilli (Wang and Gibson, 1993; Gibson and Wang, 1994; Campbell, Fahey and Wolf, 1997). Bacteroides, clostridia and fusobacteria levels seem to be reduced (Gibson *et al.*, 1995). FOS raise butyrate levels, lower pH and raise stool weight (Campbell, Fahey and Wolf, 1997). FOS stimulate substantial hydrogen production and adaptation does not diminish this (Stone-Dorshow and Levitt, 1987).

In human subjects, most (89%) of ingested FOS is not absorbed in the small intestine, but none reaches the stools and only a small amount reaches the urine (Molis *et al.*, 1996). Thus most is metabolized in the colon, giving an energy value calculated as 9.5 kJ/g.

In human subjects, FOS significantly increase stool weight by about 20% and breath hydrogen level by over threefold; but they reduce iso-butyrate and isovalerate derived from amino acids by 94% and 77% respectively (Alles *et al.*, 1997).

Enhancing bifidobacterial levels in model *in vitro* fermentation systems resulted in an initial reduction in β-glucosidase, β-glucuronidase and nitroreductase levels, but they tended to rise in the long term. FOS addition did increase bifidobacterial levels (McBain and Mac-farlane, 1997). Bouhnik *et al.* (1996) reported that FOS increased bifidobacteria levels but had no effect on stool pH, nitroreductase, azoreductase or β-glucuronidase levels, nor on bile acid or total sterol levels. It appears that addition of FOS may not affect the risk of colon cancer.

10.17.2 *trans*-Galacto-oligosaccharides (GOS)

Substances containing Gal-Gal-Gal-Gal-Gal-Glc are also not digested in the small intestine and boost growth of colonic bifidobacteria and lacto-bacilli at the expense of *Enterobacter* species. These changes reduce nitror-eductase and β-glucuronidase activity and enhance β-glucosidase activity (Hudson and Marsh, 1995). Breath hydrogen excretion is reduced and, after adaptation, acetate and lactate production increase (Bouhnik *et al.*, 1996).

It is reported that these changes diminish the conversion of the HCA IQ to the more toxic 7-hydroxyIQ, a reaction confined to the clostridia and eubacteria, which tend to be reduced by GOS consumption, but may also be due to a lower gut pH (van Tassell *et al.*, 1989). Further, administration of GOS to mice genetically at risk of colon tumours markedly reduced tumour formation (Pierre *et al.*, 1997).

10.17.3 α-Gluco-oligosaccharides

These oligosaccharides have been shown both *in vitro* and *in vivo* in rats to be used as substrates to form SCFA and gas (Djouzi and Andrieuw, 1997). *Bacteroides thetaiotaomicron* was most efficient in this process and *Clostridium butyricum* least.

10.18 IRON

Iron can act as a powerful catalyst in free radical formation. Babbs (1990), having noted high levels of hydroxyl radicals in faeces, hypothesized that a high level of iron in colon contents, perhaps chelated to bile pigments, might catalyse transformation of precursors into carcinogens. Hydroxyl or peroxyl free radicals might be generated from superoxide or hydrogen peroxide resulting from bacterial metabolism. But the preponderance of anaerobic reducing metabolism in the colon might be expected to preclude this. Since small intestinal iron absorption is very tightly controlled to prevent excessive body iron levels, in populations where iron intake is consistently generous a high proportion of dietary iron will remain in effect unabsorbed and pass into the colon. Iron

Table 10.8 Iron content of lean flesh foods (all food trimmed of visible fat; values in mg/100 g edible portion). Source: Burlingame *et al.*, 1993, except * Holland *et al.*, 1991

Beef sirloin grilled	3.8	Beef heart stewed	7.9
Beef blade stewed	4.3	Beef liver stewed	7.8
Beef kidney stewed	8.0	Lamb leg roasted	2.2
Pork leg grilled	2.7	Venison roast leg	4.2
Chicken drumstick grilled	1.9	Chicken flesh roasted	1.0
Turkey roasted	1.4	Rabbit stewed	1.9
Trout brown baked	2.0	Hoki flesh baked	0.2
Snapper flesh baked	0.7	Tuna canned	2.3
Salmon steamed*	0.8	Salmon canned*	1.4
Lobster boiled*	0.8	Crab boiled*	1.3
Shrimps canned*	5.1	Mussels boiled*	7.7
Pilchards in tomato sauce*	2.7		

levels in red meats are substantially higher than in other flesh foods (Table 10.8).

However, except for the rectum, colon cancer rates are almost the same in men and in women, yet adult women absorb more dietary iron than men and commonly men consume much larger portions of meat, so men should have a substantially greater stool iron level. The presence of free radicals in faeces does not necessarily reflect the situation in colon contents, with a very low oxygen level.

Epidemiological evidence is conflicting. Data from the NHANES studies in the USA and from other countries does provide some support for a positive relationship between body iron stores and colon cancer rates, at least in men (AIRC, 1997). Weinberg (1994) and others have suggested that a protective effect against colon cancer of phytate might be due to its ability to bind iron.

10.19 CHOLESTEROL

Two prospective cohort studies in the US (Giovannucci *et al.*, 1994; Bostick *et al.*, 1994) failed to find any relationship between dietary cholesterol intake and colon cancer (AIRC, 1997). Various case-control studies have found a weak positive relationship (Howe *et al.*, 1997). The association with egg intake may be stronger than with cholesterol intake as such, which might reflect heated protein consumption. Cholesterol feeding in animal experiments seems to enhance experimental cancers (Steinmetz and Potter, 1994).

Cholesterol is normally well absorbed from diets with usual amounts of fat. Cholesterol entering the colon tends to be metabolized by bacteria; the eubacteria convert it to coprostanol by saturation of the B ring.

10.20 HETEROCYCLIC AROMATIC AMINES (HAA)

Application of high temperatures in frying, grilling or barbecuing to flesh foods, particularly beef, produces a number of pyrolytic compounds called HAAs. HAAs are formed by the interaction of creatine or creatinine, sugars and amino acids. These are established mutagens and

Table 10.9 HAAs that induce colon cancers in rats

Glu-P-1	2-amino-6-methyl (dipyrido [1,2-a: 3′, 2′-d] imidazole
Glu-P-2	2-amino-dipyrido [1,2-a; 3′, 2′-d] imidazole
IQ	2-amino-3-methyl-3H imidazo [4,5-f] quinoline
MeIQ	2-amino-3,4-dimethylimidazo [4,5-b] quinoline
PhIP	2-amino 1-methyl-6 phenylimidazo [4,5-b] pyridine

carcinogens and have been shown to induce colon cancers in rats (Nagao and Sugimura, 1993; Table 10.9).

It has been noted that, while normal colonic bacteria convert IQ to the 7-hydroxyl form, 7-OHIQ, although mutagenic in the Ames test, is not carcinogenic (Weisburger *et al.*, 1994). Thus colonic flora might play a role in reducing the risk from these compounds.

The role of HAA in colon carcinoma remains uncertain. Epidemiological evidence often implicates consumption of red meats (beef, lamb and pork) rather than other flesh foods (poultry and fish) as significant risk factors, but the studies are conflicting, some completely failing to show any relationship. It has recently been claimed (Cox and Whichelow, 1997) that studies supporting such a link tend to come from North America, while negative studies tend to come from Europe. The American custom of charring meat is contrasted with more traditional European gentle cooking methods like stewing and casseroling. If these compounds are important risk factors, the role of further metabolism by colonic microflora is not yet elucidated.

10.21 POLYCYCLIC AROMATIC HYDROCARBONS (PAH)

Smoking, broiling and curing of foods can induce the formation of these hydrocarbon compounds. Charcoal broiling of meat, where the fat drips into the flame and is pyrolized, is a common mechanism. The main example is benzo(α)pyrene, and dibenz(a,h)anthracene and benzanthracene are commonly demonstrable in foods (Santodonato, Howard and Basu, 1981). Benzo(α)pyrene has been extensively studied as a carcinogen in animal models. However they do not seem to have been associated with colon cancer in humans (NRC, 1982).

10.22 OVERVIEW AND CONCLUSIONS

There are many other factors that may have a bearing on the risk of colon cancer, of which genetics is undoubtedly a major example. However those that do not relate directly to the colonic microflora have not been pursued here. The critical issue is: what are the relative roles of procarcinogenic factors in the environment of the colonocytes, some of which, such as the secondary bile acids, are related to the microflora, *versus* the anticarcinogenic or protective factors, some of which again, notably the short-chain fatty acids, are intimately dependent on the microflora?

While the answer to that question must at present be speculative only, evidence accumulates that, given the genetic characteristics of the individual, currently not subject to alteration, the more important factor is the provision by the microflora of a substrate environment that favours colonocyte health and normal maturation and apoptosis. If this proves indeed to be the case, this is extremely encouraging because it means that perhaps the majority of the current burden of colon cancer can be prevented by manipulation of dietary composition. If we add the possible contribution of non-specific antimalignant agents such as free-radical scavengers in foods, the outlook is bright indeed.

10.23 ADDENDUM

Cummings (1998) has recently proposed that a factor in colon cancer could be reduction of sulphates and sulphites by the colon bacteria such as *Desulphovibrio* to form sulphides. Sources would include sulphur aminoacids in meat, and sulphur dioxide and sulphites used as preservatives. H_2S would be one such compound. More quantitative date are needed to evaluate these preliminary findings.

10.24 REFERENCES

Adlercreutz, H. (1995) Phytoestrogens: epidemiology and a possible role in cancer protection. *Environmental Health Perspectives*, **103**, 103–112.

AICR (1997) *Food, Nutrition and the Prevention of Cancer: A Global Perspective*, American Institute for Cancer Research, Washington, DC.

Alberts, D. S., Ritenbaugh, C., Story, J. A. *et al.* (1996) Randomised, double-blinded, placebo-controlled study of effect of wheat bran fiber and calcium on fecal bile acids in patients with resected adenomatous colon polyps. *Journal of the National Cancer Institute*, **88**, 81–92.

Alder, R. J., McKeown-Eyssen, G. and Bright-See, E. (1993) Randomised trial of the effect of calcium supplementation on fecal risk factors for colorectal cancer. *American Journal of Epidemiology*, **138**, 804–814.

Alles, M. S., Katan, M. B., Salemans, J. M. *et al.* (1997) Bacterial fermentation of fructooligosaccharides and resistant starch in patients with an ileal pouch–anal anastomosis. *American Journal of Clinical Nutrition*, **66**, 1286-1292.

Allison, C. (1995) *Nitrogen Metabolism of the Human Large Intestinal Bacteria*, PhD thesis quoted in Macfarlane, S. and Macfarlane, G. T. Proteolysis and amino-acid fermentation, in *Human Colonic Bacteria: Role in Nutrition, Physiology and Pathology*, (eds G. R. Gibson and G. T. Macfarlane), CRC Press, Boca Raton, FL.

Antalis, T. M. and Reeder, J. A. (1995) Butyrate regulates gene expression of the plasminogen activating system in colon cancer cells. *International Journal of Cancer*, **62**, 619–626.

Atwater, W. O. and Benedict, F. G. (1899) Experiments on the metabolism and energy in the human body, *Bulletin 69*, US Department of Agriculture, Washington, DC.

Axelsen, M., Kirk, D. N., Cooley, G. *et al.* (1982) The identification of the weak oestrogen equol (7-hydroxyphenylchroman) in human urine. *Biochemical Journal*, **201**, 353–357.

Babbs, C. F. (1990) Free radicals and the etiology of colon cancer. *Free Radicals in Biology and Medicine*, **8**, 191–200.

Baghurst, P.A., Baghurst, K. I. and Record, S. J. (1996) Dietary fibre, non-starch polysaccharides and resistant starch: a review. *Food in Australia*, **48**, S3–S35.

Bancroft, T., McNally, R. A. and Phillipson, A. T. (1944) Absorption of volatile fatty acids from the alimentary tract of the sheep and other animals. *Journal of Experimental Biology*, **20**, 120–129.

Bartram, H. P., Scheppach, W., Englert, S. *et al.* (1995) Effects of deoxycholic acid and butyrate on mucosal prostaglandin E2 release and cell proliferation in the human sigmoid colon. *Journal of Parenteral and Enteral Nutrition*, **19**, 182–186.

Bartram, H. P., Gostner, A., Kelber, E. *et al.* (1996) Effects of fish oil on fecal bacterial enzymes and steroid excretion in healthy volunteers: implications for colon cancer prevention. *Nutrition and Cancer*, **25**, 71–78.

Bayerdorffer, E., Mannes, G. A., Ochsenkuhn, T. *et al.* (1995) Unconjugated secondary bile acids in the serum of patients with colorectal adenomas. *Gut*, **36**, 268–273.

Bingham, S. (1997) Meat, starch and non-starch polysaccharides, are epidemiological and experimental findings consistent with acquired genetic alterations in sporadic colorectal cancer? *Cancer Letters*, **114**, 25–34.

Block, J. B., Dietrich, M. F., Leake, R. *et al.* (1990) Fecapentaene excretion: aspects of excretion in newborn infants, children, and adult normal subjects and in adults maintained on total parenteral nutrition. *American Journal of Clinical Nutrition*, **51**, 398–404.

Bonjour, J. P. (1991) Biotin, in *Handbook of Vitamins*, 2nd edn, (ed. L. J. Machlin), Marcel Dekker, New York, pp. 393– 427.

Bostick, R. M., Potter, J. D., Lushi, L. H. *et al.* (1994) Sugar, meat and fat intake and non-dietary risk factors for colon cancer incidence in Iowa women (United States). *Cancer Causes and Control*, **5**, 38–52.

Bouhnik, Y., Flourie, B., Riottot, M. *et al.* (1996) Effects of fructo-oligosaccharides ingestion on fecal bifidobacteria and selected metabolic indexes of colon carcinogenesis in healthy humans. *Nutrition and Cancer*, **26**, 21–29.

Brunton, V. G., Ozanne, B. W., Paraskeva, C. and Frame, M. C. (1997) A role for epidermal growth factor receptor, c-Src and focal adhesion kinase in an *in vitro* model for the progression of colon cancer. *Oncogene*, **14**, 283–293.

Bugaut, M. and Bentéjac, M. (1993) Biological effects of short-chain fatty acids in nonruminant mammals. *Annual Review of Nutrition*, **13**, 217–241.

Burkitt, D. P. (1971) Epidemiology of cancer of the colon and rectum. *Cancer*, **62**, 1713–1724.

Burkitt, D. P. (1973) Epidemiology of large bowel disease: the role of fibre. *Proceedings of the Nutrition Society of the UK*, **32**, 145–149.

Burkitt, D. P. and Trowell, H. C. (1975) *Refined Carbohydrate Foods and Disease. Some Implications of Dietary Fibre*, Academic Press, London.

Burlingame, B. A., Milligan, G. C., Apamerika, D. E. and Arthur, J. M. (1993) *The Concise New Zealand Food Composition Tables*, Crop & Food, Palmerston North.

Byers, T. (1995) Dietary fiber and colon cancer risk: the epidemiological evidence, in *Dietary Fiber in Health and Disease*, (eds D. Kritchevsky and D. Bonfield), Eagan Press, Minnesota, pp. 183–190.

Calloway, D. H. and Margen, S. (1971) Variation in endogenous nitrogen excretion and dietary nitrogen utilisation as determinants of human protein requirements. *Journal of Nutrition*, **101**, 205–216.

Campbell, J. M., Fahey, G. C. and Wolf, B. W. (1997) Selected indigestible oligosaccharides affect large bowel mass, cecal and fecal short-chain fatty acids, pH and microflora in rats. *Journal of Nutrition*, **127**, 130–136.

Cassidy, A., Bingham, S. and Setchell, K. D. R. (1994) Biological effect of a diet of soy protein rich in isoflavones on the menstrual cycle of premenopausal women. *American Journal of Clinical Nutrition*, **60**, 333–340.

Cherbonnel-Lasserre, C. L., Linares-Cruz, G., Rigaut, J. P. *et al.* (1997) Strong decrease in biotin content may correlate with metabolic alterations in colorectal carcinoma. *International Journal of Cancer*, **72**, 768–775.

Christi, S. U., Bartram, H. P., Ruckert, A. *et al.* (1995) Influence of starch fermentation on bile acid metabolism by colonic bacteria. *Nutrition and Cancer*, **24**, 67–75.

Christi, S. U., Bartram, H. P., Paul, A. *et al.* (1997) Bile acid metabolism by colonic bacteria in continuous culture: effects of starch and pH. *Annals of Nutrition and Metabolism*, **41**, 45–51.

Cobiac, L. (1994) Lactose: a review of intakes and of importance to health of Australians and New Zealanders. *Food in Australia*, **46**, S3–S27.

Coward, L., Barnes, S., Setchell, K. D. R. and Barnes, S. (1993) Genistein and Diadzein and their α-glucoside conjugates: antitumour isoflavones in soybean foods from American and Asian diets. *Journal of Agricultural and Food Chemistry*, **41**, 1961–1967.

Cox, B. D. and Whichelow, M. J. (1997) Frequent consumption of red meat is not risk factor for cancer. *Lancet*, **315**, 1018.

Cummings, J. H., Pomare, E. W., Branch, W. J. *et al.* (1987) Short chain fatty acids in human large intestine, portal, hepatic and venous blood. *Gut*, **28**, 1221–1227.

Cummings, J. H., Banwell, J. G., Segal, I. *et al.* (1990) The amount and composition of large bowel contents in man. *Gastroenterology*, **98**, A408.

Cummings, J. M. (1988) *New Scientist*, 8 August.

Dang, J., Wang, Y. and Doe, W. F. (1995) Sodium butyrate inhibits expression of urokinase and its receptor mRNAs at both transcription and post-transcription levels in colon cancer cells. *FEBS Letters*, **259**, 147–150.

DeKok, T. M., van Faassen, A., Bausch-Goldbohm, R. A. *et al.* (1992) Fecapentaene excretion and fecal mutagenicity in relation to nutrient intake and fecal parameters in humans on omnivorous and vegetarian diets. *Cancer Letters*, **62**, 11–21.

Deschner, E. E., Ruperto, J., Wong, G. and Newmark, H. L. (1991) Quercetin and rutin as inhibitors of azoxymethanol-induced colonic neoplasia. *Carcinogenesis*, **12**, 1193–1196.

Djouzi, Z. and Andrieuw, C. (1997) Compared effects on three oligosaccharides in metabolism of intestinal microflora in rats inoculated with a human faecal flora. *British Journal of Nutrition*, **78**, 313–324.

Dragsted, L. O., Strube, M. and Larsen, J. C. (1993) Cancer-protective factors in fruit and vegetables; biochemical and biological background. *Pharmacology and Toxicology*, **72**(Suppl. 1), 116–135.

Eisenbrand, G., Spiegelhalder, B. and Pressman, R. (1981) Analysis of human biological specimens for nitrosamine contents, in *Gastrointestinal Cancer: Endogenous Factors, Banbury Report 7*, (eds W. P. Bruce, P. Correa, M. Lipkin *et al.*), Cold Spring Harbor Symposia, New York.

Englyst, H. N., Wiggins, H. S. and Cummings, J. H. (1982) Determination of nonstarch polysaccharides in plant foods by gas–liquid chromatography of constituent sugars as alditol acetates. *Analyst*, **107**, 307–318.

Eyssen, H. and Caenepeel, P. (1988) Metabolism of fats, bile acids and steroids, in *Role of the Gut Flora in Toxicity and Cancer*, (ed. I. R. Rowland), Academic Press, London, p. 263.

FASEB (1987) Ad Hoc Panel on Dietary Fiber, in *Physiological Effects and Health Consequences of Dietary Fiber*, (ed. S. M. Pilch), FASEB, Bethesda, MD.

Finegold, S. M., Sutter, V. L. and Mathesen, G. E. (1983) Normal indigenous intestinal flora, in *Human Intestinal Microflora in Health and Disease*, (ed. D. J. Hentges), Academic Press, London, p. 13.

Franke, A. A. and Custer, L. J. (1996) Diadzein and genistein concentrations in human milk after soy consumption. *Clinical Chemistry*, **42**, 955–964.

Garland, C., Shekelle, R. E., Barrett-Connor, E. *et al.* (1985) Dietary vitamin D and calcium and risk of colorectal cancer: a 19-year prospective study in men. *Lancet*, **i**, 307–309.

Gibson, G. R. and Macfarlane, G. T. (1995) *Human Colonic Bacteria: Role in Nutrition, Physiology and Pathology*, CRC Press, Boca Raton, FL.

Gibson, G. R. and Wang, X. (1994) Enrichment of bifidobacteria from human gut contents by oligofructose using continuous culture. *FEMS Microbiology Letters*, **118**, 121–127.

Gibson, G. R., Beatty, E. E. R., Wang, X. and Cummings, J. H. (1995) Selective stimulation of bifidobacteria in the human colon by oligofructose and inulin. *Gastroenterology*, **108**, 975–982.

Gibson, G. R., Willems, A., Reading, S. and Collins, M. D. (1996) Fermentation of non-digestible oligosaccharides by human colonic bacteria. *Proceedings of the Nutrition Society of the UK*, **55**, 899–912.

Gibson, P. R., Moeller, I., Kagelari, O. and Folino, M. (1990) Contrasting effects of butyrate on the differentiation of normal and neoplastic colonic epithelial cells *in vitro*. *Gastroenterology*, **98**, A495.

Giovannucci, E. and Goldin, B. (1997) The role of fat, fatty acids, and total energy intake in the etiology of human colon cancer. *American Journal of Clinical Nutrition*, **66**, 1564S–1571S.

Giovannucci, E., Rimm, E. B., Stampfer, M. J. *et al.* (1994) Intake of fat, meat and fiber in relation to risk of colon cancer in men. *Cancer Reviews*, **54**, 2390–2397.

Goldbohm, R. A., van den Brandt, P. A., van 't Veer, P. *et al.* (1994) A prospective cohort study on the relation between meat consumption and the risk of colon cancer. *Cancer Research*, **54**, 718–723.

Govers, M. J., Termont, D. S., Lapre, J. A. *et al.* (1996) Calcium in milk products precipitates intestinal fatty acids and secondary bile acids and thus inhibits colonic cytotoxicity in humans. *Cancer Research*, **56**, 3270–3275.

Green, L. C., Wagner, D. A., Ruiz, K. *et al.* (1981) Nitrate biosynthesis in man. *Proceedings of the National Academy of Sciences of the USA*, **78**, 7764–7768.

Groenen-van Dooren, M. M., Ronden, J. E., Soute, B. A. and Vermeer, C. (1995) Bioavailability of phylloquinone and menaquinones after oral and colorectal administration in vitamin K-deficient rats. *Biochemistry and Pharmacology*, **50**, 797–801.

Hague, A., Manning, A. M., Hanlon, K. A. *et al.* (1993) Sodium butyrate induces apoptosis in human colonic tumour cell lines in a p-53-independent pathway: implications for the possible role of dietary fibre in the prevention of large-bowel cancer. *International Journal of Cancer*, **55**, 498–505.

Hassig, C. A., Tong, J. K. and Schreiber, S. L. (1997) Fiber-derived butyrate and the prevention of colon cancer. *Chemical Biology*, **4**, 783–789.

Hata, Y., Yamamoto, M. and Nakajima, K. (1991) Effects of soybean oligosaccharides on human digestive organs: estimation of fifty percent effective dose and maximum non-effective dose based on diarrhea. *Journal of Clinical Biochemistry and Nutrition*, **10**, 135–144.

Heerdt, B. G., Houston, M. A. and Augenlicht, L. H. (1997) Short-chain fatty acid-initiated cell cycle arrest and apoptosis of colonic epithelial cells is linked to mitochondrial function. *Cell Growth and Differentiation*, **8**, 523–532.

Hennigan, T. W., Sian, M., Matthews, J. and Allen-Mersh, T. G. (1995) Protective role of lactulose in intestinal carcinogenesis. *Surgical Oncology*, **4**, 31–34.

Higginson, J. and Oettle, A. G. (1960) Cancer incidence in the Bantu and Cape coloured races in South Africa: a report of a cancer survey in the Transvaal. *Journal of the National Cancer Institute*, **24**, 589–671.

Hill, M. J. (1974) Bacteria and the etiology of colonic cancer. *Cancer*, **345**, 815–818.

Hipsley, E. H. (1953) 'Dietary fibre' and pregnancy toxaemia. *British Medical Journal*, **ii**, 420–422.

Holland, B., Welch, A. A., Unwin, I. D. *et al.* (1991) *The Composition of Foods*, Royal Society of Chemistry, Cambridge.

Hoskins, L. C. (1992) Mucin degradation in the human gastrointestinal tract and its significance to enteric microbial ecology. *European Journal of Gastroenterology and Hepatology*, **5**, 205.

Hoskins, L. C., Agustines, M., McKee, W. B. *et al.* (1985) Mucin degradation in human colon ecosystems: Isolation and properties of fecal strains that degrade ABH blood group antigens and oligosaccharides from mucin glycoproteins. *Journal of Clinical Investigation*, **75**, 944–953.

Hoverstad, T. (1986) Studies of short-chain fatty acid absorption in man. *Scandinavian Journal of Gastroenterology*, **21**, 257–260.

Howe, G. R., Aronson, K. J., Benito, E. *et al.* (1997) The relationship between dietary fat intake and risk of colorectal cancer: evidence from the combined analysis of 13 case-control studies. *Cancer Causes and Control*, **8**, 215–228.

Hu, J., Liu, Y., Yu, Y. *et al.* (1991) Diet and cancer of the colon and rectum: a case-control study in China. *International Journal of Epidemiology*, **20**, 362–367.

Hudson, M. J. and Marsh, P. D. (1995) Carbohydrate metabolism in the colon, in *Human Colonic Bacteria: Role in Nutrition, Physiology and Pathology*, (eds G. R. Gibson and G. T. MacFarlane), CRC Press, Boca Raton, FL, pp. 61–73.

Hylla, S., Gostner, A., Dusel, G. *et al.* (1998) Effects of resistant starch on the colon in healthy volunteers: possible implications for cancer prevention. *American Journal of Clinical Nutrition*, **67**, 136–142.

Jacob, R. A., Kelley, D. S., Pianalto, F. S. *et al.* (1991) Immunocompetence and oxidant defense during ascorbate depletion of healthy men. *American Journal of Clinical Nutrition*, **54**, 1302S–1309S.

Jass, J. R. (1985) Diet, butyric acid and differentiation of gastrointestinal tract tumours. *Medical Hypotheses*, **18**, 113–118.

Joannou, G. E., Kelly, G. E., Reeder, A. Y. *et al.* (1995) A urinary profile study of dietary phytoestrogens: the identification and mode of metabolism of new isoflavonoids. *Journal of Steroid Biochemistry and Molecular Biology*, **54**, 167–184.

Johnson, A. O., Semenya, J. G., Buchowski, M. S. *et al.* (1993) Correlation of lactose maldigestion, lactose intolerance and milk intolerance. *American Journal of Clinical Nutrition*, **57**, 389–401.

Kingston, D. G., van Tassell, R. L. and Wilkins, T. D. (1990) The fecapentaenes, potent mutagens from human feces. *Chemistry Research and Toxicology*, **3**, 391–400.

Kirkman, L. M., Lampe, J. W., Campbell, D. R. *et al.* (1995) Urinary lignan and isoflavanoid excretion in men and women consuming vegetable and soy diets. *Nutrition and Cancer*, **24**, 1–12.

Klurfeld, D. M. and Bull, A. W. (1997) Fatty acids and colon cancer in experimental models. *American Journal of Clinical Nutrition*, 1530S–1538S.

Koo, M. and Rao, A. V. (1991) Long-term effect of bifidobacteria and neosugar on precursor lesions of colonic cancer in CF1 mice. *Nutrition and Cancer*, **16**, 249–257.

Kritchevsky, D. (1995) Epidemiology of fibre, resistant starch and colorectal cancer. *European Journal of Cancer Prevention*, **4**, 345–352.

Lamartiniere, C. A., Moore, J. B., Holland, M. and Barnes, S. (1995) Neonatal genistein chemoprevents mammary cancer. *Proceedings of the Society for Experimental Biology and Medicine*, **208**, 120–123.

McBain, A. J. and Macfarlane, G. T. (1997) Investigations of bifidobacterial ecology and oligosaccharide metabolism in a three-stage compound continuous culture system. *Scandinavian Journal of Gastroenterology (Supplement)*, **222**, 32–40.

McBain, J. A., Eastman, A., Nobel, C. S. and Mueller, G. C. (1997) Apoptotic death in adenocarcinoma cell lines induced by butyrate and other histone deacetylase inhibitors. *Biochemistry and Pharmacology*, **53**, 1357–1368.

Macfarlane, G. T., Gibson, S. A. W. and Gibson, G. R. (1992) Proteolytic activities of the fragilis group of *Bacteroides* species, in *Medical and Environmental Aspects to Anaerobes*, (eds B. I. Duerden, J. S. Brazier, S. V. Seddon and W. G. Wade), Wrightson, Petersfield, p. 159.

Macfarlane, S. and Macfarlane, G. T. (1995) Proteolysis and amino acid fermentation, in *Human Colonic Bacteria: Role in Nutrition, Physiology and Pathology*, (eds G. R. Gibson and G. T. Macfarlane), CRC Press, Boca Raton, FL.

McMichael, A. J. and Potter, J. D. (1985) Host factors in carcinogenesis: certain bile-acid metabolic profiles that selectively increase the risk of proximal colon cancer. *Journal of the National Cancer Institute*, **75**, 185–191.

Magee, P. N. (ed.) (1981) The possible role of nitrosamines in human cancer. *Banbury Report No 12*, Cold Spring Harbor Laboratory, New York.

Massey, R. C., Key, P. E., Mallett, A. K. and Rowland, I. R. (1988) An investigation of the endogenous formation of apparent total N-nitroso compounds in conventional microflora and germ-free rats. *Food Chemistry and Toxicology*, **26**, 595–600.

Mastromarino, A., Reddy, B. S. and Wynder, E. L. (1976) Metabolic epidemiology of colon cancer: enzymatic activity of fecal flora. *American Journal of Clinical Nutrition*, **29**, 1455–1460.

Messina, M., Persky, V., Setchell, K. D. R. and Barnes, S. (1994) Soy intake and cancer risk: a review of the *in vitro* and *in vivo* data. *Nutrition and Cancer*, **21**, 113–131.

Molis, C., Flourie, B., Quarne, F. *et al.* (1996) Digestion, excretion and energy value of fructooligosaccharides in healthy humans. *American Journal of Clinical Nutrition*, **64**, 324–328.

Muir, J. G. and O'Dea, K. (1992) Measurement of resistant starch: factors affecting the amount of starch escaping digestion *in vitro*. *American Journal of Clinical Nutrition*, **56**, 123–127.

Muir, J. G., Young, G. P., O'Dea, K. *et al.* (1994) Resistant starch – the neglected 'dietary fibre'? Implications for health. *Proceedings of the Nutrition Society of Australia*, **18**, 23–32.

Nagao, M. and Sugimura, T. (1993) Carcinogenic factors in food with relevance to colon cancer development. *Mutation Research*, **290**, 43–51.

Nagengast, F. M., Grubben, M. J. and vanMunster, P. (1995) Role of bile acids in colorectal carcinogenesis. *European Journal of Cancer*, **31A**, 1067–1070.

Nakagaki, M. and Nakayama, F. (1982) Effect of female sex hormones on lithogenicity of bile. *Japanese Journal of Surgery*, **12**, 13–18.

Nakaki, M., Takikawa, H. and Yamanaka, M. (1997) Targeting immunotherapy using the avidin–biotin system for a human colon adenocarcinoma *in vitro*. *Journal of International Medical Research*, **25**, 14–23.

Nakano, K., Mizuno, T., Sowa, Y. *et al.* (1997) Butyrate activates the WAF1/Cip1 gene promoter through Sp1 sites in a p-53-negative human colon cancer cell line. *Journal of Biological Chemistry*, **272**, 22199–22206.

Narisawa, T., Magadia, N. E., Weisburger, J. H. and Wynder, E. L. (1974) Promoting effects of bile acids on colon carcinogenesis after intrarectal instillation of N-methyl-N-nitro-N-nitrosoguanidine in rats. *Journal of the National Cancer Institute*, **53**, 1093–1097.

NAS/NRC (1981) *The Health Effects of Nitrate, Nitrite and N-nitroso Compounds*, National Academy Press, Washington, DC.

Newmark, H. L., Wargovich, M. J. and Bruce, W. R. (1984) Colon cancer and dietary fat, phosphate and calcium: a hypothesis. *Journal of the National Cancer Institute*, **72**, 1323–1325.

NRC (1982) *Report of the Committee on Diet, Nutrition and Cancer*, National Academy of Sciences, Washington, DC.

Owen, R. W. (1997) Faecal steroids and colorectal carcinogenesis. *Scandinavian Journal of Gastroenterology (Supplement)*, **222**, 76–82.

Peters, J. H., Riccio, E. S., Stewart, K. R. and Reist, E. J. (1988) Mutagenic activities of fecapentaene derivatives in the Ames/*Salmonella* test system. *Cancer Letters*, **39**, 287–296.

Pierre, F., Perrin, P., Champ, M. *et al.* (1997) Short-chain fructo-oligosaccharides reduce the occurrence of colon tumours and develop gut-associated lymphoid tissue in Mm mice. *Cancer Research*, **57**, 225–228.

Pongracz, J., Clark, P., Neoptolemos, J. P. and Lord, J. M. (1995) Expression of protein kinase C isoenzymes in colorectal cancer tissue and their differential activation by different bile acids. *International Journal of Cancer*, **61**, 35–39.

Povey, A. C., Plummer, S. M., Grafstrom, R. C. and Harris, C. C. (1990) Genotoxic mechanisms of fecapentaene-12 in human cells. *Progress in Clinical Biology Research*, **347**, 155–166.

Rackis, J. J. (1981) Flatulence caused by soya and its control through processing. *Journal of the Analytical Chemistry Society*, **58**, 503–509.

Rackis, J. J., Honig, D. H., Sessa, D. J. and Steggerda, F. R. (1970a) Flavor and flatulence factors in soybean protein products. *Journal of Agricultural and Food Chemistry*, **18**, 977–982.

Rackis, J. J., Sessa, D. J., Steggerda, F. R. *et al.* (1970b) Soybean factors related to gas production by intestinal bacteria. *Journal of Food Science*, **35**, 634–639.

Roberfroid, M. B. (1997) Health benefits of non-digestible oligosaccharides. *Advances in Experimental Medicine and Biology*, **427**, 211–219.

Rose, D. P. (1993) Diet, hormones and cancer. *Annual Review of Public Health*, **14**, 1–17.

Rowland, I. R. (1995) Toxicology of the colon: role of the intestinal microflora, in *Human Colonic Bacteria: Role in Nutrition, Physiology and Pathology*, (eds G. R. Gibson and G. T. Macfarlane), CRC Press, Boca Raton, FL.

Rowland, I. R., Mallett, A. K. and Wise, A. (1985) The effect of diet on the mammalian gut flora and its metabolic activities. *Critical Review of Toxicology*, **16**, 31–103.

Rowland, I. R., Granli, T., Bockman, O. C. *et al.* (1991) Endogenous N-nitrosation in man assessed by measurement of apparent total N-nitroso compounds in faeces. *Carcinogenesis*, **12**, 1395–1401.

Santodonato, J., Howard, B. and Basu, D. (1981) Health and ecological assessment of polynuclear aromatic hydrocarbons. *Journal of Environmental Pathology and Toxicology*, **5**, 1–364.

Savage, D. C. (1986) Gastrointestinal microflora in mammalian nutrition. *Annual Review of Nutrition*, **6**, 155–178.

Scheppach, W., Bartram, H. P. and Richter, F. (1995) Role of short-chain fatty acids in the prevention of colorectal cancer. *European Journal of Cancer*, **31A**, 1077–1080.

Schiffman, M. H., van Tassell, R. L., Robinson, A. *et al.* (1989) Case-control study of colorectal cancer and fecapentaene excretion. *Cancer Research*, **49**, 1322– 1326.

Schwartz, B., Lamprecht, S. A., Polak-Charcon, S. *et al.* (1995) Induction of the differentiated phenotype in human colon cancer cell is associated with the attenuation of subcellular tyrosine phosphorylation. *Oncology Research*, **7**, 277– 287.

Scrimshaw, N. S. and Murray, E. B. (1988) Prevalence of lactose malabsorption. *American Journal of Clinical Nutrition*, **48**, 1086–1098.

Setchell, K. D. R. (1985) Naturally occurring nonsteroidal estrogens of dietary origin, in *Estrogens in the Environment: Influence on Development*, (ed. J. McLachlan), Elsevier, New York, pp. 73–106.

Setchell, K. D. R. and Adlercreutz, H. (1988) Mammalian lignans and phytoestrogens: metabolism and biological roles in health and disease, in *The Role of Gut Microflora in Toxicity and Cancer*, (ed. I. A. Rowland), Academic Press, New York, pp. 315–345.

Setchell, K. D. R., Zimmer-Nechemias, L., Cai, J. and Heubi, J. E. (1997) Exposure of infants to phytoestrogens from soy-based formula. *Lancet*, **350**, 23–27.

Shulz, T. D. and Howie, B. J. (1986) In vitro binding of steroid hormones by natural and purified fibres. *Nutrition and Cancer*, **8**, 141–147.

Southgate, D. A. T. and Durnin, J. V. G. A. (1970) Calorie conversion factors. An experimental reassessment of the factors used in the calculation of the energy value of human diets. *British Journal of Nutrition*, **24**, 517.

Steinmetz, K. A. and Potter, J. D. (1994) Egg consumption and cancer of the colon and rectum. *European Journal of Cancer Prevention*, **3**, 237–245.

Stemmermann, G. N., Nomura, A. M. and Heilbrun, L. K. (1984) Dietary fat and the risk of colorectal cancer. *Cancer Research*, **44**, 4633–4637.

Sterchi, E. E., Mills, P. R., Fransen, J. A. M. *et al.* (1990) Biogenesis of intestinal lactase phlorizin hydrolase in adults with lactose intolerance. *Journal of Clinical Investigation*, **86**, 1329–1337.

Stone-Dorshow, T. and Levitt, M. D. (1987) Gaseous response to ingestion of a poorly absorbed fructo-oligosaccharide sweetener. *American Journal of Clinical Nutrition*, **46**, 61–65.

Suttie, J. W. (1995) The importance of menaquinones in human nutrition. *Annual Review of Nutrition*, **15**, 399–417.

Taylor, P. R., Schiffman, M. H., Jones, D. Y. *et al.* (1988) Relation of changes in amount and type of dietary fat to fecapentaenes in premenopausal women. *Mutation Research*, **206**, 3–9.

Tocchi, A., Basso, L., Costa, G. *et al.* (1996) Is there a causal connection between bile acids and colorectal cancer? *Surgery Today*, **26**, 101–104.

Trowell, H. (1972) Crude fibre, dietary fibre, and atherosclerosis. *Atherosclerosis*, **16**, 138–140.

Trowell, H. (1975) Refined carbohydrate: foods and fibre, in *Refined Carbohydrate Foods and Disease*, (eds D. P. Burkitt and H. Trowell), Academic Press, London, pp. 25–41.

van der Meer, R., Lapre, J. A., Govers, M. J. and Kleibeuker, J. H. (1997) Mechanisms of the intestinal effects of dietary fats and milk products on colon carcinogenesis. *Cancer Letters*, **114**, 75–83.

van Tassell, R. L., Kingston, D. G. and Wilkins, T. D. (1990) Metabolism of dietary genotoxins by the human colonic microflora: the fecapentaenes and heterocyclic amines. *Mutation Research*, **238**, 209–221.

van Tassell, R. L., Piccariello, T., Kingston, D. G. and Wilkins, T. D. (1989) The precursors of fecapentaenes: purification and properties of a novel plasmalogen. *Lipids*, **24**, 454–459.

Velasquez, C. C., Seta, R. W., Choi, J. *et al.* (1997) Butyrate inhibits deoxycholate-induced increase in colonic mucosal DNA and protein synthesis *in vivo*. *Diseases of the Colon and Rectum*, **40**, 1368–1375.

Venitt, S. and Bosworth, D. (1988) The bacterial mutagenicity of synthetic all-trans fecapentaene-12 changes when assayed under anaerobic conditions. *Mutagenesis*, **3**, 169–173.

Visek, W. J. (1972) Effect of urea hydrolysis on cell life-span and metabolism. *Federation Proceedings*, **31**, 1178–1193.

Visek, W. J. (1978) Diet and cell growth modulation by ammonia. *American Journal of Clinical Nutrition*, **31**, S216–S220.

Wagner, D. A. and Tannenbaum, S. R. (1985) In-vivo formation of N-nitroso compounds. *Food Technology*, **39**, 89–90.

Walker, A. R. P. and Arvidsson, U. B. (1954) Fat intake, serum cholesterol concentration and atherosclerosis in the South African Bantu. *Journal of Clinical Investigation*, **33**, 1358–1365.

Walker, R. (1990) Nitrates, nitrites and N-nitroso compounds: a review of the occurrence in food and diet and the toxicological implications. *Food Additives and Contaminants*, **7**, 717–768.

Wang, X. and Gibson, G. R. (1993) Effects of the in vitro fermentation of oligofructose and inulin by bacteria growing in the human large intestine. *Journal of Applied Bacteriology*, **75**, 373–380.

Warwick, G. P. (1971) Effect of the cell cycle on carcinogenesis. *Federation Proceedings*, **30**, 1760–1765.

Watanabe, Y., Tada, M., Kawamoto, K. *et al.* (1984) A case-control study of cancer of the rectum and the colon. *Nippon Shok Gakkai Zasshi*, **81**, 185–193.

Weinberg, E. D. (1994) Association of iron with colorectal cancer. *Biometals*, **7**, 211–216.

Weisburger, J. H., Jones, R. C., Wang, C. X. *et al.* (1990) Carcinogenicity tests of fecapentaene-12 in mice and rats. *Cancer Letters*, **49**, 89–98.

Weisburger, J. H., Rivenson, A., Reinhardt, J. *et al.* (1994) Genotoxicity and carcinogenicity in rats and mice of 2-amino-3,6-dihydro-3-methyl-7H-imidazolo[4,5-f] quinolin-7-one: an intestinal bacterial metabolite of 2-amino-3-methyl-3H-imidazo[4,5-f] quinoline. *Journal of the National Cancer Institute*, **86**, 25–30.

Willett, W. C., Staempfer, M. I., Colditz, G. A. *et al.* (1990) Relation of meat, fat and fiber intake to the risk of colon cancer in a prospective study among women. *New England Journal of Medicine*, **323**, 1664–1672.

Williams, N. N., Brannigan, A., Fitzpatrick, J. M. and O'Connell, P. R. (1992) Glutamine and butyric acid metabolism measurement in biopsy specimens (ex vivo): a method of assessing treatment of inflammatory bowel conditions. *Gastroenterology*, **102**, A713.

Williams, R. D. and Olmstedt, W. (1935) A biochemical method for determining indigestible residue (crude fibre) in faeces: lignin, cellulose and non-water soluble hemicelluloses. *Journal of Biological Chemistry*, **108**, 653–666.

Wilson, R. P., Muhrer, M. E. and Bloomfield, R. A. (1968) Comparative ammonia toxicity. *Comparative Biochemistry and Physiology*, **25**, 295–301.

Wolever, T. M. S., Spadafora, P. and Eshuis, H. (1991) Interaction between colonic acetate and propionate in humans. *American Journal of Clinical Nutrition*, **53**, 681–687.

Wrong, O. M. (1988) Bacterial metabolism of protein and endogenous nitrogen compounds, in *Role of the Gut Flora in Toxicity and Cancer*, (ed. I. R. Rowland), Academic Press, London, p. 227.

Yazawa, K., Imai, K. and Tamura, Z. (1978) Oligosaccharides and polysaccharides specifically utilisable by bifidobacteria. *Chemistry and Pharmacology Bulletin (Tokyo)*, **26**, 3306–3311.

Zarkovic, M., Qin, X., Nakatsuru, Y. *et al.* (1993) Tumor promotion by fecapentaene-12 in a rat colon carcinogenesis model. *Carcinogenesis*, **14**, 1261–1264.

Zoran, D. L., Turner, N. D., Taddeo, S. S. *et al.* (1997) Wheat bran diet reduces tumor incidence in a rat model of colon cancer independent of effects on distal luminal butyrate concentrations. *Journal of Nutrition*, **127**, 2217–2225.

11

Toxicological implications of the normal microflora

Ian R. Rowland

11.1 INTRODUCTION

In relation to its role in toxicity of chemicals and in the aetiology of cancer, the intestinal microflora can have both beneficial and detrimental influences. The participation of the microflora in toxic events is often mediated by metabolism, for example the conversion of an ingested substance into a form which is more or less toxic than the parent compound, resulting in activation or detoxification respectively.

In view of the importance of intestinal bacterial metabolism in mediating toxicological events, some general points need to be made to provide an introduction to the more detailed discussions of specific reactions described below.

Little is known about the ability of the individual species that comprise the microflora to metabolize nutrients and foreign compounds. Even if it were, it would be difficult to predict whether a reaction that occurred *in vitro* with a pure culture of a gut organism would proceed when that organism was surrounded by other members of the ecosystem within the mammalian gut. Indeed, marked differences have been reported in the metabolic activity of a gut anaerobe when measured *in vitro* and when the same organism was monoassociated with a gnotobiotic rat (Cole *et al.*, 1985). Information on the reactions performed by the individual species that constitute the microflora, therefore, is usually of little use.

A more valid and useful approach for understanding the role of the microflora in toxic events in humans is to consider the microflora as an

G. W. Tannock (ed.), *Medical Importance of the Normal Microflora*, 295–311.
© 1999 *Kluwer Academic Publishers. Printed in Great Britain.*

additional 'organ' within the host, ignoring its multiorganism composition (Midtvedt, 1989; Rowland, 1989).

The reactions catalysed by the gut microflora are limited by the prevailing conditions in the lumen of the intestinal tract, notably the low redox potential and the lack of oxygen. Consequently, oxidative reactions are rare. It should be noted, however, that the physicochemical conditions, particularly in oxygen tension, in the region close to the intestinal mucosa are likely to be very different from those in the lumen and may permit very different types of reaction to occur, although these have not been studied.

An important factor to be taken into account when considering foreign compound metabolism by gut bacteria is the location of the main population in the alimentary tract. The colon is the region of the gut that harbours the greatest number of bacteria; indeed, other areas of the human gastrointestinal tract are very sparsely populated. This does not mean, however, that only poorly absorbed chemicals encounter the colonic microflora. Substances, and their metabolites, may enter the colon across the intestinal wall from the blood or may reach the colon after excretion in the bile. Thus there is ample opportunity for a wide variety of materials in the diet to encounter and be metabolized by the colonic microflora.

Metabolism of foreign or endogenously produced substances by the colonic microflora can have wide ranging implications for human health, with both beneficial and detrimental consequences. These are summarized in Table 11.1 and are considered in more detail below.

Table 11.1 Toxicological implications of gut microflora activities

1. Activation to toxicants, mutagens, carcinogens	Azo compounds
	Nitro compounds
	Plant glycosides
	IQ
2. Synthesis of carcinogens and mutagens	N-nitroso compounds
	Fecapentaenes
3. Synthesis of promoters	Bile acids
	Protein breakdown products (ammonia, phenols, cresols)
	Fecapentaenes
4. Enterohepatic circulation and deconjugation	Steroid hormones and drugs
	Carcinogens
5. Detoxification/protection	Methylmercury
	Phyto-oestrogens
	Flavonoids

11.2 ACTIVATION OF CHEMICALS TO TOXIC, MUTAGENIC AND CARCINOGENIC DERIVATIVES

11.2.1 Azo compounds

A number of colouring agents used in the production of food, cosmetics, textiles, leather and for paper printing are based on azo compounds. The intestinal microflora can reduce these compounds to varying extents to produce, ultimately, amines. The reduction products are often more toxic than the parent compound, e.g. the food dye brown FK, which after bacterial reduction causes vacuolar myopathy in rats given high doses.

Workers exposed to direct black 38, a dye used in the leather and textile industry, have an elevated risk of bladder cancer, which has been attributed to the reduction of the dye by the gut microflora to benzidine, a proven human bladder carcinogen (Cerniglia *et al.*, 1982). Incubation of direct black 38 with cultures of human intestinal bacteria results in rapid formation of benzidine, indicating that the microflora is the major site of metabolism of the dye. Studies with more sophisticated, continuous culture systems have revealed that several of the other urinary metabolites of the dye, including the *N*-acetylated derivative, can be attributed to gut microflora metabolism (Manning, Cerniglia and Federle, 1985). Since acetylation is a crucial step in the activation of aromatic amines to their ultimate carcinogenic metabolite, these studies implicate the gut microflora not only in the overall metabolism in the body of benzidine-based azo dyes but also in the generation of the reactive, carcinogenic species.

11.2.2 Nitro compounds

There is extensive human exposure, both deliberate and accidental, to nitro compounds. For example, heterocyclic and aromatic nitro compounds are extensively used in industry as important intermediates in the manufacture of consumer products and comprise an important class of drugs with uses as antibiotic, antiparasitic and radiosensitizing agents. In addition, nitroaromatics are ubiquitous environmental contaminants derived from diesel exhaust, cigarette smoke and airborne particulates. These compounds often possess toxic, mutagenic and carcinogenic activity and so may contribute to the environmental cancer risk in humans.

Reduction of the nitro group to an amine occurs via nitroso and hydroxylamino intermediates and is usually required for the pharmaceutical and toxicological activity of these compounds to be expressed. For example, reduction is a crucial step in the antitrichomonad activity and mutagenicity of metronidazole (Lindmark and Muller, 1976) and in the induction of methaemoglobinaemia by nitrobenzenes (Reddy, Pohl and Krishna, 1976). The nitro group on nitroaromatic compounds can be reduced by both hepatic- and gut-microflora-associated reductases, but in most cases the latter appear to be the more important. Evidence for the importance of bacterial reductases in the induction of methaemoglobi-naemia by nitrobenzene has been obtained by exposing conventional microflora, antibiotic-treated and germfree rats to the compound. Methaemoglobin levels of 30–40% were induced in the conventional rats, whereas no increase was detected in the animals without a gut microflora (Reddy, Pohl and Krishna, 1976). Levin and Dent (1982) demonstrated that nitrobenzene was readily reduced by caecal contents to aniline, via nitrosobenzene and phenylhydroxlamine (the presumed methaemoglobinogenic metabolites). Although aniline was also pro-duced in the presence of liver microsomes, the rate of reduction was less than 1% of that in the presence of bacteria. There is conclusive evidence, therefore, that the intestinal microflora is a major determinant of both the metabolism and the acute toxicity of nitrobenzene.

The toxicity of the important chemical intermediates dinitrotoluenes has been shown to be similarly dependent on the reductive activity of the intestinal microflora. In this case, genotoxicity and the ability to bind covalently to macromolecules (a critical early step in tumorigenesis) was lower in germfree than in conventional rats (Mirsalis *et al.*, 1982).

Studies in germfree and conventional microflora rats and in *in vitro* cultures have demonstrated the crucial role of gut-microflora-mediated reduction in the metabolism of nitrated polycyclic aromatic hydrocarbons such as 1-nitropyrene and 6-nitrochrysene (El-Bayoumy *et al.*, 1983; Man-ning *et al.*, 1988). It is significant that 6-aminochrysene is known to induce liver and lung tumours in mice, suggesting that nitroreduction by the gut microflora plays an important role in the activation and tumorigenesis by nitrochrysene.

11.2.3 Plant glycosides

Plant glycosides comprise a wide variety of low-molecular-weight sub-stances linked to sugar moieties (Brown, 1988). They are widely distrib-uted, in fruits and vegetables and in beverages, such as tea and wine, derived from plants. Estimates of human intake of plant glycosides range from 50–1000 mg per day (Brown, 1988). In their glycosidic form most of these substances are relatively harmless. It is usually assumed that they are poorly absorbed from the gut and thus pass intact into the colon; however, there is some evidence that flavonoid glycosides may be at least partially absorbed (Hollman *et al.*, 1995). Glycosides reaching the colon are subjected to the action of β-glycosidases associated with the resident microflora, which cleave the sugar moiety releasing aglycones, which have a variety of biological activities, including toxicity, mutagenicity and carcinogenicity. Amygdalin, a glycoside found in several drupes and pomes is hydrolysed to mandelonitrile, which subsequently decomposes to cyanide. After oral exposure to amygdalin, germfree rats, in contrast to rats with a conventional microflora, have lower thiocyanate levels in the blood and do not exhibit cyanide toxicity. Cyanogenic glycosides are also present in cassava, a staple crop in many countries in the tropics. In periods of food shortage, the plants are sometimes consumed without appropriate treatment for removing the glycosides, leading to cyanide exposure and acute poisoning or chronic neurological disease (Brown, 1988).

Cycasin (a glycosidic component of cycad nuts) is hydrolysed by bacterial β-glucosidase in the colon of conventional rats, releasing the aglycone methylazoxymethanol (MAM), which induces colon tumours. No such tumours are found in germfree rats fed the glycoside.

Many of the plant glycosides are mutagenic after hydrolysis by bacterial glycosidases. Some of the most common of these are the glycosides of quercetin, including rutin (quercetin-O-rutinoside) and quercetrin (quercetin-O-rhamnoside), which are found in lettuce and many other edible plants. When tested for mutagenicity using the *Salmonella*/microsome assay, mutagenic activity in these plant materials can only be detected when a faecal bacterial extract is added to the incubation mixture (Van der Hoeven *et al.*, 1983). Similarly, potent mutagenic activity can be detected in red wine and tea when the beverages are incubated with an extract of human faeces, a finding attributed to the presence of plant glycosides. Apart from cycasin, plant glycosides do not appear to be carcinogenic despite their demonstrable mutagenic activity. Quercetin has been investigated for carcinogenicity in a number of animal models and in general has been found not to exhibit tumorigenic properties (Brown, 1988). In two-stage carcinogenicity studies in rats with quercetin, no evidence was found for initiating activity for liver tumours (using phenobarbitone and partial hepatectomy as promoters) or for tumour-promoting activity (using MAM as an initiator). Similar lack of initiating and promoting activity towards rat bladder carcinogenesis has been reported. Assessment of the toxicological significance of glycoside hydrolysis by intestinal microflora is complicated by reports of potential anticarcinogenic and antimutagenic effects of flavonoid aglycones (see below).

It is clear, therefore, that hydrolysis of plant glycosides in the gut can lead, potentially, to both adverse and beneficial consequences for humans. Further studies are needed to establish the repercussions for colon cancer of flavonoid release from glycosides by β-glycosidases in the gut.

11.2.4. IQ

2-amino-3-methyl-3*H*-imidazo(4,5-*f*)quinoline (IQ) is one of several heterocyclic amine compounds formed in small quantities when meat and fish are grilled or fried. It induces tumours at various sites in rodents including the large intestine, suggesting that it may play a role in the aetiology of colon cancer in humans (Ohgaki, Takayama and Sugimura, 1991). Incubation of IQ with a human faecal suspension yields the 7-keto derivative 2-amino-3,6-dihydro-3-methyl-7*H*-imidazo(4,5-*f*)quinoline-7-one (7-OHIQ). The 7-keto metabolite has been found in faeces of individuals consuming a diet containing a high level of fried meat, indicating that the formation of 7-OHIQ can occur *in vivo* in humans (Carman *et al.*, 1988). Unlike IQ itself, the bacterial metabolite is a direct-acting and potent mutagen in *Salmonella typhimurium* and induces DNA damage in colon cells *in vitro* (Carman *et al.*, 1988; Rumney, Rowland and O'Neill, 1993; Rowland and Pool-Zobel, unpublished observation, 1996). Thus there is strong evidence for the bacterial formation in the human gut of a directly genotoxic derivative of a dietary carcinogen.

11.3 SYNTHESIS OF CARCINOGENS

11.3.1 Nitrate reduction and N-nitroso compound synthesis

Nitrate, ingested via diet and drinking water, is readily converted by the nitrate reductase activity of the intestinal microflora to its more reactive and toxic reduction product, nitrite. Nitrite reacts with nitrogenous compounds such as amines, amides and methylureas in the body to produce N-nitroso compounds, many of which are highly carcinogenic. The reaction can occur chemically in the acidic conditions prevalent in the human stomach and can also be catalysed at neutral pH by gut bacteria. To assess the significance of the gut microflora in nitrosation reactions in the body, Massey *et al.* (1988) investigated the concentration of N-nitroso compounds in tissues and gut contents of germfree and conventional rats. The compounds were detected in significant amounts only in gut contents of the conventional rats, with the largest quantities being found in the colon contents. In contrast, in the germfree rats, little or no nitroso compounds were detected in the gut contents. The enzyme involved in the nitrosation process has not been isolated, but from studies in pure cultures of bacteria, it has been suggested that nitrosation may be associated with the activity of nitrate and nitrite reductases (Leach, 1988). Subsequently, the investigations have been extended to humans. N-nitroso compounds were detectable at concentrations up to $590\,\mu g/kg$ faeces of male and female volunteers (Rowland *et al.*, 1991). The importance of ingested nitrate to the level of faecal nitroso compounds was demonstrated by providing the volunteers with nitrate-free food and water. Under these dietary conditions, faecal nitroso compound concentrations decreased to almost undetectable values over a period of 5 days. When nitrate was added to the drinking water a rapid rise in faecal nitroso compound concentration was seen, reaching a level similar to that recorded for the volunteers on their own diets.

11.3.2 Fecapentaenes

Using the Ames *Salmonella* assay, mutagenic activity has been detected in human faecal samples. Faeces from individuals excreting large amounts of mutagenic activity have been extracted and the mutagenic agents purified. Structural analysis revealed a glyceryl ether compound containing a pentaene moiety with a chain length of 12 or 14 (Bruce *et al.*, 1982; Hirai *et al.*, 1982). They have been termed fecapentaenes 12 or 14 (FP12, FP14). Both types are found in faeces although the ratio varies considerably. The gut microflora has been implicated in their synthesis by the demonstration of fecapentaene production *in vitro* by faecal suspensions under anaerobic conditions and the inhibition of that synthesis by antibiotics and heat sterilization (Hirai *et al.*, 1982).

Although FP occur in faeces of the majority of Western populations, more detailed epidemiological studies have revealed some anomalies. For example, lower FP levels were found in faeces from colorectal cancer patients than in controls and faecal excretion of FP is higher in vegetarians, a population at low risk from colon cancer (Schiffman *et al.*, 1989; De Kok *et al.*, 1992). These and other studies suggest that there is an inverse correlation between FP excretion and risk of colon cancer. This has a certain logic in that FP is a very potent direct-acting mutagen and would be expected to react extremely rapidly with DNA and other macromolecules in the colonic mucosa. Thus increased faecal excretion may reflect lower endogenous exposure to the genotoxin.

At low concentrations (0.6–10 μg/ml), fecapentaene-12 induces single strand DNA breaks, gene mutations, chromosome aberrations, sister chromatid exchanges and unscheduled DNA synthesis in human fibroblasts *in vitro* (Plummer *et al.*, 1986). Results of *in vivo* studies have yielded more equivocal results. Fecapentaene-12 increased proliferation of mucosal cells and DNA single-strand breaks when it was instilled into rat colon (Hinzman *et al.*, 1987). Similarly, intrarectal administration to mice increased mitosis, but not nuclear aberrations, in colonic crypt cells. Rodent bioassays have indicated that fecapentaene-12 does not have carcinogenic or tumour-initiating activity (Ward *et al.*, 1988; Weisburger *et al.*, 1990; Shamsuddin *et al.*, 1991). However, Zarkovic *et al.* (1993) have provided evidence for tumour-promoting activity of fecapentaene and this is consistent with some of the data described above, which indicates that the compound induces mucosal cell proliferation in the colon.

11.4 SYNTHESIS OF TUMOUR PROMOTERS

Investigations of possible tumour promoters generated by bacterial activity in the colon have focused on bile acids and protein breakdown products and have been extensively reviewed (Hill, 1988; Clinton, 1992), so a brief summary only will be provided here.

11.4.1 Secondary bile acids

The liver secretes two major bile acids – cholic acid and chenodeoxycholic acid – which are subject to extensive metabolism by the intestinal microflora (MacDonald *et al.*, 1993). Of particular importance *in vivo* is 7-α-dehydroxylation, which converts cholic to deoxycholic acid and chenodeoxycholic to lithocholic acid. These two secondary bile acids, which comprise over 80% of faecal bile acids, are postulated to play an important role in the aetiology of colon cancer by acting as promoters of the tumorigenic process. There is considerable evidence to indicate that acid steroids, in particular secondary bile acids, can exert a range of biological effects. They induce cell necrosis, hyperplasia, metabolic alterations and DNA synthesis in intestinal mucosal cells, enhance the genotoxicity of a number of mutagens in *in vitro* assays, and exhibit tumour-promoting activity in the colon (reviewed by Rowland, Mallett and Wise, 1985). There is also some recent evidence that secondary bile acids, albeit at relatively high concentrations, can induce DNA damage in colon cells in culture (Venturi *et al.*, 1997).

11.4.2 Protein and amino acid metabolites

Epidemiological studies indicate a link between high intake of protein and incidence of colon cancer. These studies, together with investigations *in vitro* and in laboratory animals, have led to the hypothesis that colonic ammonia, produced by bacterial catabolism of amino acids and other nitrogenous compounds, may participate in tumour promotion (Clinton, 1992). Increasing the protein content of the diet increases the colonic luminal ammonia concentration. Ammonia exhibits a number of effects that suggest that it may be involved in promotion, including increasing mucosal cell turnover and altering DNA synthesis. More definitively, it has been shown to increase the incidence of colon carcinomas induced by N-methyl-N-nitro-N-nitrosoguanidine in rats. Less well studied than ammonia has been the formation of phenols and cresols by bacterial deamination of certain aromatic amino acids in the colon. The generation of phenols is higher when high-protein diets are consumed and there is some evidence for tumour-promoting activity, albeit in a mouse skin assay (Hill, 1988).

11.4.3 Fecapentaenes

Evidence has been obtained that fecapentaenes may possess tumour-promoting activity in a rat colon carcinogenesis model using N-methyl-N-nitrosourea (MNU) as an initiating agent (Zarkovic *et al.*, 1993). The number of carcinoma-bearing rats and the number of carcinomas per rat were significantly higher in the animals given MNU and fecapentaene compared with those given MNU alone.

11.5 β-GLUCURONIDASE AND ENTEROHEPATIC CIRCULATION

Many toxic and carcinogenic compounds and also endogenously produced compounds such as steroids are metabolized in the liver and then conjugated by Phase II enzymes to glucuronic acid, sulphate, glutathione or amino acids before being excreted via the bile into the small intestine. In the colon, the bacterial enzyme activities such as β-glucuronidase, sulphatase and C-S lyase can hydrolyse the conjugates, releasing the parent compound, or its hepatic metabolite. Since the deconjugated compounds are usually more easily absorbed, deconjugation may result in an enterohepatic circulation as the compound is reabsorbed, returning to the liver where it can be subjected to further metabolism and conjugation. Enterohepatic circulation results in xenobiotics and steroids being retained in the body, thus potentiating their pharmacological, physiological and toxic effects. Enterohepatic circulation appears to be of particular importance in the absorption and metabolism of endogenously produced and synthetic steroids, including oestrone, oestriol, pregnanolone, tetrahydrocortisone and diethylstilboestrol. Bacterial β-glucuronidase and sulphatase acting on the conjugates of these steroids can thus affect hormonal activity and bioavailability of these steroids and their metabolites (Eyssen and Caenepeel, 1988).

A number of drugs such as chloramphenicol, diflunisal, fenclofenac, phenacetin and phenobarbitone, also undergo enterohepatic circulation, which influences their pharmaceutical and toxicological effects. For example, chloramphenicol is excreted as a glucuronide conjugate in the bile of rats and is subsequently hydrolysed in the gut and reduced to an arylamine. This is then reabsorbed and exerts a toxic effect on the thyroid (Larson, 1988).

In the case of carcinogens and mutagens, the activity of β-glucuronidase in the colon may increase the likelihood of tumour induction. For example, with the colon carcinogen 1,2-dimethylhydrazine (DMH), small amounts of procarcinogenic glucuronide conjugates of the activated metabolite (methylazoxymethanol; MAM) formed in the liver are excreted in the bile. Hydrolysis of the conjugates by colonic bacteria releases the MAM in the colon. As might be expected, germfree rats have fewer colon tumours after treatment with DMH or with MAM–glucuronic-acid conjugate than do conventional animals (Reddy *et al.*, 1974). The carcinogen benzo(α)pyrene (BaP), a common contaminant of the human diet, undergoes a similar sequence of reactions to that of DMH. Polar conjugates, including the glucuronide (about 35% of the dose), are excreted in bile following administration of BaP to rodents and are hydrolysed to BaP and BaP diols by gut bacteria (Chipman, Millburn and Brooks, 1983). These products, unlike the conjugates from which they are derived, bind to DNA and are mutagenic in *in vitro* assays. It is possible, therefore, that bacterial metabolites of BaP formed in the colon may be initiators of carcinogenesis. Other carcinogens to which β-glucuronidase activity is important for their action include benzidine, nitropyrene and dinitrotoluenes.

11.6 DETOXIFICATION AND PROTECTIVE EFFECTS OF THE MICROFLORA

11.6.1 Demethylation of methylmercury

The high neurotoxicity of methylmercury (MeHg) is due to a combination of its efficient absorption from the gut, long retention time in the body and ability to penetrate the blood–brain barrier.

Because of its rapid and virtually complete absorption from the gut, the body-burden of mercury after methylmercury exposure is determined largely by its rate of elimination, the main route of which is the faeces (Clarkson, 1979). In rats and mice given MeHg, the majority (50–90%) of the mercury in faeces is in the mercuric form; therefore, demethylation of MeHg would appear to be an important step in the excretion of the organomercurial from the body. The role of the gut microflora in this process has been reviewed recently (Rowland, 1995), so will be discussed only briefly below.

Ingested MeHg undergoes enterohepatic circulation and reaches the gut microflora as a MeHg–glutathione complex in bile. Incubation of methylmercuric chloride (or its glutathione complex) with gut contents from the rat or mouse, or with suspensions of human faeces, results in extensive metabolism of the organomercurial, with a variety of metabolites being produced including elemental mercury and mercuric ion. Studies in germfree rodents and animals treated with antibiotics to suppress their gut bacteria provides further evidence for the importance of gut bacteria in demethylation *in vivo*. Suppression or absence of the gut microflora was associated with decreased excretion of total mercury, lower amounts of mercuric ion in faeces and significantly higher body-burdens of mercury by comparison with conventional animals. The concentration of mercury in the brain after MeHg exposure was found to be

25–45% greater in germfree or antibiotic-treated animals than in controls and was associated with a higher incidence of neurotoxicity and neuropathological damage.

Changes in gut microflora with age may influence susceptibility to MeHg toxicity. Suckling mice absorb and retain the majority of an oral dose of MeHg (half-life of mercury elimination, $t^{1/2}$, greater than 100 days) whereas older mice excrete the mercury much more rapidly ($t^{1/2}$ 6–10 days). This developmental change in rate of excretion coincides with an increase in the demethylation activity of the gut microflora during weaning. Corresponding developmental changes in demethylating capacity occur in the human gut microflora. The virtual absence of demethylating capacity in faecal suspensions from unweaned human babies implies that they would absorb and retain more of an oral dose of MeHg than adults, making them more susceptible to MeHg neurotoxicity.

11.6.2 Phyto-oestrogens

Grains and other fibre-containing foods are sources of plant lignans such as secoisolariciresinol and matairesinol. When ingested, these substances are converted to enterolactone and enterodiol (the major lignans found in human urine) by hydrolysis, dehydroxylation, demethylation and oxidation reactions catalysed by the facultative anaerobes of the intestinal microflora (Setchell and Adlercreutz, 1988). Similarly, phyto-oestrogens in soy, such as glycosides of genistein, and diadzein, are modified by the intestinal microflora, with the isoflavan equol being a major metabolite, which is excreted in urine.

Lignans and isoflavones are structurally similar to the potent synthetic oestrogens such as diethylstilboestrol, and possess hormonal activity mediated by the intestinal microflora. In general, however, lignans and isoflavones are weakly oestrogenic and in animals may also exhibit anti-oestrogenic properties (Setchell and Adlercreutz, 1988). There have been a number of studies on the potential beneficial effects of phyto-oestrogens towards cancer. Epidemiological studies provide good evidence of a protective role for soy phyto-oestrogens against hormonal and other cancers, including breast, prostate and colon (Adlercreutz, 1993). In the case of lignans, fewer and less comprehensive epidemiological investigations have been carried out; nevertheless there is some support for cancer-preventing activity. Studies in laboratory animals have yielded equivocal results. Supplementation of a high-fat diet with 5% flax-seed flour, a rich source of lignans that resulted in a large increase in urinary enterolactone and enterodiol concentration, reduced cell proliferation and nuclear aberrations (considered to be early markers for carcinogenesis) in rat mammary gland, although no consistent influence on tumour incidence was detected (Serraino and Thompson, 1991a, b). Evidence that lignans may be protective against colon cancer has been obtained from a study in which rats, treated with azoxymethane as an inducer of colon tumours, were fed a basal diet with or without supplementation with flax-seed oil (Serraino and Thompson, 1992). The incidence of aberrant crypt foci (an early marker of neoplasia) in the colons of the rats after 4 weeks was reduced by about 50% in the rats fed the flax-seed oil diet.

A number of possible mechanism have been proposed for the anticarcinogenic effects of phyto-oestrogens. In hormonally dependent cancers, it is possible that the lignans may reduce the sensitivity of oestrogen receptors to oestradiol, or that the anti-oestrogenic properties of enterodiol and enterolactone may inhibit the action of oestrogen-stimulated tumour growth. Similar mechanisms could be operating in the inhibition of colon tumours, since androgen and oestrogen receptors have been identified in some primary colon tumours. However, there is also evidence for more direct, non-hormonal effects; for example, enterolactone exhibits antimitotic activity *in vitro*. There is clearly an urgent need for further studies in this area, in particular to elucidate the importance of the microbial metabolites in the beneficial effects.

11.6.3 Plant flavonoids

In addition to the capacity of flavonoid glycosides to exert toxic and genotoxic effects after hydrolysis by gut microbial enzymes (see 11.2.3 above), there is evidence that they may also afford protection against genotoxic and carcinogenic activity of other chemicals. Quercetin has been shown to antagonize tumour induction by carcinogens such as benzo(α)pyrene and 4-dimethylaminoazobenzene, and inhibits the mutagenic activity of benzo(α)pyrene and its diol-epoxide metabolite. Quercetin, myricetin and morin exhibit potent antimutagenic effects *in vitro* against a range of heterocyclic amine carcinogens in the human diet, including IQ, MeIQ and MeIQx (Alldrick, Flynn and Rowland, 1986). The mechanism of action of the flavonoids appears to be inhibition of the activity of the hepatic enzymes responsible for activating the amines to their ultimate genotoxic species. The enzymes involved, cytochrome P450 IA-dependent mixed function oxidases, as well as being present in the liver are also found in the intestinal tract. It is possible, therefore, that flavonoids released from their glycosidic forms by β-glycosidases in the gut may inhibit the bioactivation, by gut mucosal oxidases, of dietary carcinogen precursors to their reactive species. Indeed, quercetin and rutin (a glycosidic derivative) have been shown to decrease the incidence of colon tumours induced by azoxymethane in mice, although the effect appeared to be on the promotional phase of the carcinogenic process, possibly by decreasing gut mucosal cell proliferation (Deschner *et al.*, 1991).

11.6.4 Binding of carcinogens

There are a number of reports describing the binding, by freeze-dried preparations of intestinal bacteria and particularly lactic acid bacteria, of a variety of dietary carcinogens including the heterocyclic amines formed during cooking of meat, the fungal toxin aflatoxin B_1 (AFB_1) and the food contaminant AF2 (Morotomi and Mutai, 1986). The extent of the binding was dependent on the mutagen and on the bacterial strain used. Significant binding ability was observed towards the heterocyclic amines 3-amino-1,4-dimethyl-5H-pyrido (4,3-b) indole (Trp-P-1), 3-amino-1-methyl-5H-pyrido (4,3-b) indole (Trp-P-2), and the least binding was observed with AF2 and AFB_1. In some studies, binding was associated with a concomitant decrease in mutagenic activity *in vitro* (Morotomi and Mutai, 1986; Orrhage *et al.*, 1994). The binding of carcinogens was reported to be pH-dependent and to occur with dead cells. These results indicate that the binding is a physical phenomenon, mostly due a cation exchange mechanism involving cell wall peptidoglycans and polysaccharides (Zhang and Ohta, 1991).

Carcinogen binding has important implications for carcinogenesis, since it could lead, in theory, to a decrease in bioavailability of ingested carcinogens in the gut, reducing their capacity to damage the intestinal mucosa and decreasing their absorption into the blood and hence their interaction with other tissues in the body. There is some experimental evidence in rats that administration of lactic acid bacteria decreases the amount of orally administered carcinogens reaching the blood (Zhang and Ohta, 1993). However, the animals were probably in an abnormal physiological state, having been starved for 4 days. More recently, Bolognani, Rumney and Rowland (1997) reported no effect of consumption of lactic acid bacteria on absorption or genotoxic activity *in vivo* of carcinogens in normally fed rats.

The results of the latter study suggest that, although LAB may bind carcinogens *in vitro*, this has little biological significance *in vivo* either for absorption and distribution of carcinogens in the body, or for their genotoxic activity in the liver.

11.7 FACTORS AFFECTING GUT MICROFLORA METABOLISM AND TOXICITY

A large number of factors can influence both metabolism of the gut microflora and its involvement in toxic events. Such factors include age, drugs, diet and species differences and have been reviewed previously (Mallett and Rowland, 1988)

11.7.1 Species differences in gut microflora metabolism

Although laboratory animals are very convenient for studying metabolism and toxic events and enable monitoring of the microflora in different gut regions as well as faeces, their use as models for the human faecal microflora and its metabolism is questionable. Differences exist in the bacterial composition of the gut microfloras of animals and humans (Drasar, 1988) and, more importantly, in gut microbial enzymic activity (Rowland *et al.*, 1986). Marked differences are apparent in azoreductase, nitroreductase, nitrate reductase, β-glucuronidase and β-glucosidase activities in gut contents of rat, mouse, hamster, guinea pig, rabbit, marmoset and humans, with no animal species providing a similar enzyme profile to that of humans. Consequently, although they are more difficult and expensive to perform, studies using human volunteers or germfree animals associated with a complete human microflora provide more relevant data (Rumney and Rowland, 1992)

11.7.2 The effect of age on microbial metabolism

During weaning, a major change in microbial population occurs in the mammalian gut and there is a concomitant change in various xenobiotic-metabolizing enzymes (Brennan-Craddock *et al.*, 1992). In mice aged 2–12 weeks, nitrate reductase activity tended to decrease with age, while nitroreductase and β-glucuronidase activities increased sharply between 2 and 4 weeks. Increases in the rate of bacterial demethylation of methylmercury at the time of weaning have also been reported, which may have important implications for the susceptibility of infants to neurotoxic damage by the organomercurial.

11.8 REFERENCES

Adlercreutz, H. (1993) Dietary lignans and isoflavonoid phytoestrogens and cancer. *Kliinisk Laboratorie*, **2a**, 4–12.

Alldrick, A. J., Flynn, J. and Rowland, I. R. (1986) Effect of plant derived flavonoids and polyphenolic acids on the activity of mutagens from cooked food. *Mutation Research*, **163**, 225–232.

Bolognani, F., Rumney, C. J. and Rowland, I. R. (1997) Influence of carcinogen binding by lactic acid-producing bacteria on tissue distribution and *in vivo* mutagenicity of dietary carcinogens. *Food Chemistry and Toxicology*, **6**, 535–545.

Brennan-Craddock, W. E., Mallett, A. K., Rowland, I. R. and Neale, S. (1992) Developmental changes to gut microflora metabolism in mice. *Journal of Applied Bacteriology*, **73**, 163–167.

Brown, J. P. (1988) Hydrolysis of glycosides and esters, in *Role of the Gut Flora in Toxicity and Cancer*, (ed. I. R. Rowland), Academic Press, London, pp. 109–144.

Bruce, W. R., Baptista, J., Che, T. *et al.* (1982) General structure of 'fecapentaenes', the mutagenic substances in human feces. *Naturwissenschaften*, **69**, 557–558.

Carman, R. J., van Tassell, R. L., Kingston, D. G. I. *et al.* (1988) Conversion of IQ, a dietary pyrolysis carcinogen, to a direct-acting mutagen by normal intestinal bacteria of human. *Mutation Research*, **206**, 335–342.

Cerniglia, C. E., Freeman, J. P., Franklin, W. and Pack, L. D. (1982) Metabolism of benzidine and benzidine-congener based dyes by human, monkey and rat intestinal bacteria. *Biochemical and Biophysical Research Communications*, **107**, 1224–1229.

Chipman, J. K., Millburn, P. and Brooks, T. M. (1983) Mutagenicity and *in vivo* disposition of biliary metabolites of benzo(α)pyrene. *Toxicology Letters*, **17**, 233–240.

Clarkson, T. W. (1979) General principles underlying the toxic action of metals, in *Handbook on Toxicology of Metals*, (eds L. Friburg, G. F. Nordbery and V. B. Vouk), Elsevier, Amsterdam, pp. 99–117.

Clinton, S. K. (1992) Dietary protein and carcinogenesis, in *Nutrition, Toxicity, and Cancer*, (ed. I. R. Rowland), CRC Press, Boca Raton, FL, pp. 455–475.

Cole, C. B., Fuller, R., Mallett, A. K. and Rowland, I. R. (1985) The influence of the host on expression of intestinal microbial enzyme activities involved in metabolism of foreign compounds. *Journal of Applied Bacteriology*, **58**, 549–553.

De Kok, T. M. C. M., van Faasen, A., ten Hoor, F. and Kleinjans, J. C. S. (1992) Fecapentaene excretion and fecal mutagenicity in relation to nutrient intake and fecal parameters in humans on omnivorous and vegetarian diets. *Cancer Letters*, **62**, 11.

Deschner, E. D., Ruperto, J., Wong, G. and Newmark, H. L. (1991) Quercetin and rutin as inhibitors of azoxymethanol-induced colonic neoplasia. *Carcinogenesis*, **12**, 1193.

Drasar, B. S. (1988) The bacterial flora of the intestine, in *Role of the Gut Flora in Toxicity and Cancer*, (ed. I. R. Rowland), Academic Press, London, pp. 23–38.

El-Bayoumy, K., Sharma, C., Louis, Y. M. *et al.* (1983) The role of intestinal microflora in the metabolic reduction of 1-nitropyrene to 1-aminopyrene in conventional and germfree rats and in humans. *Cancer Letters*, **19**, 311.

Eyssen, H. and Caenepeel, P. (1988) Metabolism of fats, bile acids and steroids, in *Role of the Gut Flora in Toxicity and Cancer*, (ed. I. R. Rowland), Academic Press, London, pp. 263–286.

Hill, M. J. (1988) Gut flora and cancer in humans and laboratory animals, in *Role of the Gut Flora in Toxicity and Cancer*, (ed. I. R. Rowland), Academic Press, London, pp. 461–502.

Hinzman, M. J., Novotny, C., Ullah, A. and Shamsuddin, A. M. (1987) Fecal mutagen fecapentaene-12 damages mammalian colon epithelial DNA. *Carcinogenesis*, **8**, 1475.

Hirai, N., Kingston, D. G. I., van Tassell, R. L. and Wilkins, T. D. (1982) Structure elucidation of a potent mutagen from human feces. *Journal of the American Chemistry Society*, **104**, 6149–6150.

Hollman P. C. H., de Vries, J. M. H., van Leeuwen, S. D. *et al.* (1995). Absorption of dietary quercetin glycosides and quercetin in healthy ileostomy volunteers. *American Journal of Clinical Nutrition*, **62**, 1276–1282.

Larsen, G. L. (1988) Deconjugation of biliary metabolites by microfloral β-glucuronidases, sulphatases and cysteine conjugate β-lyases and their subsequent enterohepatic circulation, in *Role of the Gut Flora in Toxicity and Cancer*, (ed. I. R. Rowland), Academic Press, London, pp. 79–107.

Leach, S. (1988) Mechanisms of endogenous N nitrosation, in *Nitrosamines Toxicology and Microbiology*, (ed. M. J. Hill), Ellis Horwood, Chichester, pp. 69–87.

Levin, A. A. and Dent, J. G. (1982) Comparison of the metabolism of nitrobenzene by hepatic microsomes and cecal microflora from Fischer-344 rats *in vitro* and the relative importance of each *in vivo*. *Drug Metabolism and Disposition*, **10**, 450–454.

Lindmark, D. G. and Muller, M. (1976) Antitrichomonad action, mutagenicity, and reduction of metronidazole and other nitroimidazoles. *Antimicrobial Agents and Chemotherapy*, **10**, 476.

MacDonald, I. A., Bokkenheuser, V. D., Winter, J. *et al.* (1993) Degradation of steroids in the human gut. *Journal of Lipid Research*, **24**, 675–700.

Mallett, A. K. and Rowland, I. R. (1988) Factors affecting the gut microflora, in *Role of the Gut Flora in Toxicity and Cancer*, (ed. I. R. Rowland), Academic Press, London, pp. 348–382.

Manning, B. W., Cerniglia, C. E. and Federle, T. W. (1985) Metabolism of the benzidine-based azo dye Direct Black 38 by human intestinal microbiota. *Applied and Environmental Microbiology*, **50**, 10.

Manning, B. W., Campbell, W. L., Franklin, W. *et al.* (1988) Metabolism of 6-nitro-chrysene by intestinal microflora. *Applied and Environmental Microbiology*, **54**, 197.

Massey, R. C., Key, P. E., Mallett, A. K. and Rowland, I. R. (1988) An investigation of the endogenous formation of apparent total N-nitroso compounds in conventional microflora and germ-free rats. *Food Chemistry and Toxicology*, **26**, 595–600.

Midtvedt, T. (1989) Monitoring the functional state of the microflora, in *Recent Advances in Microbial Ecology*, (eds T. Hattori, Y. Ishida, Y. Maruyama *et al.*), Japan Scientific Societies Press, Tokyo, pp. 515–519.

Mirsalis, J. C., Hamm, T. E., Sherrill, J. M. and Butterworth, B. E. (1982) Role of gut flora in genotoxicity of dinitrotoluene. *Nature*, **295**, 322–323.

Morotomi, M. and Mutai, M. (1986) In vitro binding of potent mutagenic pyrolyzates to intestinal bacteria. *Journal of the National Cancer Institute*, **77**, 195–201.

Ohgaki, H., Takayama, S. and Sugimura, T. (1991) Carcinogenicities of heterocyclic amines. *Mutation Research*, **259**, 399–411.

Orrhage, K., Sillerstrom, E., Gustafsson, J. A. *et al.* (1994) Binding of mutagenic heterocyclic amines by intestinal and lactic acid bacteria. *Mutation Research*, **311**, 239–248.

Plummer, S. M., Grafstom, R. C., Yang, L. L. *et al.* (1986) Fecapentaene-12 causes DNA damage and mutations in human cells. *Carcinogenesis*, **7**, 1607–1609.

Reddy, B. G., Pohl, L. R. and Krishna, G. (1976) The requirement of the gut flora in nitrobenzene-induced methemoglobinemia in rats. *Biochemistry and Pharmacology* **25**, 1119–1122.

Reddy, B. S., Weisburger, J. H., Narisawa, T. and Wynder, E. L. (1974) Colon carcinogenesis in germfree rats with 1,2-dimethylhydrazine and N-methyl-N-nitro-N-nitrosoguanidine. *Cancer Research*, **74**, 2368–2372.

Rowland, I. R. (1989) Metabolic profiles of intestinal floras, in *Recent Advances in Microbial Ecology*, (eds T. Hattori, Y. Ishida, Y. Maruyama *et al.*), Japan Scientific Societies Press, Tokyo, pp. 510–514.

Rowland, I. R. (1995). Interaction of the gut flora with metal compounds, in *Role of Gut Bacteria in Human Toxicology and Pharmacology*, (ed. M. J. Hill), Taylor & Francis, London, pp 197–211.

Rowland, I. R., Mallett, A. K. and Wise, A. (1985) The effect of diet on the mammalian gut flora and its metabolic activities. *CRC Critical Reviews in Toxicology*, **16**, 31–103.

Rowland, I. R., Mallett, A. K., Bearne, C. A. and Farthing, M. J. G. (1986) Enzyme activities of the hindgut microflora of laboratory animals and man. *Xenobiotica*, **16**, 519–523.

Rowland, I. R., Granli, T., Bockman, O. C. *et al.* (1991) Endogenous N-nitrosation in man assessed by measurement of apparent total N-nitroso compound in faeces. *Carcinogenesis*, **12**, 1395–1401.

Rumney, C. J. and Rowland, I. R. (1992) *In vivo* and *in vitro* models of the human colonic flora. *Critical Reviews in Food Science and Nutrition*, **31**, 299–331.

Rumney, C. J., Rowland, I. R. and O'Neill, I. K. (1993) Conversion of IQ to 7-OHIQ by gut microflora. *Nutrition and Cancer*, **19**, 67–76.

Schiffman, M. H., van Tassell, R. L., Robinson, A. *et al.* (1989) Case control study of colorectal cancer and fecapentaene excretion. *Cancer Research*, **49**, 1322.

Serraino, M. and Thompson, L. U. (1991a) The effect of flaxseed supplementation on early risk markers for mammary carcinogenesis. *Cancer Letters*, **60**, 135.

Serraino, M. and Thompson, L. U. (1991b) The effect of flaxseed supplementation on the initiation and promotional stages of mammary tumorigenesis. *Nutrition and Cancer*, **17**, 153.

Serraino, M. and Thompson, L. U. (1992) Flaxseed supplementation and early markers of colon carcinogenesis. *Cancer Letters*, **63**, 159.

Setchell, K. D. R. and Adlercreutz, H. (1988) Mammalian lignans and phyto-oestrogens. Recent studies on their formation, metabolism and biological role in health and disease, in *Role of the Gut Flora in Toxicity and Cancer*, (ed. I. R. Rowland), Academic Press, London, pp. 315–345.

Shamsuddin, A. M., Ullah, A., Baten, A. and Hale, E. (1991) Stability of fecapentaene-12 and its carcinogenicity in F344 rats. *Carcinogenesis*, **12**, 601.

Van der Hoeven, J. C., Lagerweij, W. J., Bruggeman, I. M. *et al.* (1983) Mutagenicity of extracts of some vegetables commonly consumed in the Netherlands. *Journal of Agricultural and Food Chemistry*, **31**, 1020.

Venturi, M., Hambly, R. J., Glinghammar, B. *et al.* (1997) Genotoxic activity in human faecal water and the role of bile acids: a study using the alkaline comet assay. *Carcinogenesis*, **18**, 2353.

Ward, J. M., Anjo, T., Ohannesian, L. *et al.* (1988) Inactivity of fecapentaene-12 as a rodent carcinogen or tumor initiator. *Cancer Letters*, **42**, 49.

Weisburger, J. H., Jones, R. C., Wang, C. X. *et al.* (1990) Carcinogenicity tests of fecapentaene-12 in mice and rats. *Cancer Letters*, **49**, 89.

Zarkovic, M., Qin, X., Nakatsuru, Y. *et al.* (1993) Tumor promotion by fecapentaene-12 in a rat colon carcinogenesis model. *Carcinogenesis*, **14**, 1261.

Zhang X. B. and Ohta Y. (1991) Binding of mutagens by fractions of the cell wall skeleton of lactic acid bacteria on mutagens. *Journal of Dairy Science*, **74**, 1477–1481.

Zhang X. B. and Ohta Y. (1993) Microorganisms in the gastrointestinal tract of the rat prevent absorption of the mutagen-carcinogen 3-amino-1,4-dimethyl-5H-pyrido(4,3-b)indole. *Canadian Journal of Microbiology*, **39**, 841–845.

12

Chemical transformations of bile salts by the intestinal microflora

J. Van Eldere

12.1 INTRODUCTION

Bile acids are synthesized as primary bile acids in the liver from choles-terol. In humans and all other mammals, bile acid synthesis is the pre-dominant pathway for elimination of cholesterol. After conjugation with glycine or taurine the bile acids are secreted in the bile and via the bile into the intestinal tract where they assist in the digestion and absorption of fat. The largest part of the bile acids are reabsorbed and via the portal blood return to the liver. A small fraction escapes from the enterohepatic circulation and is excreted in the faeces. In the intestinal tract, micro-organisms that are part of the intestinal microflora can transform the primary bile acids into secondary bile acids. These secondary bile acids can also be reabsorbed and return to the liver where they can again be transformed into the so-called tertiary bile acids before they are again excreted into the bile. In this chapter the transformations of bile acids by the intestinal microflora and the impact of these bile acid transformations on the host will be discussed.

12.2 BILE ACID BIOSYNTHESIS

Humans, in common with the higher mammals, synthesize primarily cholic and chenodeoxycholic acid as primary bile acids. The synthesis of the primary bile acids from cholesterol includes the introduction of a hydroxyl group in the 7α-position, the saturation of the 5,6 double bond,

G. W. Tannock (ed.), *Medical Importance of the Normal Microflora*, 312–337.
© 1999 *Kluwer Academic Publishers. Printed in Great Britain.*

the epimerization of the 3β-hydroxyl group to a 3α-position, introduction of supplementary hydroxyl groups, mostly in the 12α-position and the oxidative shortening of the C-27 cholesterol side chain to a C-24 carboxyl group (Fig. 12.1).

The 7α-hydroxylation is the first and rate-limiting step in this bile acid biosynthesis pathway. This reaction is controlled by a homeostatic negative feedback system consisting of the bile acids that return to the liver via the enterohepatic circulation. In the steady state, faecal excretion of bile acids is compensated by biosynthesis of primary bile acids from cholesterol. Bile acid synthesis and excretion are affected, however, by age, diet,

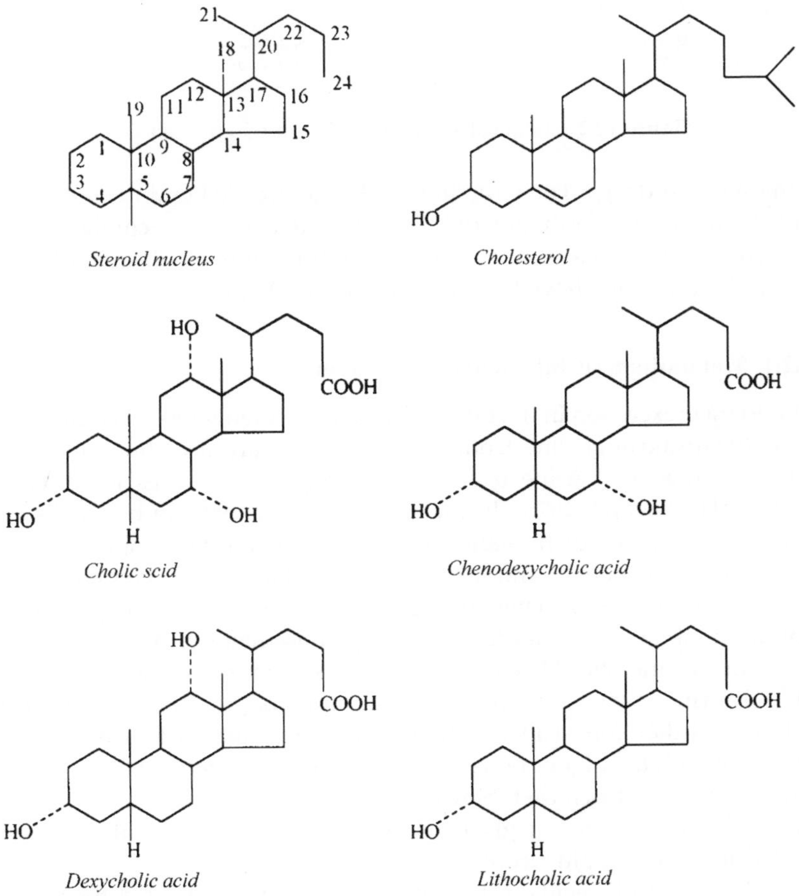

Figure 12.1 Molecular structure of cholesterol and the major bile acids.

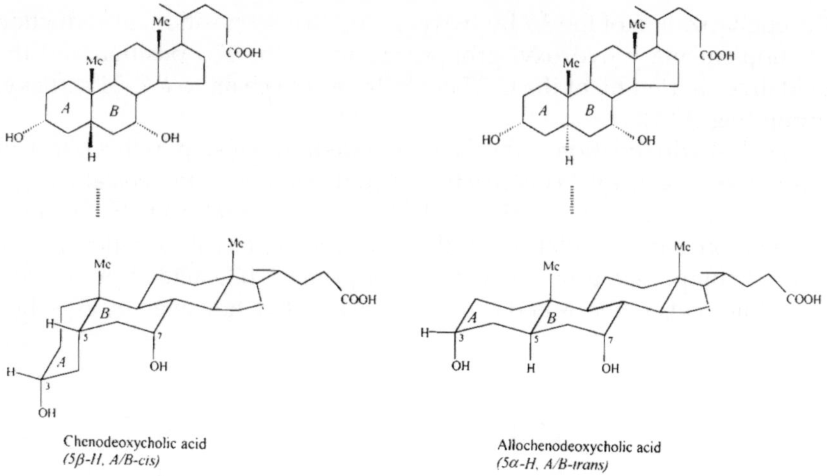

Chenodeoxycholic acid
(5β-H, A/B-cis)

Allochenodeoxycholic acid
(5α-H, A/B-trans)

Figure 12.2 Chemical structure of 5α and 5β bile acids.

hormones and drugs. The reduction of the 5,6 double bond in cholesterol leads to the specific formation of bile acids with a 5α- or 5β-configuration. In humans, bile acids have a 5β-configuration; only small quantities of 5α- or allo-bile acids have been identified (Fig. 12.2).

12.2.1 Metabolism of bile acids in the liver

Prior to their excretion in the bile, bile acids are conjugated at the carboxyl group to taurine or glycine. Conjugated bile acids are also often called bile salts. The ratio of taurine over glycine conjugation varies in different animals (Haslewood, 1978). In the rat for example, 90% of bile acids are taurine conjugated. In humans the amount of taurine conjugated bile acids varies between 15 and 50%. The major factor that determines the ratio of taurine over glycine conjugation in humans is the hepatic concentration of taurine (Hardison, 1978). Consequently, a high dietary intake of taurine should lead to an increased tauroconjugation of bile acids. Taurine is found in all tissues of animal origin; diets rich in meat and fat will therefore lead to a higher dietary intake of taurine. This is consistent with the higher percentage of tauroconjugated bile acids found in those on a meat rich diet (Sjövall, 1959).

In addition to tauro- or glycoconjugation, bile acids can also be conjugated to sulphate, glucuronic acid or glucose.

Sulphation of bile acids in humans was first demonstrated for microbially formed lithocholic acid in bile (Palmer and Bolt, 1971). In healthy people, the toxic lithocholic acid is the only bile acid that is significantly sulphated: up to 60% per liver-passage. It was shown that sulphation of lithocholic acid plays an important role in the metabolism and elimination of this bile acid in humans (Cowen *et al.*, 1975). Subsequently it was established that bile acid sulphates are produced in a variety of hepatobiliary diseases (Makino *et al.*, 1975; Stiehl, 1974; van Berge Henegouwen *et al.*, 1976). In humans, sulphated cholic and chenodeoxycholic acid were recovered from the serum and urine of patients suffering from cholestasis but also from the meconium of newborn babies. Sulphation of bile acids increases their water-solubility and their critical micellar concentration. As a result, the intestinal reabsorption of sulphated bile acids is impaired and their faecal elimination is increased (Eyssen *et al.*, 1985). In this way sulphation promotes the elimination from the body of potentially toxic substances such as lithocholic acid. Increased sulphation of di- and trihydroxy-bile acids is found in cholestasis and significantly increases the urinary elimination of these bile acids (Makino *et al.*, 1975; van Berge Henegouwen *et al.*, 1976).

Glucuronidated bile acids were identified in bile, serum and urine, but again only in significant quantities in patients with cholestasis (Back, Spaczynski and Gerok, 1974; Fröhling *et al.*, 1977; Takikawa *et al.*, 1983). Hyodeoxycholic acid is the only bile acid that is glucuronidated to a significant extent in healthy people and is as such eliminated via the urine (Sacquet *et al.*, 1983). However, in healthy people, hyodeoxycholic acid is only present in very small quantities (Almé *et al.*, 1977). Glucuronidation can take place at the C-3 and C-6 position of bile acids (Radominska-Pyrek *et al.*, 1986, 1987; Sacquet *et al.*, 1983). Glucuronidation, like sulphation, is a means for detoxification of bile acids or elimination of bile acids in cholestasis (Back and Gerok, 1977).

Glucose-conjugated bile acids have been identified in small quantities in the urine of healthy people. The physiological role of glycosylation is probably the same as that of sulphation or glucuronidation of bile acids (Marshall *et al.*, 1987).

12.2.2 The enterohepatic circulation of bile acids

The enterohepatic circulation is a physiological and not an anatomical entity, although it takes place within well defined anatomical boundaries. It is an efficient feedback mechanism that enables the body to use its bile acid pool optimally but at the same time it is the basis for the interaction of the intestinal microflora with the bile acids.

The driving forces behind the enterohepatic circulation are the transport systems of the hepatocytes that extract the bile acids from the portal blood, the intestinal peristalsis that propels the bile acids towards the terminal ileum where reabsorption is most efficient, and the transport systems in the enterocytes in the terminal ileum that absorb the bile acids from the intestinal lumen. The ileal bile acid transport system is homeostatically regulated by feedback inhibition of the bile acids that are reabsorbed. In addition to this active absorption in the terminal ileum, bile acids can also be passively reabsorbed from the intestinal tract (Wilson, 1981; Fricker *et al.*, 1987; Barbara *et al.*, 1984). The intestinal reabsorption and the hepatic extraction are very efficient. As a consequence, only a small percentage of the intestinal bile acids is excreted in the faeces, and an even smaller percentage is extracted from the general circulation by the kidneys. It is estimated that, per day, approximately 1 g of bile acids is excreted in the faeces and 1 mg in the urine (De Barros *et al.*, 1982; van der Werf, van Berge Henegouwen and van den Broeck, 1985). This implies that, per cycle of enterohepatic circulation, only 2% of bile acids are excreted in the faeces. However, per intestinal passage, only approximately 75% of bile acids are reabsorbed in the ileum. Consequently, it is estimated that, per intestinal passage, between 130 and 650 mg of bile acids reach the colon where they are exposed to the enzymatic activity of the intestinal microflora.

12.2.3 The composition, localization and stability of the human intestinal microflora

The bacteria that reside in the intestinal tract constitute an enormously complex but relatively stable ecosystem that interacts in various ways with the physiology and metabolism of the healthy host (Finegold, Sutter and Mathesen, 1983).

In between meals, the upper half of the small intestine contains few, relatively acid-resistant, bacteria such as streptococci and lactobacilli. Gastric acid, intestinal peristalsis, secretory immunoglobulins, lysozyme, pancreatic fluid but also bile acids limit bacterial growth. The distal ileum constitutes a transition zone; the bacterial count gradually increases and the composition of the microflora changes from Gram-positive and aerobic to mainly Gram-negative and anaerobic. The microflora in the distal ileum, therefore, resembles that of the colon and transformations of bile acids also take place in the ileum (Raedsch *et al.*, 1983).

The colon is the most densely populated part of the intestinal tract and contains approximately 10^{10} microorganisms/g and more than 400 different species. It is therefore not surprising that bacterial transformations of bile acids are most intense in the colon. Anaerobic species outnumber aerobic species 1000 to 1. In humans the predominant species are *Bacteroides*, *Bifidobacterium*, *Eubacterium*, anaerobic cocci and to a lesser degree *Clostridium*, *Lactobacillus* and *Streptococcus*. The 'hold' function of the colon allows the multiplication of these bacteria in the colon. Several studies also report as many as 10^{8} microorganisms/g in the mucous layer, attached to the epithelium or in the crypts of the colon. It is not yet clear whether the mucosal microflora is identical to the luminal flora or whether it is distinct from the luminal microflora and if it has its own particular metabolic activity.

In spite of the enormous diversity of species in the intestinal microflora and in spite of the constant challenge from ingested bacteria, the composition of the intestinal microflora is fairly stable. The specificity and stability of the intestinal microflora are due on the one hand to the immunological interactions between the intestinal microflora and the host and on the other hand to the metabolic interactions between the different species that constitute the intestinal microflora. Secretory IgA and epithelial receptors are means by which the host can affect intestinal adherence and multiplication of certain species of bacteria. Metabolic interdependence between species, competition for limited sources of nutrients and production of growth-inhibiting substances such as bacteriocins or short-chain fatty acids are mechanisms by which the intestinal microflora itself can stabilize its composition and discourage the multiplication of transient exogenous species.

Examples of this metabolic interdependence can also be found with regards to bile acids. We isolated from human faeces a binary culture of two strains that together were able to 7α-dehydroxylate tauroconjugated cholic acid (Van Eldere *et al.*, 1996). One strain deconjugated taurocholic acid rapidly and reduced the deconjugated taurine rapidly to H_2S. This strain, however, did not 7α-dehydroxylate cholic acid. In fact this strain, a *Bacteroides* sp. strain termed R1, did not grow unless taurine or another suitable reducible sulphur source was present. In addition, the H_2S produced by this *Bacteroides* sp. strain was a necessary growth factor for the second strain, a *Clostridium* sp. strain termed 9/1 that was able to 7α-dehydroxylate cholic acid only after it had been deconjugated.

All these mechanisms together result in an intestinal microflora of which the composition is stable due to an equilibrium between its different composing species. The resistance of the intestinal microflora against colonization of the intestinal tract by transient and possibly pathogenic species that do not belong to the microflora is called colonization resistance (van der Waaij, Berghuis de Vries and Lekkerkerk van der Wees, 1971). Consequently the composition of the intestinal microflora is insensitive to all but strong exogenous influences such as drastic changes in diet or in particular, antibiotics.

However, although the composition of the intestinal microflora might be quite stable, certain aspects of the metabolic activity of the intestinal microflora are labile and are clearly influenced by even mild external factors such as modifications of the diet. Interesting examples of this are, for example, the effect of an increased meat and fat intake on various oxidoreduction activities of the intestinal microflora such as the 7α-dehydroxylation reaction of bile acids and steroids. More emphasis is therefore put on changes in the intestinal metabolic activity, the influence of this metabolic activity on the host and the factors that cause these changes rather than on the changes in the composition of the intestinal microflora.

12.3 TRANSFORMATION OF BILE ACIDS BY THE INTESTINAL MICROFLORA

The intestinal microflora of humans can transform the primary bile acids into a variety of between 15 and 20 metabolites (Hayakawa, 1973; Midvedt, 1974). The major bile acid transformations performed by the intestinal microflora are: hydrolysis of conjugated bile acids into free bile acids, dehydroxylation of the 7α-hydroxyl group of the steroid nucleus and epimerization of hydroxyl groups. Other reactions are desulphation, deglucuronidation and dehydrogenation of the steroid nucleus. Reductive reactions are favoured over the other reactions because the intestinal environment is strictly anaerobic.

Biliary bile acids are almost completely conjugated. Conjugated bile acids are actively absorbed from the ileum and returned to the liver via the enterohepatic circulation. However, a sizeable portion of the bile acids escapes this reabsorption and reaches the terminal ileum and particularly the colon, where they are exposed to the biotransforming activity of the intestinal microflora. The initial reaction in the transformation of the bile acids by the intestinal microflora is the deconjugation of the bile acids. In fact, the bile acids recovered in faeces in humans are almost all deconjugated (Ali, Kuksis and Beveridge, 1966). Once the bile acids are deconjugated, they can then be transformed into many other different metabolites by the intestinal microflora.

The primary bile acids constitute between 60% and 90% of the biliary excreted bile acids; the remainder are conjugated secondary or tertiary bile acids. Secondary bile acids arise through bacterial transformation of the primary bile acids and are reabsorbed via the circulation through passive diffusion in the proximal colon. Tertiary bile acids are the result of further metabolic transformations of these absorbed secondary bile acids in the liver. Thus the biliary excreted bile acids are in fact a complex mixture of primary, secondary and tertiary bile acids as a result of liver biosynthesis and bacterial transformation and reabsorption. In the faeces the predominant bile acids are the secondary bile acids deoxycholic and lithocholic acid.

The transformation of the bile acids by the intestinal microflora changes the physicochemical properties, the biological activities and the enterohepatic circulation of the bile acids. It is therefore not surprising that certain bile acids have been reported to play important roles in the pathogenesis of some diseases of the gastrointestinal tract.

12.3.1 Deconjugation of bile acids

In humans, bile acids are excreted in the bile as taurine or glycine conjugates. Tauro- and glycodeconjugation is the initial transformation of the bile acids by the intestinal microflora. Germfree animals without an intestinal microflora excrete only conjugated bile acids in their faeces (Eyssen and Van Eldere, 1984). Deconjugation of bile acids is normally limited to the large bowel and the terminal ileum. Under pathological conditions, e.g. in the blind loop syndrome or in gastrectomized patients, the numbers of deconjugating bacteria tend to increase in the more proximal parts of the intestinal tract, leading to abnormal fat absorption and steatorrhoea.

The ability to hydrolyse the peptide bond of conjugated bile acids is found in many intestinal bacteria belonging to widely different genera such as *Bacteroides* (Midtvedt and Norman, 1967; Shimada, Bricknell and Finegold, 1969), *Bifidobacterium* (Shimada, Bricknell and Finegold, 1969; Ferrari, Pacini and Canzi, 1980), *Fusobacterium* (Shimada, Bricknell and Finegold, 1969; Dickinson, Gustafsson and Norman, 1971), *Clostridium* (Midtvedt and Norman, 1967; Masuda, 1981), *Lactobacillus* (Gilliland and Speck, 1977), *Peptostreptococcus* (Dickinson, Gustafsson and Norman, 1971; Kobashi *et al.*, 1978) and *Streptococcus* (Shimada, Bricknell and Finegold, 1969; Kobashi *et al.*, 1978). There is, however, considerable variation regarding the presence and extent of this activity. Most bacteria with deconjugating activity deconjugate both amino-acid conjugates of the same bile acids (Shimada, Bricknell and Finegold, 1969; Masuda, 1981). Some, however, exhibit specificity for either taurine or glycine (Midvedt, 1974; Kobashi *et al.*, 1978) or for the steroid portion of the bile acid (Masuda, 1981).

The effect of microbial deconjugation of bile acids on faecal bile acid excretion has been studied by Kellogg, Knight and Wostmann (1970) in germfree rats selectively associated with a bile-acid-deconjugating strain of *Clostridium perfringens*. Although in these animals the faecal bile acids were deconjugated, the quantities excreted were not increased compared to germfree rats. This is not surprising because, in the normal animal, deconjugation mainly takes place in the large intestine, whereas the vast majority of bile acids are absorbed from the distal small intestine. This might also apply to the observation that microbial deconjugation of bile acids in the large intestine does not seem to interfere significantly with fat absorption in normal animals. However, when a complex deconjugating microflora becomes established in the small intestine, as for example in the contaminated small bowel syndrome, malabsorption of fat becomes a major effect.

12.3.2 Dehydroxylation of bile acids

Undoubtedly the most important transformation of the free bile acids by the intestinal flora is the 7α-dehydroxylation reaction that transforms the primary bile acids cholic and chenodeoxycholic acid into the secondary bile acids deoxycholic and lithocholic acid. It has been estimated that the colon contains 10^3–10^5 bacteria per gram wet weight of faeces capable of performing this reaction (Ferrari, Pacini and Canzi, 1980). The majority of presently known strains that possess a stable bile acid 7α-dehydroxylation activity are anaerobic Gram-positive rods that belong to the genera *Clostridium* and *Eubacterium*. The 7α-dehydroxylation reaction is irreversible. Based on the isolation of bile acid intermediates and the conversion of these intermediates to 7α-dehydroxylated products, the 7α-dehydroxylation reaction consists of the following steps. The 7α-hydroxy bile acid is first conjugated to ADP or coenzyme A and oxidized in two steps to yield a 3-oxo-4-cholenic acid. This intermediate is then thought to undergo a trans-elimination of water across the C6–C7 bond to give a 3-oxo-4,6-choldienoic acid. Three successive reductive reactions then follow, resulting in the production of the corresponding 7α-dehydroxylated bile acid (Björkhem *et al.*, 1989; Coleman *et al.*, 1987).

The presence of bacterial 7α-dehydroxylation of chenodeoxycholic acid is important because of the hepatotoxicity of its 7α-dehydroxylation reaction product, lithocholic acid. Deoxycholic acid, the 7α-dehydroxylation product of cholic acid, has been implicated in the pathogenesis of colon cancer, albeit that the exact role of bile acids in colon cancer pathogenesis has not yet been determined.

Dehydroxylation of bile acid hydroxyl groups in the C-3 and C-12 position has also been reported (Borriello and Owen, 1982; Edenharder, 1984).

12.3.3 Oxidation–reduction reactions of hydroxyl groups

Bile acid dehydrogenase or hydroxysteroid dehydrogenase activity leads to the epimerization of the hydroxyl groups of different bile acids. This is because the keto bile acids that are formed by these enzymes are, in the strictly anaerobic environment of the large intestine, again reduced by the same or different bacteria to α- as well as β-epimers of the hydroxyl groups. Bacteria have been isolated that can reversibly oxidize the hydroxyl groups at C-3, C-7 and C-12. The presence of a bile acid with a 3β-hydroxyl group, for instance, is a fingerprint usually left by microbial enzymatic activity. Hydroxysteroid dehydrogenase activity is not widely distributed among anaerobic intestinal bacteria. It is, however, found in genera that are major constituents of the intestinal microflora such as *Bacteroides* (Hylemon and Sherrod, 1975; Edenharder, Stubenrauch and Slemrova, 1976) and *Eubacterium* (Macdonald *et al.*, 1979; Hirano and Masuda, 1981a). It has also been demonstrated in *Clostridium* (Dickinson, Gustafsson and Norman, 1971; Edenharder and Deser, 1981), *Bifidobacterium* (Aries and Hill, 1970; Ferrari, Pacini and Canzi, 1980), *Lactobacillus* and *Peptostreptococcus* (Dickinson, Gustafsson and Norman, 1971; Hirano and Masuda, 1981b) and *Fusobacterium* and *Escherichia* (Midtvedt and Norman, 1967).

12.3.4 3α-hydroxysteroid dehydrogenase activity

Epimerization of the 3α-hydroxyl group has been described in species of the genera *Clostridium* and *Eubacterium* (Midtvedt and Norman, 1967; Dickinson, Gustafsson and Norman, 1971). *Clostridium perfringens*, for example, has been shown to produce 3-keto and 3β-hydroxy bile acids from 3α-hydroxy bile acids. The extent of epimerization is low in growing cultures but increases in resting cell suspensions. Epimerization was strongly favoured by anaerobic conditions and aeration of resting cell suspensions or cultures promoted 3α-hydroxysteroid dehydrogenase activity but prevented the reduction of 3-keto intermediates to 3β-hydroxy bile acids. Another bacterial species that has also been shown to be capable of forming 3-keto and 3β-hydroxy derivatives of 3α-hydroxy bile acids is *Eubacterium lentum* (Midtvedt and Norman, 1967; Macdonald *et al.*, 1979; Hirano and Masuda, 1981a). As in *Clostridium perfringens*, anaerobic conditions enhanced epimerization of the 3α-hydroxyl group whereas aeration of resting cell suspensions resulted in 3-keto bile acid formation. In time course experiments, *Eubacterium lentum* cultures oxidized cholic and chenodeoxycholic acid to the 3-keto forms during the exponential growth phase and subsequently reduced the 3-keto to the 3α- and 3β-hydroxy bile acids as growth proceeded. *Eubacterium lentum* also contains 7α- and 12α-hydroxysteroid dehydrogenase activities. Both the *Clostridium perfringens* and the *Eubacterium lentum* enzymes exhibited broad substrate specificity against 3α-hydroxyconjugated bile acids, free bile acids and neutral steroids.

12.3.5 6β-hydroxysteroid dehydrogenase activity

This activity has been demonstrated in anaerobic intestinal strains isolated from rats. In rats and pigs, the β-muricholic bile acid, which contains a 6β-hydroxy group, is one of the major bile acids (Eyssen and Parmentier, 1979).

12.3.6 7-hydroxysteroid dehydrogenase activity

The 7α- and 7β-hydroxysteroid dehydrogenase enzymes are the most common among the hydroxysteroid dehydrogenases. The combined activities of a 7α-hydroxysteroid dehydrogenase and a 7β-hydroxysteroid dehydrogenase are required for epimerization of the 7-hydroxyl group. An additional problem in the evaluation of 7-hydroxy-epimerization is that it competes with the irreversible 7α-dehydroxylation of bile acids *in vivo*. Epimerization of 7β-hydroxy bile acids prior to 7α-dehydroxylation by the intestinal flora is probably the major metabolic pathway of 7β-hydroxy bile acids.

7α-dehydroxylation of bile acids has been found in *Eubacterium lentum*, *Escherichia coli* and several *Bacteroides* species.

Eubacterium lentum forms 7-keto bile acids as major transformation products (Midtvedt and Norman, 1967; Hirano and Masuda, 1981a). This 7α-hydroxysteroid dehydrogenase activity is oxygen sensitive. *Escherichia coli* forms 7-keto derivates from free and conjugated bile acids during growth (Midtvedt and Norman, 1967); its 7α-hydroxysteroid dehydrogenase was NAD-dependent and was active on a broad range of free and conjugated bile acids (Aries and Hill, 1970; Macdonald, Williams and Mahony, 1973). *Bacteroides* species form only 7-keto bile acids from cholic and deoxycholic acid during growth or as resting cell suspensions (Hylemon and Stellwag, 1976; Edenharder, Stubenrauch and Slemrova, 1976). Oxidation of the 7β-hydroxy bile acids ursocholic and ursodeoxycholic acid by anaerobic bacteria has been documented in cell extracts of *Clostridium absonum*, *Eubacterium aerofaciens* and *Peptostreptococcus productus* (Macdonald, Hutchison and Forrest, 1981; Macdonald *et al.*, 1982; Hirano and Masuda, 1982). In *Clostridium absonum* – a soil bacterium that has not yet been isolated from human faeces – both 7β- and 7α-hydroxysteroid dehydrogenase activities were found.

12.3.7 12α-hydroxysteroid dehydrogenase activity

The presence of 12α-hydroxysteroid dehydrogenase activity in human intestinal bacteria is derived from the fact that incubation of cholic acid with human faecal cultures leads to the formation of 12-keto bile acids in low concentrations (Hirano *et al.*, 1981). 12α-hydroxysteroid dehydrogenase activity has indeed been demonstrated in species of the genera *Eubacterium* and *Clostridium* (Midtvedt and Norman, 1967; Macdonald *et al.*, 1979).

12.3.8 Desulphation of bile acids

The percentage of sulphated bile acids in normal human faeces was measured as lower than 5% (Breuer *et al.*, 1984; Islam, Raicht and Cohen, 1981), but increased significantly after treatment with the antibiotic oxytetracycline (Korpela, Fotsis and Adlercreutz, 1986). This effect was hypothesized to be due to the elimination of the desulphating bacteria from the intestinal tract by the antibiotic treatment. Although less than 1% of the bile acids are sulphated prior to biliary excretion, the concentration of these sulphated bile acids will have increased severalfold when they reach the large intestine. This is because bile acid sulphates are very poorly absorbed from the small intestine (Palmer and Bolt, 1971; Low-Beer, Tyor and Lack, 1969; Cowen *et al.*, 1975). Studies with labelled lithocholic acid sulphate confirmed that desulphation of bile acid sulphates took place in humans in the distal part of the intestinal tract, i.e. the colon (Cowen *et al.*, 1975). Desulphation of bile acid sulphates could also be demonstrated *in vitro*, in mixed faecal anaerobic cultures (Borriello and Owen, 1982; Huijghebaert, Parmentier and Eyssen, 1984). In these *in vitro* experiments, only bile acid 3-sulphates were desulphated. However, sulphation of bile acids in humans is almost exclusively at the 3-hydroxyl group.

The bile acid desulphating ability is not widespread among intestinal bacteria. Imperato, Wong and Chen (1977) reported the first isolation of a bile-acid-desulphating species from human faeces. The *Pseudomonas aeruginosa* strain described by Imperato, Wong and Chen transformed lithocholic acid sulphate into lithocholic acid but only in aerobic conditions and in the absence of any other source of sulphur. It remains to be proven that this strain actually contributes to the desulphating activity in the strictly anaerobic, sulphur-rich environment of the colon.

Huijghebaert and Eyssen (1982) and Robben, Parmentier and Eyssen (1986) isolated two bile-acid-sulphatase-producing *Clostridium* sp. strains, termed S-1 and S-2, from rat faecal flora. These bacteria produced enzymes capable of desulphation of 3α-sulphates of 5β-bile acids, 3β-sulphates of 5β- and 5α-bile acids (Huijghebaert and Eyssen, 1982) and 3α-sulphates of 5β- and 5α-bile acids (Robben, Parmentier and Eyssen, 1986). A *Peptococcus niger* strain that was originally isolated from human faeces for its steroid desulphating activity (Van Eldere, De Pauw and Eyssen, 1987) was later found also to contain enzymes capable of desulphating the 3β-sulphates of 5α- and 5β-sulphated bile acids and to a lesser extent also the 3α-sulphates of 5α- and 5β-sulphated bile acids. Studies in ex-germfree rats selectively associated with bile-acid-desulphating bacteria have shown that desulphation stimulated the enterohepatic circulation of sulphated bile acids and reduced their faecal elimination (Eyssen and Van Eldere, 1984; Eyssen *et al.*, 1985).

12.3.9 Deglucuronidation of bile acids

Deconjugation of β-glucuronidated steroids including bile acids can also take place through the activity of the mucosal β-glucuronidases. Therefore the impact and importance of intestinal bacterial deglucuronidation is more difficult to assess and most probably it is also less important for the metabolism of bile acids. Nevertheless, the intestinal microflora is considered to be a major source of β-glucuronidase activity because the hydrolysis of many xenobiotic glucuronides is dramatically reduced in germfree or antibiotic-treated animals (Grantham *et al.*, 1970). β-glucuronidase activity has been detected in many strictly anaerobic strains belonging to many different genera found in the intestinal tact. However the substrate specificity of these enzymes for bile acid glucuronides has not been determined. In our laboratory, we have isolated a cholic acid 3-glucuronide-deconjugating *Peptostreptoccus productus* from human intestinal microflora (Van Eldere, unpublished).

12.4 CHEMICAL TRANSFORMATION OF BILE ACIDS BY MEMBERS OF THE NORMAL MICROFLORA AND ITS SIGNIFICANCE TO HUMAN HEALTH

The normal intestinal microflora interacts with its host in many different ways. In many of these interactions the transformation of bile acids most probably plays no role. Examples of this kind of interaction are the protection of the host against colonization of the intestinal tract by exogenous pathogenic microorganisms or the stimulation of the host's immune system.

However, chemical transformation by the intestinal microflora of bile acids does play a role in the following interactions between host and intestinal microflora: the pathogenesis of colon cancer, the enterohepatic circulation of bile acids and the contaminated small bowel syndrome.

12.4.1 Contaminated small bowel syndrome and transformation of bile acids

(a) Introduction

'Contaminated small bowel syndrome' or 'small bowel bacterial overgrowth' is a condition in which the number of bacteria in the small intestine is at least 10–100 times higher than in the healthy individual. In contaminated small bowel syndrome the number of bacteria in the fasting state is equal to or greater than 10^5 CFU/ml or the number of anaerobic bacteria is equal to or greater than 10^4 CFU/ml. This also implies that the bacterial microflora in the small intestine in the contaminated small bowel syndrome is not that normally isolated from the healthy individual. Instead it resembles closely the microflora normally found in the large intestine.

There are many possible causes of the contaminated small bowel syndrome (Gracey, 1983). The main defences against bacterial overgrowth in the small intestine are peristalsis, gastric acid secretion and local immunity. It is therefore not surprising that, in patients with contaminated small bowel syndrome, there are often anatomical abnormalities of the gut that lead to obstruction or fistulas. In other cases there have been surgical interventions leading to stasis of bowel contents. Disordered motility, reduced gastric acidity or infection of the biliary tract are clinical settings in which the contaminated small bowel syndrome may also occur (Rolfe, 1989; Rudell *et al.*, 1980; Gracey, 1983). Finally, antibiotics or immune deficiency disorders that impair local intestinal immunity can also disrupt the intestinal microbial ecosystem and therefore lead to this syndrome.

(b) Clinical manifestations and pathogenesis of the contaminated small bowel syndrome and the role of bile acid transformations

The clinical symptoms that accompany this syndrome may be minor or include a wide range of non-specific gastrointestinal manifestations such as abdominal pain and bloating, excessive gas production and weight loss. The nutritional consequences include carbohydrate malabsorption, diarrhoea with electrolyte loss or steatorrhoea, vitamin B_{12} deficiency and protein malnutrition (Neale *et al.*, 1972; Gracey, 1983).

The pathogenesis of the symptoms associated with the contaminated small bowel syndrome is most probably multifactorial. There is no single particular bacterial species that is responsible for the pathological symptoms nor is there evidence of invasion of bacteria into the intestinal tissue. However, there are clear indications that the abnormally numerous anaerobic microflora found in the small bowel in the contaminated small bowel syndrome is involved in the pathogenesis of this syndrome (Kinsella, Hennessy and George, 1961; Tabaqchali *et al.*, 1966; Polter *et al.*, 1968).

The pathologically profuse microflora associated with the contaminated small bowel syndrome induces morphological and functional changes in the gastrointestinal tract. These changes are highlighted by the differences that are found between small intestinal mucosa in germfree and conventional animals. In germfree animals, the intestinal epithelium is more regular, with tall slender villi, a wide brush border and shallow crypts (Abrams, Bauer and Sprinz, 1963). In conventional animals, there is an increase in surface area of the small intestine (Gordon and Bruckner-Kardoss, 1961) and in epithelial cell turnover. Broadening and shortening of small intestinal villi have also been demonstrated in patients with contaminated small bowel syndrome (King and Toskes, 1976). These morphological changes are accompanied by functional changes that may be of physiological importance. Indeed, in conventional animals there is a relative lack of activities of such brush border enzymes as alkaline phosphatase (Yolton, Stanely and Savage, 1971) and disaccharidases (Reddy and Wostmann, 1966). In addition, there is some evidence that absorption of sugars, vitamins and minerals may be more efficient in the absence of a luminal microflora (Ford and Coates, 1971; Heneghan, 1963; Reddy, Pleasants and Wostmann, 1969).

What causes the morphological abnormalities in the contaminated small bowel syndrome? There is little evidence of bacterial invasion to explain them, but there is some evidence to support a role for toxic bacterial metabolic byproducts. There is also evidence supporting a causative role of deconjugated bile acids. Experiments have shown that deconjugated bile acids can reproduce contaminated small bowel syndrome lesions in animals (Gracey, Papadimitriou and Burke, 1973) and reversibly inhibit uptake of nutrients such as sugars, amino acids, peptides and fatty acids (Baraona *et al.*, 1968; Gracey, Burke and Oshin, 1971; Gracey *et al.*, 1971; Burke *et al.*, 1975; Shimoda, O'Brien and Saunders, 1974). Nevertheless it is clear that factors other than deconjugated bile acids are also involved in the pathogenesis of mucosal injury in contaminated small bowel syndrome (Bloch *et al.*, 1973; Giannella, Rout and Toskes, 1974).

(c) Pathogenic mechanisms involved in steatorrhoea

Steatorrhoea is the clinical hallmark of the contaminated small bowel syndrome and many studies have addressed the role of bile acids in its pathogenesis. Many anaerobic bacteria that belong to the normal colonic microflora are able to deconjugate bile acids. When these bacteria are present in large numbers in the small intestine, as in contaminated small bowel syndrome, they will lead to the production of free bile acids that are poorly water-soluble at the pH of the upper small intestinal lumen. If the bacterial contamination causes sufficiently extensive deconjugation, the luminal conjugated bile acid concentration will fall below the minimal concentration for mixed micelle formation and lead to steatorrhoea (Tabaqchali, Hatzioannou and Booth, 1968). Other studies have shown an inhibitory effect of unconjugated bile acids on fatty acid uptake and esterification in rat jejunum (Dawson and Isselbacher, 1960), and a direct effect of deconjugated deoxycholate on fat uptake. This observation is suggestive of a direct toxic effect of deconjugated bile acids (Shimoda, O'Brien and Saunders, 1974). It must also be remembered that intestinal bacteria are capable of transforming bile acids in many more ways that might also interfere with lipid digestion. Other possible mechanisms involved in the pathogenesis of steatorrhoea are excessive production of hydroxy fatty acids and volatile fatty acids (Kim and Spritz, 1968).

(d) Carbohydrate malabsorption

A possible link between carbohydrate malabsorption and bile acid deconjugation lies in the damaging effect of free bile acids on the metabolic functions of the enterocyte (Gracey, Burke and Oshin, 1971). Others have suggested that intraluminal bacterial consumption of sugars causes carbohydrate malabsorption (Goldstein *et al.*, 1970). Finally there are also data suggestive of a negative effect of toxic bacterial metabolic byproducts on intestinal monosaccharide absorption (Gracey, 1981).

(e) Hypoproteinaemia

Several, not necessarily mutually exclusive, pathogenic mechanisms have also been proposed for the hypoproteinaemia in contaminated small bowel syndrome. Intraluminal bacterial breakdown of amino acids (Tomkin and Weir, 1972; King and Toskes, 1976) and increased ammonia production (Jonas *et al.*, 1968) have been documented. These metabolic processes have also been used to devise breath tests for contaminated small bowel syndrome (King and Toskes, 1976; Thomas *et al.*, 1982). Increased intestinal protein loss due to enterocyte damage (Nygaard and Rootwelt, 1968), impaired mucosal digestion due to reduced peptidase and enterokinase activity (Rutgeerts *et al.*, 1974) and reduced mucosal uptake (Baraona *et al.*, 1968; Giannella, Rout and Toskes, 1974) have all been documented and suggested as pathogenic mechanisms contributing to the hypoproteinaemia.

As to deconjugation of bile acids, it has been shown that deconjugated bile acids inhibit amino acid and dipeptide uptake (Pope, Parkinson and Olson, 1966; Hajjar, Khuri and Bikhazi, 1975; Burke *et al.*, 1975).

(f) Vitamin deficiency

Vitamin B_{12} deficiency in contaminated small bowel syndrome is due to bacterial binding of the vitamin (Donaldson, 1962) and the production of competitive inhibitors of vitamin B_{12} uptake (Giannella, Broitman and Zamchek, 1971). A role for bile acids in the pathogenesis of vitamin B_{12} deficiency has not been suggested. Vitamin K and D deficiency, which might be expected in contaminated small bowel syndrome because absorption of these vitamins depends on the dispersion of normal intraluminal bile acids, is, however, rare. A deficiency of vitamin A and β-carotene has been reported in a patient with contaminated small bowel syndrome (Levy and Toskes, 1974).

(g) Water and electrolyte loss

Fluid diarrhoea with electrolyte loss is a possible manifestation of contaminated small bowel syndrome. Deconjugated bile acids may contribute to this manifestation because they alter water, sodium and electrolyte transport in rat (Forth, Rummel and Glasner, 1966) and humans (Mekhjian, Phillips and Hofmann, 1968). Cell-free broth cultures of a wide range of human intestinal bacteria can, however, also cause fluid secretion into the lumen, presumably through the action of toxic products such as alcohols, short-chain fatty acids and bacterial exotoxins (Baraona, Pirola and Lieber, 1974; Klipstein *et al.*, 1973).

(h) Diagnosis and treatment

Microbiological examination of aspirated small intestinal fluid provides the most convincing evidence of bacterial overgrowth, but is not usually done in a routine setting. A breath test for determination of altered bile acid metabolism has been used for the diagnosis of contaminated small bowel syndrome but has been shown to be insufficiently discriminative (Taylor *et al.*, 1981). Other breath tests such as the xylose breath test and the fasting breath hydrogen test also do not offer sufficient sensitivity (Riordan *et al.*, 1995). It has been reported that the glucose H_2 breath test (Bjorneklett, Fausa and Midtvedt, 1983) may be a sufficiently discriminative test. There are also reports that determination of the proportion of unconjugated to total cholic acid in peripheral venous blood may be useful as a simple screening test for detection of contaminated small bowel syndrome (Einarsson *et al.*, 1992).

There are several different approaches to the treatment of contaminated small bowel syndrome (Gracey, 1983). Anatomical abnormalities, if present, can be surgically corrected. Second, antibiotic therapy directed against anaerobes is also effective in many patients. Thirdly, nutritional support, e.g. vitamin supplementation and low fat diet, also frequently leads to long-lasting remission of symptoms in some patients.

12.4.2 Role of bile acids and transformations of bile acids by the intestinal microflora in the pathogenesis of intestinal cancer

A complex interaction exists between the diet, the intestinal microflora and intestinal cancer, particularly colon cancer. Epidemiological studies have shown a higher incidence of colon cancer associated with environmental factors and in particular the characteristic 'Western' diet with its high content of meat and fat and its low fibre content (Willett *et al.*, 1990).

Several studies have examined the composition of the intestinal microflora in relation to a possible link between diet and colon cancer. The overall conclusion from these studies, however, was that the composition of the faecal microflora, if influenced at all by the diet, was only modified to a minor extent by differences in diet. The existence of a specific 'Western-type' diet microflora, associated with a higher risk for colon cancer, could not be demonstrated in these studies. In one more recent study, however, comparison of the faecal microfloras of groups considered to be at high risk of colon cancer (polyp patients and Japanese Hawaiians) *versus* groups documented to be at low risk (rural native

Japanese and rural native Africans) did show some differences. Total concentrations of *Bacteroides* sp. and *Bifidobacterium* sp. were generally positively associated with high risk of colon cancer. Some *Lactobacillus* sp. and *Eubacterium aerofaciens* showed the strongest association with low risk of colon cancer (Moore and Moore, 1995).

There is evidence suggesting that a possible causal link between the intestinal microflora and colon cancer may be found in the intestinal bacterial metabolism of exogenous and endogenous compounds. Although some chemicals are capable of directly inducing colon cancer in animals, it has become apparent that most carcinogens are modified by the intestinal microflora before they become active. Several bacterial enzymes have been implicated in the generation of mutagens, carcinogens and various tumour promoters, e.g. β-glucuronidase, β-glucosidase, β-galactosidase, nitroreductase, azoreductase, 7α-steroid dehydroxylase and 7α-hydroxysteroid dehydrogenase. It is therefore no coincidence that the most common site for the development of cancer in the intestinal tract, the large bowel, also harbours a metabolically highly active microflora.

Studies have also shown that this bacterial metabolic activity is labile and can be influenced by diet. For example measurements in both humans and rats of intestinal bacterial 7α-hydroxylase, nitroreductase, azoreductase and β-glucuronidase activity levels have shown these to be higher in groups on a mixed 'Western-type' meat-rich diet than in groups on a diet poor in meat (Goldin and Gorbach, 1976; Mastromarino, Reddy and Wynder, 1976; Reddy, Weisburger and Wynder, 1974). Changing from a diet poor in meat and fat to a meat-rich diet increased the levels of activity of these enzymes significantly.

What chemicals are responsible for colon cancer and how does the intestinal microflora interact with these compounds? Some procarcinogens may occur naturally in our environment whereas others may be ingested as food preservatives, dyes, additives or pollutants. There are carcinogens that are glucuronidated in our liver prior to biliary excretion (N-hydroxy-N-2-fluor-enylacetamide, diethylstilboestrol) and are retoxified by intestinal bacteria through their β-glucuronidase activity (Fischer *et al.*, 1966). Many natural plant products occur as glycosides; hydrolysis of the glycosidic linkages renders some of these compounds mutagenic (Tamura *et al.*, 1980; Macdonald *et al.*, 1984). Aromatic nitro compounds and azo compounds are widely distributed in our environment. The bacterial nitro- and azoreductase enzymes reduce these compounds, leading to the formation of potentially carcinogenic substances such as nitrosamines, N-hydroxy compounds, substituted phenyl- and napthylamines.

Secondary bile acids are also cancer promoting substances (Hill *et al.*, 1975; Reddy and Wynder, 1973; Barnes and Powrie, 1982). Deoxycholic acid in particular can act as a cancer promoter and cocarcinogen (Wilpart *et al.*, 1983; Narisawa *et al.*, 1974). Bile acids have been shown to have various effects on gastrointestinal mucosal cells, such as hyperplasia, metabolic alterations, cell-membrane disruptions and stimulation of DNA synthesis. In this way secondary bile acids might affect the frequency and the rate of progression of colon carcinogenesis. High faecal concentrations of secondary bile acids, and more precisely deoxycholic acid, are correlated with a Western-style, high-beef diet (Reddy and Wynder, 1973). In addition, population groups at high risk of developing colon cancer were reported to excrete larger amounts of secondary bile acids in their faeces (Hill and Aries, 1971; Reddy and Wynder, 1973; Reddy, 1981). Case-control studies, however, have yielded conflicting results (Murray *et al.*, 1980; Mudd *et al.*, 1980; Breuer *et al.*, 1985). As already mentioned, the secondary bile acids deoxycholic and lithocholic acid are formed from the primary bile acids cholic and chenodeoxycholic acid through the activity of the bacterial 7α-dehydroxylase. It has been observed that the activity of this bacterial 7α-dehydroxylase is affected by the type of diet. A 'Western-type' diet increases the intestinal bacterial 7α-dehydroxylase activity (Goldin *et al.*, 1980). We have shown recently that tauroconjugation of cholic acid also increases 7α-dehydroxylation compared to glycoconjugation (Van Eldere *et al.*, 1996). This observation fits in with the reports on the effect of a 'Western-type' diet on the 7α-dehydroxylase activity because a 'Western-type' diet will increase the level of bile acid tauroconjugation.

Finally, other endogenous compounds such as ketosteroids and glycerol ethers linked to long-chain polyunsaturated alcohols (fecapentaenes) were also shown to act as cancer inducers or promoters (Gupta *et al.*, 1984).

12.5 REFERENCES

Abrams, G. D., Bauer H. and Sprinz, H. (1963) Influence of the normal flora on mucosal morphology and cellular renewal in the ileum: a comparison of germ-free and conventional mice. *Laboratory Investigation*, **12**, 355–364.

Ali, S. S., Kuksis, A. and Beveridge, J. M. (1966) Excretion of bile acids by three men on a fat-free diet. *Canadian Journal of Biochemistry*, **4**, 957–969.

Almé, B., Bremmelgaard, A., Sjövall, J. *et al.* (1977) Analysis of metabolic profiles of bile acids in urine using a lipophilic anion exchanger and computerized gas–liquid chromatography–mass spectometry. *Journal of Lipid Research*, **18**, 339–362.

Aries, V. and Hill, M. J. (1970) Degradation of steroids by intestinal bacteria. II. Enzymes catalyzing the oxidoreduction of the 3α-, 7α, and 12α-hydroxyl groups in cholic acid and the dehydroxylation of the 7-hydroxyl group. *Biochimica et Biophysica Acta*, **202**, 535–543.

Back, P. and Gerok, W. (1977) Differences in renal excretion between glycoconjugates, tauroconjugates, sulfoconjugates and glucuronoconjugates of bile acids in cholestasis, in *Bile Acid Metabolism in Health and Disease*, (eds G. Paumgartner and A. Stiehl), MTP Press, Lancaster, pp. 93–100.

Back, P., Spaczynski, K. and Gerok, W. (1974) Bile-salt glucuronides in urine. *Hoppe-Seyler's Zeitschrift für Physiologische Chemie*, **355**, 749–752.

Baraona, E., Pirola, R. C. and Lieber, C. S. (1974) Small intestinal damage and changes in cell population produced by ethanol ingestion in the rat. *Gastroenterology*, **66**, 226–234.

Baraona, E., Palma, R., Navia, E. *et al.* (1968) The role of unconjugated bile salts in the malabsorption of glucose and tyrosine by everted sacs of jejunum of rats with the 'blind loop syndrome'. *Acta Physiologica Latinoamerica*, **18**, 291–297.

Barbara, L., Festi, D., Bazzoli, F. *et al.* (1984) The influence of gallbladder and intestinal motility on serum bile acid levels, in *The Enterohepatic Circulation of Bile Acids and Sterol Metabolism*, (eds G. Paumgartner, A. Stiehl and W. Gerok), MTP Press, Lancaster, pp. 139–142.

Barnes, W. S. and Powrie, W. D. (1982) Clastogenic activity of bile acids and organic fractions in human feces. *Cancer Letters*, **15**, 317–327.

Björkhem, I., Einarsson, K., Melone, P. *et al.* (1989) Mechanism of intestinal formation of deoxycholic acid from cholic acid in humans: evidence for a 3-oxo-Δ4-steroid intermediate. *Journal of Lipid Research*, **30**, 1033–1039.

Bjorneklett, A., Fausa, O. and Midtvedt, T. (1983). Small-bowel bacterial overgrowth in the postgastrectomy syndrome. *Scandinavian Journal of Gastroenterology*, **18**, 277–287.

Bloch, R., Menge, H., Lorenz-Meyer, M. *et al.* (1973) Morphologische und biochemische Veränderungen der Dunndarmschleimhaut beim Blind-sack Syndrom. *Verhandlungen der deutschen Gesellschaft für interne Medizin*, **79**, 853–856.

Borriello, S. P. and Owen, R. W. (1982) The metabolism of lithocholic acid and lithocholic acid-3-α-sulfate by human fecal bacteria. *Lipids*, **17**, 477–482.

Breuer, N., Dommes, P., Tandon, R. *et al.* (1984) Isolierung und Quantifizierung nicht-sulfatierter und sulfatierter Gallensäuren im Stuhl. *Journal of Clinical Chemistry and Clinical Biochemistry*, **22**, 623–631.

Breuer, N. F., Dommes, P., Jaekel, S. *et al.* (1985) Fecal bile acid excretion pattern in colonic cancer patients. *Digestive Diseases and Sciences*, **30**, 852–859.

Burke, V., Gracey, M., Thomas, J. *et al.* (1975) Inhibition of intestinal amino acid absorption by unconjugated bile salts *in vivo*. *Australian and New Zealand Journal of Medicine*, **5**, 430–432.

Coleman, J. P., White, W. B., Egestad, B. *et al.* (1987) Biosynthesis of a novel bile acid nucleotide and mechanism of 7α-dehydroxylation by an intestinal *Eubacterium* species. *Journal of Biological Chemistry*, **262**, 4701–4707.

Cowen, A. E., Korman, M. G., Hofmann, A. F. *et al.* (1975) Metabolism of lithocholate in healthy man. I. Biotransformation and biliary excretion of intravenously administered lithocholate, lithocholylglycine, and their sulfates. *Gastroenterology*, **69**, 59–66.

Dawson, A. M. and Isselbacher, K. J. (1960) Studies on lipid metabolism in the small intestine with observations on the role of bile salts. *Journal of Clinical Investigation*, **39**, 730–740.

De Barros, S. G., Balistreri, W. F., Soloway, R. D. *et al.* (1982) Response of total and individual serum bile acids to endogenous and exogenous bile acid input to the enterohepatic circulation. *Gastroenterology*, **82**, 647–652.

Dickinson, A. B., Gustafsson, B. E. and Norman, A. (1971) Determination of bile acid conversion potencies of intestinal bacteria by screening *in vitro* and subsequent establishment in germ-free rats. *Acta Pathologica et Microbiologica Scandinavica Section B*, **79B**, 691–698.

Donaldson, R. M. Jr (1962) Malabsorption of Co60-labeled cyanocobalamin in rats with intestinal diverticula. I. Evaluation of possible mechanisms. *Gastroenterology*, **43**, 271–281.

Edenharder, R. (1984) Dehydroxylation of cholic acid at C12 and epimerization at C5 and C7 by *Bacteroides* species. *Journal of Steroid Biochemistry*, **21**, 413–420.

Edenharder, R. and Deser, H. J. (1981) The significance of the bacterial steroid degradation for the etiology of large bowel cancer. VIII. Transformation of cholic, chenodeoxycholic and deoxycholic acid by lecithinase-lipase-negative Clostridia. *Zentralblatt für Bakteriologie, Parasitenkunde, Infektionskrankheiten und Hygiene Abt 1: orig., reihe B*, **174**, 91–104.

Edenharder, R., Stubenrauch, S. and Slemrova, J. (1976) Die Bedeutung des bakteriellen Steroidabbaus für die Aetiologie des Dickdarmkrebs. V. Metabolismus von Chenodesoxycholsaure durch saccharolytische *Bacteroides*-arten. *Zentralblatt für Bakteriologie, Parasitenkunde, Infektionskrankheiten und Hygiene Abt 1: orig., reihe B*, **162**, 506–518.

Einarsson, K., Bergstrom, M., Eklof, R. *et al.* (1992). Comparison of the proportion of unconjugated to total serum cholic acid and the (^{14}C)-xylose breath test in patients with suspected small intestinal bacterial overgrowth. *Scandinavian Journal of Clinical and Laboratory Investigation*, **52**, 425–430.

Eyssen, H. J. and Parmentier, G. G. (1979) Influence of the microflora of the rat on the metabolism of fatty acids, sterols and bile salts in the intestinal tract, in *Clinical and Experimental Gnotobiotics*, (eds T. M. Fliedner, H. Heit, D. Niethammer and H. Pflieger), *Zentralblatt für Bakteriologie*, Suppl. 7. Gustav Fischer Verlag, Stuttgart, pp. 39–44.

Eyssen, H. and Van Eldere, J. (1984) Metabolism of bile acids, in *The Germ-Free Animal in Biomedical Research*, (eds M. E. Coates and. B. E. Gustafsson), Laboratory Animals, London, pp. 291–316.

Eyssen, H., Van Eldere, J., Parmentier, G. *et al.* (1985) Influence of microbial bile salt desulfation upon the fecal excretion of bile salts in gnotobiotic rats. *Journal of Steroid Biochemistry*, **22**, 547–554.

Ferrari, A., Pacini, N. and Canzi, E. (1980) A note on bile acid transformations by strains of *Bifidobacterium*. *Journal of Applied Bacteriology*, **49**, 193–197.

Finegold, S. M., Sutter, V. L. and Mathesen, G. E. (1983) Normal indigenous intestinal flora, in *Human Intestinal Microflora in Health and Disease*, (ed. D. J. Hentges), Academic Press, New York, pp. 3–31.

Fischer, L. J., Millburn, P., Smith, R. L. *et al.* (1966) The fate of ^{14}C-stilboestrol in the rat. *Biochemical Journal*, **100**, 69–72.

Ford, D. J. and Coates, M. E. (1971) Absorption of glucose and vitamins of the B complex by germfree and conventional chicks. *Proceedings of the Nutrition Society*, **30**, 10A.

Forth, W., Rummel, W. and Glasner, H. (1966) Zur resorptionhemmenden Wirkung von Gallsauren. *Naunyn-Schmeidebergs Archiv der Pharmakologie*, **254**, 364–380.

Fricker, G., Schneider, S., Gerok, W. *et al.* (1987) Identification of different transport systems for bile salts in sinusoidal and canalicular membranes of hepatocytes *Biological Chemistry Hoppe-Seyler*, **368**, 1143–1150.

Fröhling, W., Stiehl, A., Czygan, P. *et al.* (1977) Induction of bile acid glucuronide formation in children with intrahepatic cholestasis, in *Bile Acid Metabolism in Health and Disease*, (eds G. Paumgartner and A. Stiehl), MTP Press, Lancaster, pp. 101–104.

Giannella, R. A., Broitman, S. A. and Zamchek, N. (1971) Vitamin B_{12} uptake by intestinal micro-organisms: mechanism and relevance to syndromes of intestinal bacterial overgrowth. *Journal of Clinical Investigation*, **50**, 1100–1107.

Giannella, R. A., Rout, W. R. and Toskes, P. P. (1974) Jejunal brush border injury and impaired sugar and amino-acid uptake in the blind loop syndrome. *Gastroenterology*, **67**, 965–974.

Gilliland, S. E. and Speck, M. L. (1977) Deconjugation of bile acids by intestinal lactobacilli. *Applied and Environmental Microbiology*, **33**, 15–18.

Goldin, B. R. and Gorbach, S. L. (1976) The relationship between diet and fecal bacterial enzymes implicated in colon cancer. *Journal of the National Cancer Institute*, **57**, 371–375.

Goldin, B. R., Swenson, L., Dwyer, J. *et al.* (1980) Effect of diet and *Lactobacillus acidophilus* supplements on human fecal bacterial enzymes. *Journal of the National Cancer Institute*, **64**, 255–261.

Goldstein, F., Karacadag, S., Wirts, C. W. *et al.* (1970) Intraluminal small-intestinal utilization of D-xylose by bacteria: a limitation of the D-xylose absorption test. *Gastroenterology*, **59**, 380–386.

Gordon, H. A. and Bruckner-Kardoss, E. (1961) Effect of normal microbial flora on intestinal surface area. *American Journal of Physiology*, **201**, 175–178.

Gracey, M. (1981) Nutrition, bacteria and the gut. *British Medical Bulletin*, **37**, 71–75.

Gracey, M. (1983) The contaminated small bowel syndrome, in *Human Intestinal Microflora in Health and Disease*, (ed. D. J. Hentges), Academic Press, New York, pp. 495–515.

Gracey, M., Papadimitriou, J. and Burke, V. (1973) Effects on small-intestinal function and structure induced by feeding a deconjugated bile salt. *Gut*, **14**, 519–528.

Gracey, M., Burke, V. and Oshin, A. (1971) Reversible inhibition of intestinal active sugar transport by deconjugated bile salt *in vitro*. *Biochimica et Biophysica Acta*, **225**, 308–314.

Gracey, M., Burke, V., Oshin, A. *et al.* (1971) Bacteria, bile salts and intestinal monosaccharide malabsorption. *Gut*, **12**, 683–692.

Grantham, P. H., Horton, R. E., Weisburger, E. K. *et al.* (1970) Metabolism of the carcinogen N-2-fluorenylacetamide in germ-free and conventional rats. *Biochemical Pharmacology*, **19**, 163–171.

Gupta, I., Suzuki, K., Bruce, W. R. *et al.* (1984) A model study of fecapentaenes: mutagens of bacterial origin with alkylating properties. *Science*, **225**, 521–523.

Hajjar, J. J., Khuri, R. N. and Bikhazi, A. B. (1975) Effect of bile salts on amino acid transport by rabbit intestine. *American Journal of Physiology*, **229**, 518–523.

Hardison, W. G. M. (1978) Hepatic taurine concentration and dietary taurine as regulators of bile acid conjugation with taurine. *Gastroenterology*, **75**, 71–75.

Haslewood, G. A. D. (1978). The biological importance of bile salts, in *Frontiers of Biology*, vol. 47, (eds A. Neuberger and E. L. Tatum), North-Holland, Amsterdam, pp. 206.

Hayakawa, S. (1973) Microbiological transformation of bile acids. *Advances in Lipid Research*, **11**, 143–192.

Heneghan, J. B. (1963) Influence of microbial flora on xylose absorption in rats and mice. *American Journal of Physiology*, **205**, 417–420.

Hill, M. J. and Aries, V. C. (1971) Faecal steroid composition and its relationship to cancer of the large bowel. *Journal of Pathology*, **104**, 129–139.

Hill, M. J., Drasar, B. S., Williams, R. E. *et al.* (1975) Faecal bile-acids and clostridia in patients with cancer of the large bowel. *Lancet*, **i**, 535–539.

Hirano, S. and Masuda, N. (1981a) Transformation of bile acids by *Eubacterium lentum*. *Applied and Environmental Microbiology*, **42**, 912–915.

Hirano, S. and Masuda, N. (1981b) Epimerization of the 7-hydroxyl group of bile acids by the combination of two kinds of microorganisms with 7α- and 7β-hydroxysteroid dehydrogenase activity, respectively. *Journal of Lipid Research*, **22**, 1060–1068.

Hirano, S. and Masuda, N. (1982) Characterization of the NADP-dependent 7β-hydroxysteroid dehydrogenase from *Peptostreptococcus productus* and *Eubacterium aerofaciens*. *Applied and Environmental Microbiology*, **43**, 1057–1063.

Hirano, S., Masuda, N., Oda, H. *et al.* (1981) Transformation of bile acids by mixed microbial cultures from human feces and bile acid transforming activities of isolated bacterial strains. *Microbiology and Immunology*, **25**, 271–282.

Huijghebaert, S. M. and Eyssen, H. J. (1982) Specificity of bile salt sulfatase activity from *Clostridium* sp. strain S1. *Applied and Environmental Microbiology*, **44**, 1030–1034.

Huijghebaert, S., Parmentier, G. and Eyssen, H. (1984) Specificity of bile salt sulfatase activity in man, mouse and rat intestinal microflora. *Journal of Steroid Biochemistry*, **20**, 907–912.

Hylemon, P. B. and Sherrod, J. A. (1975) Multiple forms of 7α-hydroxysteroid dehydrogenase in selected strains of *Bacteroides fragilis*. *Journal of Bacteriology*, **122**, 418–422.

Hylemon, P. B. and Stellwag, E. J. (1976) Bile acid transformation rates of selected gram-positive and gram-negative intestinal bacteria. *Biochemical and Biophysical Research Communications*, **69**, 1088–1094.

Imperato, T. J., Wong, C. G., Chen, J. L. (1977) Hydrolysis of lithocholate sulfate by *Pseudomonas aeruginosa*. *Journal of Bacteriology*, **130**, 545–547.

Islam, M. A., Raicht, R. F. and Cohen, B. I. (1981) Isolation and quantification of sulfated and unsulfated steroids in human feces. *Analytical Biochemistry*, **112**, 371–377.

Jonas, E. A., Craigie, A., Tavill, A. S. *et al.* (1968) Protein metabolism in the intestinal stagnant loop syndrome. *Gut*, **9**, 466–469.

Kellogg, T. F., Knight, P. L. and Wostmann, B. S. (1970) Effect of bile acid deconjugation on the fecal excretion of steroids. *Journal of Lipid Research*, **11**, 362–366.

Kim, Y. S. and Spritz, N. (1968) Metabolism of hydroxy fatty acids in dogs with steatorrhea secondary to experimentally produced intestinal blind loops. *Journal of Lipid Research*, **9**, 487–491.

King, C. E. and Toskes, P. P. (1976) Malabsorption following gastric secretion, in *Postgastrectomy Syndromes, Major Problems in Clinical Surgery, vol. 20*, (eds F. L. Bushkin and E. R. Woodward), W. B. Saunders, Philadelphia, PA, pp. 129–146.

Kinsella, V. J., Hennessy, W. B. and George, E. P. (1961) Studies on postgastrectomy malabsorption: the importance of bacterial contamination of the upper small intestine. *Medical Journal of Australia*, **2**, 257–261.

Klipstein, F. A., Holdeman, L. V., Corcino, J. J. *et al.* (1973) Enterotoxigenic intestinal bacteria in tropical sprue. *Annals of Internal Medicine*, **79**, 632–641.

Kobashi, K., Nishizawa, I., Yamada, T. *et al.* (1978) A new hydrolase specific for taurine conjugates of bile acids. *Journal of Biochemistry (Tokyo)*, **84**, 495–497.

Korpela, J. T., Fotsis, T. and Adlercreutz, H. (1986) Multicomponent analysis of bile acids in faeces by anion exchange and capillary column gas–liquid chromatography: application in oxytetracycline treated subjects. *Journal of Steroid Biochemistry*, **25**, 277–284.

Levy, N. S. and Toskes, P. P. (1974) Fundus albipunctatus and vitamin A deficiency. *American Journal of Ophthalmology*, **150**, 219–238.

Low-Beer, T. S., Tyor, M. P. and Lack, L. (1969) Effects of sulfation of taurocholic and glycolithocholic acids on their intestinal transport. *Gastroenterology*, **56**, 721–726.

Macdonald, I. A., Hutchison, D. M. and Forrest, T. P. (1981) Formation of urso- and ursodeoxycholic acids from primary bile acids by *Clostridium absonum*. *Journal of Lipid Research*, **22**, 458–466.

Macdonald, I. A., Williams, C. N. and Mahony, D. E. (1973) 7α-Hydroxysteroid dehydrogenase from *Escherichia coli* B. Preliminary studies. *Biochimica et Biophysica Acta*, **309**, 243–253.

Macdonald, I. A., Jellett, J. F., Mahoney, D. E. *et al.* (1979) Bile salt 3α- and 12α-hydroxysteroid dehydrogenases from *Eubacterium lentum* and related strains. *Applied and Environmental Microbiology*, **37**, 992–1000.

Macdonald, I. A., Rochon, Y. P., Hutchison, D. M. *et al.* (1982) Formation of ursodeoxycholic acid from chenodeoxycholic acid by a 7β-hydroxysteroid dehydrogenase-elaborating *Eubacterium aerofaciens* strain co-cultured with 7α-hydroxysteroid dehydrogenase-elaborating organisms. *Applied and Environmental Microbiology*, **44**, 1187–1195.

Macdonald, I. A., Bussard, R. G., Hutchinson, D. M. *et al.* (1984) Rutin-induced β-glucosidase activity in *Streptococcus faecium* VGH-1 and *Streptococcus* strain FRP-17 isolated from human feces: formation of the mutagen quercetin, from rutin. *Applied and Environmental Microbiology*, **47**, 350–355.

Makino, I., Hashimoto, H., Shinozaki, K. *et al.* (1975) Sulfated and nonsulfated bile acids in urine, serum, and bile of patients with hepatobiliary diseases. *Gastroenterology*, **68**, 545–553.

Marshall, H. U., Egestad, B., Matern, H. *et al.* (1987) Evidence for bile acid glucosides as normal constituents in human urine. *FEBS Letters*, **213**, 411–414.

Mastromarino, A., Reddy, B. S. and Wynder E. L. (1976) Metabolic epidemiology of colon cancer: enzymatic activity of the fecal flora. *American Journal of Clinical Nutrition*, **29**, 1455–1460.

Masuda, N. (1981) Deconjugation of bile salts by *Bacteroides* and *Clostridium*. *Microbiology and Immunology*, **25**, 1–11.

Mekhjian, H. S., Phillips, S. F. and Hofmann, A. F. (1968) Conjugated bile salts block water and electrolyte transport by the human colon. *Gastroenterology*, **54**, 1256 (abstr.).

Midtvedt, T. (1974) Microbial bile acid transformation. *American Journal of Clinical Nutrition*, **27**, 1341–1347.

Midtvedt, T. and Norman, A. (1967) Bile acid transformations by microbial strains belonging to genera found in intestinal contents. *Acta Pathologica et Microbiologica Scandinavica*, **71**, 629–638.

Moore, W. E. C. and Moore, L. H. (1995) Intestinal flora of populations that have a high risk of colon cancer. *Applied and Environmental Microbiology*, **61**, 3202–3207.

Mudd, D. G., McKelvey, S. T., Norwood, W. *et al.* (1980) Faecal bile acid concentrations of patients with carcinoma or increased risk of carcinoma in the large bowel. *Gut*, **21**, 587–590.

Murray, W. R., Blackwood, A., Trotter, J. M. *et al.* (1980) Faecal bile acids and clostridia in the aetiology of colorectal cancer. *British Journal of Cancer*, **41**, 923–928.

Narisawa, T., Magadia, N. E., Weisburger, J. H. *et al.* (1974) Promoting effect of bile acids on colon carcinogenesis after intrarectal instillation of N-methyl-N'-nitro-N-nitrosoguanidine in rats. *Journal of the National Cancer Institute*, **53**, 1093–1097.

Neale, G., Gompertz, D., Schjonsby, H. *et al.* (1972) The metabolic and nutritional consequences of bacterial overgrowth in the small intestine. *American Journal of Clinical Nutrition*, **25**, 1409–1417.

Nygaard, L. and Rootwelt, K. (1968) Intestinal protein loss in rats with blind segments on the bowel. *Gastroenterology*, **54**, 52–55.

Palmer, R. H. and Bolt, M. G. (1971) Bile acid sulfates. I. Synthesis of lithocholic acid sulfates and their identification in human bile. *Journal of Lipid Research*, **12**, 671–679.

Polter, D. E., Boyle, J. D., Miller, L. G. *et al.* (1968) Anaerobic bacteria as a cause of the blind loop syndrome. *Gastroenterology*, **54**, 1148–1154.

Pope, J. L., Parkinson, T. M. and Olson, J. A. (1966) Action of bile salts on the metabolism and transport of water-soluble nutrients by perfused rat jejunum *in vitro*. *Biochimica et Biophysica Acta*, **130**, 218–232.

Radominska-Pyrek, A., Zimniak, P., Chari, M. *et al.* (1986) Glucuronides of mono-hydroxylated bile acids: specificity of microsomal glucuronyltransferase for the glucuronidation site, C-3 configuration and side chain length. *Journal of Lipid Research*, **27**, 89–101.

Radominska-Pyrek, A., Zimniak, P., Irshaid, Y. M. *et al.* (1987) Glucuronidation of 6α-hydroxy bile acids by human liver microsomes. *Journal of Clinical Investigation*, **80**, 234–241.

Raedsch, R., Stiehl, A., Rudolph, G. *et al.* (1983) On the ileal output and composition of bile acids in man. *Gastroenterology*, **84**, 1391.

Reddy, B. S. (1981) Diet and excretion of bile acids. *Cancer Research*, **41**, 3766–3768.

Reddy, B. S. and Wostmann, B. S. (1966) Intestinal disaccharidase activities in the growing germfree and conventional rats. *Archives of Biochemistry and Biophysics*, **113**, 609–616.

Reddy, B. S. and Wynder, E. L. (1973) Large bowel carcinogenesis: fecal constituents of populations with diverse incidence rates of colon cancer. *Journal of the National Cancer Institute*, **50**, 1437–1442.

Reddy, B. S., Pleasants, J. B. and Wostmann, B. S. (1969) Effect of intestinal microflora on iron and zinc metabolism, and on activities of metalloenzymes in rats. *Journal of Nutrition*, **102**, 101–108.

Reddy, B. S., Weisburger, J. and Wynder E. L. (1974) Fecal beta-glucuronidase: control by diet. *Science*, **183**, 416–417.

Riordan, S. M., McIver, C. J., Duncombe, V. M. *et al.* (1995) Factors influencing the 1-g ^{14}C-D-xylose breath test for bacterial overgrowth. *American Journal of Gastroenterology*, **90**, 1455–1460.

Robben, J., Parmentier, G. and Eyssen, H. (1986) Isolation of a rat intestinal *Clostridium* strain producing 5α- and 5β-bile salt 3α-sulfatase activity. *Applied and Environmental Microbiology*, **51**, 32–38.

Rolfe, R. D. (1989) Role of anaerobic bacteria in other bowel pathology in *Anaerobic Infections in Humans*, (eds S. M. Finegold and W. L. George), Academic Press, San Diego, CA, pp. 679–690.

Ruddell, W. S., Axon, A. T., Findlay, J. M. *et al.* (1980) Effect of cimetidine on the gastric bacterial flora. *Lancet*, **i**, 672–674.

Rutgeerts, L., Mainguet, P., Tytgat, G. *et al.* (1974) Enterokinase in contaminated small-bowel syndrome. *Digestion*, **10**, 249–254.

Sacquet, E., Parquet, M., Riottot, M. *et al.* (1983) Intestinal absorption, excretion, and biotransformation of hyodeoxycholic acid in man. *Journal of Lipid Research*, **24**, 604–613.

Shimada, K., Bricknell, K. S. and Finegold, S. M. (1969) Deconjugation of bile acids by intestinal bacteria: review of the literature and additional studies. *Journal of Infectious Diseases*, **119**, 273–281.

Shimoda, S. S., O'Brien, T. K. and Saunders, D. R. (1974) Fat absorption after infusing bile salts into the human small intestine. *Gastroenterology*, **67**, 7–18.

Sjövall, J. (1959) Dietary glycine and taurine on bile acid conjugation in man. Bile acids and steroids 75. *Proceedings of the Society for Experimental Biology and Medicine*, **100**, 676–678.

Stiehl, A. (1974) Bile salt sulphates in cholestasis. *European Journal of Clinical Investigation*, **4**, 59–63.

Tabaqchali, S., Hatzioannou, J. and Booth, C. C. (1968) Bile salt deconjugation and steatorrhoea in patients with the stagnant loop syndrome. *Lancet*, **ii**, 12–16.

Tabaqchali, S., Okubadejo, O. A., Neale, G. *et al.* (1966) Influence of abnormal bacterial flora on small intestine function. *Proceedings of the Royal Society of Medicine*, **59**, 1244–1246.

Takikawa, H., Otsuka, H., Beppu, T. *et al.* (1983) Serum concentrations of bile acid glucuronides in hepatobiliary diseases. *Digestion*, **27**, 189–195.

Tamura, G., Gold, C., Ferro-Luzzi, A. *et al.* (1980) Fecalase: a model for activation of dietary glycosides to mutagens by intestinal flora. *Proceedings of the National Academy of Sciences of the USA*, **77**, 4961–4965.

Taylor, R. H., Avgerinos, A., Taylor, A. J. *et al.* (1981) Bacterial colonisation of the jejunum: an evaluation of five diagnostic tests (abstract). *Gut*, **22**, A442–A443.

Thomas, G. P., Nakagaki, M., Hofmann, A. F. *et al.* (1982) A systemic marker for the rapid detection of experimental bacterial overgrowth: the ureolysis breath test (abstract). *Gastroenterology*, **82**, 1197.

Tomkin, G. H. and Weir, D. G. (1972) Indicanuria after gastric surgery. *Quarterly Journal of Medicine*, **41**, 191–203.

van Berge Henegouwen, G. P., Brandt, K. H., Eyssen, H. *et al.* (1976) Sulphated and unsulphated bile acids in serum, bile and urine of patients with cholestasis. *Gut*, **17**, 861–869.

van der Waaij, D., Berghuis de Vries, J. M. and Lekkerkerk van der Wees, J. E. C. (1971) Colonisation resistance of the digestive tract in conventional and anti-biotic-treated-mice. *Journal of Hygiene (Cambridge)*, **69**, 405–411.

van der Werf, S. D. J., van Berge Henegouwen, G. P. and van den Broeck, W. (1985) Estimation of bile acid pool sizes from their spillover into systemic blood. *Journal of Lipid Research*, **26**, 168–174.

Van Eldere, J., De Pauw, G. and Eyssen, H. J. (1987) Steroid sulfatase activity in a *Peptococcus niger* strain from the human intestinal microflora. *Applied and Environmental Microbiology*, **53**, 1655–1660.

Van Eldere, J., Celis, P., De Pauw G. *et al.* (1996) Tauroconjugation of cholic acid stimulates 7α-dehydroxylation by fecal bacteria. *Applied and Environmental Microbiology*, **62**, 656–661.

Willett, W. C., Stampfer, M. J., Colditz, B. A. *et al.* (1990) Relation of meat, fat, and fiber intake to the risk of colon cancer in a prospective study among women. *New England Journal of Medicine*, **323**, 1664–1672.

Wilpart, M., Mainguet, R., Maskens, A. *et al.* (1983) Structure-activity relationship among biliary bile acids showing co-mutagenic activity towards 1,2-dimethyl-hydrazine. *Carcinogenesis*, **4**, 1239–1241.

Wilson, F. A. (1981) Intestinal transport of bile acids. *American Journal of Physiology*, **241**, G83–G92.

Yolton, D. P., Stanely, C. and Savage, D. C. (1971) Influence of the indigenous gastrointestinal microbial flora or duodenal alkaline phosphatase activity in mice. *Infection and Immunity*, **3**, 768–773.

13

Translocation of microbes from the intestinal tract

Rodney D. Berg

13.1 INTRODUCTION

As early as 1900, viable bacteria were cultured from the livers, spleens, and kidneys of various animals and humans (Ford, 1900, 1901). There was not much interest in this phenomenon, however, until Ellis and Dragstedt (1930) confirmed the report of Wolbach and Saiki (1909) that the livers of presumably healthy dogs contain viable anaerobes. Further studies in the 1950s suggested that these bacteria pass from the gastrointestinal (GI) tract in dogs to the portal blood and liver (Chau, Goldbloom and Gurd, 1951; Schatten, 1954). Thus there were scattered early reports of the presence of bacteria of GI origin in extraintestinal sites, but no systematic investigations were undertaken.

Various terms, such as absorption, persorption, transmural migration and translocation, have been used to describe the passage of various substances, including bacteria, viruses and even non-viable particles, from the GI tract across the intestinal epithelial barrier (Table 13.1).

Translocation seems the most appropriate of these terms since it simply describes the relocation of bacteria from one site (GI tract) to another (extraintestinal site) without implying anything concerning the mechanisms involved (Berg and Garlington, 1979).

Indigenous bacteria are continuously translocating across the intestinal mucosa at low rates in normal healthy mice, but then are killed by the host immune defenses during their transit through the lamina propria and *in situ* in reticuloendothelial organs, such as the mesenteric lymph

G. W. Tannock (ed.), *Medical Importance of the Normal Microflora*, 338–370.

Table 13.1 Terms describing the passage of particles and microorganisms across the intestinal epithelium

Term	Reference
Absorption	Wilson and Walzer, 1935
Transmural migration	Frank, Seligman and Fine, 1948
Persorption	Volkheimer, 1974
Phage translocation	Keller and Engley, 1958 Hildebrand and Wolochow, 1962
Bacterial translocation	Wolochow, Hildebrand and Lamanna, 1966 Fuller and Jayne-Williams, 1970 Berg and Garlington, 1979

node complex (MLN). On rare occasions, translocating indigenous bacteria that have temporarily escaped killing by the host immune defenses may be cultured from the MLN of healthy adult mice (Berg and Garlington, 1979). These 'spontaneously' translocating bacteria are usually indigenous lactobacilli and enterococci that normally populate the GI tract at fairly high levels (10^{7-9}/g intestinal contents). Indigenous obligately anaerobic bacteria are only rarely detected in the MLN of healthy mice, even though these obligate anaerobes populate the GI tract at very high levels (10^{10-11}/g intestinal contents). It is assumed that the obligate anaerobes physically cross the intestinal barrier (i.e. translocate) as readily as do the facultative bacteria but the obligate anaerobes are probably killed more easily in extraintestinal sites because of their sensitivity to oxygen. In normal mice, it also is rare to detect spontaneous translocating bacteria in sites other than the MLN, such as the spleen, liver, kidney or blood.

The epithelial cells lining the GI tract can be thought of as 'non-professional' phagocytes that readily engulf any particle or bacterium that comes in close contact. Consequently, in animal models that do not exhibit increased intestinal permeability or actual physical damage to the intestinal mucosa, the major route of translocation of indigenous bacteria is engulfment by and passage through the intestinal epithelial cells, i.e. an intracellular route (Berg, 1981a, 1985; Fig. 13.1).

Even overtly pathogenic microorganisms, such as *Salmonella typhimurium* (Takeuchi, 1967), enteropathogenic *Escherichia coli* (Staley, Jones and Corly, 1969) and *Candida albicans* (Alexander *et al.*, 1990), are seen by microscopy to translocate from the GI tract primarily by an intracellular route rather than an intercellular route between the intestinal epithelial cells. Once across the intestinal epithelium, translocating bacteria travel via the lymphatics to the MLN. Translocating bacteria can be

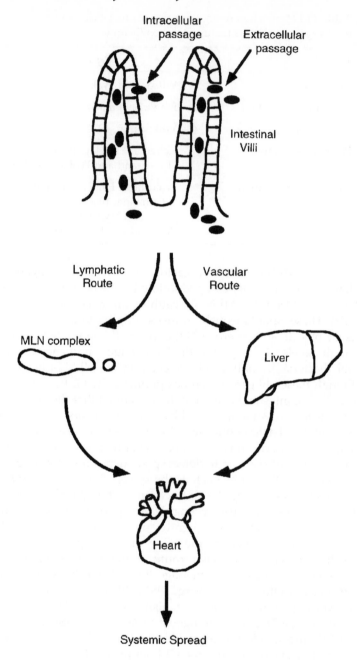

Figure 13.1 Routes of bacterial translocation from the GI tract.

seen microscopically free in the lymph and engulfed by macrophages in the lymph. However, if the intestinal epithelium or the tight junction between epithelial cells is physically damaged, as readily occurs during endotoxic shock, hemorrhagic shock, or thermal injury, then bacteria can translocate by the intercellular route through the spaces left by denuded epithelial cells and through ulcerated areas of the intestinal epithelium (Berg, 1983a). Under these conditions, translocating bacteria can enter the blood directly and travel to the liver without passing through the lymph.

Different segments of the MLN complex drain lymph from particular regions of the GI tract (Carter and Collins, 1964; Fig. 13.2). Segment-1, the small portion of the MLN complex that is separated from the main body of the MLN, drains lymph from the distal ileum, cecum and ascending colon. Segment-2, one half of the large body of the MLN closest to segment-1, drains lymph from the proximal ileum. Segment-3,

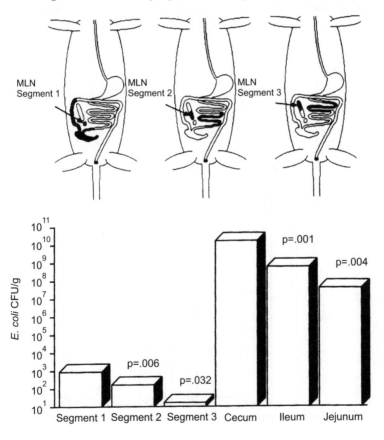

Figure 13.2 Bacterial translocation to different segments of the MLN complex.

the other half of the large body of the MLN, drains lymph from the jejunum. Interestingly, indigenous bacteria translocate to each of these three MLN segments in direct relation to their population levels in the region of the intestines supplying lymph to the particular MLN segment (Gautreaux, Deitch and Berg, 1994a). For example, indigenous *E. coli* population levels are highest in the cecum and ascending colon, and *E. coli* translocates at the highest rate to segment-1 draining lymph from this region of the GI tract. *E. coli* populations are lowest in the jejunum and *E. coli* translocates at the lowest rate to segment-3 draining lymph from the jejunum. These results demonstrate the importance of GI population levels in the promotion of translocation of indigenous bacteria from the GI tract.

Even though Peyer's patches have been suggested as conduits for the passage of bacteria and particles from the GI tract, Peyer's patches are not required for the translocation of indigenous bacteria. *E. coli* translocates to the MLN equally well from ligated ileal segments not containing visible

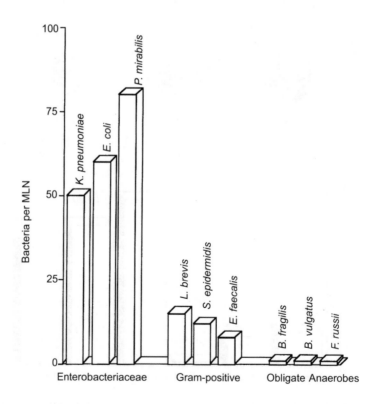

Figure 13.3 Translocation efficiencies of various indigenous bacteria from the GI tract to the MLN after 1 week in monoassociated gnotobiotic mice.

Peyer's patches as from ligated ileal segments containing visible Peyer's patches (Maejima and Berg, unpublished observations). Pathogenic *S. typhimurium* also translocates equally well to the MLN when inoculated into ligated ileal segments with or without Peyer's patches, even though *S. typhimurium* is thought to exhibit predilection for Peyer's patches. Therefore, it appears that indigenous microorganisms can translocate across the intestinal epithelium by routes other than through Peyer's patch M-cells.

The various species of indigenous bacteria do not all translocate at the same efficiency from the GI tract to the MLN (Steffen and Berg, 1983). *Pseudomonas aeruginosa, Enterococcus faecalis,* and the Gram-negative facultatively anaerobic *Enterobacteriaceae,* such as *E. coli, Klebsiella pneumoniae* and *Proteus mirabilis,* translocate at the greatest rate from the GI tract to the MLN in ex-germfree mice monoassociated for 1 week with each of these bacterial types (Fig. 13.3).

The Gram-positive oxygen-tolerant bacteria, such as *Staphylococcus epidermidis,* translocate at an intermediate level. Interestingly, the obligately anaerobic bacteria, such as *Bacteroides fragilis, Bacteroides vulgatus* and *Fusobacterium russii,* are the least effective translocators, even though they colonize the GI tract at the highest population levels (10^{10-11}/g cecum). Thus, the *Enterobacteriaceae* are the bacterial types that translocate most readily from the GI tract in animal models of intestinal bacterial overgrowth and these same bacterial types most commonly cause septicemia in hospitalized patients.

This same pattern of bacterial translocation efficiency is observed *in vitro* across human Caco-2 tissue culture monolayers in double-chambered Costar TranswellX™ filter units (Cruz *et al.,* 1994). Caco-2, a transformed human colonic carcinoma cell line, forms polarized monolayers of differentiated small intestinal enterocytes exhibiting both tight junctions and brush border microvilli. Bacterial species translocating across Caco-2 monolayers exhibit the following order of frequency: *E. coli, Citrobacter freundii, Enterococcus faecalis, Enterobacter aerogenes, Staphylococcus aureus* and finally *B. fragilis.* Thus, the *Enterobactericeae* translocate at a higher rate than do the obligate anaerobes, similar to observations in animal models of translocation. Direct comparisons cannot be made between the animal models and Caco-2 monolayers since the *in vitro* tissue culture monolayer lacks mucus, secretory IgA, phagocytic cells, T cells and normal intestinal peristalsis. Nonetheless, it is important to note that indigenous *Enterobacteriaceae* translocate more efficiently than other bacterial types both *in vivo* and *in vitro,* and *Enterobacteriaceae* are the most common cause of opportunistic infections originating from the GI tract in hospitalized patients.

The three primary mechanisms promoting translocation of microorganisms from the GI tract are: (1) disruption of the GI microecology allowing

1. Intenstinal Bacterial Overgrowth

- Mice given oral antibiotics
- Monoassociated gnotobiotic (ex-germfree) mice
- Starvation
- Protein malnutrition
- Parenteral alimentation
- Enteral alimentation
- Bowel obstruction
- Bile duct ligation
- Streptozotocin-induced diabetes

2. Impaired Host Immune Defenses

- Mice injected with immunosuppressive agents
- Thymectomy
- Athymic (nu/nu) mice
- CD4/CD8 T-cell depletion
- Beige/nude (bg/bg;nu/nu) mice
- Lymphoma
- Leukemia
- Streptozotocin-induced diabetes
- Endotoxemia
- Thermal injuiry
- Hemorrhagic shock

3. Physical Damage to Intestinal Mucosa

- Oral ricinoleic acid (castor oil)
- Endotoxemia
- Injection of zymosan
- Thermal injury
- Hemorrhagic shock

Figure 13.4 Animal models demonstrating the three major mechanisms promoting bacterial translocation from the GI tract.

intestinal overgrowth by certain indigenous microorganisms and/or colonization by more pathogenic exogenous microorganisms; (2) increased intestinal permeability or physical disruption of the intestinal epithelial barrier; and (3) compromised host immune defenses (reviewed in Berg, 1981a, 1983a, 1992a, 1995). Various animal models demonstrating bacterial translocation are categorized in Fig. 13.4 according to the primary translocation promotion mechanism. As can be seen, more than one mechanism often is operating to promote bacterial translocation in certain animal models and in various clinical situations.

13.2 MECHANISMS PROMOTING BACTERIAL TRANSLOCATION FROM THE GI TRACT

13.2.1 Intestinal bacterial overgrowth

Berg and Garlington (1979) reported that indigenous *E. coli*, *E. faecalis*, *K. pneumoniae* and *P. mirabilis* readily translocate from the GI tract to the MLN of ex-germfree (gnotobiotic) mice inoculated intragastrically (IG) with the whole cecal flora from specific pathogen-free (SPF) mice and tested after 1 week, whereas SPF mice receiving a similar cecal inoculum do not exhibit bacterial translocation to the MLN. Bacterial translocation occurs in the gnotobiotic mice and not the SPF mice because, during this first week after IG inoculation, the GI microecology in the gnotobiotic mice has not had time to stabilize and certain indigenous bacteria are present in various regions of the GI tract at abnormally high levels. The whole cecal inoculum has no effect on the GI microecology of SPF mice since their GI ecological niches are already occupied by stable populations of indigenous bacteria. Thus, intestinal overgrowth by certain species of indigenous bacteria readily promotes their translocation from the GI tract.

E. coli translocates to 100% of the MLN of *E. coli*-monoassociated gnotobiotic mice or antibiotic-decontaminated SPF mice monoassociated with *E. coli* (Berg and Owens, 1979). SPF mice were antibiotic-decontaminated with oral penicillin plus streptomycin *ad libitum* to eliminate their indigenous GI tract microflora. The intestinal lamina propria is much thicker in SPF mice than in germfree mice that have not been antigenically stimulated by an indigenous GI microflora. It is therefore somewhat surprising that *E. coli* translocates with equal efficiency from the GI tract to the MLN in *E. coli*-monoassociated SPF mice with a normal-thickness intestinal lamina propria compared with *E. coli*-monoassociated gnotobiotic mice with a very thin intestinal lamina propria. Apparently, the abnormally high GI tract population levels of *E. coli* (10^{9-10}/g cecum) in *E. coli*-monoassociated SPF mice overcomes any limitations imposed by their thicker intestinal lamina propria.

E. coli establishes a 'steady-state' pattern of translocation in *E. coli*-monoassociated gnotobiotic mice since the numbers of *E. coli* cultured from the MLN at 2 days of monoassociation are similar to the numbers of *E. coli* cultured from the MLN after 100 days of monoassociation (Berg and Owens, 1979). It seems likely that the host immune defenses over time would become better able to eliminate the translocating *E. coli*, especially after 100 days of *E. coli* monoassociation. *E. coli* maintained high GI population levels (10^{9-10}/g intestinal contents) throughout the entire 100-day test period.

E. coli translocation to the MLN ceases within 24 hours, however, when *E. coli*-monoassociated gnotobiotic mice are given oral penicillin to eliminate

E. coli from their GI tracts. Thus, translocating *E. coli* do not establish a permanent 'infection' in the MLN, but rather *E. coli* in the GI tract continuously 'seed' the MLN and are then detected as translocating *E. coli*.

A 'cause and effect' relationship is demonstrated between *E. coli* GI population levels and *E. coli* translocation to the MLN by comparing gnotobiotic mice monoassociated with *E. coli* with gnotobiotic mice colonized with *E. coli* plus an indigenous GI microflora (Berg and Owens, 1979). In *E. coli*-monoassociated gnotobiotes, *E. coli* cecal populations remain greater than 10^{11}/g during the entire 1-week test period and *E. coli* translocates to 100% of the MLN. In gnotobiotes colonized with *E. coli* plus a whole indigenous GI tract microflora, *E. coli* cecal populations decrease 1000-fold within 48 hours as a result of antagonism by the indigenous microflora and *E. coli* translocation to the MLN decreases concomitantly. Repeating these experiments with antibiotic-decontaminated SPF mice monoassociated with *E. coli* produces similar results (Berg, 1980a). Furthermore, neither *E. coli* cecal populations nor *E. coli* translocation is decreased in *E. coli*-monoassociated gnotobiotic mice inoculated IG with heat-treated, formalin-treated or filter-sterilized cecal contents from SPF mice (Berg, 1980a). Thus, viable indigenous anaerobes antagonize indigenous *E. coli* intestinal populations and provide defense against *E. coli* translocation from the GI tract.

The direct relationship between cecal bacterial population levels and bacterial translocation to the MLN is dramatically demonstrated by comparing the cecal populations of three different species of *Enterobacteriaceae* in triassociated gnotobiotic mice and the translocation of these three species to the MLN (Steffen, Berg and Deitch, 1988; Table 13.2).

Germfree mice were triassociated with indigenous *E. coli*, *K. pneumoniae* and *P. mirabilis* for 1, 3, 5 and 12 weeks. The proportion of each of the bacterial species in the total cecal populations was compared to the proportion of each bacterial species in the total number of bacteria translocating to the MLN. For example, at 1 week following triassociation, the

Table 13.2 Percentage of each bacterial species in the total populations of bacteria in the cecum and MLN of triassociation gnotobiotic mice

Triassociation (weeks)	E. coli		P. mirabilis		K. pneumoniae	
	Cecum (%)	MLN (%)	Cecum (%)	MLN (%)	Cecum (%)	MLN (%)
1	26	34	6	10	68	56
3	21	22	9	23	70	55
5	28	20	7	20	63	61
12	25	32	10	14	66	54

cecal populations consisted of 26% *E. coli*, 6% *P. mirabilis* and 68% *K. pneumoniae* and the translocating bacteria in the MLN were 34% *E. coli*, 10% *P. mirabilis* and 56% *K. pneumoniae*. Statistically, 26% *E. coli* in the cecum is the same as 34% *E. coli* in the MLN. The same is true for the percentages of *K. pneumoniae* or *P. mirabilis*. Thus, the proportion of each bacterial species in the cecum is statistically the same as the proportion of each bacterial species translocating to the MLN, demonstrating a direct relationship between cecal populations and bacterial translocation from the GI tract.

Patients treated with oral antibiotics often exhibit the adverse side effect of intestinal overgrowth by certain antibiotic-resistant indigenous bacteria or intestinal colonization by antibiotic-resistant non-indigenous bacteria. To mimic this clinical condition, mice were given either penicillin, clindamycin or metronidazole orally for 4 days, but tested for bacterial translocation to the MLN for the next 32 days (Berg, 1981b). Within 2 days of antibiotic treatment, the numbers of indigenous obligate anaerobes in the cecum decrease 1000-fold and the numbers of indigenous *Enterobacteriaceae* increase 10 000-fold from 10^{4-5}/g to more than 10^{9-10}/g cecum. Oral treatment with a single antibiotic does not remove all the 400–500 bacterial species normally present in the GI tract but does remove the portion of the indigenous microflora that is normally antagonistic to indigenous *Enterobacteriaceae*. The subsequent increase in *Enterobacteriaceae* cecal populations promotes Enterobacteriaceae translocation to the MLN. The translocating Enterobactericeae in order of frequency are *Enterobacter cloacae*, *E. aerogenes*, *K. pneumoniae*, *E. coli*, *Proteus morganii* and *P. mirabilis*. When the oral antibiotic is discontinued on day 4, the obligate anaerobe cecal populations increase back to their normal high levels and the indigenous Enterobacteriaceae are again antagonized to their normal low levels of 10^{4-5}/g cecum. However, it takes 30 days after oral antibiotic is discontinued for the normal GI microecology to re-establish, i.e. for the various indigenous *Enterobacteriaceae* to return completely to normal cecal levels and for *Enterobacteriaceae* translocation to cease to the MLN.

These results suggest that patients receiving oral antibiotic therapy are also at increased risk of bacterial translocation and opportunistic infections originating from their GI tract, even after their antibiotic therapy has been discontinued. Many clinical situations lead to intestinal bacterial overgrowth and subsequent bacterial translocation from the GI tract, such as parenteral alimentation, enteral alimentation, protein malnutrition and endotoxic shock (Berg, 1995).

The intestinal bacterial overgrowth model has several advantages over other animal models in the study of bacterial translocation. In the intestinal bacterial overgrowth model, bacterial challenge is by a natural oral route rather than by the artificial intravenous (i.v.) or intraperitoneal (i.p.) injection of bacteria. Most importantly, promotion of bacterial translocation in the intestinal overgrowth model is due directly to bacterial overgrowth and not confounded by variables, such as increased intestinal permeability or physical injury to the intestinal epithelium, that are present in more complex models of bacterial translocation, such as the endotoxic shock model.

Theoretically, a single translocating bacterium will produce a positive MLN culture. The culture of viable bacteria from the MLN has proved a more sensitive method of detecting bacterial translocation than even the use of fluorescein-labeled or radiolabeled challenge bacteria. Thus bacterial translocation to the MLN and other organs is an extremely sensitive indicator of bacterial infection originating from the GI tract.

13.2.2 Physical disruption of intestinal mucosal barrier

The intact intestinal mucosa provides a physical or mechanical barrier preventing bacteria colonizing the GI tract from translocating to extraintestinal sites. Thus, the integrity of the gut mucosa is an important host defense mechanism against bacterial translocation. Indigenous GI bacteria are not normally cultured from the MLN or other extraintestinal sites in healthy adult animals (Berg and Garlington, 1979). Indigenous bacteria readily translocate from the GI tract, however, when the intestinal epithelium is physically damaged or exhibits increased permeability. For example, ricinoleic acid (12-hydroxy-9-octadecanoic acid), the pharmacologically active constituent of castor oil, given once i.g. to mice severely damages the intestinal mucosa and readily promotes bacterial translocation (Morehouse *et al.*, 1986). Massive exfoliation of columnar epithelial cells in the proximal small intestine occurs within 2 hours after ricinoleic acid administration and both facultative and obligate anaerobic bacteria translocate to the MLN, spleen and liver. The incidence of bacterial translocation is greatest at 4 days following ricinoleic acid administration but ceases by 7 days when the intestinal epithelium has healed.

The intestinal mucosa also is severely damaged during endotoxic shock and allows various species of indigenous bacteria to translocate from the GI tract (Berg, 1992b, c). Even though the large populations of indigenous Gram-negative bacteria in the GI tract continuously release endotoxin during cell growth, and particularly upon cell death and lysis, only very small amounts of endotoxin are normally absorbed from the healthy intestines and this absorbed endotoxin is easily detoxified by liver Kupffer cells (Nolan, 1981). Severely ill or injured patients, however, often exhibit increased intestinal permeability or even damaged intestinal mucosa and, in these situations, bacteria and endotoxin readily cross the mucosal barrier to gain access to extraintestinal tissues and the systemic circulation.

SPF mice injected once i.p. with *E. coli* 026:B6 endotoxin exhibit translocating indigenous bacteria in their MLN within 24 hours (Deitch and Berg, 1987). Endotoxin-challenged mice exhibit edematous ileal and cecal lamina propria with separation of the epithelium from the lamina propria at the villus tips. To determine which components of the endotoxin structure are involved in promoting bacterial translocation, mice were injected with endotoxin from six R-mutant strains of *Salmonella* (Ra, Rb, Rc, Rd, Re and lipid A), all differing in their endotoxin composition (Deitch *et al.*, 1989a). Injection of intact *Salmonella* whole endotoxin (wild type), Ra endotoxin fragment or the Rb endotoxin fragment promotes bacterial translocation to the MLN, whereas injection of *Salmonella* endotoxin fragments Rc, Rd, Re or lipid A does not. It appears that endotoxin must contain the terminal three sugars of the core polysaccharide in order to induce intestinal mucosal damage and thereby promote bacterial translocation.

Both endotoxin and lipid A produce their toxic manifestations by stimulating host cells, especially macrophages, to release mediator substances that then act as second messengers to disrupt various homeostatic systems. Oxygen-free radicals generated by xanthine oxidase are released during the ischemia/reperfusion that occurs during endotoxic shock (Parks *et al.*, 1982). Injection of the translocation-promoting Ra endotoxin fragment increases ileal xanthine oxidase and xanthine dehydrogenase activities, whereas injection of the non-promoting Rc and Re endotoxin fragment does not increase these enzymatic activities (Deitch *et al.*, 1989b). Inhibition of xanthine oxidase activity in endotoxin-shocked animals by: (1) pretreatment with allopurinol, a xanthine oxidase inhibitor; (2) dimethylsulfoxide, a hydroxyl scavenger; or (3) deferoxamine, an iron chelator, reduces the intestinal damage and does not allow bacterial translocation from the GI tract (Deitch *et al.*, 1988, 1989c). Intestinal damage and subsequent bacterial translocation also are inhibited in animals that have been fed a tungsten-supplemented, molybdenum-free diet to deplete intestinal xanthine oxidase for 2 weeks prior to endotoxin injection. These results suggest that the promotion of bacterial translocation during endotoxic shock is due, in part at least, to intestinal mucosal damage caused by xanthine-generated hydroxyl radicals.

Hemorrhagic shock and the shock of thermal injury also produce extensive intestinal mucosal damage and promote bacterial translocation (Maejima, Deitch and Berg, 1984; Baker *et al.*, 1988). Rats submitted to hemorrhagic shock for 90 minutes exhibit necrosis of the ileal mucosa and increased bacterial translocation to the MLN. Intestinal mucosal damage in the hemorrhagic shock and thermal injury models also is mediated by oxidants generated by the xanthine oxidase system and the use of xanthine oxidase inhibitors lessens the damage and decreases bacterial translocation (Deitch *et al.*, 1988, 1989a).

Interestingly, cecal populations of *Enterobacteriaceae*, such as *E. coli*, increase 100-fold following i.p. endotoxin injection (Deitch and Berg, 1987). The mechanisms whereby endotoxin injection increases cecal populations of certain indigenous bacteria are unknown, although endotoxin reduces GI motility and this stasis could account for intestinal overgrowth by certain indigenous bacteria. Zymosan, a yeast cell wall product unrelated to endotoxin, but inflammatory because of its complement activating activity, also increases cecal populations of indigenous *Enterobacteriaceae* when injected i.p. and promotes *Enterobacteriaceae* translocation to the MLN (Mainous *et al.*, 1991). Thus physical damage to the intestinal mucosa, intestinal bacterial overgrowth and compromised host immune defenses all are operating in these complex shock models to promote bacterial translocation from the GI tract.

Endotoxin shock, hemorrhagic shock, bacterial infection and host immunodeficiency all are clinical conditions that increase gut mucosal permeability resulting in the increased translocation of bacteria and bacterial products (e.g. endotoxin) from the GI tract to the portal and systemic circulations. Translocated endotoxin then activates various plasma protein cascades, resident macrophages and circulating neutrophils to release monokines and proteins that, in turn, further increase gut mucosal permeability. It is hypothesized that gut barrier failure allowing absorption of endotoxin and bacterial translocation, in conjunction with hepatic dysfunction, promotes or potentiates multiple organ failure syndrome, a common final pathway leading to death in a variety of patients (Carrico *et al.*, 1986).

13.2.3 Deficiencies in host immune defenses

Most studies of bacterial translocation have concentrated on intestinal bacterial overgrowth and physical disruption of the mucosal barrier as the primary mechanisms promoting bacterial translocation rather than on the role of decreased host immune defenses. The practical definition of bacterial translocation, however, includes not only the physical passage of bacteria across the intestinal epithelial barrier but also the survivability of translocating bacteria in transit through the lamina propria and *in situ* in reticuloendothelial organs, such as the MLN. Even though translocating bacteria physically cross the intestinal epithelium and 'migrate' to the MLN, cultures of the MLN will be negative if the translocating bacteria are killed *en route* from the GI tract or *in situ* in the MLN by resident macrophages, and bacterial translocation by definition will not have taken place (Berg, 1980b, 1992b). The host immune defenses are therefore integral components in the pathogenesis of bacterial translocation.

All compartments of the host immune system, such as mucosal immunity, cell-mediated immunity and serum immunity, are likely involved in the prevention of opportunistic extraintestinal infections caused by translocating bacteria. Mice injected once i.p. with an immunosuppressive agent such as cyclophosphamide, prednisolone, methotrexate, 5-fluorouracil or cytosine arabinoside exhibit translocation of various indigenous GI bacteria to their MLN, spleen and liver (Berg, 1983b). Immunocompromised host defenses also play a part in promoting bacterial translocation from the GI tract in animal models of diabetes (Berg, 1985), leukemia (Penn, Maca and Berg, 1986), endotoxemia (Deitch and Berg, 1987), thermal injury (Maejima, Deitch and Berg, 1984) and hemorrhagic shock (Baker *et al.*, 1988). These complex animal models demonstrate the importance of host immune defenses in preventing bacterial translocation but do not delineate the relative roles of the major immune compartments or identify the immune mechanisms actually responsible for killing the translocating microorganisms.

(a) Serum and secretory immunity

There is very little information concerning the role of serum immunity in the defense against bacterial translocation, although it is assumed that antibacterial serum antibodies act as opsonins to enhance phagocytosis and killing of translocating bacteria by polymorphonuclear leukocytes (PMNs) and macrophages. *E. coli* translocation to the spleen and liver but not the MLN is decreased in *E. coli*-monoassociated SPF mice vaccinated i.p. with killed *E. coli* to induce anti-*E.-coli* IgG and IgM serum antibodies (Gautreaux, Deitch and Berg, 1990). Thus serum immunity is not effective in defending against the initial translocation of bacteria from the GI tract to the MLN, but is effective in reducing the subsequent spread of translocating bacteria from the MLN to other sites, such as liver, spleen and blood stream.

It seems likely that secretory IgA on mucosal surfaces would inhibit the close association of a bacterium with intestinal epithelial cells that must occur prior to the translocation of any bacteria across the intestinal barrier. In fact, 'spontaneous' bacterial translocation from the GI tract observed in athymic nude mice may be due in part to lack of T-cell-dependent secretory IgA in their intestinal secretions (Owens and Berg, 1980). There is little concrete information to date, however, on the specific role of secretory IgA in the defense against bacterial translocation from the GI tract.

(b) Macrophages

Bacteria translocating from the GI tract are always cultured first in the MLN prior to their appearance in organs such as the liver or spleen (Berg, 1981a, 1983a, 1992a). MLN-resident macrophages, therefore, are strategically located to prevent translocating bacteria from surviving in the MLN and to inhibit their spread from the MLN to other sites. Since many different bacterial species translocate from the GI tract, it would be desirable to non-specifically activate macrophages and PMNs to more efficiently engulf and kill a variety of bacterial types. Consequently, three immunomodulators known to non-specifically activate macrophages, namely yeast glucan, muramyl dipeptide and formalin-killed *Propionibacterium acnes* (formerly classified as *Corynebacterium parvum*), were tested for their abilities to inhibit bacterial translocation to the MLN (Fuller and Berg, 1985).

Mice were antibiotic-decontaminated, monoassociated with indigenous *E. coli*, *P. mirabilis* or *E. cloacae*, injected once i.p. with either of the immunomodulators and tested for bacteria translocation to the MLN 1 week following vaccination. All three immunomodulators induced a lymphoreticular response demonstrated by splenomegaly, a commonly used indicator of macrophage activation. However, only *P. acnes* vaccination decreased translocation of these bacteria to the MLN. Plastic-adherent spleen or MLN cells (predominantly macrophages) from *P.-acnes*-vaccinated mice adoptively transferred to non-vaccinated recipients also inhibit bacterial translocation, whereas non-adherent control cells (predominantly lymphocytes) do not (Gautreaux, Deitch and Berg, 1990).

These results indicate that macrophages are important immune effector cells in the host defense against bacterial translocation. Harmsen *et al.* (1985), however, demonstrated that fluorescent microspheres could be transported by macrophages from the lungs to tracheobronchial lymph nodes, suggesting that macrophages also might promote bacterial translocation. Wells, Maddus and Simmons (1987) subsequently reported that PMNs engulf indigenous bacteria in the GI tract and transport them to experimental abdominal abscesses. Bacterial translocation from the GI tract to the MLN and liver, however, is neither increased nor decreased in op/op mice, genetically deficient in CSF-1-dependent macrophage populations (Feltis *et al.*, 1994). However, only a few op/op mice were tested. It still seems likely that macrophages and PMNs are involved in the immune defense against bacterial translocation. However, more investigation is required to determine the special situations in which phagocytic cells might also promote bacterial translocation.

Interestingly, *P. acnes* vaccination readily reduces *E. coli* translocation to the MLN of antibiotic-decontaminated SPF mice monoassociated with *E. coli* but not to the MLN of gnotobiotic mice monoassociated with *E. coli* (Fuller and Berg, 1985). Gnotobiotic mice are immunologically stimulated by *P. acnes* vaccination since they exhibit even greater splenomegaly than *P.-acnes*-vaccinated, antibiotic-decontaminated SPF mice. *E. coli* also populates the GI tract to the same high levels (10^9/g cecum) in both *E.-coli*-monoassociated gnotobiotic and *E.-coli*-monoassociated

antibiotic-decontaminated SPF mice. The only difference between gnotobiotic and antibiotic-decontaminated SPF mice is that gnotobiotic mice,
unlike antibiotic-decontaminated SPF mice, have not been previously
exposed to an indigenous GI microflora. Thus, it appears that the
mouse immune system must be 'primed' by antigens of the indigenous
GI microflora before a second stimulation by *P. acnes* vaccination activates
macrophages sufficiently to kill translocating bacteria.

To test this hypothesis, adult germfree mice were colonized for 8 weeks
with a whole cecal microflora from SPF mice (400–500 bacterial species) to
'prime' their immune system. These gnotobiotes were then decontamin-
ated with oral antibiotics, monoassociated with *E. coli* and vaccinated
with killed *P. acnes* as previously. Surprisingly, *P. acnes* vaccination still
did not reduce *E. coli* translocation to the MLN in these adult gnotobiotes
exposed to the indigenous GI microflora for 8 weeks (Berg and Itoh,
1986). Further experimentation revealed that germfree mice must be
exposed to the whole indigenous GI microflora within 1 week after
birth in order for *P. acnes* vaccination at 8 weeks of age to decrease *E.
coli* translocation to the MLN (Berg and Itoh, 1986). This is yet another
example of the profound influence of the indigenous GI microflora on the
host's immunological development (Berg, 1983c, 1996).

(c) T-cell-mediated immunity

Most studies attempting to delineate host immune defenses against bacterial translocation from the GI tract have concentrated on T-cell-
mediated immunity. Thymectomy of neonatal mice promotes the translocation of indigenous GI bacteria to 46% of MLN, spleen, liver and
kidneys compared with only 5% positive organs in sham-thymectomized
control mice (Owens and Berg, 1982). Athymic (nu/nu) mice also exhibit
'spontaneous' translocation of aerobic, facultative and obligately anaerobic bacteria to 50% of their MLN, spleen, liver and kidneys compared
with 5% positive organs in control euthymic (nu/+) mice (Owens and
Berg, 1980). Grafting neonatal thymuses from nu/+ donors to nu/nu
recipients decreases spontaneous bacterial translocation in the recipients to 8% of these organs, statistically similar to the 5% positive
organ cultures noted in control nu/+ mice. Thus, T-cell-mediated
immunity contributes to the host defense against bacterial translocation and is especially effective in preventing the spread of translocating bacteria from the MLN to other sites, such as the spleen and
liver.

It is not surprising that athymic mice are more susceptible to the
systemic spread of translocating bacteria since athymic mice are more
susceptible than euthymic mice to a variety of bacterial, mycotic and viral
infections. Subsequent studies by Maddus *et al.* (1988), however, did not

detect spontaneous bacterial translocation in athymic nu/nu mice nor did they observe an increase in *E. coli* translocation in *E.-coli*-monoassociated nu/nu mice compared with *E.-coli*-monoassociated nu/+ mice. Furthermore, they did not detect increased bacterial translocation after i.p. injection of antithymocyte globulin or anti-CD4+ T cell monoclonal antibodies (GK 1.5).

Because of these conflicting results, we also depleted mice of CD4+ and/or CD8+ T cells with injections of specific anti-T-cell monoclonal antibodies and tested for increased bacterial translocation (Gautreaux *et al.*, 1994b). Mice were antibiotic-decontaminated with oral penicillin plus streptomycin and then monoassociated with streptomycin-resistant *E. coli* C25. Neither antibiotic-decontamination or *E. coli* monoassociation changed T cell phenotypic profiles in the intestinal epithelium, lamina propria, MLN or spleen. Mice were depleted of CD4+ T cells with two i.p. injections of monoclonal antibody GK1.5 (rat anti-mouse CD4) and/or depleted of CD8+ T cells with two i.p. injections of monoclonal antibody 2.43 (rat anti-mouse CD8). Flow cytofluorometric analyses demonstrated 100% depletion of CD4+ and CD8+ T cells in the intestinal epithelium, lamina propria, MLN and spleen. *E. coli* translocation from the GI tract to the MLN was increased in these mice depleted of either CD4+ T cells, CD8+ T cells or both CD4+ and CD8+ T cells. Depletion of CD4+ and/or CD8+ T cells also increased translocation of indigenous bacteria to the MLN in mice not antibiotic-decontaminated but harboring their full complement of indigenous GI bacteria.

These results are in agreement with our previous studies with athymic mice, thymus-grafted nu/nu mice and thymectomized mice, demonstrating the importance of T-cell-mediated immunity in the host defense against bacterial translocation. Maddus *et al.* (1988) may have obtained different results because their T cell depletion regimen only reduced T cells by 75% in the peripheral lymph nodes whereas our methodology achieved 100% depletion of T cells in the intestinal epithelium, lamina propria, MLN and spleen.

To further confirm the role of T-cell-mediated immunity in the defense against bacterial translocation, CD4+ and/or CD8+ T cells harvested from donor mice were adoptively transferred to mice that had been depleted of T cells by thymectomy plus i.p. injections of anti-CD4+ and CD8+ T cell monoclonal antibodies (Gautreaux *et al.*, 1995). These thymectomized mice depleted of CD4+ and/or CD8+ T cells were antibiotic-decontaminated and monoassociated with *E. coli*. As previously, T-cell depletion increased *E. coli* translocation to the MLN. Adoptive transfer of CD4+ and/or CD8+ T cells from MLNs or spleens of donor mice then reduced this *E. coli* translocation. Migration of the adoptively transferred T cells to the MLN was demonstrated by using donor Thy 1.1 T cells and recipient mice expressing Thy 1.2 T cells.

The observation that either CD4$^+$ or CD8$^+$ adoptively transferred T cells are effective in reducing *E. coli* translocation suggests an effector function common to both subsets of T cells. CD4$^+$ and CD8$^+$ T cells secrete γ-interferon and granulocyte/macrophage colony-stimulating factor, both of which activate phagocytic cells. Since T cells do not kill bacteria by themselves, the ultimate immune effector cells are probably phagocytic cells, such as macrophages. In fact, as presented earlier, *E. coli* translocation to the MLN is reduced in mice adoptively transferred with *P.-acnes*-stimulated macrophages (Gautreaux, Deitch and Berg, 1990). Quantitation of the types of cytokine produced by the adoptive CD4$^+$ and CD8$^+$ T cells *in situ* in the intestinal epithelium, lamina propria and MLN of recipient mice would provide additional insights into the cell-mediated mechanisms.

Since secretory immunity is T-cell-dependent, it also is possible that the lack of secretory IgA on the intestinal mucosal surfaces of athymic mice allows increased bacterial translocation from the GI tract. The relative contributions have not been elucidated of mucosal immunity (secretory IgA), cell-mediated immunity (macrophages and T cells) and serum immunity (IgG and IgM) in the host immune defense against bacterial translocation. This information is required in order to devise immunological strategies to decrease bacterial translocation and opportunistic infections originating from the GI tract.

13.3 PROMOTION OF BACTERIAL TRANSLOCATION BY MULTIPLE MECHANISMS

More than one mechanism is operating to promote bacterial translocation from the GI tract in complex animal models such as endotoxic shock, hemorrhagic shock, thermal injury and streptozotocin-induced diabetes (Berg, 1983a, 1992a, 1995). To determine whether these mechanisms might operate synergistically to promote bacterial translocation, mice were given a combination of an oral antibiotic (penicillin, clindamycin or metronidazole) to induce intestinal bacterial overgrowth together with an immunosuppressive agent (cyclophosphamide or prednisolone) to compromise host immune defenses (Berg, Wommack and Deitch, 1988). Mice given an oral antibiotic alone exhibit translocating indigenous bacteria in their MLN but the translocating bacteria do not spread systemically to other organs or sites. Mice injected with an immunosuppressive agent also exhibit bacterial translocation to the MLN but to a lesser degree than do mice given only an oral antibiotic. The combination of an oral antibiotic plus an immunosuppressive agent, however, not only promotes bacterial translocation to the MLN, but the translocating bacteria spread systemically to the peritoneal cavity and blood stream. In fact, mice given the combination of an antibiotic plus an immunosuppressive agent all die within 2 weeks of septicemia caused by indigenous bacteria translocating from their GI tract. The oral antibiotic/immunosuppressive agent combination therefore synergistically promotes bacterial translocation from the GI tract.

The synergistic promotion of bacterial translocation by multiple mechanisms has also been demonstrated in other animal models. For example, protein malnutrition alone produces histological atrophy of the mucosa of the small bowel and cecum, but remarkably the mucosal barrier to bacterial translocation remains intact (Deitch and Berg, 1987). The combination of protein malnutrition plus one i.p. injection of endotoxin, however, produces intestinal ulcerations with concomitant bacterial translocation from the GI tract (Ma *et al.*, 1989). Similarly, a 30% total body surface area burn plus one i.p. endotoxin injection synergistically promotes intestinal mucosal damage and bacterial translocation (Deitch and Berg, 1987). Studies are in progress to delineate the relationships among increased intestinal permeability, intestinal mucosal injury, bacterial translocation and lethal sepsis in these complex animal models.

13.4 HOST AND BACTERIAL FACTORS INFLUENCING TRANSLOCATION

13.4.1 Host genetics

Different strains of mouse vary in their immune response to various infectious agents and these immune reactions are genetically determined. It is possible that various strains of mouse might also exhibit different patterns of bacterial translocation, since host immune competence is a major determining factor in the pathogenesis of bacterial translocation. Translocation of indigenous microflora bacteria from the GI tract to the MLN, however, did not occur to any significant extent in any of 14 inbred mouse strains examined, suggesting that genetic differences in these mouse strains have no effects on the 'spontaneous' translocation of indigenous bacteria (Maejima, Shimoda and Berg, 1984).

Immunodeficiency is not very effective in initiating bacterial translocation from the GI tract to the MLN (Berg, 1992b). Consequently, any effect of host genetics in promoting spontaneous bacterial translocation from the GI tract to the MLN may be too slight to be recognized in this model. Intestinal bacterial overgrowth, on the other hand, is the most effective mechanism promoting initial bacterial translocation from the GI tract to the MLN (Berg, 1992b). Consequently, an influence of host genetics on bacterial translocation might be observed in the intestinal overgrowth model in which bacteria are already translocating to the MLN. Ten inbred mouse strains were antibiotic-decontaminated, monoassociated with streptomycin-resistant *E. coli* C25 and tested 2 days later for *E. coli* C25 translocation to the MLN (Maejima, Shimoda and Berg, 1984; Table 13.3).

Table 13.3 Translocation of *E. coli* to the MLN of various genetic strains of SPF mice antibiotic-decontaminated and monoassociated with *E. coli*

Mouse strain	E. coli/g cecum (Mean $\log_{10}$)	E. coli/MLN (Mean $\log_{10}$)
BALB/cCrSlc	8.9	1.17
C3H/HeSlc	9.1	1.13
A/JSlc	9.2	1.40
C57BL/6CrSlc	8.7	1.24
C57BL/10SnSlc	9.0	1.77
B10.A/SgSnSlc	8.8	1.77
B10.BR/SgSnSlc	9.1	1.86
B10.D2/nSnSlc	8.9	1.70
Slc:ICR	9.0	1.23
DBA/2CrSlc	8.9	1.47

As expected, *E. coli* C25 reached similar high cecal populations in all ten mouse strains and translocated at low levels to the MLN. However, there were no significant differences among mouse strains as to either the incidence of *E. coli* translocation to the MLN or the numbers of translocating *E. coli* per MLN, again suggesting that host genetics do not appreciably influence the rate of bacterial translocation from the GI tract.

13.4.2 Bacterial virulence determinants

Much research has been devoted to the identification and characterization of virulence determinants involved in directing the passage of non-indigenous pathogenic bacteria, such as *Shigella* and *Salmonella*, from the GI lumen into host tissues. Nothing is known, however, concerning the attributes required of the indigenous GI tract bacteria enabling them to translocate the intact intestinal mucosal barrier and survive in the hostile environment of the MLN and extraintestinal tissues.

E. coli is the most common indigenous bacteria translocating from the GI tract in various animal models of translocation and is also the most common cause of human urinary tract infections, pyelonephritis, bacteremia and septicemia (Donnenberg *et al.*, 1994). The GI tract is the reservoir for these opportunistic *E. coli* infections. A few studies have compared indigenous *E. coli* isolates from the feces of healthy adults with *E. coli* isolates causing infection in extraintestinal sites of patients as to their complement of virulence determinants (Table 13.4).

Table 13.4 Frequency of *E. coli* isolates exhibiting putative virulence determinants (%) isolated from patients (in extraintestinal sites *versus* healthy subjects in feces)

Virulence determinant	Healthy subjects	Patients	Reference
Hemolytic activity	18	ND	Smith, 1963
	11	ND	Cooke, 1968
	5	35 – blood	Minshew *et al.*, 1978
		59 – misc. site	
	12	39 – blood	Johnson, 1991
		43 – misc. site	
	11	ND	Siitonen, 1992
Aerobactin	34	69 – blood	Carbonetti *et al.*, 1986
Colicin V	0	12 – blood	Minshew *et al.*, 1978
		16 – misc. site	
	ND	18 – blood	Smith, 1963
Capsule	23	ND	Johnson, 1991
	62	ND	Siitonen, 1992
	21	ND	Wullenweber *et al.*, 1993
Type 1C fimbriae	7	ND	Siitonen, 1992
P fimbriae mannose-	11	ND	Siitonen, 1992
resistant	15	59 – misc. sites	Minshew *et al.*, 1978
hemagglutination			
Adhesin operon	ND	52 – blood	Maslow *et al.*, 1993
(*pap, sfa, afa, hly*)		(exhibit at least	
		one operon)	

As can be seen, there is no clear association of one or even a set of *E. coli* virulence determinants with any extraintestinal disease caused by *E. coli* originating from the GI tract, even *E. coli* septicemia. For example, only 5–18% of *E. coli* isolates from the feces of healthy subjects possess hemolytic activity and this percentage is increased to only 35–39% for *E. coli* strains isolated from the blood of septicemic patients. Even more surprising, only 52% of *E. coli* blood isolates exhibit *pap, sfa, afa* or *hly* operons, suggesting that these virulence determinants are not absolute requirements for the development of *E. coli* septicemia originating from the GI tract.

We hypothesized that known non-indigenous pathogenic *E. coli* strains would be better able to translocate from the GI tract to the MLN and extraintestinal sites than the indigenous *E. coli* strains. Consequently, *E. coli* translocation was quantified in groups of antibiotic-decontaminated SPF mice monoassociated for 4 days with *E. coli* strains indigenous to animals, *E. coli* strains pathogenic to animals, or *E. coli* strains pathogenic to humans (Fig. 13.5).

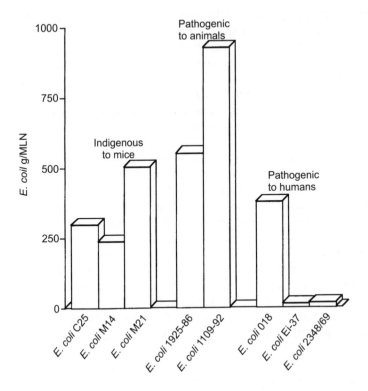

Figure 13.5 Translocation efficiencies to the MLN of various indigenous and pathogenic *E. coli* strains in monoassociated antibiotic-decontaminated SPF mice.

All the *E. coli* strains readily colonized antibiotic-decontaminated SPF mice at similar high population levels (10^{10}/g cecum). The *E. coli* strains indigenous to mice (*E. coli* C25, *E. coli* M14, *E. coli* M21) translocate in only low numbers to the MLN. *E. coli* 1925–86, a domestic cat pathogen, and *E. coli* 1109–92, a rat pathogen, also translocate at only low numbers to the MLN. Unexpectedly, *E. coli* O18, a serotype known to be a common cause of human neonatal meningitis/septicemia, translocates to the MLN in only low numbers, similar to those of indigenous *E. coli* strains. Even more surprisingly, *E. coli* EI-37, a prototype human enteroinvasive strain (EIEC), and *E. coli* 2348/69, a prototype human enteropathogenic strain (EPEC), translocate to the MLN at levels tenfold lower than any of the *E. coli* strains tested, including the indigenous *E. coli* strains.

Enteroinvasive *E. coli* (EIEC) are particularly adept at penetrating the human intestinal mucosa. Enteropathogenic *E. coli* (EPEC) interact with target intestinal epithelial cells causing attaching and effacing (A/E) lesions and are a leading cause of infantile diarrhea in developing countries. It is therefore unexpected that these enteropathogenic and enteroinvasive *E. coli* strains, known to possess multiple virulence determinants, do not translocate from the GI tract to the MLN at a rate at least as high as that of the indigenous *E. coli* strains. More investigation is required to draw any firm conclusions, but these results question the role of bacterial virulence determinants in *E. coli* translocation from the GI tract to the MLN.

Since specific adhesins or other bacterial virulence determinants do not appear to be essential in the translocation of indigenous or even pathogenic strains of *E. coli*, we also tested whether non-viable plastic particles, completely lacking virulence factors, might translocate from the GI tract. Normal and antibiotic-decontaminated SPF mice were inoculated i.g. three times per day with 2.8×10^7 fluorescein-labeled latex particles (polystyrene beads) $1.0\,\mu m$ or $5.0\,\mu m$ in size. Latex particles were detected microscopically in the MLN as early as 24 hours following i.g. inoculation and peak latex translocation occurred at 4 days. Unexpectedly, latex particles translocate at the same rate to the MLN in SPF mice harboring their entire complement of indigenous GI bacteria (400–500 species) as in antibiotic-decontaminated SPF mice without an indigenous GI microflora (Table 13.5).

Thus, the enormous numbers of indigenous obligately anaerobic bacteria normally layered on the intestinal mucosa of SPF mice do not physically hinder the close contact of latex particles with intestinal epithelial cells that must occur prior to latex particle translocation. Also, $1\,\mu m$ latex particles, approximately the size of *E. coli*, translocate at a higher rate to the MLN in both normal and antibiotic-decontaminated SPF mice than do $5\,\mu m$ latex particles, approximately the size of *Candida albicans*.

Table 13.5 Translocation of latex particles 1 μm *versus* $5\,\mu m$ in size from the GI tract to the MLN

Latex particle size (μm)	Mice*	Latex inoculum†	Latex particles (Mean/g MLN)
1	Normal	2.78×10^7/ml	2462
5	Normal	3.34×10^7/ml	563
1	Antibiotic-decontaminated	2.78×10^7/ml	3420
5	Antibiotic-decontaminated	3.34×10^7/ml	252

* Mice antibiotic-decontaminated with streptomycin plus penicillin-G in drinking water for 3 days † 0.5 ml 3× i.g. per day for 3 days. Mice examined for latex translocation on day 4

Latex particles 1 μm in size translocate to the MLN in higher numbers in this model than either indigenous or pathogenic *E. coli* strains. The latex challenge (10^7) is at least 1000-fold less than the population levels reached by *E. coli* in monoassociated SPF mice (10^{10}/g cecum) and the concentration of latex in the GI lumen begins to decrease immediately following i.g. inoculation as a result of normal intestinal peristalsis. Thus, latex challenge is much less than *E. coli* challenge and latex particles are continuously eliminated from the GI tract without resupply. In contrast, *E. coli* multiplies in the GI tract and maintains very high population (challenge) levels.

Translocating *E. coli*, however, are probably killed *en route* to the MLN and *in situ* in the MLN by phagocytic cells, and the number of viable *E. coli* cultured from the MLN represents the number of *E. coli* crossing the intestinal epithelium minus the number of *E. coli* killed *en route*. Latex particles, on the other hand, are not 'killed' by the host immune system and, as discussed previously, phagocytic cells may engulf latex particles and transport them to the MLN thereby actually promoting rather than inhibiting their translocation. Consequently, the kinetics of translocation between inert latex particles and viable *E. coli* cannot be compared directly in this *in vivo* model.

These preliminary results, nonetheless, pose questions concerning the role of specific virulence determinants, such as bacterial adhesins, in the pathogenesis of bacterial translocation from the GI tract. There is not as yet enough information to conclude whether bacterial virulence determinants promote initial translocation across the intestinal barrier and/or protect translocating bacteria in the hostile environment of extraintestinal sites. The study of translocation in more controlled *in vitro* tissue culture models should provide more definitive information concerning these important questions.

13.5 BACTERIAL TRANSLOCATION IN HUMANS

Indirect evidence of the importance of bacterial translocation in the pathogenesis of human opportunistic infections originating from the GI tract is provided by comparing the population levels of certain bacterial biotypes/serotypes present at the highest level in the feces of immunosuppressed patients and the particular bacterial biotypes/serotypes subsequently causing septicemia in these patients. The bacterial biotype/serotype present at the highest level in the patient's feces is also the biotype/serotype most likely to cause septicemia in that patient (Tancrede and Andremont, 1985; Wells *et al.*, 1987). Indigenous bacteria have also been cultured directly from the lymph nodes of certain types of patients. For example, patients with increased intestinal permeability, such as those with Crohn's disease, ulcerative colitis or inflammatory bowel disease, exhibit higher percentages of MLN cultures positive for indigenous GI bacteria than do control patients (Ambrose *et al.*, 1984; Peltzman, Udekwu and Ochoa, 1991; Sedman *et al.*, 1994; Table 13.6). Indigenous GI tract bacteria also are cultured from the MLN of patients with bowel obstruction or bowel cancer (Deitch, 1989; Sedman *et al.*, 1994; Vincent *et al.*, 1988).

In agreement with results from animal models, hemorrhagic shock in humans promotes bacterial translocation from the GI tract to the MLN. Moore *et al.* (1991, 1992) cultured indigenous bacteria from the MLN of hemorrhagic shock patients in two separate studies and Rush *et al.* (1988) demonstrated indigenous GI bacteria in the blood of patients when their systolic blood pressure decreased to less than 80 mmHg. *Candida albicans* was cultured in large numbers from the blood and urine of a volunteer who ingested a large quantity of viable *C. albicans* (Krause, Matheis and Wulf, 1969).

Interestingly, Brathwaite *et al.* (1993) detected bacteria by electron microscopy in 12/16 MLN of trauma patients but could culture indigenous bacteria from only 4/16 MLN. Reed *et al.* (1994) detected *E. coli* β-galactosidase by immunofluorescence in macrophages from 20/20 MLN from trauma patients but could culture bacteria from only 1/20 MLN. It is assumed that a portion of the bacteria translocating from the intestinal lumen to the MLN are killed *en route* by the host immune defenses, and

Table 13.6 Translocation of indigenous bacteria from GI tract to MLN in human patients; bacteria in MLN of patients identified in order of frequency as *E. coli*, *Enterobacter*, *Klebsiella*, *Enterococcus*, *Staphylococcus* and *Bacteroides*

Patient type	Patients (% positive MLN)	Controls (% positive MLN)	Reference
Crohn's disease	33	2	Ambrose *et al*, 1984
Ulcerative colitis	75	0	Peltzman, Udekwu and Ochoa, 1991
Inflammatory bowel disease	16	ND	Sedman *et al.*, 1994
Bowel obstruction	59	4	Deitch, 1989
	39	ND	Sedman *et al.*, 1994
Bowel cancer	65	30	Vincent *et al.*, 1988
	11	ND	Sedman *et al.*, 1994
Hemorrhagic shock	56		Rush *et al.*, 1988
	33	ND	Moore *et al.*, 1991
	15		Moore *et al.*, 1992
Multiple organ donors	79	ND	Van Goor *et al.*, 1994
Trauma patients	81	ND	Brathwaite *et al.*, 1993
	100	ND	Reed *et al.*, 1994

that the number of viable bacteria cultured from the MLN is only a fraction of the total number of translocating bacteria. However, examination by fluorescent microscopy of frozen sections of the lamina propria and MLN from animal models exhibiting bacterial translocation has not revealed large numbers of non-viable translocating bacteria *en route* to the MLN or even in the MLN (Berg, unpublished observations). Thus, more investigation is required to determine what percentage of the total bacteria translocating the intestinal mucosal barrier are eventually cultured from the MLN.

Evidence is accumulating that bacterial translocation from the GI tract is an important early step in the pathogenesis of certain opportunistic infections of debilitated patients. Bacterial translocation from the GI tract is strongly suspected in the pathogenesis of human septicemia, multiple organ failure and acute respiratory death syndrome (Berg, 1992b). Selective antibiotic decontamination of the GI tract may eventually prove to be beneficial in certain debilitated patients to eliminate particular bacterial types, such as the *Enterobacteriaceae*, that are especially efficient in translocating from the GI tract.

13.6 CONCLUSION

The results from the various animal models of bacterial translocation suggest that the pathogenesis of indigenous bacterial translocation from the GI tract in both animals and humans occurs in several discrete stages (Fig. 13.6).

In the healthy host, 'spontaneous' or 'natural' translocation of indigenous bacteria to the MLN probably occurs continuously at a low rate, but

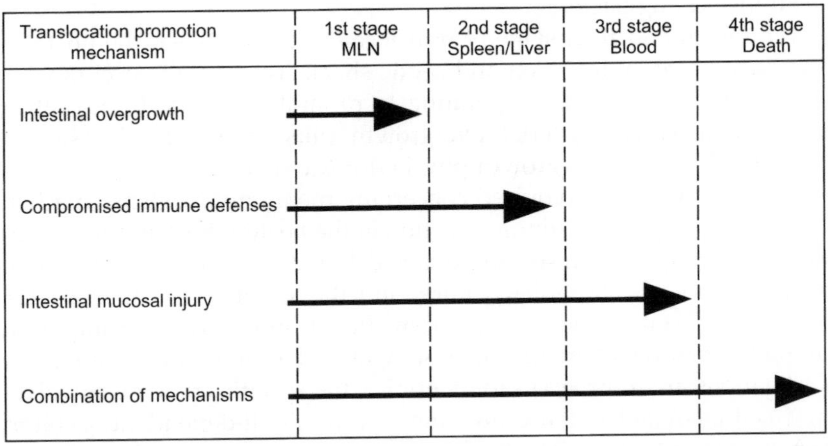

Translocation promotion mechanism	1st stage MLN	2nd stage Spleen/Liver	3rd stage Blood	4th stage Death
Intestinal overgrowth	→			
Compromised immune defenses		→		
Intestinal mucosal injury			→	
Combination of mechanisms				→

Figure 13.6 Stages in the pathogenesis of bacterial translocation from the GI tract.

the translocating bacteria are killed *en route* or *in situ* by the host immune defenses. This low level of spontaneous translocation of indigenous bacteria to the MLN may actually be beneficial in stimulating the host immune system to respond more rapidly and more effectively to more pathogenic non-indigenous microorganisms.

In the first stage of translocation, as occurs during oral antibiotic administration, intestinal bacterial overgrowth readily promotes translocation of indigenous bacteria to the MLN. In fact, intestinal bacterial overgrowth is more effective than physical disruption of the intestinal epithelium or host immunodeficiency in initiating bacterial translocation from the GI tract. Fortunately, in this first stage of translocation the host immune defenses usually are able to control the numbers of translocating bacteria in the MLN and translocating bacteria do not spread from the MLN to other organs or extraintestinal sites. However, if the host is colonized by more virulent non-indigenous bacteria, then the first stage of translocation can have more serious consequences for the host.

The second stage of translocation occurs when translocating bacteria spread from the MLN to other organs, such as the liver, spleen or kidney. This second stage readily occurs when the host immune system is compromised. Depending upon the degree of immunodeficiency and the virulence characteristics of the translocating bacteria, the host still may be able to confine this infection to these particular organs. In some cases, the infection will spread systemically to the third stage of translocation.

The third stage occurs when translocating bacteria spread systemically to the peritoneal cavity and blood stream. The host may still recover from this systemic infection depending on the degree of immunosuppression, the amount of intestinal mucosal damage and the virulence properties of the translocating bacteria.

The fourth and final stage of translocation occurs when the host succumbs to bacterial infection and septic shock. This usually occurs when more than one process is operating to promote bacterial translocation, such as intestinal bacterial overgrowth plus immunosuppression or intestinal bacterial overgrowth plus endotoxic shock.

More knowledge is required concerning the immune defense mechanisms preventing bacterial translocation in the healthy host and how these defenses are compromised in certain debilitated patients. Studies with athymic mice, T-cell-depleted mice, and the adoptive transfer of $CD4^+$ and $CD8^+$ T cells prove that T-cell-mediated immunity is an important immune component in the defense against bacterial translocation. Investigations are in progress to determine the specific cytokines involved in T-cell-mediated defense and the role of T-cell-dependent secretory IgA in preventing initial translocation of bacteria across the intestinal barrier.

Macrophages and PMNs are probably the immune effector cells that kill translocating bacteria. *P. acnes* vaccination non-specifically activates macrophages to inhibit the translocation of various *Enterobacteriaceae* from the GI tract to the MLN. The non-specific activation of macrophages is of practical significance, since it cannot be predicted which of the 400–500 species of indigenous bacteria normally populating the GI tract will translocate in any given clinical situation.

Bacterial translocation has been studied extensively in animal models but evidence is rapidly accumulating that bacterial translocation from the GI tract is also an important initial event in humans in the pathogenesis of opportunistic infections originating from the GI tract. In animal models of bacterial translocation, the bacterial species most likely to translocate from the GI tract – *Pseudomonas aeruginosa*, *E. faecalis*, and especially the *Enterobacteriaceae* – are also the most common causes of septicemia in hospitalized patients. Selective antibiotic decontamination to remove indigenous *Enterobacteriaceae* but leave intact the indigenous obligate anaerobes that exert colonization resistance against more pathogenic non-indigenous bacteria may eventually prove beneficial in certain types of patients. Oral and even systemic antibiotics, however, must be employed with caution, since intestinal bacterial overgrowth is the most efficient mechanism of initiating bacterial translocation from the GI tract. The use of probiotics, such as *Lactobacillus acidophilus* or *Saccharomyces boulardii*, as an alternative to antibiotic therapy also requires investigation.

Maintaining a stable ecological balance in the GI tract is a major defense mechanism preventing intestinal bacterial overgrowth and subsequent bacterial translocation. Consequently, much more information is required concerning the very complex ecological inter-relationships between the host and its indigenous GI microflora if we are to understand the pathogenesis of indigenous bacterial translocation. The mechanisms whereby host immune defenses prevent bacterial translocation from the GI tract, and especially inhibit the systemic spread translocating bacteria, must also be delineated in order to devise rational immunological therapeutic approaches. Bacterial translocation from the GI tract has become even more relevant with the dramatic rise in the numbers of patients with compromised immune systems and increased intestinal permeability, such as the elderly, trauma and shock patients, and those with cancer, diabetes, transplants, invasive devices or AIDS.

13.7 REFERENCES

Alexander, J. W., Boyce, S. T., Babcock, G. F. *et al.*. (1990) The process of microbial translocation. *Archives of Surgery*, **212**, 496–512.

Ambrose, M. S., Johnson, M., Burdon, D. W. and Keighley, M. R. B. (1984) Incidence of pathogenic bacteria from mesenteric lymph nodes and ileal serosa during Crohn's disease surgery. *British Journal of Surgery*, **71**, 623–625.

Baker, J. W., Deitch, E. A., Ma, L. and Berg, R. (1988) Hemorrhagic shock promotes the systemic translocation of bacteria from the gut. *Journal of Trauma*, **28**, 896–906.

Berg, R. D. (1980a) Inhibition of *Escherichia coli* translocation from the gastrointestinal tract by normal cecal flora in gnotobiotic or antibiotic-decontaminated mice. *Infection and Immunity*, **29**, 1073–1081.

Berg, R. D. (1980b) Mechanisms confining indigenous bacteria to the gastrointestinal tract. *American Journal of Clinical Nutrition*, **33**, 2472–2484.

Berg, R. D. (1981a) Factors influencing the translocation of bacteria from the gastrointestinal tract, in *Recent Advances in Germfree Research*, (eds S. Sasaki, A. Ogawa and K. Hashimoto), Tokai University Press, Tokyo, pp. 411–418.

Berg, R. D. (1981b) Promotion of the translocation of enteric bacteria from the gastrointestinal tracts of mice by oral treatment with penicillin, clindamycin, or metronidazole. *Infection and Immunity*, **33**, 854–861.

Berg, R. D. (1983a) Translocation of the indigenous bacteria from the intestinal tract, in *Human Intestinal Microflora in Health and Disease*, (ed. D. J. Hentges), Academic Press, New York, pp. 333–352.

Berg, R. D. (1983b) Bacterial translocation from the gastrointestinal tracts of mice receiving immunosuppressive chemotherapeutic agents. *Current Microbiology*, **8**, 285–292.

Berg, R. D. (1983c) The immune response to indigenous intestinal bacteria, in *Human Intestinal Microflora in Health and Disease*, (ed. D. J. Hentges), Academic Press, New York, pp. 101–126.

Berg, R. D. (1985) Bacterial translocation from the intestines. *Experimental Animals*, **34**, 1–16.

Berg, R. D. (1992a) Translocation and the indigenous gut flora, in *Probiotics. The Scientific Basis*, (ed. R. Fuller), Chapman & Hall, London, pp. 55–85.

Berg, R. D. (1992b) Translocation of enteric bacteria in health and disease, in *Gut-Derived Infectious-Toxic Shock (GITS). A Major Variant of Septic Shock*, (eds H. Cottier and R. Kraft), Current Studies in Hematology and Blood Transfusion 59, S. Karger, Basel, pp. 44–65.

Berg, R. (1992c) Bacterial translocation from the gastrointestinal tract. *Journal of Medicine*, **23**, 217–244.

Berg, R. D. (1995) Bacterial translocation from the gastrointestinal tract. *Trends in Microbiology*, **3**, 149–154.

Berg, R. D. (1996) The indigenous gastrointestinal microflora. *Trends in Microbiology*, **4**(11), 430–435.

Berg, R. D. and Garlington, A. W. (1979) Translocation of certain indigenous bacteria from the gastrointestinal tract to the mesenteric lymph nodes and other organs in a gnotobiotic mouse model. *Infection and Immunity*, **23**, 403–411.

Berg, R. D. and Itoh, K. (1986) Bacterial translocation from the gastrointestinal tract – immunologic aspects. *Microecology and Therapy*, **16**, 131–145.

Berg, R. D. and Owens, W. E. (1979) Inhibition of translocation of viable *Escherichia coli* from the gastrointestinal tract to the mesenteric lymph nodes and other organs in a gnotobiotic mouse model. *Infection and Immunity*, **25**, 820–827.

Berg, R. D., Wommack, E. and Deitch, E. A. (1988) Immunosuppression and intestinal bacterial overgrowth synergistically promote bacterial translocation. *Archives of Surgery*, **123**, 1359–1364.

Brathwaite, C. E. M., Ross, S. E., Nagele, R. *et al.* (1993) Bacterial translocation occurs in humans after traumatic injury: evidence using immunofluorescence. *Journal of Trauma*, **34**, 586–590.

Carbonetti, N. H., Boonchal, S., Parry, S. H. *et al.* (1986) Aerobactin-mediated iron uptake by *Escherichia coli* isolates from human extraintestinal infections. *Infection and Immunity*, **51**, 966–968.

Carrico, C. J., Meakins, J. L., Marshall, J. C. *et al.* (1986) Multiple organ failure syndrome. *Archives of Surgery*, **121**, 196–208.

Carter, P. B. and Collins, F. M. (1964) The route of enteric infection in normal mice. *Journal of Experimental Medicine*, **139**, 1189–1203.

Chau, A. Y. S., Goldbloom, V. C. and Gurd, F. N. (1951) Clostridia infection as a course of death after ligation of the hepatic artery. *Archives of Surgery*, **63**, 390–402.

Cooke E. M. (1968) Properties of strains of *Escherichia coli* isolated from the faeces of patients with ulcerative colitis, patients with acute diarrhoea and normal persons. *Journal of Pathology and Bacteriology*, **95**, 101–113.

Cruz, N., Qi, L., Alvarez, X. *et al.* (1994) The CaCo-2 cell monolayer system as an *in vitro* model for studying bacterial-enterocyte interactions and bacterial translocation. *Journal of Burn Rehabilitation*, **15**, 207–212.

Deitch, E. A. (1989) Simple intestinal obstruction causes bacterial translocation in man. *Archives of Surgery*, **123**, 699–701.

Deitch, E. A. and Berg, R. D. (1987) Endotoxin but not malnutrition promotes bacterial translocation of the gut flora in burned mice. *Journal of Trauma*, **27**, 161–166.

Deitch, E. A., Bridges, W., Baker, J. *et al.* (1988) Hemorrhagic shock-induced bacterial translocation is reduced by xanthine oxidase inhibition or inactivation. *Surgery*, **104**, 191–198.

Deitch, E. A., Ma, L., Ma, J.-W. *et al.* (1989a) Inhibition of endotoxin-induced bacterial translocation in mice. *Journal of Clinical Investigation*, **84**, 36–42.

Deitch, E. A., Ma, J.-W., Ma, L. *et al.* (1989b) Endotoxin-induced bacterial translocation. A study of mechanisms. *Surgery*, **106**, 292–300.

Deitch, E. A., Taylor, M., Grisham, M. *et al.* (1989c) Endotoxin induces bacterial translocation and increases xanthine oxidase activity. *Journal of Trauma*, **29**, 1679–1683.

Donnenberg, M. S., Newman, B., Utsalo, S. J. *et al.* (1994) Internalization of *Escherichia coli* into human kidney epithelial cells: comparisons of fecal and pyelonephritis-associated strains. *Journal of Infectious Disease*, **169**, 831–838.

Ellis, J. C. and Dragstedt, L. R. (1930) Liver autolysis *in vivo*. *Archives of Surgery*, **20**, 8–16.

Feltis, B. A., Jechorek, R. P., Erlandsen, S. L. and Wells, C. L. (1994) Bacterial translocation and lipopolysaccharide-induced mortality in genetically macrophage-deficient op/op mice. *Shock*, **2**, 29–33.

Ford, W. W. (1900) The bacteriology of healthy organs. *Transactions of the Association of American Physicians*, **15**, 389–415.

Ford, W. W. (1901) On the bacteriology of normal organs. *Journal of Hygiene*, **1**, 277–284.

Frank, H. A., Seligman, A. M. and Fine, J. (1948) Further experiences with peritoneal irrigation for acute renal failure. *Annals of Surgery*, **128**, 561–608.

Fuller, K. G. and Berg, R. D. (1985) Inhibition of bacterial translocation from the gastrointestinal tract by nonspecific immunostimulation, in *Germfree Research: Microflora Control and its Application to the Biomedical Sciences*, (eds B. S. Wostmann, J. R. Pleasants, M. Pollard *et al.*), Alan R. Liss, New York, pp. 195–198.

Fuller, R. and Jayne-Williams, D. J. (1970) Resistance of the fowl (*Gallus domesticus*) to invasion by its intestinal flora. II. Clearance of translocated intestinal bacteria. *Research of Veterinary Science*, **11**, 368–374.

Gautreaux, M. D., Deitch, E. A. and Berg, R. D. (1990) Immunological mechanisms preventing bacterial translocation from the gastrointestinal tract. *Microecology and Therapy*, **20**, 31–34.

Gautreaux, M., Deitch, E. and Berg, R. D. (1994a) Bacterial translocation from the gastrointestinal tract to various segments of the mesenteric lymph node complex. *Infection and Immunity*, **62**, 2132–2134.

Gautreaux, M., Deitch, E. and Berg, R. D. (1994b) T-cells in the host defense against bacterial translocation from the gastrointestinal tract. *Infection and Immunity*, **62**, 2874–2884.

Gautreaux, M., Gelder, F., Jennings, S. *et al.* (1995) Adoptive transfer of T-cells to T-cell depleted mice inhibits bacterial translocation from the gastrointestinal tract. *Infection and Immunity*, **63**, 3827–3834.

Harmsen, A. G., Muggenburg, B. A., Snipes, M. B. and Bice, D. E. (1985) The role of macrophages in particle translocation from lungs to lymph nodes. *Science*, **230**, 1277–1280.

Hildebrand, G. J. and Wolochow, H. (1962) Translocation of bacteriophage across the intestinal wall of the rat. *Proceedings of the Society of Experimental Biology and Medicine*, **109**, 183–185.

Johnson, J. R. (1991) Virulence factors in *Escherichia coli* urinary tract infections. *Clinical Microbiology Reviews*, **4**, 80–128.

Keller, R. and Engley, F. B. Jr (1958) Fate of bacteriophage particles introduced into mice by various routes. *Proceedings of the Society of Experimental Biology and Medicine*, **98**, 577–580.

Krause, W., Matheis, H. and Wulf, K. (1969) Fungaemia and funguria after oral administration of *Candida albicans*. *Lancet*, **i**, 598–600.

Ma, L., Specian, R. D., Berg, R. D. and Deitch, E. A. (1989) Effects of protein malnutrition and endotoxin on the intestinal mucosal barrier to the translocation of indigenous flora in mice. *Journal of Parenteral and Enteral Nutrition*, **13**, 572–578.

Maddus, M. A., Wells, C. L., Platt, J. L. *et al.* (1988) Effect of T cell modulation on the translocation of bacteria from the gut and mesenteric lymph node. *Annals of Surgery*, **207**, 387–398.

Maejima, K., Deitch, E. A. and Berg, R. D. (1984) Bacterial translocation from the gastrointestinal tract of rats receiving thermal injury. *Infection and Immunity*, **43**, 6–10.

Maejima, K., Shimoda, K. and Berg, R. D. (1984) Assessment of mouse strain on bacterial translocation from the gastrointestinal tract. *Experimental Animals*, **33**, 345–349.

Mainous, M., Tso, P., Berg, R. D. and Deitch, E. A. (1991) Studies of the route, magnitude, and time course of bacterial translocation in a model of systemic inflammation. *Archives of Surgery*, **126**, 33–37.

Maslow, J. N., Mulligan, M. E., Adams, K. S. *et al.* (1993) Bacterial adhesins and host factors: role in the development and outcome of *Escherichia coli* bacteremia. *Clinical Infectious Diseases*, **17**, 89–97.

Minshew, B. H., Jorgensen, J., Swanstrum, M. *et al.* (1978) Some characteristics of *Escherichia coli* strains isolated from extraintestinal infections of humans. *Journal of Infectious Disease*, **137**, 648–654.

Moore, F. A., Moore, E. E., Poggetti, R. *et al.* (1991) Gut bacterial translocation via the portal vein: a clinical perspective with major torso trauma. *Journal of Trauma*, **31**, 629–638.

Moore, F. A., Moore, E. E., Poggetti, R. S. and Read, R. A. (1992) Postinjury shock and early bacteremia. *Archives of Surgery*, **127**, 893–898.

Morehouse, J., Specian, R., Stewart, J. and Berg, R. D. (1986) Promotion of the translocation of indigenous bacteria of mice from the GI tract by oral ricinoleic acid. *Gastroenterology*, **91**, 673–682.

Nolan, J. P. (1981) Endotoxin, reticuloendothelial function, and liver injury. *Hepatology*, **1**, 458–465.

Owens, W. E. and Berg, R. D. (1980) Bacterial translocation from the gastrointestinal tract of athymic (nu/nu) mice. *Infection and Immunity*, **24**, 308–312.

Owens, W. E. and Berg, R. D. (1982) Bacterial translocation from the gastrointestinal tracts of thymectomized mice. *Current Microbiology*, **7**, 169–174.

Parks, D. A., Bulkley, G. B., Granger, D. N. et al. (1982) Ischemic injury in the cat small intestine: role of superoxide radicals. *Gastroenterology*, **82**, 9–15.

Peltzman, A. B., Udekwu, A. O. and Ochoa, J. (1991) Bacterial translocation in trauma patients. *Journal of Trauma*, **31**, 1083–1087.

Penn, R. L., Maca, R. D. and Berg, R. D. (1986) Leukemia promotes the translocation of indigenous bacteria from the gastrointestinal tract to the mesenteric lymph nodes and other organs. *Microecology and Therapy*, **15**, 85–91.

Reed, L. L., Martin, M., Manglano, R. et al. (1994) Bacterial translocation following abdominal trauma in humans. *Circulatory Shock*, **42**, 1–6.

Rush, B. F., Sori, A. J., Murphy, T. F. et al. (1988) Endotoxemia and bacteremia during hemorrhagic shock: the link between trauma and sepsis? *Annals of Surgery*, **207**, 549–554.

Schatten, W. F. (1954) The role of intestinal bacteria in liver necrosis following experimental excision of the hepatic arterial supply. *Surgery*, **36**, 256–269.

Sedman, P. C., Macfie, J., Sagar, P. et al. (1994) The prevalence of gut translocation in humans. *Gastroenterology*, **107**, 643–649.

Siitonen, A. (1992) *Escherichia coli* in fecal flora of healthy adults: serotypes, P and Type 1C fimbriae, non-P mannose-resistant adhesins, and hemolytic activity. *Journal of Infectious Disease*, **166**, 1058–1065.

Smith, H. W. (1963) The hemolysins of *Escherichia coli*. *Journal of Pathology and Bacteriology*, **85**, 197–211.

Staley, T. E., Jones, E. W. and Corly, L. D. (1969) Attachment and penetration of *E. coli* into epithelium of the ileum in newborn pigs. *American Journal of Pathology*, **56**, 371–392.

Steffen, E. K. and Berg, R. D. (1983) Relationship between cecal population levels of indigenous bacteria and translocation to the mesenteric lymph nodes. *Infection and Immunity*, **39**, 1252–1259.

Steffen, E. K., Berg, R. D. and Deitch, E. A. (1988) Comparison of the translocation rates of various indigenous bacteria from the gastrointestinal tract to the mesenteric lymph nodes. *Journal of Infectious Disease*, **157**, 1032–1037.

Takeuchi, A. (1967) Electron microscope studies of experimental *Salmonella* infection. *American Journal of Pathology*, **50**, 109–136.

Tancrede, C. H. and Andremont, A. O. (1985) Bacterial translocation and gram-negative bacteremia in patients with hematological malignancies. *Journal of Infectious Disease*, **152**, 99–103.

Van Goor, H., Rosman, C., Grond, J. et al. (1994) Translocation of bacteria and endotoxin in organ donors. *Archives of Surgery*, **129**, 1063–1066.

Vincent, P., Colombel, J. F., Lescut, D. et al. (1988) Bacterial translocation in patients with colorectal cancer. *Journal of Infectious Disease*, **158**, 1395–1396.

Volkheimer, G. (1974) Passage of particles through the wall of the gastrointestinal tract. *Environmental Health Perspective*, **9**, 215–225.

Wells, C. L., Maddus, M. A. and Simmons, R. L. (1987) Role of the macrophage in the translocation of intestinal bacteria. *Archives of Surgery*, **122**, 48–53.

Wells, C. L., Ferrieri, P., Weisdorf, D. J. and Rhame, F. S. (1987) The importance of surveillance stool cultures during episodes of severe neutropenia. *Infection Control*, **8**, 317–319.

Wilson, S. J. and Walzer, M. (1935) Absorption of undigested proteins in human beings. IV. Absorption of unaltered egg protein in infants and in children. *American Journal of Diseases of Children*, **50**, 49–54.

Wolbach, S. B. and Saiki, T. (1909) A new anaerobic spore-bearing bacterium commonly present in the livers of healthy dogs, and believed to be responsible for many changes attributed to aseptic autolysis of liver tissue. *Journal of Medical Research*, **21**, 267–278.

Wolochow, G., Hildebrand, G. J. and Lamanna, C. (1966) Translocation of microorganisms across the intestinal wall of the rat: Effects of microbial size and concentration. *Journal of Infectious Disease*, **116**, 523–528.

Wullenweber, M., Beutin, L., Zimmermann, S. and Jonas, C. (1993) Influence of some bacterial and host factors on colonization and invasiveness of *Escherichia coli* K1 in neonatal rats. *Infection and Immunity*, **61**, 2138–2144.

14

The normal microflora and antibiotic-associated diarrhoea and colitis

S. P. Borriello and C. Roffe

14.1 INTRODUCTION

Normal gut structure and function is the end-product of a complex set of interactions between the host and the complex microflora that inhabits the gastrointestinal tract. This balanced interaction is subjected both directly and indirectly to environmental influences, such as dietary factors, which can affect gut function directly through effects on the host or indirectly by effects on the gut microflora. In many cases these are an interactive loop of cause and effect whereby changes to the host alter the gut microflora, which in turn influences gut function. The 'environmental' factor that is most likely to have a significant direct impact on the gut microflora is antibiotic treatment.

There are a number of ways in which antibiotics may cause diarrhoea. Firstly, disruption of the equilibrium between the host and gut microflora may cause diarrhoea by facilitating accumulation of secretagogues or inhibition of absorption, and secondly, antibiotics may predispose to overgrowth or colonization *de novo* of diarrhoea-causing pathogens. The best established pathogen in this respect is *Clostridium difficile*. Thirdly, some cases may be due to direct action of antibiotics on the host. However, most of the evidence on this third aspect of host–antibiotic interaction relates to induction of nausea, abdominal cramps and vomiting. The interested reader is directed to a review that covers these considerations

G. W. Tannock (ed.), *Medical Importance of the Normal Microflora*, 371–387.

(Borriello, 1992). This chapter will first deal with antibiotic-associated diarrhoea and colitis due to causes other than *C. difficile*, and then with disease caused by *C. difficile*.

14.2 NON-*C. DIFFICILE* ANTIBIOTIC-ASSOCIATED DIARRHOEA

The majority of cases of antibiotic-associated diarrhoea are not due to *C. difficile*. We know little about the cause of most of these. Some are due to other pathogens, some are likely to be due to other pathogens and some may be due to disruption of the normal host–gut-microflora equilibrium.

14.2.1 Diarrhoea due to normal microflora disruption

The gut microflora influences the structure and function of the gut. This interaction has been reviewed in detail (Borriello, 1984a). The most striking evidence for this comes from a comparison of the intestines and state of gut contents of germfree and conventional animals. The major differences are presented in Table 14.1. In addition to those listed, germfree animals have fewer plasma cells and Peyer's patches, thinner villi, shallower crypts, reduced mucosal surface area and reduced cellular turnover.

It is evident from such observations that the germfree state is associated with increased hydration and decreased transit time of the gut contents. The gut microflora most probably contributes, therefore, to the reduced hydration and increased transit time seen in conventional counterpart animals. Different antibiotics will have differing effects on the gut microflora and exhibit differences in the degree to which the activities of the gut microflora are compromised. The same antibiotics may also have different effects in different individuals because of the non-uniformity of the composition of the gut microflora among different individuals.

Table 14.1 Comparison of the features of the gut of germfree and conventional animals

Elevated in germfree state	*Elevated in conventional state*
Degree of hydration of gut contents	Chloride and bicarbonate ion concentrations
Caecum size	Intestinal wall thickness and cellularity
Transit time	Intestinal motility
Nutrient uptake	Gastric emptying
Contactile activity	Cholecystokinin
pH	
Redox potential	

How the microbial microflora of the gut contributes to the hydration state and transit time is not fully understood. However, there is some evidence for a role in controlling secretion and absorption, as well as motility, with some indicators of mechanisms involved.

(a) *Effects on absorption and secretion*

The caecum of a germfree animal can represent up to 50% of the total body weight, the contents of which consist of a hypotonic fluid low in chloride and bicarbonate as well as other diffusable anions. Administration of certain antibiotics can induce a similarly dramatic effect. A possible explanation for these observations is the presence of osmotically active compounds that would normally be metabolized by gut bacteria (Gracey, Papadimitriou and Bower, 1974). Such compounds would include mucin and dietary carbohydrates. Antibiotics that reduce this metabolizing capacity of the gut microflora and inhibit reduction in the level of these high-molecular-weight osmotically active compounds may predispose to osmotic diarrhoea. There is much evidence in the literature for such an impact of antibiotics, e.g. the decreased colonic fermentation observed in antibiotic-associated diarrhoea by measurement of short-chain fatty acids (Clausen *et al.*, 1991).

Obviously, increased hydration of gut contents and possible diarrhoea is not only dependent on antibiotic-induced decrease in absorption. Any indirect effect on increasing secretion will also contribute. Such compounds include the dihydroxy bile acids (chenodeoxycholate and deoxycholate) and fatty acids, and have been shown to cause secretion in the gut in humans and other animals (Mekhjian, Phillips and Hofmann, 1971; Binder and Rawlins, 1973; Bright-Asare and Binder, 1973). The overall effect of gut bacterial metabolism is to reduce the levels of the secretion inducing primary and secondary bile acids. Based on this, it has been proposed that some antibiotic-associated diarrhoea is due to increased faecal levels of the primary bile acid chenodeoxycholate following disruption of bile acid metabolism (Bright-Asare and Binder, 1973). Further, indirect evidence of the involvement of bile acids in secretion comes from the observation that ileal resection frequently causes diarrhoea which can be controlled by administration of cholestyramine (Hofmann, 1969).

There are a number of long-chain faecal hydroxy fatty acids (James, Webb and Kellock, 1961), the major one, 10-hydroxy-stearic acid, having similarity to ricinoleic acid. It was postulated that this fatty acid may have similar cathartic action (James, Webb and Kellock, 1961). It was later shown in large intestine perfusion studies in volunteers that 10-hydroxy-stearic acid inhibited water and electrolyte absorption (Amon and Phillips, 1973). Bacterial metabolism of dietary lipids contributes to the levels of faecal long-chain fatty acids. It is possible that selective inhibitory effects of some antibiotics could lead to an accumulation of such compounds (i.e. does not prevent their formation but inhibits their degradation).

(b) Normal microflora, antibiotics and motility

Although it has been known for many years that the peristaltic activity of the small bowel is stimulated by the presence of bacteria (Abrams and Bishop, 1996), little is known of the mechanism(s) by which this is achieved. A possible contributor to the differing transit times between germfree and conventional animals is the octapeptide of cholecystokinin. This compound, which can induce repetitive bursts of action potentials in the gut, is present in lower concentrations in germfree animals (Pen and Welling, 1981). Bile acids may also play a role in influencing transit time. Both primary and secondary bile acids increase colonic motility in the rabbit and humans (Kirwan *et al.*, 1975), with primary being more potent than secondary bile acids. The increase in primary bile acids in the germfree gut due to absence of bacterial metabolism should serve to decrease transit time. As transit time is increased, other factors must have more influence. However, it remains possible that antibiotic-associated reduction in gut microflora and its metabolism has the potential to increase motility because of a shift towards higher concentrations of primary bile acids as well as reducing the capacity to degrade secondary bile acids.

14.2.2 Antibiotic-associated diarrhoea due to pathogens other than *C. difficile*

Animal and volunteer challenge studies demonstrate that in certain circumstances infection with enteric pathogens can be established with a lower challenge dose following pre-exposure with antibiotics. This potential antibiotic-induced higher susceptibility to colonization with enteric pathogens may account for some cases of antibiotic-associated diarrhoea. This was shown to be the case for enterotoxigenic *Clostridium perfringens* (Borriello *et al.*, 1984; Williams *et al.*, 1985). The features of this infection are a little different to those of classical food poisoning (Larson and Borriello, 1988), and an epidemiological study showed cross-infection to be the route of infection and failed to implicate food as a source (Borriello *et al.*, 1985b). This does not exclude food as a possible source in other cases, and the possibility that this may account for some antibiotic-associated diarrhoea in the community can not be excluded.

There have been a few other sporadic case reports of specific pathogens implicated in antibiotic-associated diarrhoea, some of which have been reviewed by Borriello and Carman (1988), but no major common pathogen has been implicated.

14.3 *C.-DIFFICILE*-ASSOCIATED DIARRHOEA

C. difficile represents the most common identifiable bacterial cause of cases of nosocomial diarrhoea. The vast majority of these cases are associated with antibiotics. What makes the pathogenesis of this infection unique is the apparent dependence on antibiotic-induced susceptibility.

Much has been learnt about pathogenesis of the infection over the last 20 years since the discovery of the role of *C. difficile* in antibiotic-associated diarrhoea and pseudomembranous colitis. This overview will not comment on virulence factors of *C. difficile*, host responses, mechanisms of action of the toxins or epidemiology. The interested reader is directed to two books (Borriello, 1984b; Rolfe and Finegold, 1988) and three reviews (Borriello, 1990, 1998; Lyerly, Krivan and Wilkins, 1988). Instead it will comment only on antibiotic-induced disruption of colonization resistance that predisposes to infection, and studies on attempts to reconstitute colonization resistance to prevent (bacterioprophylaxis) or treat (bacteriotherapy) infection.

14.3.1 The normal microflora and colonization resistance

The barrier effect to infection exerted by the normal microflora (colonization resistance) is extremely effective against *C. difficile* and readily disrupted by antibiotics. This interaction of antibiotic-induced predisposition to infection and disease following exposure to *C. difficile* was clearly demonstrated in the hamster model of disease. Hamsters challenged with toxigenic *C. difficile* suffered no ill effects unless pretreated with an antibiotic, and further, even antibiotic-pretreated hamsters remained healthy if protected against exposure to *C. difficile* (Larson, Price and Borriello, 1980). Neither antibiotic exposure nor challenge with *C. difficile* alone was sufficient to cause disease; both were required. This barrier effect to *C. difficile* colonization can also be demonstrated *in vitro* with human faecal or hamster caecal microflora (Borriello and Barclay, 1986). This *in vitro* model consists of faecal or caecal emulsions prepared in sterile distilled water. The crucial importance of the presence of viable normal microbial components of the microflora in this system was demonstrated by the ability of *C. difficile* to grow and produce toxin in heat- or filter-sterilized emulsions from healthy animals or volunteers, but not in untreated emulsions. The degree of inhibition in emulsions prepared from patients on antibiotics differed and was indicative of susceptibility to colonization in some cases (i.e. untreated emulsions supported growth of *C. difficile*). The predictive capability of such a model to indicate susceptibility to infection was determined in the hamster model (Borriello, Barclay and Welch, 1988). In these experiments, different antibiotics were given to sets of hamsters; half were killed to prepare caecal emulsions and half were challenged with C. *difficile*. The ability or inability of the emulsions to suppress the growth of C. *difficile* was predictive of outcome in the hamster challenge experiments.

Additional work with the hamster model of disease demonstrated that for those antibiotics which induce susceptibility to infection, the degree of susceptibility induced is similar with respect to the challenge dose of C. *difficile* needed to cause disease. The major difference between the antibiotics examined was the longevity of the susceptibility induced (Larson and Borriello, 1990).

The mechanism(s) by which the normal microflora inhibits growth of C. *difficile* and prevents implantation are not understood. A number of mechanisms could contribute to the competitive exclusion exhibited by the normal microflora. These include competition for nutrients, presence of antibacterial substances and/or availability of mucosal receptors. There is some experimental evidence for all three mechanisms.

Wilson and Perini (1988) showed that C. *difficile* had a requirement for exogenous N-acetylgucosamine in a continuous-flow chemostat model of mouse gut microflora resistance to C. *difficile* establishment. It was postulated that competition for this nutrient, which C. *difficile* could not cleave from mucosal glycoproteins, contributed to colonization resistance against C. *difficile*. However, later work by Seddon, Hemingway and Borriello (1990) demonstrated that C. *difficile* could yield this substrate as an end-product of its hydrolysis of the connective tissue hyaluronic acid.

Most work on inhibitory metabolites has been on the role of volatile fatty acids. For example, it has been shown that coincidental with development of colonization resistance in the baby hamster gut is an increase in volatile fatty acids to levels that are inhibitory to C. *difficile*. (Rolfe, 1984).

More recently, it has been shown that exclusion of C. *difficile* in the mouse by minimum re-colonization of the gut with C. *indolis*, C. *cocleatum* and a *Eubacterium* sp. may be due to competition for mucosal receptors (Boureau *et al.*, 1994).

Complicating the situation further is the existence of protective mechanisms against colonization/disease in neonates. Babies are frequently colonized with C. *difficile* and may have both toxins present (Larson *et al.*, 1982; Libby, Donta and Wilkins, 1983), yet very rarely succumb to disease. It has been postulated that protection is due to the absence of receptors in the gut for the toxins or the organism (Borriello, 1989). As colonization of the gut progresses and conventional competitive exclusion becomes established, the early life protective mechanism(s) are lost. In hamsters it has been shown that, within 25 days of birth, 25% of animals in the experiment were completely reliant on competitive exclusion for protection, and that this loss of the early-life protective mechanism and reliance on an established gut microflora was complete by 5 months of age (Borriello *et al.*, 1985a).

14.3.2 Reconstitution of colonization resistance in animals

There have been many attempts to establish colonization resistance to *C. difficile* in animals using complex microflora derived from faecal or caecal homogenates from healthy animals of the same species or from different species (Table 14.2).

These bacterioprophylactic approaches have been very successful. This contrasts with the outcome when attempts are made to identify minimum components of the microflora that will have the same protective effect (Table 14.3).

The work of Wilson and Freter (1986) clearly demonstrates the complexity of the problem. In this study 150 isolates, from the predominant

Table 14.2 Reconstitution of colonization resistance using complete microflora from a donor animal (Figures compiled from Dubos *et al.*, 1984, Ducluzeau *et al.*, 1981, Itoh *et al.*, 1987, Raibaud *et al.*, 1980, Wilson, Silva and Fekety, 1981, Wilson *et al.*, 1986 and Wilson and Freter, 1986).

Host	Source of bacterial flora	Protection
Mouse	Mouse	22/23
Hamster	Hamster	24/30
Hare	Hare	49/49
Mouse	Hamster	22/28
Mouse	Hare	10/10
Mouse	Human	6/10

Table 14.3 Reconstitution of colonization resistance using components of the normal mouse microflora

Source of bacteria	Species	Protection	Reference
Mouse	Lactobacilli (3) Bacteroides (37) Clostridia (46)	0/7	Itoh *et al.*, 1987
	Total spore formers	7/7	
	Peptostreptococcus micros	13/13	Carlstedt-Duke, unpublished
Hamster	150 (not speciated)	0/8	Wilson and Freter, 1986
	149 (not speciated)	0/4	
	167 (not speciated)	0/6	
Hare	14 (not speciated) Clostridia (3)	0/8	Dubos *et al.*, 1984
	Eubacteria (2) Bacteroides (1)	0/16	

Table 14.4 Prevention and treatment of C. *difficile* gastrointestinal infection (DBPC = double-blind placebo-controlled; RPC = randomized placebo-controlled)

Study	Study size	Study design	Treatment	Indication	Eventual outcome
Gorbach et al., 1987	5	Open	Lactobacillus GG	Relapse	Resolution in all five
Seal et al., 1987	2	Open	Non-toxigenic C. difficile	Relapse	Resolution in both patients
Surawicz et al., 1989a	13	Open	Saccharomyces boulardii	Relapse	Resolution in 11/13
Surawicz et al., 1989b	180	DBPC	S. boulardii	Prevention	No reduction in acquisition of C. difficile
Cano, Chapoy and Corthier, 1989	2	Open	S. boulardii	First episode treatment	Resolution in both patients
Kimmey et al., 1990	1	Open	S. boulardii	Relapse	Resolution
	1	Open	S. boulardii	Relapse	Resolution
Siitonen et al., 1990	16	RPC	Lactobacillus GG	Prevention	No acquisition of C. difficile
Buts, Corthier and Delmee, 1993	19	Open	S. boulardii	First episode treatment	Resolution in 18/19
McFarland et al., 1994	64	DBPC	S. boulardii	First episode treatment	No significant effect on rate of relapse
	60	DBPC	S. boulardii	Relapse	Further relapses reduced ($p = 0.04$)

Reference	N	Design	Organism	Indication	Outcome
Schellenberg *et al.*, 1994	3	Open	*Saccharomyces cerevisiae*	Relapse	Resolution in all three patients
Chia, Chan and Goldstein, 1995	4	Open	*S. cerevisiae*	Relapse	Resolution in all four patients
Biller *et al.*, 1995	4	Open	*Lactobacillus* GG	Relapse	Resolution in all four patients
Lewis, Potts and Barry, 1998	72	DBPC	*S. boulardii*	Prevention	No reduction in acquisition of *C. difficile*

climax stage microflora of hamster gut microflora in a chemostat, reduced by a factor of ten the numbers of *C. difficile* in gnotobiotic mice precolonized with *C. difficile*. This was improved to a 1000-fold reduction with 67 isolates from a preclimax stage microflora followed by 100 isolates from the climax stage microflora. However, this still failed to eradicate the majority of resident *C. difficile*.

Studies have been undertaken to use human faecal microflora components in animal models to try to identify components important in colonization resistance. This has been tried with mixed human faecal microflora in the mouse with some success (Raibaud *et al.*, 1980; Table 14.2), and with full success in rats (Borriello and Fuller, unpublished). Corthier, Dubos and Raibaud (1985) examined eight different human faecal isolates in mice di-associated in turn with one of the species and *C. difficile*. Four of them afforded some protection (*Escherichia coli*, *Bifidobacterium bifidum*, *Eubacterium ventriosum* and *Enterococcus faecalis*). Protection was based on a combination of a reduction in the numbers of *C. difficile* and the levels of cytotoxin. However, how reflective these experiments are of complex interactions in the gut of antibiotic-treated patients, who still, in most cases, harbour a qualitatively and quantitatively complex gut microflora, is difficult to determine.

14.3.3 Bacteriotherapy and bacterioprophylaxis in humans

The most commonly used agents in the treatment of *C.-difficile*-mediated antibiotic diarrhoea are vancomycin and metronidazole (Keighley *et al.*, 1978; Cherry *et al.*, 1982). Unfortunately, relapses occur, the frequency of which can be as high as 25%, and a given patient may suffer multiple relapses (Bartlett *et al.*, 1980; Wilcox and Spencer, 1992).

These problems have stimulated an interest in the use of live organisms for prophylaxis or therapy of initial diarrhoea or (more commonly) relapses (Table 14.4). The approaches taken include use of mixed faecal microflora, components of the faecal microflora, single species of the gut microflora, non-toxigenic *C. difficile* or organisms (e.g. yeasts) that are not normal components of the gut microflora.

(a) Use of undefined mixed faecal microflora

The use of complete undefined faecal microflora, administered as enemas, has been successful in the treatment of pseudomembranous colitis. Most of these cases preceded the discovery of *C. difficile* as the aetiological agent responsible (Eisemann *et al.*, 1958; Bowden, Mansberger and Lykins, 1981), although one of the 18 patients in the report of treatment experience over 16 years by Bowden and colleagues (1981), was sufficiently recent to have been tested for *C. difficile* toxin. This was one of the

13 patients who responded positively. The best documented case is that of Schwan and colleagues (1984) where two faecal enemas prepared from a relative were administered 2 days apart to treat a patient who had suffered multiple relapses. The diarrhoea settled after the first enema and the patient recovered.

(b) Use of defined faecal microflora

An attempt to refine this approach was made a few years later (Tvede and Rask-Maden, 1989). In this study of six patients with relapsing C. *difficile* diarrhoea, a cocktail of ten faecal organisms consisting of three *Bacteroides* spp., three *Clostridium* spp., two *Escherichia coli*, *Enterococcus faecalis* and *Peptostreptococcus productus* were given by rectal infusion in five patients with success. One patient received a crude faecal enema and also recovered.

Animal experimentation had shown that non-toxigenic strains of C. *difficile* could prevent infection from toxigenic strains in susceptible hamsters (Wilson and Sheagren, 1983; Borriello and Barclay, 1985). This approach to treatment of two multiple relapse cases in elderly patients was used successfully to prevent further relapse (Seal *et al.*, 1987).

A human isolate of lactobacillus (*Lactobacillus rhamnosus* GG) was used to treat five patients with relapsing C. *difficile* diarrhoea (Gorbach, Chang and Goldin, 1987). This was administered in milk for 7–10 days. All five of these patients responded immediately; one of them relapsed once. More recently, a 2-week course of L. *rhamnosus* GG was given to four children who had a history of multiple relapses (Biller *et al.*, 1995). All four responded to this treatment, although two relapsed. However, both responded to re-treatment. L. *rhamnosus* GG has also been used as a component of yoghurt to determine its ability to protect against erythro-mycin-induced gut symptoms in volunteers (Siitonen *et al.*, 1990). None of the 16 volunteers had toxin present in the stools. However, prevention of C. *difficile* diarrhoea was not the main hypothesis being tested in this experiment, erythromycin is not commonly associated with C. *difficile* diarrhoea, and the volunteers were not in an environment at high risk of infection with C. *difficile*.

(c) Use of Saccharomyces *species*

Saccharomyces boulardii is a non-pathogenic yeast first used in France in the 1950s to treat diarrhoea (reviewed by McFarland and Bernasconi, 1993). It is resistant to antibiotics and survives undigested into the lower gastrointestinal tract after oral ingestion. It is not a normal consti-tuent of the human intestinal microflora and does not multiply in the gut. Early work in hamsters (Toothaker and Elmer, 1984) and mice (Corthier,

Dubos and Ducluzeau, 1986) provided evidence that this yeast could protect against *C. difficile* infection. A few years later work was undertaken to assess the potential of *S. boulardii* to prevent and treat *C. difficile* infection in humans, and as yet results are still equivocal. While there is some evidence that *S. boulardii* may prevent the development of diarrhoea in antibiotic-treated patients (Surawicz *et al.*, 1989b), it did not reduce the acquisition of *C. difficile* or toxin production in two double-blind placebo controlled studies including some 250 patients (Surawicz *et al.*, 1989b; Lewis, Potts and Barry, 1998). Two small open studies report resolution of *C.-difficile*-associated pseudomembranous colitis in four out of four adults (Cano, Chapuy and Corthier, 1989) and of chronic diarrhoea or colic in 18 of 19 children (Buts, Corthier and Delmee, 1993) treated with *S. boulardii*.

Relapses are a common problem in *C. difficile* diarrhoea, occurring in about 20% of initially successfully treated patients. *S. boulardii* failed to prevent relapse when given in addition to conventional treatment with metronidazole or vancomycin in a randomized controlled study of 94 patients with *C. difficile* diarrhoea (McFarland *et al.*, 1994), but did reduce relapses by almost 50% (from 65% to 34%, $p = 0.04$) in a study of 60 patients with established relapsing *C. difficile* diarrhoea. This symptomatic response was associated with a reduction in toxin excretion but no change in *C. difficile* carrier status.

Other *Saccharomyces* species have been used in small open studies to treat relapsing *C. difficile* diarrhoea. Chia, Chan and Goldstein (1995) report resolution of relapses in four patients after treatment with baker's yeast. The therapeutic effect may not depend on viability of the yeast since brewer's yeast, a non-viable preparation of yeast sold as a food supplement, was also found to be effective in an open study of three patients with relapsing *C. difficile* diarrhoea (Schellenberg *et al.*, 1994).

Published evidence so far suggests that *S. boulardii* reduces the incidence of *C. difficile* diarrhoea after antibiotic treatment and the number of relapses in established *C. difficile* colitis. However, all positive evidence for this treatment comes from studies performed by one group of authors, and studies from other research groups to corroborate the results would help to confirm the findings. The observed clinical response is not due to prevention of *C. difficile* infection or accelerated clearance from the bowel. Some of the studies suggest a reduction of toxin excretion. Recent experimental work in a rat ileal loop model indicates that *S. boulardii* activity against *C.-difficile*-induced disease may be mediated via a serine protease that accelerates degradation of rat ileal receptors for the toxin (Pothoulakis *et al.*, 1993). However, although there is some evidence for degradation of toxin receptors, the same group have since generated better data that more strongly implicate degradation of toxin A as the main protective mechanism (Castagliuolo *et al.*, 1996).

14.4 CONCLUSIONS

Although there is no doubt that competitive exclusion against C. *difficile* is a feature of the normal gut microflora, there is as yet little information on the mechanisms involved or the minimum key components of the gut microflora involved. An understanding of this might help to better identify patients at risk of infection and/or relapse.

There is also little evidence on which to draw conclusions for the efficacy of individual proposed biotherapeutic agents. However, it would appear that there is no compelling evidence for their use for prophylaxis, but some evidence for their use as an adjunct to treatment with vancomycin. Of these agents, S. *boulardii* has been most used. Unfortunately, most of the studies with this yeast have been conducted by the same group of researchers. More studies by other groups are needed.

14.5 REFERENCES

Abrams, G. D. and Bishop, J. E. (1966) Effect of the normal microbial flora on the resistance of the small intestine to infections. *Journal of Bacteriology*, **92**, 1604–1608.

Amon, H. V. and Phillips, S. F. (1973) Inhibition of colonic water and electrolyte absorption by fatty acids in man. *Gastroenterology*, **65**, 744–749.

Bartlett, J. G., Tedesco, F. J., Shull, S. *et al.* (1980) Symptomatic relapse after oral vancomycin therapy of antibiotic-associated pseudomembranous colitis. *Gastroenterology*, **78**, 431–434.

Biller, J. A., Katz. A. J., Flores, A. F. *et al.* (1995) Treatment of recurrent *Clostridium difficile* colitis with *Lactobacillus* GG. *Journal of Pediatric Gastroenterology and Nutrition*, **21**, 224–226.

Binder, H. J. and Rawlins, C. L. (1973) Effect of conjugated bile salts on electrolyte transport in rat colon. *Journal of Clinical Investigation*, **52**, 1460–1466.

Borriello, S. P. (1984a) Bacteria and gastrointestinal secretion and motility. *Scandinavian Journal of Gastroenterology*, **19**(Suppl. 93), 115–121.

Borriello, S. P. (ed.) (1984b) *Antibiotic-Associated Diarrhoea and Colitis*. Martinus Nijhoff, Boston, MA.

Borriello, S. P. (1989) Influence of the normal gut flora of the gut on *Clostridium difficile*, in *The Regulatory and Protective Role of the Normal Microflora*, (eds R. Grubb, T. Midtvedt and E. Norrin), Macmillan, Basingstoke, pp 239–251.

Borriello, S. P. (1990) Pathogenesis of *Clostridium difficile* infection of the gut. *Journal of Medical Microbiology*, **33**, 207–215.

Borriello, S. P. (1992) Possible mechanisms of action of antimicrobial agent-associated gastrointestinal symptoms. *Postgraduate Medical Journal*, **68**(Suppl. 3), S38–S42.

Borriello, S. P. (1998) Pathogenesis of *Clostridium difficile* infection. *Journal of Antimicrobial Chemotherapy*, in press.

Borriello, S. P. and Barclay, F. E. (1985) Protection of hamsters against *Clostridium difficile* ileocaecitis by prior colonisation with non-pathogenic strains. *Journal of Medical Microbiology*, **19**, 339–349.

Borriello, S. P. and Barclay, F. E. (1986) An in-vitro model of colonisation resistance to *Clostridium difficile* infection. *Journal of Medical Microbiology*, **21**, 299–309.

Borriello, S. P. and Carman, R. J. (1988) Other clostridial causes of diarrhoea and colitis in man and animals, in *Clostridium difficile: Its Role in Intestinal Disease*, (eds R. D. Rolfe and S. M. Finegold), Academic Press, San Diego, CA, pp. 65–98.

Borriello, S. P., Barclay, F. E. and Welch, A. R. (1988) Evaluation of the predictive capability of an in-vitro model of colonisation resistance to *Clostridium difficile* infection. *Microbiology Ecology in Health and Disease*, **1**, 61A.

Borriello, S. P., Larson, H. E., Welch, A. R. *et al.* (1984) Enterotoxigenic *Clostridium perfringens:* a possible cause of antibiotic-associated diarrhoea. *Lancet,* **i**, 305–307.

Borriello, S. P., Barclay, F. E., Welch, A. R. *et al.* (1985a) Host and microbial determinants of the spectrum of *Clostridium difficile* mediated gastrointestinal disorders. *Microecology and Therapy*, **15**, 231–236.

Borriello, S. P., Barclay, F. E., Welch, A. R. *et al.* (1985b) Epidemiology of diarrhoea caused by enterotoxigenic *Clostridium perfringens*. *Journal of Medical Microbiology*, **20**, 363–372.

Boureau, H., Salanon, C., Decaaens, C. and Bourlioux, P. (1994) Caecal localisation of the specific microbiota resistant to *Clostridium difficile* colonisation in gnotobiotic mice. *Microbial Ecology in Health and Disease*, **7**, 111–117.

Bowden, T. A., Mansberger, A. R. and Lykins, L. E. (1981) Pseudomembranous enterocolitis: mechanism of restoring flora homeostasis. *American Surgeon*, **47**, 178–183.

Bright-Asare, P. and Binder, H. J. (1973) Stimulation of colonic secretion of water and electrolytes by hydroxy fatty acids. *Gastroenterology*, **64**, 81–88.

Buts, J.-P., Corthier, G. and Delmee, M. (1993) *Saccharomyces boulardii* for *Clostridium difficile* associated enteropathies in infants. *Journal of Pediatric Gastroenterology and Nutrition*, **16**, 419–425.

Cano, N., Chapoy, P. and Corthier, G. (1989) *Saccharomyces boulardii:* un traitment des colites pseudomembraneuses? *Presse Medicale*, **18**, 1299.

Castagliuolo, I., LaMont, J. T., Nikulasson, S. T. and Pothoulakis, C. (1996) *Saccharomyces boulardii* protease inhibits *Clostridium difficile* toxin A effects in the rat ileum. *Infection and Immunity*, **64**, 5225–5232.

Cherry, R. D., Portnoy, D., Jabbari, M. *et al.* (1982) Metronidazole: an alternative therapy for antibiotic-associated colitis. *Gastroenterology*, **82**, 849–851.

Chia, J. K. S., Chan, S. M. and Goldstein, H. (1995) Baker's yeast as adjunctive therapy for relapses of *Clostridium difficile* diarrhoea. *Clinical Infectious Diseases*, **20**, 1581.

Clausen, M. R., Bronnen, H., Tvede, M. and Mortensen, P. B. (1991) Colonic fermentation to short-chain fatty acids is decreased in antibiotic-associated diarrhoea. *Gastroenterology*, **101**, 1497–1504.

Corthier, G. Dubos, F. and Ducluzeau, R. (1986) Prevention of *Clostridium difficile* mortality in gnotobiotic mice by *Saccharomyces boulardii*. *Canadian Journal of Microbiology*, **32**, 894–896.

Corthier, G. Dubos, F. and Raibaud, P. (1985) Modulation of cytotoxin production by *Clostridium difficile* in the intestinal tracts of gnotobiotic mice inoculated with various human intestinal bacteria. *Applied and Environmental Microbiology*, **49**, 250–252.

Dubos, F., Martinet, L., Dabard, J. and Ducluzeau, R. (1984) Immediate postnatal inoculation of a microbial barrier to prevent neonatal diarrhoea induced by *Clostridium difficile* in young conventional and gnotobiotic hares. *American Journal of Veterinary Research*, **45**, 1241–1244.

Ducluzeau, R., Dubos, F., Hudault, S. *et al.* (1981) Microbial barriers against enteropathogenic strains in the digestive tract of gnotoxenic animals. Application to the treatment of *Clostridium difficile* diarrhoea in the young hare, in *Recent*

Advances in Germ-free Research, (eds A. Sasaki, K. Ozawa and K. Hashimoto), Tokai University Press, Tokyo.

Eisemann, B., Silem, W., Bascomb, W. S. and Kanvor, A. J. (1958) Fecal enema as an adjunct in the treatment of pseudomembranous enterocolitis. *Surgery*, **44**, 854–858.

Gorbach, S. L., Chang, T.-W. and Goldin, B. (1987) Successful treatment of relapsing *Clostridium difficile* colitis with *Lactobacillus* GG. *Lancet*, **ii**, 1519.

Gracy, M., Papadimitriou, J. and Bower, G. (1974) Ultrastructural changes in the small intestine of rats with self-filling blind loops. *Gastroenterology*, **67**, 646–651.

Hofmann, A. F. (1967) The syndrome of ileal disease and the broken enterohepatic circulation: cholerheic enteropathy. *Gastroenterology*, **52**, 752–757.

Itoh, K., Lee, W. K., Kawamura, H. *et al.* (1987) Intestinal bacteria antagonistic to *Clostridium difficile* in mice. *Laboratory Animal*, **21**, 20–25.

James, AT., Webb, J. P. W. and Kellock, T. D. (1961) The occurrence of unusual fatty acids in faecal lipids from human beings with normal and abnormal fat absorption. *Biochemical Journal*, **78**, 333–339.

Keighley, M. R. B., Burdon, D. W., Arabi, Y. *et al.* (1978) Randomised controlled trial of vancomycin for pseudomembranous colitis and post-operative diarrhoea. *British Medical Journal*, **ii**, 1667–1669.

Kimmey, M. B., Elmer, G. W., Surawicz, C. M. and McFarland, L. V. (1990) Prevention of further recurrences of *Clostridium difficile* colitis with *Saccharomyces boulardii*. *Digestive Diseases and Sciences*, **35**, 897–901.

Kirwan, W. O., Smith, A. N., Mitchell, W. D. *et al.* (1975) Bile acids and colonic motility in the rabbit and the human. Part 1. The rabbit. *Gut*, **16**, 894–902.

Larson, H. E. and Borriello, S. P. (1988) Infectious diarrhoea due to *Clostridium perfringens*. *Journal of Infectious Diseases*, **157**, 390–391.

Larson, H. E. and Borriello, S. P. (1990) Quantitative study of antibiotic induced susceptibility to *Clostridium difficile* enterocolitis in hamsters. *Antimicrobial Agents and Chemotherapy*, **34**, 1348–1353.

Larson, H. E., Price, A. B. and Borriello, S. P. (1980) Epidemiology of experimental enterocecitis due to *Clostridium difficile*. *Journal of Infectious Diseases*, **142**, 408–413.

Larson, H. E., Barclay, F. E., Honour, P. and Hill I. D. (1982) Epidemiology of *Clostridium difficile* in infants. *Journal of Infectious Diseases*, **146**, 727–733.

Lewis, S. J., Potts, L. F. and Barry, R. E. (1998) The therapeutic effects of *Saccharomyces boulardii* in the prevention of antibiotic related-diarrhoea in elderly patients. *Journal of Hospital Infection*, **36**, in press.

Libby, J. M., Donta, S. T. and Wilkins, T. D. (1983) *Clostridium difficile* toxin A in infants. *Journal of Infectious Diseases*, **148**, 606.

Lyerly, D. M., Krivan, H. E. and Wilkins, T. D. (1988) *Clostridium difficile*: its disease and toxins. *Clinical Microbiology Reviews*, 1–18.

McFarland, L. V. and Bernasconi, P. (1993) *Saccharomyces boulardii*: a review of an innovative biotherapeutic agent. *Microbial Ecology in Health and Disease*, **6**, 157–171.

McFarland, L. V., Surawicz, C. M., Greenberg, R. N. *et al.* (1994) A randomized placebo-controlled trial of *Saccharomyces boulardii* in combination with standard antibiotics for *Clostridium difficile* disease. *Journal of the American Medical Association*, **271**, 1913–1918.

Mekhjian, H. S., Phillips, S. F. and Hofmann, A. F. (1971) Colonic secretion of water and electrolytes induced by bile acids: perfusion studies in man. *Journal of Clinical Investigation*, **50**, 1569–1577.

Pen, J. and Welling, G. W. (1981) The concentration of cholecystokinin in the intestinal tract of germ-free and control mice. *Antonie van Leeuwenhoek*, **47**, 84–85.

Pothoulakis, C., Kelly, C. P., Joshi, M. A. *et al.* (1993) *Saccharomyces boulardii* inhibits *Clostridium difficile* toxin A binding and enterotoxicity in rat ileum. *Gastroenterology*, **104**, 1108–1115.

Raibaud, P., Ducluzeau R., Dubos, F. *et al.* (1980) Implantation of bacteria from the digestive tract of man and various animals into gnotobiotic mice. *American Journal of Clinical Nutrition*, **33**, 2440–2447.

Rolfe, R. D. (1984) Role of volatile fatty acids in colonisation resistance to *Clostridium difficile*. *Infection and Immunity*, **45**, 185–191.

Rolfe, R. D. and Finegold, S. M. (eds) (1988) *Clostridium difficile: Its Role in Intestinal Disease*. Academic Press, San Diego, CA.

Schellenberg, D., Bonington, A., Champion, C. M. *et al.* (1994) Treatment of relapsing *Clostridium difficile* diarrhoea with brewer's yeast. *Lancet*, **343**, 171–172.

Schwan, A., Sjolin, S., Trottestam, U. and Aronsson, B. (1984) Relapsing *Clostridium difficile* enterocolitis cured by rectal infusion of normal faeces. *Scandinavian Journal of Infectious Diseases*, **16**, 211–215.

Seal, D. V., Borriello, S. P., Barclay, F. *et al.* (1987) Treatment of relapsing *Clostridium difficile* diarrhoea by administration of a non-toxigenic strain. *European Journal of Clinical Microbiology*, **6**, 51–53.

Seddon, S. V., Hemingway, I. and Borriello, S. P. (1990) Hydrolytic enzyme production by *Clostridium difficile* and its relationship to toxin production and virulence in the hamster model. *Journal of Medical Microbiology*, **31**, 169–174.

Surawicz, C. M., Elmer, G. M., Speelman, P. *et al.* (1989a) Prevention of antibiotic-associated diarrhea by *Saccharomyces boulardii*: a prospective study. *Gastroenterology*, **96**, 981–988.

Surawicz, C. M., McFarland, L. V., Elmer, G. W. and Chinn, J. (1989b) Treatment of recurrent *Clostridium difficile* colitis with vancomycin and *Saccharomyces boulardii*. *American Journal of Gastroenterology*, **84**, 1285–1287.

Siitonen, S., Vapaatalo, H., Salminen, S. *et al.* (1990) Effect of *Lactobacillus* GG yoghurt in prevention of antibiotic associated diarrhoea. *Annals of Medicine*, **22**, 57–59.

Toothaker, R. D. and Elmer, G. W. (1984) Prevention of clindamycin-induced mortality in hamsters by *Saccharomyces boulardii*. *Antimicrobial Agents and Chemotherapy*, **26**, 552–556.

Tvede, M. and Rask-Maden, J. (1989) Bacteriotherapy for chronic relapsing *Clostridium difficile* diarrhoea in six patients. *Lancet*, **i**, 1156–1160.

Wilcox, H. M. and Spencer, R. C. (1992) *Clostridium difficile* infection: responses, relapses and re-infections. *Journal of Hospital Infection*, **22**, 85–92.

Williams, R., Piper, M., Borriello, S. P. *et al.* (1985) Diarrhoea due to enterotoxigenic *Clostridium perfringens*: clinical features and management of a cluster often cases. *Age and Ageing*, **14**, 296–302.

Wilson, K. H. and Freter, R. (1986) Interaction of *Clostridium difficile* and *Escherichia coli* with microfloras in continuous-flow cultures and gnotobiotic mice. *Infection and Immunity*, **54**, 354–358.

Wilson, K. H. and Perini, F. (1988) Role of competition for nutrients in suppression of *Clostridium difficile* by the colonic microflora. *Infection and Immunity*, **56**, 2610–2614.

Wilson, K. H. and Sheagren, J. N. (1983) Antagonism of toxigenic *Clostridium difficile* by non-toxigenic *C. difficile*. *Journal of Infectious Diseases*, **147**, 733–736.

Wilson, K. H., Silva, J. and Fekety, R. F. (1981) Suppression of *Clostridium difficile* by normal hamster cecal flora and prevention of antibiotic-associated cecitis. *Infection and Immunity*, **34**, 626–628.

Wilson, K. H., Sheagren, J. N., Freter, R. *et al.* (1986) Gnotobiotic models for study of the microbial ecology of *Clostridium difficile* and *Escherichia coli*. *Journal of Infectious Diseases*, **153**, 547–551.

15

The normal microflora as a reservoir of antibiotic resistance determinants

Gerald W. Tannock

The complex communities of bacteria that are resident in various body sites are potential crucibles in which the evolution and transfer of genetic determinants encoding antibiotic resistance could occur. Bacterial cells with relatively short generation times, representing many diverse species, coexist in habitats that may be exposed to antibiotics, which act as selective agents for cells containing appropriate genetic determinants of resistance. Antibiotics may be administered to the host for a number of medical or veterinary reasons: (1) to treat an infection, (2) as a prophylactic measure, (3) as a growth promotion measure (farm animals).

The evolution of new DNA constructs encoding antibiotic resistance is most likely to occur under conditions in which an antibiotic persists in the habitat for some time at a concentration that is sublethal to the bacterial cells. These conditions are unlikely to occur during most treatment regimes. They are usually of short duration and the patient is exposed to relatively high concentrations of antibiotic. Bacteria susceptible to the antibiotic at these concentrations will be eliminated from the habitat and conditions will not be conducive to the emergence of novel antibiotic resistance determinants (Salyers, 1995). Cells that are already resistant to the antibiotic will, however, survive and will proliferate under the selective conditions provided by the antibiotic. Infections are often caused by a single bacterial strain that has established in a body site that is normally sterile. The pathogen has little opportunity, at the

G. W. Tannock (ed.), *Medical Importance of the Normal Microflora*, 388–404.
© 1999 *Kluwer Academic Publishers. Printed in Great Britain.*

infected site, of acquiring antibiotic resistance from other bacterial cells. Pathogens may sometimes pass through regions of the body that are inhabited by a normal microflora, however, and may acquire antibiotic resistance determinants at this stage. Bacterial pathogens may also be coresident in body sites with the normal microflora, for long or short periods, in hosts that are usually termed 'carriers'. Of course, as other chapters in this book recount, many infections are caused by actual members of the normal microflora. Thus the conditions under which antibiotic resistance arises and by means of which such resistances can be transferred among the members of the normal microflora and to pathogens is of considerable medical importance. It is likely that these conditions relate to the inappropriate use of antibiotics in human and veterinary medicine, and to the use of these drugs as growth promoters in the husbandry of farm animals.

In human medicine, antibiotics have sometimes been prescribed inappropriately for infections of non-bacterial aetiology and have been widely used as topical applications, so that conditions that are selective for resistance have been applied unnecessarily to bacterial populations. Poor patient compliance has also contributed to the selection of resistant strains. Failure of patients to follow prescription directions exactly results in subtherapeutic concentrations of antibiotic in the body, which serve to select for antibiotic-resistant strains (Levy, 1992). The emergence of multiple-resistant strains of *Mycobacterium tuberculosis* is partly a consequence of poor patient compliance (Ellner *et al.*, 1993; Kaufmann and van Embden, 1993). Self-prescribing of antibiotics is also a problem. Failure of patients to complete the prescribed course of antibiotic results in leftover drug being stored in the home and used occasionally, and inadequately, when family members have colds or influenza. In some countries, this problem is compounded because the purchase of antibiotics without a doctor's prescription is possible. Antibiotics for veterinary use are, in some parts of the world, obtainable without a prescription, even though the same drugs are regulated for human use. Since antibiotic products for veterinary use are cheaper than those used in human medicine, alternative sources for antibiotic abuse are available (Levy, 1992).

Perhaps the practice that provides the greatest risk of selection of antibiotic-resistant bacteria among the normal microflora is the prophylactic use of these agents in human and veterinary medicine. In the latter case, in particular, the antibiotics are added to animal feeds so that the whole farm environment becomes permeated with antibiotic-containing dust from the food. In a recent study (Bateup *et al.*, 1998), the majority of *Lactobacillus* isolates obtained from pigs on a farm were found to be resistant to two antibiotics. Resistance to erythromycin and tetracycline was widespread among the gastrointestinal isolates, presumably because the macrolides lincomycin and tylosin, and oxytetracycline, were used as prophylactic drugs administered in the animals' feed. The nature or location of the determinants of antibiotic resistance in the bacterial cells was not investigated, but previous studies have demonstrated that erythromycin and tetracycline resistance in lactobacilli can be plasmid-associated (Tannock *et al.*, 1994; Tannock and Luchansky, unpublished). Interestingly, not all the isolates were resistant to the antibiotics, even though they were cultured from animals exposed to the drug at the time of isolation. Isolates that were resistant to both antibiotics were also recovered from 6-day-old piglets that had not yet been administered lincomycin or tetracycline, suggesting that a *Lactobacillus* microflora existed on the farm that was essentially antibiotic-resistant and to which all the animals were exposed. It should be noted that lincomycin and tetracycline were used for relatively brief periods in animal husbandry, yet the determinants encoding erythromycin and tetracycline resistance were maintained in the lactobacilli long after the animals were administered the drugs. For example, strains resistant to erythromycin and tetracycline were detected in animals 129 days after their last exposure to lincomycin, and 79 days after exposure to oxytetracycline (Bateup *et al.*, 1998).

Even more menacing may be the practice of adding subtherapeutic concentrations of antibiotics to animal feeds in order to promote the growth of the animals. Soon after the Second World War, farmers observed that farm animals fed mashes obtained as a byproduct of antibiotic manufacture gained weight more rapidly than animals fed conventional feeds. The mashes had been used to cultivate antibiotic-producing microbes. Even after extraction of the antibiotic from the mash, low concentrations were present, so that farm animals were receiving subtherapeutic concentrations of the antimicrobial substances (Bird, 1969). Trials confirmed that the addition of antibiotics (40–50 ppm) to animal feeds increased the weight gain and improved feed efficiency (the gain in weight in relation to the amount of food consumed). Antibiotic supplementation of feeds did not affect the growth rate of germfree animals; therefore it was concluded that the drugs must have been influencing the normal microflora or pathogens causing subclinical infections (Feighner and Dashkevicz, 1987). Inexplicably, the bacterial populations that are modified by the administration of antibiotics in animal feeds have never been identified. Enterococci and clostridia have been proposed as candidates for the causation of growth suppression, as have bile-salt-hydrolysing lactobacilli, but conclusive evidence as to the nature of the responsible bacterial species and their mode of action in growth depression has not been produced (Elam *et al.*, 1954; Eyssen and de Somer, 1967; Feighner and Dashkevicz, 1987, 1988; Fuller, Coates and Harrison, 1979; Houghton, Fuller and Coates, 1981).

Despite the lack of a detailed scientific basis for the administration of antibiotics in animal feeds, the practice continues to be widespread in many countries. About 50% of the production of antibiotics in the USA, for example, is destined for use in animal feeds (Levy, 1992). The continuous exposure of the normal microflora of farm animals to low concentrations of antibiotics must surely pose a serious public health problem. Selection of bacterial cells harbouring the determinants of antibiotic resistance increases the risk of transfer of these determinants to bacterial pathogens also harboured by farm animals. If these pathogens are capable of infecting humans as well as the farm animals, these animal husbandry practices could affect human health. Evidence from California has already demonstrated the spread of an antibiotic-resistant pathogen (*Salmonella newport*) from farm animals to humans (Holmberg, Wells and Cohen, 1984; Holmberg *et al.*, 1984; Spika *et al.*, 1987). Transmission of antibiotic-resistant members of the normal microflora of farm animals to humans could result in the transfer of resistance determinants to the microflora of the new host. The determinants could then be transferred to bacteria that are specifically pathogens of humans (Anderson, 1968; Swartz, 1989). Of particular concern is the emergence of vancomycin-resistant enterococci, first reported in Europe in 1988, which are seen as a major threat to public health because of the possibility that the resistance determinants could be transferred from enterococci to methicillin-resistant staphylococci (Uttley *et al.*, 1988). Strains of *Enterococcus faecium* that are resistant to multiple antibiotics, including vancomycin, are themselves aetiological agents of blood stream infections of humans. Vancomycin-resistant enterococci have been detected in environmental samples (waste water, sewage), food samples (chicken, pork) and livestock faeces. The recovery of vancomycin-resistant enterococci from farm samples has suggested the possibility that the use of a glycopeptide growth promoter, avoparcin, is associated with the emergence of these antibiotic-resistant bacteria. Bacteria that are resistant to avoparcin are also resistant to related glycopeptides such as vancomycin. Genetic fingerprinting (ribotyping, pulsed field gel electrophoresis) has detected the presence of similar strains of vancomycin-resistant enterococci in farm animals, uncooked meats and human patients, which strongly suggests that interspecies transfer of these bacteria can occur and may contribute to colonization and infection of humans (McDonald *et al.*, 1997).

It is perhaps worth revising, at this stage, the mechanisms by which resistance to antibiotics may be acquired by bacterial cells. Bacteria can become resistant to an antibiotic by a mutation occurring in one of their genes. The mutation can alter the molecule or structure that the antibiotic targets in the bacterial cell. Changes as a result of mutations to the amino acid sequence of the RNA polymerase subunit, for example, decrease the binding of rifampicin to the enzyme and thus increase the resistance of the bacterial cell to this antibiotic. Mutations affecting the porins of the outer membrane of Gram-negative bacterial cells can impede the transport of β-lactam antibiotics to the sites of peptidoglycan synthesis, hence increasing the resistance of the cell to these drugs. Finally, mutations can activate genes that encode resistance mechanisms but are not expressed in the absence of the mutation or other chromosomal alteration (cryptic resistance).

Bacterial cells can acquire resistance determinants from other bacteria by three processes (transformation, transduction, conjugation) all of which require the transfer of DNA from one cell to another. The genes acquired in this way can encode enzymes that alter the target molecules affected by the antibiotic (e.g. *erm* genes: erythromycin resistance methylase), or encode proteins that pump the antibiotic out of the bacterial cell so that it does not reach a toxic level (e.g. tetracycline efflux proteins), or enzymes that inactivate the antibiotic molecule by structural modification (e.g. β-lactamases). Transformation usually involves the uptake of small fragments of linear DNA by bacterial cells from the extracellular environment. The transforming DNA has presumably originated from the lysis of other cells. The heterologous DNA is incorporated into the chromosome of the recipient cell, which may thus be transformed to antibiotic resistance should the DNA fragment contain an appropriate gene. Transformation is well known in laboratory studies, but its significance as a source of genetic variation *in vivo* is not clear.

Gene transfer can also be mediated by bacteriophages (transduction). In both generalized and specialized transduction, bacterial DNA can be packaged in a virion along with the viral DNA. The heterologous DNA can thus be transferred by a bacteriophage infecting a bacterial cell. While transfer of antibiotic resistance has been shown to be mediated under laboratory conditions, the *in vivo* situation has not been studied (Raya and Klaenhammer, 1992). Many bacteriophages can be demonstrated by electron microscopy of gastrointestinal samples so phage-mediated transfer of antibiotic resistance determinants between members of the normal microflora may occur (Klieve and Bauchop, 1988; Havelaar, 1993).

The best-described DNA transfer mechanism by which antibiotic resistance determinants can be acquired by bacterial cells is conjugation. During close contact of the donor and recipient cells, conjugal transfer elements, either plasmids or transposons, are transferred from one cell to the other. R-plasmids encode resistance to one or more antibiotics. Recipient bacterial cells can acquire a conjugative R-plasmid encoding resistance to several drugs and thus become multiply resistant as a result of a single conjugal event. The use of any one of the antibiotics to which the cell is resistant selects for all the resistances encoded by the R-plasmid. Transposons are DNA elements that can integrate into DNA molecules (bacterial chromosome, plasmid, bacteriophage genome). They are composed of insertion sequences that flank additional DNA that may encode resistance to antibiotics. Conjugative transposons found in bacterial species associated with humans (e.g. *Bacteroides* sp., *Enterococcus faecalis*) often carry antibiotic resistance genes, notably some that encode tetracycline resistance. As the name suggests, the conjugative transposons encode mechanisms that mediate their transfer from one bacterial cell to another during bacterial conjugation. Both conjugative plasmids and conjugative transposons can mobilize other DNA elements (other plasmids or fragments of chromosome) that are not self-transmissible. The potential for the widespread acquisition of antibiotic resistance by bacterial species is therefore enormous, but whether the phenomena observed *in vitro* occur *in vivo* must be considered.

Although there are numerous publications concerning the mechanisms and transfer of antibiotic resistance determinants between cells, only selected examples will be discussed in the remainder of this chapter. These selected publications provide adequate evidence that the administration of antibiotics selects for resistant bacteria, especially in the intestinal tract, and that the transfer of antibiotic resistance determinants occurs among the normal microflora. They show, too, that antibiotic-resistant bacteria can be transmitted between different species of animals, including to human hosts.

As reviewed by Linton (1976), research in several countries has demonstrated a close correlation between the extent of antimicrobial usage and the incidence of antibiotic-resistant *Escherichia coli* in the normal microflora of the intestinal tract of farm animals. The continuous administration to farm animals of subtherapeutic concentrations of antibiotics in their food produced the highest incidence of resistance, increased the frequency of multiple resistance (resistance to antibiotics in addition to the one administered to the animals) and lengthened the period in which the microflora contained resistant bacteria after cessation of antibiotic administration. In an experiment carried out by Linton (1976) in support of these general findings, the incidence of antibiotic-resistant *E. coli* in the faeces of two calves was determined. One calf served as a control while the other received 100 ppm of tetracycline in the food for 30 days. The proportion of antibiotic-resistant *E. coli* in the faeces of both calves was high at the beginning of the experiment, but declined gradually with time in the calf that did not receive tetracycline-containing food. In the calf administered the antibiotic, susceptible bacteria diminished in numbers and the *E. coli* population was soon composed entirely of resistant bacteria. The resistant population persisted for 24 days after antibiotic administration was stopped prior to slaughter of the animals.

In a recent study, analysis (genetic fingerprinting of isolates) of the strain composition of enterobacterial populations in the faeces of two human subjects revealed thought-provoking observations concerning the microbial ecology of the intestinal tract in relation to the administration of an antibiotic (McBurney and Tannock, unpublished).

The enterobacterial microflora was monitored over a 12-month period in which monthly faecal samples from two human subjects were obtained and analysed. In the case of Subject 1, although quite marked variation occurred in the total number of enterobacteria present in the faecal samples, a single *E. coli* strain predominated throughout the 12-month period. Only one isolate of *E. coli* that was resistant to an antibiotic (tetracycline) was detected during the 12-month period. Two *E. coli* strains were predominant in the faecal samples obtained from Subject 2 prior to the administration of amoxycillin for the treatment of a respiratory tract infection. These *E. coli* strains were susceptible to ampicillin, doxycycline, trimethoprim and nalidixic acid. Following amoxycillin administration (1 g/d for 7 days), however, the microbial ecology of the intestinal tract was severely altered. This perturbation was not obvious from examination of total enterobacterial numbers because they had always fluctuated widely in this individual, but was apparent when the strain composition and antibiotic sensitivity of the enterobacterial population was determined. *Escherichia coli* strains that exhibited multiple resistances (plasmid-encoded) predominated in the faecal samples for a period of about 13 weeks following administration of the antibiotic. The multiple-drug-resistant strains were then no longer detectable but were replaced by a complex community containing ampicillin-resistant *Klebsiella* and antibiotic-susceptible *Enterobacter* and *Serratia* strains. It was striking that antibiotic-resistant strains continued to predominate in the enterobacterial microflora for almost 6 months, even though amoxycillin was administered for a standard treatment period of 7 days.

The origin of the antibiotic-resistant strains could not be determined, but there were several possible sources. First, they may have been subdominant members of the intestinal microflora of Subject 2 prior to antibiotic administration. The method of analysis of bacterial populations using genetic fingerprinting to differentiate between strains, unfortunately, only permits detection of the numerically predominant types. Second, the plasmids encoding the multiple drug resistance were self-transmissible, so they may have been transmitted from an allochthonous organism to indigenous but numerically subdominant *E. coli* strains, which then became dominant under the selective conditions provided by the antibiotic. Third, the antibiotic-resistant strains may all have been allochthonous bacteria that were able to establish in the intestinal tract under the conditions imposed by antibiotic treatment. The persistence of the antibiotic-resistant strains for many weeks was surprising as the amoxycillin was administered for only 7 days and is an antibiotic that is excreted relatively rapidly from the body (Simon, Stille and Wilkinson, 1985). The reasons for the persistence of the antibiotic-resistant strains and for the multiple-drug-resistant strains being superseded by ampicillin-resistant bacteria were not known. It was reassuring to observe that the resistant strains were eventually replaced by a susceptible strain of *E. coli*, albeit one that had not been detected prior to antibiotic adminis-tration. The results of this study showed that even a standard treatment regime administered to a compliant patient resulted in profound changes in the composition of the enterobacterial population of the intestinal tract, including the selection of antibiotic-resistant strains.

Molecular analysis of the genetic elements encoding antibiotic resist-ances in disparate bacterial groups has provided indirect evidence of the spread of resistance determinants among members of the normal micro-flora. There are 12 classes (*tetA–E*, *tetK–M*, *tetO–Q*, *tetX*) of gene encoding resistance to tetracycline (Salyers, Speer and Shoemaker, 1990; Speer, Shoemaker and Salyers, 1992). The *tet* classes have been defined on the basis of DNA–DNA hybridizations. *tetA–E* have been detected in Gram-negative bacteria, whereas *tetK*, *L* and *P* were present in Gram-positive bacteria. *tetM* has been detected in Gram-negative, Gram-positive and cell-wall-free bacteria (Table 15.1), many of which were species compris-ing the normal microflora.

These observations concerning *tetM* provide evidence that transfer of antibiotic resistance determinants occurs among the members of the nor-mal microflora. Since *tetM* has also been detected in pathogenic species of bacteria, exchange of DNA between normal microflora and pathogens is strongly suggested (Morse *et al.*, 1986; Speer, Shoemaker and Salyers, 1992).

Table 15.1 Members of the normal microflora in which *tetM* has been detected (Sources: Courvalin, 1994; Jorgensen *et al.*, 1990; Knapp *et al.*, 1988; Mullany *et al.*, 1990; Roberts and Hillier, 1990; Roberts and Kenny, 1986; Roberts and Moncla, 1988; Roberts *et al.*, 1985; Salyers, 1995)

- *Actinomyces* sp.
- *Bacteroides ureolyticus*
- *Bifidobacterium* sp.
- *Eikenella corrodens*
- *Enterococcus faecalis*
- *Fusobacterium nucleatum*
- *Gardnerella vaginalis*
- *Haemophilus* spp.
- *Kingella denitrificans*
- *Mycoplasma hominis*
- *Neisseria perflava–N. sicca*
- *Peptostreptococcus anaerobius*
- *Staphylococcus* sp.
- *Streptococcus agalactiae*
- *Ureaplasma urealyticum*
- *Veillonella parvula*

At least 13 classes of *erm* genes (erythromycin resistance) encoding coresistance to macrolides (erythromycin), lincosamides and streptogramin B have been described (Table 15.2). A plasmid (pGT633) encoding *ermGT* replicates in a wide range of Gram-positive bacterial hosts in which the resistance gene is also expressed (Table 15.3). Similarly, the broad-host-range conjugative plasmid pAMβ1 transfers, replicates and expresses *ermAM* in several Gram-positive species (Table 15.4). These experimental observations demonstrate the potential for the spread of antibiotic resistance determinants in natural habitats.

Direct evidence of the transfer of *ermAM* between members of the normal microflora was obtained in a study conducted by McConnell, Mercer and Tannock (1991). *Lactobacillus*-free mice were inoculated with a strain of *Lactobacillus reuteri* that harboured plasmid pAMβ1. The strain colonized the digestive tract of the mice. The mice were administered

Table 15.2 Classes of *erm* genes

Class	Reference
A	Murphy, 1985
AM	Martin *et al.*, 1987
B	Horinouchi, Byeon and Weisblum, 1983
C	Horinouchi and Weisblum, 1982
CD	Serworld-Davis and Groman, 1988
D	Gryczan *et al.*, 1984
E	Uchiyama and Weisblum, 1985
F	Rasmussen, Odelson and Macrina, 1986
FS	Smith, 1987
GT	Tannock *et al.*, 1994
M	Lampson and Parisi, 1986
X	Trieu-Cuot, Arthur and Courvalin, 1987
Z	Hachler, Berger-Bachi and Kayer, 1987

Table 15.3 Host range of plasmid pGT633
(Source: Tannock *et al.*, 1994)

- *Lactobacillus reuteri*
- *Lactobacillus gasseri*
- *Lactobacillus fermentum*
- *Lactobacillus salivarius* subsp. *salicinius*
- *Lactobacillus delbrueckii*
- *Enterococcus faecalis*
- *Staphylococcus aureus*
- *Bacillus subtilis*
- *Streptococcus sanguis* (*gordonii*)

Table 15.4 Host range of plasmid pAMβ1

Host	Reference
Enterococcus faecalis	LeBlanc and Lee, 1984
Enterococcus faecium	McConnell, Mercer and Tannock, 1991
Butyrovibrio fibrosolvens	Hespell and Whitehead, 1991
Lactobacillus acidophilus	Vescovo *et al.*, 1983
Lactobacillus plantarum	Shrago, Chassy and Dobrogosz, 1986
Lactobacillus reuteri	Tannock, 1987
Lactobacillus fermentum	Tannock, 1987
Lactobacillus murinus	Tannock, 1987
Lactobacillus casei	Iwata, 1988
Lactobacillus salivarius	Vescovo *et al.*, 1983
Lactococcus lactis	Gasson and Davies, 1980
Bacillus subtilis	Van der Lelie and Venema, 1987
Clostridium acetobutylicum	Oultram and Young, 1985
Staphylococcus aureus	Engel *et al.*, 1980
Streptococcus sanguis	LeBlanc and Hassell, 1976

lincomycin (19 mg/l of drinking water, equivalent to a dietary level of 25 ppm). Plasmid transfer did not occur in the digestive tract of adult mice, but *Enterococcus faecium* harbouring pAMβ1 were detected in the intestinal tract of infant mice from 8 days of age. Transfer occurred in the infant mouse intestine because, in these animals, the enterococci had achieved a population size suitable for conjugation events to occur between potential donor (lactobacilli) and potential recipient (enterococci). *Lactobacillus* numbers were relatively high in the caecum of adult mice (10^{7-8}/g) but enterococcal numbers were low (10^{3-4}). Since the donor efficiency for the *L. reuteri* strain, determined *in vitro*, was poor (about 1×10^{-8}), a large potential recipient population would have been necessary for conjugal transfer of pAMβ1 to have been detected. Transfer of pAMβ1 in the infant mouse intestinal tract was possible because at days 8, 10 or 12 after birth large, approximately equal-sized populations of lactobacilli and enterococci were present. In other experiments, transfer of pAMβ1 from *E. faecium* to *Enterococcus faecalis* and to *Lactobacillus fermentum* was demonstrated in the intestinal tract of adult mice.

Lincomycin was necessary for the maintenance of pAMβ1 in the *L. reuteri* host. Without antibiotic administration, the *Lactobacillus* continued to colonize the mice, but pAMβ1 was not retained by the bacterial cells. Erythromycin-resistant lactobacilli could not usually be detected in the faeces of mice later than 24 days after inoculation of animals to which lincomycin was not administered. In a group of mice to which the antibiotic had been administered for 13 weeks, however, pAMβ1 was retained by the lactobacilli after cessation of antibiotic use. Modification of pAMβ1 in the *L. reuteri* cells inhabiting these animals had occurred, resulting in an increased concentration of plasmid: approximately seven times more plasmid DNA could be extracted from intestinal isolates of *Lactobacillus* compared to the stock strain used to inoculate the mice. The increased *in vivo* stability of the antibiotic-resistant phenotype was presumably due to the increased copy number of molecules of pAMβ1 in the *Lactobacillus* cells: the higher copy number may have ensured that partitioning of pAMβ1 between daughter cells occurred more reliably (better segregational stability) than was the case when plasmid concentration was lower.

Erythromycin-resistant colonies of *E. coli* were detected in cultures from the caecum of a mouse during the course of the study. The increase in resistance compared to that of strains usually encountered in the mice was about fourfold, suggesting that the antibiotic resistance resulted from a mutation rather than by acquisition of the *ermAM* of pAMβ1. DNA–DNA hybridization did not occur between the resistant *E. coli* isolate and a probe composed of *ermAM*. Isolates of erythromycin-resistant *Eubacterium* species were obtained from the caecum of lincomycin-administered mice. These isolates were about 20 000 times more resistant to erythromycin than were similar isolates from mice not administered antibiotic, strongly suggesting that *ermAM* had been transferred to the eubacteria. The presence of *ermAM* in these cells could not, however, be detected by DNA–DNA hybridization. It was shown subsequently (Roberts and Tannock, unpublished) that the erythromycin-resistant eubacteria harboured *ermFS*, first detected in *Bacteroides fragilis* (Smith, 1987). This gene must have been harboured by a small proportion of the eubacteria inhabiting the digestive tract of the mice prior to antibiotic treatment. The administration of lincomycin at subtherapeutic level resulted in the detection of the antibiotic-resistant eubacteria as a result of their replication under the selective conditions provided by the antibiotic. This study provided direct evidence that: (1) transfer of antibiotic resistance encoded by a conjugative plasmid can occur between members of the normal microflora of the digestive tract; (2) modification (evolution) of an R plasmid can occur under conditions of the administration of a subtherapeutic concentration of an antibiotic; (3) selection of a bacterial mutant resistant to an antibiotic can occur in the digestive tract; and (4) selection of an occult bacterial population that harbours a resistance determinant can occur during a period of antibiotic administration.

There is little doubt that intestinal lactobacilli are able gene jockeys in their own right. Evidence for this was provided by a study conducted by Heng and Tannock (unpublished) using recombinant plasmid pAzGKV5 (Fig. 15.1).

This plasmid was harboured by a strain of *L. reuteri* and encoded two antibiotic resistance genes (erythromycin resistance, *ermC*, and

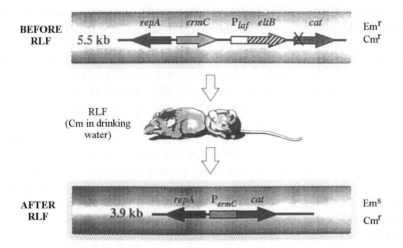

Figure 15.1 Diagrammatic summary of the deletion event that occurred in recombinant plasmid pAzGKV5 while harboured by lactobacilli inhabiting the digestive tract of mice administered chloramphenicol. Key: RLF = lactobacillus-free mice; repA = lactococcal plasmid (pGKV210) replication gene; *ermC* = erythromycin resistance gene; Plaf = lactacin F promoter; *eltB* = LT-B gene; *cat* = promoterless chloramphenicol resistance gene. The *Lactobacillus* strain used to inoculate the mice was slightly resistant to chloramphenicol (MIC 10 µg/ml) as a result of the activity of Plaf. The bacteria harbouring the deleted form of the plasmid were 16 times more resistant to chloramphenicol than was the stock strain (MIC 160 µg/ml).

chloramphenicol resistance, *cat*). *Lactobacillus* cells containing this plasmid were resistant to erythromycin but were only marginally resistant to chloramphenicol because the *cat* gene lacked its promoter. Its expression in this recombinant plasmid was weak because it was controlled by the lactacin F promoter (Fig. 15.1). The recombinant *Lactobacillus* strain was used to inoculate previously lactobacillus-free mice that were administered chloramphenicol in their drinking water. A few days later, lactobacilli that were resistant to chloramphenicol but not to erythromycin were isolated from the faeces of the mice. Analysis of the plasmid harboured by these bacteria showed that the promoter of the erythromycin gene was now fused to the chloramphenicol gene so that the latter was expressed efficiently (Fig. 15.1). Hence the lactobacilli had become, as a result of a deletion (1.6 kb of DNA) and fusion of plasmid DNA, well equipped for life in a milieu containing chloramphenicol.

In a study conducted by Marshall, Petrowski and Levy (1990), the spread of antibiotic-resistant *E. coli* was demonstrated to occur in a farm environment in the absence of antibiotic administration. A spontaneously occurring nalidixic-acid-resistant mutant of *E. coli* isolated from a heifer was transformed to multiple antibiotic resistance using plasmid pSL222–1. The *E. coli* strain was then used to inoculate a bull. The bacterial strain was excreted in the bull's faeces and, within a few days, two farm workers who were involved in husbandry of the bull began to excrete the antibiotic-resistant *E. coli*. The strain was also isolated from inhabitants of the barn in which the bull was housed (mice, pigs, poultry, and flies) at intervals during a period of 70 days. When a pig was dosed with a nalidixic-acid-resistant *E. coli*, mice and flies excreted the bacterial strain during a 20-day period. That the isolates recovered from other animal species were identical to that used to inoculate the bull or pig was confirmed by comparing patterns produced by restriction endonuclease digestion of genomic DNA from the *E. coli* strains, and by analysis of plasmid DNA. Plasmid pSL222–1 was not transferred to members of the normal microflora of any of the animals.

In a similar experiment using chickens, Levy, Fitzgerald and Macone (1976) demonstrated the transmission of an *E. coli* strain that was resistant to multiple antibiotics (including tetracycline) from the inoculated birds to other members of the flock. Transfer to other birds occurred only if the potential recipient chickens were fed tetracycline-containing food. Transmission between flocks held in cages situated about 16 m apart was demonstrated. Recovery of *E. coli* with characteristics compatible with those of the strain used to inoculate the chickens was reported from several human subjects who had contact with the birds. These experiments clearly demonstrated intra-and interspecies transfer of antimicrobial-resistant *E. coli* in the farm environment.

The apparent ease by which antibiotic-resistant strains of bacteria have arisen in response to the widespread use of antibiotics has produced a sense of alarm among medical scientists. The existence of multiple-antibiotic-resistant *M. tuberculosis*, penicillin-resistant *Streptococcus pneumoniae*, methicillin-resistant *Staphylococcus aureus* and vancomycin-resistant enterococci are of particular concern (Bloom, 1992). Indirect and direct evidence shows that the members of the normal microflora have a role in the maintenance and evolution of DNA elements that encode resistance determinants. From an optimistic viewpoint, however, there are clearly limits to the extent to which the spread of antibiotic resistance determinants can occur. Mutations and the expression of resistance genes probably impose a biological cost on the bacterial cell because resistant strains may be less fit for life in a competitive environment in the absence of antibiotic (Gillespie and McHugh, 1997). While broad-host-range conjugative plasmids exist, even they are limited in the extent to which they can spread. Expression of resistance determinants is also limited because codon usage differs between bacterial groups and because specific gene promoters are not universally recognized. Nevertheless, increased vigilance of antibiotic prescribing practices in hospitals, nursing homes and in the community, and curtailment of the availability of antibiotics for use in animal husbandry, would seem to be wise measures if the situation is not to deteriorate. The use of antibiotics with pharmacological properties that would avoid exposure of the normal microflora, especially that of the intestinal tract (most antibiotics are administered by the oral route), to subtherapeutic levels of the drug would doubtless be helpful. Practitioners should remember that not only pathogens are affected by antibiotic administration: the body is home to the normal microflora and appropriate or inappropriate therapy will probably result in exposure of some of the indigenous bacteria to the effects of the drug.

15.1 REFERENCES

Anderson, E. S. (1968) The ecology of transferable drug resistance in the enterobacteria. *Annual Review of Microbiology*, **22**, 131–180.

Bateup, J., Dobbinson, S., McConnell, M. A. *et al.* (1998) Molecular analysis of *Lactobacillus* populations inhabiting the stomach and cecum of pigs. *Microbial Ecology in Health and Disease*, in press.

Bird, H. R. (1969) Biological basis for the use of antibiotics in poultry feeds, in *The Use of Drugs in Animal Feeds*, National Academy of Sciences, Washington, DC, pp. 31–41.

Bloom (1992) Back to a frightening future. *Nature*, **358**, 538–539.

Courvalin, P. (1994) Transfer of antibiotic resistance genes between gram-positive and gram-negative bacteria. *Antimicrobial Agents and Chemotherapy*, **38**, 1447–1451.

Elam, J. F., Jacobs, R. L., Fowler, J. and Couch, J. R. (1954) Effect of dietary clostridia upon growth-promoting responses of penicillin. *Proceedings of the Society for Experimental Biology and Medicine*, **85**, 645–648.

Ellner, J. J., Hinmans, A. R., Dooley, S. W., Fischl, M. A. *et al.* (1993) Tuberculosis symposium: emerging problems and promise. *Journal of Infectious Diseases*, **168**, 537–551.

Engel, H. W. B., Soedirman, N., Rost, J. A. *et al.* (1980) Transferability of macrolide, lincomycin, and streptogramin resistances between group A, B, and D streptococci, *Streptococcus pneumoniae*, and *Staphylococcus aureus*. *Journal of Bacteriology*, **142**, 407–443.

Eyssen, H. and de Somer, P. (1967) Effects of *Streptococcus faecalis* and a filterable agent on growth and absorption in gnotobiotic chicks. *Poultry Science*, **46**, 323–333.

Feighner, S. D. and Dashkevicz, M. P. (1987) Subtherapeutic levels of antibiotics in poultry feeds and their effects on weight gain, feed efficiency, and bacterial cholyltaurine hydrolase activity. *Applied and Environmental Microbiology*, **53**, 331–336.

Feighner, S. D. and Dashkevicz, M. P. (1988) Effect of dietary carbohydrates on bacterial cholyltaurine hydrolase in poultry intestinal homogenates. *Applied and Environmental Microbiology*, **54**, 337–342.

Fuller, R., Coates, M. E. and Harrison, G. F. (1979) The influence of specific bacteria and a filterable agent on the growth of gnotobiotic chicks. *Journal of Applied Bacteriology*, **46**, 335–342.

Gasson, M. J. and Davies, F. L. (1980) Conjugal transfer of the drug resistant plasmid pAMβ1 in the lactic streptococci. *FEMS Microbiology Letters*, **7**, 51–53.

Gillespie, S. H. and McHugh, T. D. (1997) The biological cost of antimicrobial resistance. *Trends in Microbiology*, **5**, 337–339.

Gryczan, T., Israeli-Reches, M., Del Blue, M. and Dubnau, B. (1984) DNA sequence and regulation of *ermD*, a macrolide–lincosamide–streptogramin B resistance element from *Bacillus licheniformis*. *Molecular and General Genetics*, **194**, 349–356.

Hachler, H., Berger-Bachi, B. and Kayer, F. H. (1987) Genetic characterization of a *Clostridium difficile* erythromycin–clindamycin resistance determinant that is transferrable to *Staphylococcus aureus*. *Antimicrobial Agents and Chemotherapy*, **31**, 1039–1045.

Havelaar, A. H. (1993) Bacteriophages as models of human enteric viruses in the environment. *American Society for Microbiology News*, **58**, 614–619.

Hespell, R. B. and Whitehead, T. R. (1991) Conjugal transfer of Tn916, Tn916E, and pAMβ1 from *Enterococcus faecalis* to *Butyrivibrio fibrisolvens* strains. *Applied and Environmental Microbiology*, **57**, 2703– 2709.

Holmberg, S. D., Wells, J. G. and Cohen, M. L. (1984) Animal-to-man transmission of antimicrobial-resistant *Salmonella*: investigation of US outbreaks, 1971–1983. *Science*, **225**, 833–835.

Holmberg, S. D., Osterholm, M. T., Senger, K. A. and Cohen, M. L. (1984) Drug-resistant *Salmonella* from animals fed antimicrobials. *New England Journal of Medicine*, **311**, 617–622.

Horinouchi, S. and Weisblum, B. (1982) Nucleotide sequence and functional map of pE194, a plasmid that specifies inducible resistance to macrolide, lincosa-mide, and streptogramin B type antibiotics. *Journal of Bacteriology*, **150**, 804–814.

Horinouchi, S., Byeon, W.-H. and Weisblum, B. (1983) A complex attenuator regulates inducible resistance to macrolides, lincosamides, and streptogramin type B antibiotics in *Streptococcus sanguis*. *Journal of Bacteriology*, **154**, 1252–1262.

Houghton, S. B., Fuller, R. and Coates, M. E. (1981) Correlation of growth depres-sion of chicks with the presence of *Streptococcus faecium* in the gut. *Journal of Applied Bacteriology*, **51**, 113–120.

Iwata, M. (1988) Characterization of a pAMβ1 deletion derivative isolated from *Lactobacillus casei* after conjugation. *Biochimie*, **70**, 553–558.

Jorgensen, J. H., Doern, G. V., Maher, L. A. *et al.* (1990) Antimicrobial resistance among respiratory isolates of *Haemophilus influenzae*, *Moraxella catarrhalis*, and *Streptococcus pneumoniae* in the United States. *Antimicrobial Agents and Che-motherapy*, **34**, 2075–2080.

Kaufmann, S. H. E. and van Embden, J. D. A. (1993) Tuberculosis: a neglected disease strikes back. *Trends in Microbiology*, **1**, 2–5.

Klieve, A. V. and Bauchop, T. (1988) Morphological diversity of ruminal bacter-iophages from sheep and cattle. *Applied and Environmental Microbiology*, **54**, 1637–1641.

Knapp, J. S., Johnson, S. R., Zenilman, J. M. *et al.* (1988) High-level tetracycline resistance resulting from *tetM* in strains of *Neisseria* spp., *Kingella denitrificans*, and *Eikenella corrodens*. *Antimicrobial Agents and Chemotherapy*, **32**, 765–767.

Lampson, B. C. and Parisi, J. T. (1986) Nucleotide sequence of the constitutive macrolide–lincosamide–streptogramin B resistance plasmid pNE131 from *Sta-phylococcus epidermidis* and homologies with *Staphylococcus aureus* plasmids pE194 and pSN2. *Journal of Bacteriology*, **167**, 888–892.

LeBlanc, D. J. and Hassell, F. P. (1976) Transformation of *Streptococcus sanguis* Challis by plasmid deoxyribonucleic acid from *Streptococcus faecalis*. *Journal of Bacteriology*, **128**, 347–355.

LeBlanc, D. J. and Lee, L. N. (1984) Physical and genetic analysis of streptococcal plasmid pAMβ1 and cloning of its replication region. *Journal of Bacteriology*, **157**, 445–453.

Levy, S. B. (1992) *The Antibiotic Paradox. How Miracle Drugs are Destroying the Miracle*, Plenum Press, New York.

Levy, S. B., Fitzgerald, G. B. and Macone, A. B. (1976) Spread of antibiotic-resistance plasmids from chicken and from chicken to man. *Nature*, **260**, 40–42.

Linton, A. H. (1976) Antibiotics, animals and man – an appraisal of a contentious subject, in *Antibiotics and Antibiosis in Agriculture*, (ed. M. Woodbine), Butterworths, London, pp. 315–343.

McConnell, M. A., Mercer, A. A. and Tannock, G. W. (1991) Transfer of plasmid pAMβ1 between members of the normal microflora inhabiting the murine digestive tract and modification of the plasmid in a *Lactobacillus reuteri* host. *Microbial Ecology in Health and Disease*, **4**, 343–355.

McDonald, L. C., Kuehnert, M. J., Tenover, F. C. and Jarvis, W. R. (1997) Vancomycin-resistant enterococci outside the health-care setting: prevalence, sources, and public health implications. *Emerging Infectious Diseases*, **3**, 311– 317.

Marshall, B., Petrowski, D. and Levy, S. B. (1990) Inter- and intraspecies spread of *Escherichia coli* in a farm environment in the absence of antibiotic usage. *Proceedings of the National Academy of Sciences of the USA*, **87**, 6609–6613.

Martin, B., Alloing, G., Mejean, V. and Claverys, J-P. (1987) Constitutive expression of erythromycin resistance mediated by the *ermAM* determinant of plasmid pAMβ1 results from deletion of 5′ leader peptide sequences. *Plasmid*, **18**, 250–253.

Morse, S. A., Johnson, S. R., Biddle, J. W. and Roberts, M. C. (1986) High-level tetracycline resistance in *Neisseria gonorrhoeae* is result of acquisition of streptococcal *tetM* determinant. *Antimicrobial Agents and Chemotherapy*, **30**, 664–670.

Mullany, P., Wilks, M., Lamb, I. *et al.* (1990) Genetic analysis of a tetracycline resistance element from *Clostridium difficile* and its conjugal transfer to and from *Bacillus subtilis*. *Journal of General Microbiology*, **136**, 1343–1349.

Murphy, E. (1985) Nucleotide sequence *ermA*, a macrolide–lincosamide–streptogramin B determinant in *Staphylococcus aureus*. *Journal of Bacteriology*, **162**, 633–640.

Oultram, J. D. and Young, M. (1985) Conjugal transfer of plasmid pAMβ1 from *Streptococcus lactis* and *Bacillus subtilis* to *Clostridium acetobutylicum*. *FEMS Microbiology Letters*, **27**, 129–134.

Rasmussen, J. L., Odelson, D. A. and Macrina, F. L. (1986) Complete nucleotide sequence and transcription of *ermF*, a macrolide–lincosamide–streptogramin B resistance determinant from *Bacteroides fragilis*. *Journal of Bacteriology*, **168**, 523–533.

Raya, R. R. and Klaenhammer, T. R. (1992) High-frequency transduction by *Lactobacillus gasseri* bacteriophage φadh. *Applied and Environmental Microbiology*, **58**, 187–193.

Roberts, M. C. and Hillier, S. L. (1990) Genetic basis of tetracycline resistance in urogenital bacteria. *Antimicrobial Agents and Chemotherapy*, **34**, 261–264.

Roberts, M. C. and Kenny, G. E. (1986) Dissemination of the *tetM* tetracycline resistance determinant to *Ureaplasma urealyticum*. *Antimicrobial Agents and Chemotherapy*, **29**, 350–352.

Roberts, M. C. and Moncla, B. (1988) Tetracycline resistance and *tetM* in oral anaerobic bacteria and *Neisseria perflava–N. sicca*. *Antimicrobial Agents and Chemotherapy*, **32**, 1271–1273.

Roberts, M. C., Koutsky, L. A., Holmes, K. K. *et al.* (1985) Tetracycline-resistant *Mycoplasma hominis* strains contain streptococcal *tetM* sequences. *Antimicrobial Agents and Chemotherapy*, **28**, 141–143.

Salyers, A. A. (1995) *Antibiotic Resistance Transfer in the Mammalian Intestinal Tract: Implications for Human Health, Food Safety and Biotechnology*, Molecular Biology Intelligence Unit, R. G. Landes Company, Austin, TX.

Salyers, A. A., Speer, B. S. and Shoemaker, N. B. (1990) New perspectives in tetracycline resistance. *Molecular Microbiology*, **4**, 151–156.

Serworld-Davis, T. M. and Groman, M. B. (1988) Identification of a methylase gene for erythromycin resistance within the sequence of a spontaneously deleting fragment of *Corynebacterium diphtheriae* plasmid pNG2. *FEMS Microbiology Letters*, **56**, 7–14.

Shrago, A. W., Chassy, B. M. and Dobrogosz, W. J. (1986) Conjugal plasmid transfer (pAMβ1) in *Lactobacillus plantarum*. *Applied and Environmental Microbiology*, **52**, 574–576.

Simon, C., Stille, W. and Wilkinson, P. J. (1985). *Antibiotic Therapy in Clinical Practice*, F. K. Schattauer Verlag, Stuttgart.

Smith, C. J. (1987) Nucleotide sequence analysis of Tn4551: use of *ermFS* operon fusions to detect promoter activity in *Bacteroides fragilis*. *Journal of Bacteriology*, **169**, 4589–4596.

Speer, B. S., Shoemaker, N. B. and Salyers, A. A. (1992) Bacterial resistance to tetracycline: mechanisms, transfer, and clinical significance. *Clinical Microbiology Reviews*, **5**, 387–399.

Spika, J. S., Waterman, S. H., Soo Hoo, G. W. *et al.* (1987) Chloramphenicol-resistant *Salmonella newport* traced through hamburger to dairy farms. *New England Journal of Medicine*, **316**, 566–570.

Swartz, M. N. (chairman) (1989) *Report of a Study: Human Health Risks with the Subtherapeutic Use of Penicillin or Tetracyclines in Animal Feed*, National Academy Press, Washington, DC.

Tannock, G. W. (1987) Conjugal transfer of plasmid pAMβ1 in *Lactobacillus reuteri* and between lactobacilli and *Enterococcus faecalis*. *Applied and Environmental Microbiology*, **53**, 2693–2695.

Tannock, G. W., Luchansky, J. B., Miller, L. *et al.* (1994) Molecular characterization of a plasmid-borne (pGT633) erythromycin resistance determinant (*ermGT*) from *Lactobacillus reuteri* 100-63. *Plasmid*, **31**, 60–71.

Trieu-Cuot, P., Arthur, M. and Courvalin, P. (1987) Origin, evolution and dissemination of antibiotic resistance genes. *Microbiological Sciences*, **4**, 263–266.

Uchiyama, H. and Weisblum, B. (1985) N-methyl transferase of *Streptomyces erythraeus* that confers resistance to the macrolide–lincosamide–streptogramin B antibiotics: amino acid sequence and its homology to cognate R-factor enzymes from pathogenic bacilli and cocci. *Genetics*, **38**, 103–110.

Uttley, A. H. C., Collins, C. H., Naidoo, J. and George, R. C. (1988) Vancomycin-resistant enterococci. *Lancet*, **i**, 57–58.

Van der Lelie, E. and Venema, G. (1987) *Bacillus subtilis* generates a major specific deletion in pAMβ1. *Applied and Environmental Microbiology*, **53**, 2458–2463.

Vescovo, M., Morelli, L., Bottazzi, V. and Gasson, M. J. (1983) Conjugal transfer of broad-host-range plasmid pAMβ1 into enteric species of lactic acid bacteria. *Applied and Environmental Microbiology*, **46**, 753–755.

16

The development of microbial biofilms on medical prostheses

Gregor Reid

16.1 INTRODUCTION

Each year in many countries around the world more and more medical prostheses are used in clinical practice and in the management of body functions (Reid, 1994). For example, in the specialty of urology alone, devices such as incontinence pads, catheters, ureteral stents, drainage lines, penile prostheses, prostatic stents and combinations with tissues such as bowel for bladder replacement are used in patient care (Light, Lapin and Vohra, 1995; Reid *et al.*, 1995b; Reid, Tieszer and Bailey, 1995; Herschorn and Ordorica, 1995; Wilson and Delk, 1995). The enormity of the field is illustrated by the fact that over 25 million surgical procedures carried out in the USA alone each year, require the permanent or temporary use of biomaterials. Many examples of biofilm-associated infections on medical devices can be found, such as those in central venous catheters (Kowalewska-Grochowska *et al.*, 1991; Elliott and Faroqui, 1992), a total artificial heart (Jarvik, 1981), contact lenses and intraocular devices (Elder *et al.*, 1995), voice prostheses (Neu *et al.*, 1992) as well as a vast number of prosthetic implants, reviewed in full elsewhere (Dankert, Hogt and Feijen, 1986).

Interestingly, this extensive usage of medical devices by patients in developed countries leads to what might be regarded as relatively few complications or side-effects overall. Although a lack of biocompatability can lead to problems, e.g. due to toxins released when a material decomposes, it is microbial infection, often fatal, that is regarded as the main

G. W. Tannock (ed.), *Medical Importance of the Normal Microflora*, 405–422.
© 1999 *Kluwer Academic Publishers. Printed in Great Britain.*

impediment to further use of medical devices. Moreover, it is the formation of organisms into microcolonies termed 'biofilms' that represents the key component in disease causation. Such biofilms were first described in the mammalian intestine (Savage, 1977), but their clinical importance has been most often associated with the use of medical devices.

16.2 THE STAGES OF BIOFILM FORMATION

A microbial biofilm has been defined as an accumulation of microorganisms and their extracellular products to form a structured community on a surface (Costerton *et al.*, 1987). More recently, the term has been broadened to include biofilms some distance away from a surface and those that exist in dense as well as single layers of cells (Reid, Tieszer and Bailey, 1995).

The question might be asked: 'Why do organisms form biofilms?' There appear to be many reasons, including better access to nutrients, faster mass transfer due to shorter diffusion distances, metabolic synergism and formation of cooperative consorts, a means to resist host and chemical attack, and a method of avoiding being flushed from a microniche (Geesey, Stupy and Bremer, 1992). As an experimental example, Fletcher (1986) showed that attached organisms assimilated two to five times more glucose than free-living cells, and the surface-associated cells respired at a greater rate.

The formation of an infectious biofilm on biomaterials consists of several sequential steps, and includes:

- adsorption of host conditioning film on to the device;
- deposition of the infectious microorganisms;
- adhesion of the organisms;
- anchoring by exopolymer production;
- growth and spread of the organisms.

Immediately after insertion of a device into the body, the material surface comes into contact with body fluids such as blood, saliva or urine. Macromolecular components from these body fluids adsorb to form a conditioning film, which can then act as the substrate to which approaching organisms attach. The actual film need not necessarily be uniform over the whole device at the time the first organism arrives. Thus, microorganisms may adhere to relatively unchanged polymers on a bare biomaterial surface, or via components of the conditioning film, such as proteins.

The deposition of conditioning films has been demonstrated on orthopedic devices (Gristina, 1987), urinary catheters (Reid *et al.*, 1992b) and ureteral stents (Reid, Davidson and Denstedt, 1994). The film composition, as one would expect, varies from patient to patient, and to some extent from device to device. Studies using X-ray photoelectron spectroscopy and energy dispersive X-ray analysis, have demonstrated adsorption to urinary devices from urine of elements including nitrogen, carbon, oxygen, calcium, sodium and phosphorus (Fig. 16.1). These elements can, in time, cause dense encrustation deposits to adsorb on to the prosthesis (Fig. 16.2).

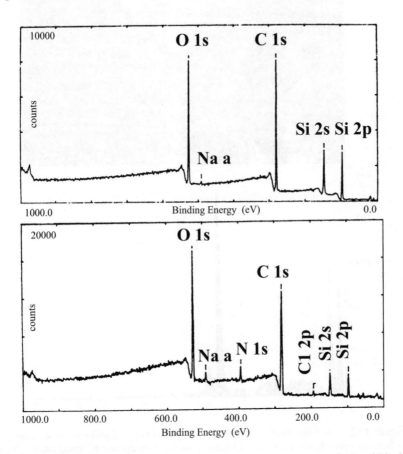

Figure 16.1 X-ray photoelectron spectroscopy broad-spectrum analysis of Black Beauty unused stent (top) and one that had been inserted into a patient for 12 days (bottom). The deposition of oxygen, nitrogen and chlorine, and the reduced peaks of silicone (due to overlay of conditioning film covering detection of silicone) demonstrate conditioning film deposition. Source: reproduced from *Colloids and Surfactants B: Biointerfaces*, **5**, Reid, G., Tieszer, C., Denstedt, J. and Kingston, D. Examination of bacterial and encrustation deposition on ureteral stents of differing surface properties, after indwelling in humans, pp. 171–179, 1995, with kind permission of Elsevier Science – NL, Sara Burgerhartstraat 25, 1055 KV Amsterdam, The Netherlands.

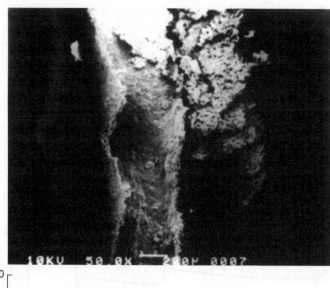

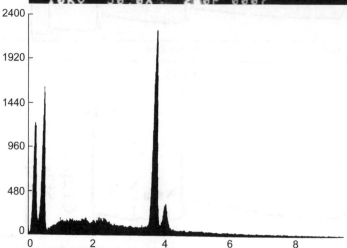

Figure 16.2 Scanning electron micrograph and energy dispersive X-ray analysis of SofFlex stent from a patient. Extensive encrustations were found on the inner portion of the stent. The analysis showed the deposits to be calcium oxalate. Source: reproduced from *Colloids and Surfactants B: Biointerfaces*, **5**, Reid, G., Tieszer, C., Denstedt, J. and Kingston, D. Examination of bacterial and encrustation deposition on ureteral stents of differing surface properties, after indwelling in humans, pp. 171–179, 1995, with kind permission of Elsevier Science – NL, Sara Burgerhartstraat 25, 1055 KV Amsterdam, The Netherlands.

Many examples exist of protein adsorption to substrata. It has been shown that within minutes, proteins such as serum albumin, fibrinogen, collagen and fibronectin absorb to polymer surfaces and influence wettability, cell-spreading (Schakenraad *et al.*, 1989; Busscher, Stokroos and Schakenraad, 1991) and subsequent bacterial adhesion (Wadström, 1989; Brokke *et al.*, 1991; Chamberlain, 1992).

The importance of the conditioning film and the initially adhering microorganisms are only now being realized. Most investigations have concentrated upon the subsequent growth of the organisms, their biofilm formation and the clinical disease end-point. However, the first link in the chain of events leading to the formation of a mature biofilm is the anchor from which further events take place. This bond represents the link with the growing biofilm. If it was possible to break this link, e.g. under the influence of fluctuating shear forces or using electric current or enzymes, it could be feasible for the entire biofilm to detach, thereby aiding the eradication of infection.

Given the importance of the conditioning film and the initially adhering organisms, it has been proposed (Busscher, Bos and van der Mei, 1995; Reid, van der Mei and Busscher, 1998) that three parts of a biofilm should be distinguished (Plate 1): the linking film, the base film and the surface film.

In situ techniques, such as Fourier transform infrared spectroscopy (FTIR) and confocal scanning laser microscopy (CSLM), have indicated that the surface film is pivotal to the transport of nutrients and the interference with transport of antimicrobial substances (Nivens *et al.*, 1993; Caldwell, 1995).

16.3 PHYSICOCHEMICAL FACTORS AND STRUCTURE OF BIOFILMS

The movement of organic material present in aqueous solution from the bulk phase to surfaces is primarily due to rapid (within 15 minutes – Bryers, 1987) molecular diffusion (Koch, 1991). The resultant film, containing proteins, glycoproteins, electrolytes and perhaps carbohydrates, provides reactive binding sites to which additional solutes, cell debris and bacterial adhesins can bind. The bare surface thus acquires a net negative charge, and its surface free energy and other specific properties (e.g. hydrogel coat) tend to become 'neutralized' or at least altered (Korber *et al.*, 1995). It has been shown that bacterial appendages can extend through the free energy barrier and connect with the organic matrix to form reversible short-range electrostatic, covalent or hydrogen bonds (Plate 2). In *Staphylococcus epidermidis* a polysaccharide-based hemagglutinin has been associated with adhesion and biofilm formation on devices (Rupp *et al.*, 1995).

Many *in vitro* studies have demonstrated that pathogen adhesion to surfaces involves a complicated interplay of electrostatic interactions, hydrophobicity, van der Waals forces, the presence or absence of structural features on the microbial cell surface and many more factors.

The physicochemical characteristics of medical device surfaces are more commonly being designed to resist encrustation, coagulation and bacterial adhesion. At relatively short distances from the surface ($\leqslant 1$–10 nm), attractive and repulsive forces (DVLO theory: London van der Waals forces) influence adhesion (van Loosdrecht, Norde and Zehnder, 1987; Absolom *et al.*, 1983). The theory defines a primary minimum where small distances ($\leqslant 1$ nm) separate the substratum and approaching bacterium, and a secondary minimum (5–10 nm) where larger distances separate two surfaces (Characklis, Turakhia and Zelver, 1990). The DVLO theory also predicts that when a bacterial radius decreases, the repulsive electrostatic energy barrier decreases (Sjollema *et al.*, 1990). This might partly explain why Gram-positive coccal human pathogens, which have a small radius, have a propensity to adhere to medical devices.

The critical surface tension of the substratum, suspending fluid and bacteria all play a role in determining whether or not an organism is likely to adhere (Absolom *et al.*, 1983; Reid *et al.*, 1991; Lappin-Scott and Costerton, 1989). In this way, if the total free energy of a system (ΔF^{adh}) is reduced by bacterial cell contact with a surface, then adhesion will occur (Fletcher and Marshall, 1982; Absolom, 1988; Plate 3). In static systems, such as the interface between the skin and an intravenous line, other factors such as Brownian motion, gravity and cell motility help organisms overcome electrostatic repulsive forces to move from the bulk phase to retention on surfaces.

While thermodynamic studies indicate an ability of hydrophobic organisms to adhere better to hydrophobic surfaces (Absolom *et al.*, 1983), this does not appear to be the case for experiments done in the presence of physiologically relevant suspending fluids (Reid *et al.*, 1991, 1993a). The exception could be the oral cavity, where hydrophobic materials have been reported to attract fewer biofilms (Quirynen *et al.*, 1994). In relation to uropathogens, however, studies have shown that the possession of various fimbriae by uropathogenic *Escherichia coli* does not alter whole cell hydrophilicity nor adhesion ability to materials (Reid *et al.*, 1996). This is not to say that hydrophobic components on the cell are not involved in adhesion, but rather that their role is negligible for biomaterials.

Individual uropathogens are capable of adhering *in vitro* to urological devices and subsequently forming biofilms. *In vivo* data on ureteral stents supports this adhesion capacity (Reid *et al.*, 1992a). However, the situation becomes a little more complex when experiments are performed in the presence of multiple species. In this scenario, the role of cell and material surface hydrophobicity is even less clear, as there is presumably competition among the organisms for space to adhere. The net result can be the reduction in adhesion to a given surface for one or more of the organisms being tested (Reid and Tieszer, 1993, 1995; Reid *et al.*, 1995a).

Diffusion within biofilms is important for microbial viability. It is caused by the random, thermally energetic movement of particles from regions of high to low concentration, and it is a process that drives motion (Koch, 1991). Once the biofilm forms, encased in extracellular glycocalyx (composed of lipopolysaccharides, proteins, polysaccharides) matrix, diffusion governs growth and mass transport, with solute flux being a constant proportional to migration distance. The biofilm thickness is influenced by factors such as heat transfer efficiency, fluid frictional resistance and transfer of nutrients such as oxygen (Christensen and Characklis, 1990).

Bacteria appear able to transfer genetic information within biofilms, and communicate with each other and their external environment. For example, plasmid transfer from *E. coli* to Gram-positive bacteria has been demonstrated (Trieu-Cuot *et al.*, 1987), implying that this could occur within biofilms (Plate 4) as shown to occur in a marine biofilm system (Angles, Marshall and Goodman, 1993). Other plasmid transfer reactions among Gram-positive organisms have been reported (Schrago, Chassy and Dobrogosz, 1986; Tannock, 1987).

Signal transduction is involved in communication with the external environment. There are two families of proteins, the so-called signal transduction proteins, whereby one senses the environment and the second responds to this message by relaying the information to chromosomal elements (Stock, Stock and Mottonen, 1990; Goodman and Marshall, 1995). It is thought that *E. coli* may have over 50 pairs of these regulatory proteins, each responsible for a specific reaction to a stimulus. In this way, pathogens can coordinate expression of their virulence determinants (Miller, Mekalanos and Falkow, 1989).

16.4 THE NATURE OF INFECTIONS AND PREDISPOSING FACTORS

Implanted medical devices, such as heart valves and artificial veins and joints, are especially vulnerable to microbial biofilm formation (Gristina, 1987). These surfaces are relatively unprotected by host defenses and provide a focal point from which infection arises. Indeed, closed implants are more frequently associated with life-threatening situations, with *Staphylococcus epidermidis* and *Staphylococcus aureus* being the chief culprits (Christensen *et al.*, 1989). From a bacteriology perspective, it is intriguing that a member of the normal skin flora (*S. epidermidis*) could become so pathogenic to the host. Indeed, for many years, the finding of coagulase-negative staphylococci in urine and prostate specimens was viewed as being a normal microflora contaminant. It has been suggested that humoral host defense alterations and functional failure of polymorphs in the vicinity of an implant can lead to *S. epidermidis* infections (Vaudaux, Lew and Waldvogel, 1989).

It should be pointed out that infection need not always arise, even in the presence of bacterial biofilms. Two examples illustrate this finding, namely bacterial colonization of intrauterine devices (Reid *et al.*, 1988), and Gram-positive coccal colonization of peritoneal dialysis catheters and exit site polymers (Dasgupta *et al.*, 1987). The explanation for these organisms not infecting the host remains unclear. However, it appears that some normal microflora constituents are actually potential pathogens, depending upon their access to internal host sites, availability of nutrients and degree of host defense response to their presence.

Other host factors can predispose a patient to infection. Penile prostheses have been implanted successfully in clinical practice (Herschorn and Ordorica, 1995); however infection rates, ranging from 8% in non-diabetics to 18% in diabetics, have been reported in larger studies (Wilson and Delk, 1995). It would seem likely that advances in polymer technology could enhance the performance of penile prostheses, but infection by organisms in biofilms will need a fairly radical approach.

Three devices open to the external environment of the patient are commonly used in urological practice: peritoneal dialysis catheters, urethral or suprapubic urinary catheters and nephrostomy tubes. The open nature of these devices means that the patients are exposed to a high risk of infection. Indeed, in urethral catheterized patients, urinary tract infection is inevitable without antimicrobial coverage and patients become systemically ill with bacteremia and sepsis. In these open systems, access to the prosthesis is less of a barrier for the organisms.

For *E. coli* biofilms, cell growth alters when nutrients are lacking, in that cells grow shorter in length, surrounded by thicker extracellular matrix (Dewanti and Wong, 1995). Furthermore, their adhesive ability varies depending upon the surface energy of the substratum (Rittle *et al.*, 1990). In *S. epidermidis* a specific polysaccharide antigen functions to accumulate and stabilize the biofilm (Mack *et al.*, 1996).

Intact biofilms have been examined from devices recovered from critically ill patients (Inglis *et al.*, 1995). It was discovered that often more than one species of bacteria was present per site, and many neutrophil polymorphonuclear cells were dispersed throughout. This implies an integral part for host inflammatory cells in certain biofilms.

Polymorphonuclear neutrophils (PMN) are prominent in chronic biofilm-associated infections, but their oxidation burst response is almost half that obtained against planktonic cells for opsonized and non-opsonized bacteria (Jensen *et al.*, 1990, 1992). Although oxygen radicals are released by the macrophages and polymorphs, they do not appear able to effectively attack the biofilm organisms (Plate 5). Complement activation, via peptidoglycan in Gram-positives and lipopolysaccharides in Gram-negatives, in combination with antibody responses and immune complex formation can worsen chronic biofilm-associated infection (Berger, 1991). The subsequent destruction of tissues and repeated failure of antibiotics to clear the infection nidus, makes it imperative that immune responses to biofilms be better understood for future therapeutics to be effective (Hoiby *et al.*, 1995).

In the urinary tract, voiding mechanisms, including removal of sloughed uroepithelial cells, normally constitute a means to prevent infection, along with cellular and humoral immune responses (Hopkins, Uehling and Balish, 1987) and possibly a glycosaminoglycan layer on the tissue mucosa (McLean, Nickel and Olson, 1995). However, when residual urine, calculi or polymers such as a catheter are present, the patient becomes significantly more likely to develop infection.

16.5 PREVENTION AND TREATMENT OF BIOFILMS

The development of materials which resist bacterial colonization is a continuous process in polymer science. Various surface and device modifications have been made, e.g. the creation of hydrophilic outer layers, antimicrobial coated surfaces, low surface energy and carbon rich materials, highly biocompatible substances, biodegradable materials and cell or protein grafted surfaces (Ratner, 1980; Reid, van der Mei and Busscher, 1998; Strachan, 1995). However, clinically, success has been scant and short-term at best. Organisms have been shown to survive even in the presence of daily dosing with antibiotics (Fig. 16.3).

In urology, many different materials are being developed, to prevent vesicoureteral reflux (e.g. Teflon, collagen, macroplastique; Schulman, 1995), to treat urinary incontinence (e.g. bioglass, collagen, autologous fat, Teflon; Williams, 1995a), and to manage prostate disease such as hypertrophy and strictures (various materials; Williams, 1995b). Evidence to date indicates that infections do not arise in large numbers, indicating that progress has been made in the field. The search for materials that are completely resistant to microorganisms will continue, but microorganisms will still prove a difficult challenge to overcome.

In order for a device to work effectively in a biocompatible way with the host, the material must not be carcinogenic, cytotoxic or overly immunogenic or allergenic, and it should resist microbial colonization. The recognition that polymer biocompatability is important in integration of a material into the host's tissues and in avoiding attack by host defenses has led to the testing of mesh-like materials through which host cells can grow. Another procedure receiving interest is the combined use of a prosthesis with host tissue, such as bowel, for bladder reconstruction (Light, Lapin and Vohra, 1995). The development of such techniques is needed when one considers that over 3000 cystectomies are carried out each year in the US alone, and high (up to 50%) infection rates have been reported.

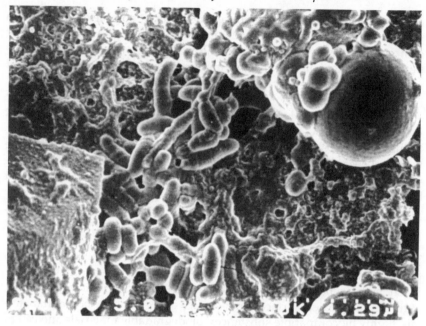

Figure 16.3 Scanning electron micrograph showing the incorporation of bacteria deep within the crystal structure on the luminal surface of a SofFlex stent from a patient, 10 days after insertion. Antibiotic therapy with trimethoprim did not kill the organisms, and sonication for 5 minutes failed to dislodge them from the encrustations.

The rationale for antimicrobial coatings is to provide a period of protection against infection, which either allows the patient to recover from an illness and no longer require the device (e.g. urinary catheter in postsurgical patient for 5 days) or allows the biomaterial to integrate with the host tissue while a wound heals (e.g. hip joint). Certain antimicrobial agents, such as silver and fluoroquinolones, have produced promising results with respect to reducing the ability of pathogens to colonize the device and potentially infect the host (Johnson *et al.*, 1990; Ozaki *et al.*, 1993; Reid *et al.*, 1993b; Reid *et al.*, 1994; Preston *et al.*, 1996). Nevertheless, even fluoroquinolones may have difficulty eradicating certain highly adherent, mucus-encased organisms such as *Pseudomonas aeruginosa* (Evans *et al.*, 1991; Gilbert and Brown, 1995; Plate 6).

One hypothesis that is being explored in my laboratory is that oral intake of antibiotics leads to adsorption of the drug on to a device surface, thereby enhancing prevention against biofilm formation. Results with 12 devices to date support this theory and show that differences probably exist between device types and between antibiotics (unpublished data). Further studies are under way.

In considering susceptibility, it must be appreciated that there are different types of biofilms: namely defined as 'young' (<2 days old) and 'old' ($\geqslant 7$ days old) biofilms (Anwar and Strap, 1992). Studies have shown that young biofilms can be eradicated by antibiotics such as $10\,\mu g$ tobramycin with $200\,\mu g$ piperacillin, whereas old biofilms were completely resistant (Anwar and Costerton, 1990). This 'age' factor should be considered when analysing new antimicrobial agents for the ability to eradicate biofilms.

The clinical problem of antibiotic failure against biofilms is potentially very serious. For example, *S. epidermidis* is a common infectant in medical-device-associated disease and vancomycin ($250\,mg$ i.v. every $12\,h$) is recommended. However, this approach (often given for several weeks) leads to the very real possibility of emergence of Gram-positive cocci resistant to vancomycin and every available antibiotic. In a hospital setting the emergence of such bacteria could be devastating. Already, in hospitals in London, Ontario, we have had cases of illness due to multi-drug-resistant pathogens.

There is some data to suggest that combination antimicrobial therapy can prevent colonization by slime-forming staphylococci, Gram-negative organisms and yeasts (Anglen *et al.*, 1994; Raad *et al.*, 1995; Teichman *et al.*, 1994; Yasuda *et al.*, 1994). For prostatitis, effective eradication may require new antibiotic classes and possibly use of alternative methods of administration, such as direct injection into the prostate, urethra or anal mucosa (Nickel, 1996).

With respect to treatment of a device-associated infection, antimicrobial therapy is invariably ineffective, and devices have to be removed to increase the chances of curing the patient (Khardori and Yassien, 1995). However, there are some data to indicate that infection can be repressed and replacement devices reimplanted successfully (Parsons, 1996). Of course, device replacement is not necessarily simple: urinary catheters can be replaced easily while removal of peritoneal dialysis catheters and penile implants requires invasive surgery. Therefore, there is a financial cost as well as a morbidity factor for the patient.

Various theories have been proposed to explain the mechanisms of bacterial biofilm resistance to antibiotic and chemical treatments. The nature of the bacterial glycocalyx as an array of fine fibers in a thick, hydrated and polyanionic matrix influences access of antibiotics to bacterial surfaces (Plate 1; Costerton, Irwin and Cheng, 1981). It is now appreciated that the type and concentration of antibiotic, the binding capacity of the glycocalyx towards it, the bacterial growth rate and environmental factors such as pH influence this interaction (Gilbert and Brown, 1995).

Nutrient limitation can arise at the surface, base and linking regions of a biofilm. In time, as the organisms in the linkage area consume nutrients, so the supply diminishes, waste products are produced and physiological growth rate is reduced. Changes also occur in the cell wall and envelope, in porin protein composition, lipopolysaccharide and cation content. These alterations result in different phenotypes being present within the biofilm, leading to increased or decreased susceptibility to antimicrobial agents, depending upon the nature of the agents themselves (Gilbert and Brown, 1995). Changes in bacterial surface porins have been shown to alter sensitivity of *P. aeruginosa* to gentamicin, polymyxin and EDTA (Nicas and Hancock, 1980). Penicillin binding proteins in the cytoplasmic membrane have been shown to decrease susceptibility to β-lactam antibiotics (Brown and Williams, 1985).

Bacterial growth rate (ranging from rapid cell division of two to three per hour to much slower rates) experiments have shown that slower-growing cells are especially recalcitrant to antimicrobials (Brown, 1975; Tuomanen, 1986). In an attempt to explain clinical findings of biofilm resistance to antibiotics, experiments were carried out on biofilm organisms, with the result that differences in growth rates were found in relation to drug resistance (Brown, Allison and Gilbert, 1988). In slow-growing *E. coli*, penicillin binding proteins are poorly expressed, hence the antibiotics ceftoxidine and ceftriaxone have little effect against the organisms, irrespective of growth-limiting nutrients (Cozens *et al.*, 1986). Thus, organisms within a biofilm can use various means to prevent their eradication by antimicrobial agents.

Another approach to treatment is the use of biocide chemicals, which are believed to be taken up by a cell, passaged to target sites where they damage the organism. The response of the bacteria to biocides depends upon the nature of the chemical agent and the organism itself, as well as environmental factors such as pH, presence of organic matter, temperature, redox potential and other factors. Mechanisms of resistance to biocides can be intrinsic or acquired. One problem with application to humans is that some biocides are toxic. Of those that are safe, cetrimide, triclosan and chlorhexidine have shown some effect against Gram-positive and Gram-negative pathogens (Russell, 1995). However, even these biocides do not eradicate organisms such as *P. aeruginosa* and multidrug-resistant staphylococci.

In conclusion, biofilm formation on medical prostheses is a widespread occurrence, it is complicated in nature and it provides many challenges for scientists wishing to understand the microbial environment and physicians wishing to selectively prevent and treat infection.

16.6 SUMMARY

Polymer materials have made a significant contribution to human health through their incorporation into medical devices that are used, in a multi-

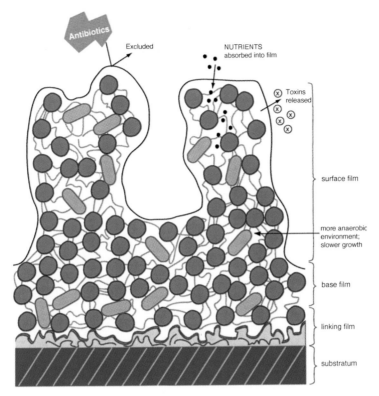

Plate 1 Illustration of the various layers which comprise a bacterial biofilm adherent to a prosthesis. The organisms in deeper sections of the biofilm tend to be microaerophilic or anaerobic. Antibiotics are often excluded and nutrients enter and diffuse through the biofilm. Toxin and bacterial by-product release can occur and be problematic for the patient.

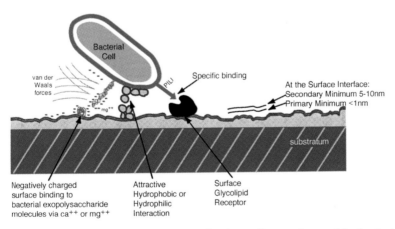

Plate 2 This illustration depicts three mechanisms that can be used by bacteria to first adhere to a prosthesis and/or its conditioning film. These are: charge interactions, hydrophilic-hydrophilic or hydrophobic-hydrophobic interactions, and specific adhesin-receptor interactions where the receptor molecule has adsorbed to the device in the conditioning film. Produced with reference to material presented by Gristina, 1987, and Reid and Busscher, 1998.

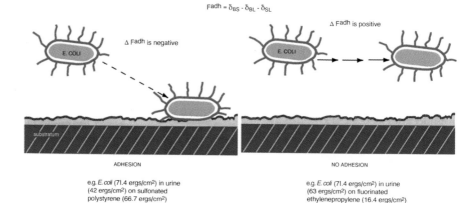

$$Fadh = \delta_{BS} - \delta_{BL} - \delta_{SL}$$

Δ Fadh is negative

Δ Fadh is positive

E. COLI

E. COLI

substratum

ADHESION

NO ADHESION

e.g. *E.coli* (71.4 ergs/cm²) in urine (42 ergs/cm²) on sulfonated polystyrene (66.7 ergs/cm²)

e.g. *E.coli* (71.4 ergs/cm²) in urine (63 ergs/cm²) on fluorinated ethylenepropylene (16.4 ergs/cm²)

Plate 3 Adhesion of bacteria to surfaces occurs when the net free energy of adhesion is negative, but not when it is positive. Produced from data reported by Absolom et al. 1983, and Reid et al. 1991.

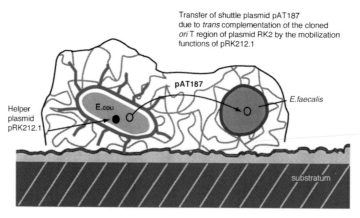

Transfer of shuttle plasmid pAT187 due to *trans* complementation of the cloned *ori* T region of plasmid RK2 by the mobilization functions of pRK212.1

pAT187

E.faecalis

Helper plasmid pRK212.1

E.COLI

substratum

Plate 4 This illustrates transfer of genetic information between two different species, depicted here within a biofilm. The diagram is an original using some of the findings reported by Trieu-Cuot et al. 1987.

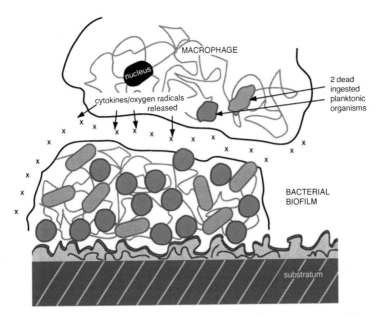

Plate 5 Although immune cells, in this case a macrophage, can come into contact with biofilms and release toxic oxygen radicals, they are quite ineffective at eradicating the organisms within the sessile population.

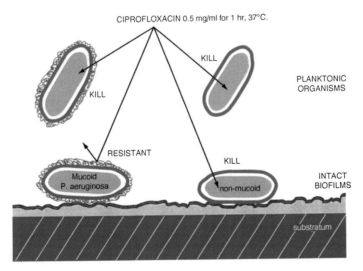

Plate 6 This figure illustrates findings of Evans et al. 1991 and Gilbert and Brown 1996, which describe how ciprofloxacin can kill planktonic organisms and adherent ones which do not have a capsule, but it is less able to kill encapsulated *Pseudomonas aeruginosa*.

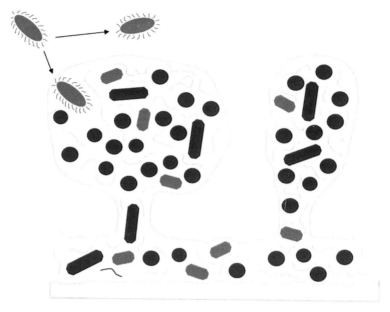

Plate 7 Illustrates fimbriated *E. coli* entering the urogenital microenvironment, where they either become members of the biofilm adherent to cells or ascend straight to the urinary tract. The dark circles represent Gram-positive cocci; the dark rods represent lactobacilli; and the open rods represent Gram-negative coliforms. The biofilms are encased in glycocalyx strands.

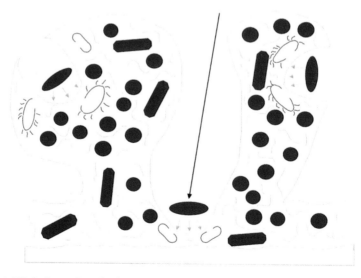

Plate 8 We believe that the biosurfactant lactobacilli (illustrated as round-edged dark rods) can enter the urogenital biofilms and disrupt the Gram-negative coliforms, thereby changing the biofilm composition.

tude of forms, shapes and sizes, worldwide on a daily basis. For the most part, these devices do not pose significant problems. However, in a small but significant number of cases, infections and often death can ensue as a consequence of microorganisms colonizing the devices and infecting the host. In many such instances the organisms accumulate in structured communities termed biofilms. This chapter represents a review of biofilm formation, structure and function, and the outcome in terms of disease. While the stages of biofilm formation are reasonably well understood, there has been less progress in preventing their deposition on medical devices and in eradicating them over long periods of instillation into the body. The innate resistance of biofilms to antimicrobial agents compounds the problem. Future management of medical prostheses will require new approaches to therapy when organisms interfere with the ability of the device to operate, and when the patient's defenses fail to fight off the attack by pathogens.

16.7 ACKNOWLEDGMENTS

The financial support of Ortho-McNeil and Bayer is appreciated. I acknowledge the input of my many colleagues whose research and ideas have helped my understanding of this area. Thanks to Ms Louise Gadbois for the graphics and Ms Christina Tieszer for the electron micrographs.

16.8 REFERENCES

Absolom, D. R. (1988) The role of hydrophobicity in infection: bacterial adhesion and phagocytic ingestion. *Canadian Journal of Microbiology*, **34**, 287–298.

Absolom, D. R., Lamberti, F. V., Policova, Z. *et al.* (1983) Surface thermodynamics of bacterial adhesion. *Applied and Environmental Microbiology*, **46**, 90–97.

Anglen, J. O., Apostoles, S., Christensen, G. and Gainor, B. (1994) The efficacy of various irrigation solutions in removing slime-producing *Staphylococcus. Journal of Orthopedics and Trauma*, **8**, 390–396.

Angles, M. L., Marshall, K. C. and Goodman, A. E. (1993) Plasmid transfer between marine bacteria in the aqueous phase and biofilms in reactor microcosms. *Applied and Environmental Microbiology*, **59**, 843–850.

Anwar, H. and Costerton, J. W. (1990) Enhanced activity of combination of tobramycin and piperacillin for eradication of sessile biofilm cells of *Pseudomonas aeruginosa. Antimicrobial Agents and Chemotherapy*, **34**, 1666–1671.

Anwar, H. and Strap, J. L. (1992) Changing characteristics of aging biofilms. *International Biodeterioration and Biodegradation*, **30**, 177–186.

Berger, M. (1991) Inflammation in the lung in cystic fibrosis – a vicious cycle that does more harm than good. *Clinical Review of Allergy*, **9**, 119–142.

Brokke, P., Dankert, J., Carballo, J. and Feijen, J. (1991) Adherence of coagulase-negative staphylococci onto polyethylene catheters *in vitro* and *in vivo*: a study on the influence of various plasma proteins. *Journal of Biomaterials Applied*, **5**, 204–226.

Brown, M. R. W. (1975) The role of the envelope in resistance, in *Resistance of Pseudomonas aeruginosa*, (ed. M. R. W. Brown), John Wiley, Chichester, pp. 71–107.

Brown, M. R. W. and Williams, P. (1985) The influence of environment on envelope properties affecting survival of bacteria in infections. *Annual Review of Microbiology*, **39**, 527–556.

Brown, M. R. W., Allison, D. G. and Gilbert, P. (1988) Resistance of bacterial biofilms to antibiotics: a growth related effect? *Journal of Antimicrobial Chemotherapy*, **22**, 777–783.

Bryers, J. D. (1987) Biologically active surfaces: processes governing the formation and persistence of biofilms. *Biotechnology Progress*, **3**, 57–68.

Busscher, H. J., Bos, R. and van der Mei, H. C. (1995) Initial microbial adhesion is a determinant for the strength of biofilm adhesion. *FEMS Microbiology Letters*, **128**, 229–234.

Busscher, H. J., Stokroos, I. and Schakenraad, J. M. (1991) Two dimensional, spatial arrangement of fibronectin adsorbed to biomaterials with different wettabilities. *Cells and Materials*, **1**, 49–57.

Caldwell, D. E. (1995) Cultivation and study of biofilm communities, in *Microbial Biofilms*, (eds H. M. Lappin-Scott and J. W. Costerton), Cambridge University Press, Cambridge, pp. 64–79.

Chamberlain, A. H. L. (1992) The role of adsorbed layers in bacterial adhesion, in *Biofilms – Science and Technology*, (ed. L. F. Melo), Kluwer Academic, Dordrecht, pp. 59–67.

Characklis, W. G., Turakhia, M. H. and Zelver, N. (1990) Transfer and interfacial transport phenomena, in *Biofilms*, (eds W. G. Characklis and K. C. Marshall), John Wiley, New York, pp. 265–340.

Christensen, B. E. and Characklis, W. G. (1990) Physical and chemical properties of biofilms, in *Biofilms*, (eds W. G. Characklis and K. C. Marshall), John Wiley, New York, pp. 93–130.

Christensen, G. D., Baddour, L. M., Hasty, D. L. *et al.* (1989) Microbial and foreign body factors in the pathogenesis of medical device infections, in *Infections Associated with Indwelling Medical Devices*, (eds A. L. Bisno and F. A. Waldvogel), American Society for Microbiology, Washington, DC, pp. 27–59.

Costerton, J. W., Irwin, R. T. and Cheng, K.-T. (1981) The bacterial glycocalyx in nature and disease. *Annual Review of Microbiology*, **35**, 399–424.

Costerton, J. W., Cheng, K.-J., Geesey, G. G. *et al.* (1987) Bacterial biofilms in nature and disease. *Annual Review of Microbiology*, **41**, 435–464.

Cozens, R. M., Tuomanen, E., Tosch, W. *et al.* (1986) Evaluation of the bactericidal activity of β-lactam antibiotics upon slowly growing bacteria cultured in the chemostat. *Antimicrobial Agents and Chemotherapy*, **29**, 797–802.

Dankert, J., Hogt, A. H. and Feijen, J. (1986) Biomedical polymers: bacterial adhesion, colonization, and infection. *CRC Critical Reviews of Biocompatability*, **2**, 219–301.

Dasgupta, M. K., Bettcher, K. B., Ulan, R. A. *et al.* (1987) Relationship of adherent bacterial biofilms to peritonitis in chronic ambulatory peritoneal dialysis. *Peritoneal Dialysis Bulletin*, **7**, 168–173.

Dewanti, R. and Wong, A. C. L. (1995) Influence of culture conditions on biofilm formation by *Escherichia coli* 0157:H7. *International Journal of Food Microbiology*, **26**, 147–164.

Elder, M. J., Stapleton, F., Evans, E. and Dart, J. K. (1995) Biofilm-related infections in ophthalmology. *Eye*, **9**, 102–109.

Elliott, T. S. J. and Faroqui, M. H. (1992) Infections and intravascular devices. *British Journal of Hospital Medicine*, **48**, 496–503.

Evans, D. J., Allison, D. G., Brown, M. R. W. and Gilbert, P. (1991) Susceptibility of *Pseudomonas aeruginosa* and *Escherichia coli* biofilms towards ciprofloxacin: effect of specific growth rate. *Journal of Antimicrobial Chemotherapy*, **27**, 177–184.

Fletcher, M. (1986) Measurement of glucose utilization by *Pseudomonas fluorescens* that are free-living and that are attached to surfaces. *Applied and Environmental Microbiology*, **52**, 672–676.

Fletcher, M. and Marshall, K. C. (1982) Bubble contact angle method for evaluating substratum interfacial characteristics and its relevance to bacterial attachment. *Applied and Environmental Microbiology*, **44**, 184–192.

Geesey, G. G., Stupy, M. W. and Bremer, P. J. (1992) The dynamics of biofilms. *International Biodeterioration and Biodegradation*, **30**, 135–154.

Gilbert, P. and Brown, M. R. W. (1995) Mechanisms of the protection of bacterial biofilms from antimicrobial agents, in *Microbial Biofilms*, (eds H. M. Lappin-Scott and J. W. Costerton), Cambridge University Press, Cambridge, pp. 118–130.

Goodman, A. E. and Marshall, K. C. (1995) Genetic responses of bacteria at surfaces, in *Microbial Biofilms*, (eds H. M. Lappin-Scott and J. W. Costerton), Cambridge University Press, Cambridge, pp. 81–98.

Gristina, A. G. (1987) Biomaterial-centered infection: microbial adhesion versus tissue integration. *Science*, **237**, 1588–1595.

Herschorn, S. and Ordorica, R. C. (1995) Penile prosthesis insertion with corporeal reconstruction with synthetic vascular graft material. *Journal of Urology*, **154**, 80–84.

Hoiby, N., Fomsgaard, A., Jensen, E. T. *et al.* (1995) The immune response to bacterial biofilms, in *Microbial Biofilms*, (eds H. M. Lappin-Scott and J. W. Costerton), Cambridge University Press, Cambridge, pp. 233–250.

Hopkins, W. J., Uehling, D. T. and Balish, E. (1987) Local and systematic antibody responses accompany spontaneous resolution of experimental cystitis in cynomolgus monkeys. *Infection and Immunity*, **55**, 1951–1956.

Inglis, T. J. J., Tit-Meng, L., Mah-Lee, N. *et al.* (1995) Structural features of tracheal tube biofilm formed during prolonged mechanical ventilation. *Chest*, **108**, 1049–1052.

Jarvik, R. K. (1981) The total artificial heart. *Scientific American*, **244**, 74.

Jensen, E. T., Kharazmi, A., Lam, K. *et al.* (1990) Human polymorphonuclear leukocyte response to *Pseudomonas aeruginosa* grown in biofilm. *Infection and Immunity*, **58**, 2383–2385.

Jensen, E. T., Kharazmi, A., Hoiby, N. and Costerton, J. W. (1992) Some bacterial parameters influencing the neutrophil oxidative burst response to *Pseudomonas aeruginosa* biofilms. *Acta Pathologica Microbiologica et Immunologica Scandinavica*, **100**, 727–733.

Johnson, J. R., Roberts, P. L., Olsen, R. J. *et al.* (1990) Prevention of catheter-associated urinary tract infection with a silver oxide-coated urinary catheter: clinical and microbiologic correlates. *Journal of Infectious Diseases*, **162**, 1145–1150.

Khardori, N. and Yassien, M. (1995) Biofilms in device-related infections. *Journal of Industrial Microbiology*, **15**, 141–147.

Koch, A. L. (1991) Diffusion: the crucial process in many aspects of the biology of bacteria, in *Advances in Microbial Ecology 8*, (ed. K. C. Marshall), Plenum Press, New York, pp. 37–70.

Korber, D. R., Lawrence, J. R., Lappin-Scott, H. M. and Costerton, J. W. (1995) Growth of microorganisms on surfaces, in *Microbial Biofilms*, (eds H. M. Lappin-Scott and J. W. Costerton), Cambridge University Press, Cambridge, pp. 15–45.

Kowalewska-Grochowska, K., Richards, R., Moysa, G. L. *et al.* (1991) Guidewire catheter change in central venous catheter biofilm formation in a burn population. *Chest*, **100**, 1090–1095.

Lappin-Scott, H. M. and Costerton, J. W. (1989) Bacterial biofilms and surface fouling. *Biofouling*, **1**, 323–342.

Light, J. K., Lapin, S. and Vohra, S. (1995) Combined use of bowel and the artificial urinary sphincter in reconstruction of the lower urinary tract: infectious complications. *Journal of Urology*, **153**, 331–333.

Mack, D., Fischer, W., Krokotsch, A. *et al.* (1996) The intracellular adhesin involved in biofilm accumulation of *Staphylococcus epidermidis* is a linear β-1,6-linked glucosaminoglycan: purification and structural analysis. *Journal of Bacteriology*, **178**, 175–183.

McLean, R. J. C., Nickel, J. C. and Olson, M. E. (1995) Biofilm associated urinary tract infections, in *Microbial Biofilms*, (eds H. M. Lappin-Scott and J. W. Costerton), Cambridge University Press, Cambridge, pp. 261–273.

Miller, J. F., Mekalanos, J. J. and Falkow, S. (1989) Coordinate regulation and sensory transduction in the control of bacterial virulence. *Science*, **243**, 916–922.

Neu, T. R., Dijk, F., Verkerke, G. J. *et al.* (1992) Scanning electron microscopy study of biofilms on silicone voice prostheses. *Cells and Materials*, **2**, 261–269.

Nicas, T. and Hancock, R. E. W. (1980) Outer membrane protein H1 of *Pseudomonas aeruginosa*: involvement in adaptive and mutational resistance to ethylene-diaminetetracetate, polymyxin B, and gentamicin. *Journal of Bacteriology*, **50**, 715–731.

Nickel, J. C. (1996) Prostatitis, in *Antibiotic Therapy in Urology*, (ed. S. G. Mulholland), pp. 57–69.

Nivens, D. E., Chambers, J. Q., Anderson, T. R. *et al.* (1993) Monitoring microbial adhesion and biofilm formation by attenuated total reflection/Fourier Transform Infrared Spectroscopy. *Journal of Microbiological Methods*, **17**, 199–213.

Ozaki, C. K., Phaneuf, M. D., Bide, M. J. *et al.* (1993) *In vivo* testing of an infection-resistant vascular graft material. *Journal of Surgical Research*, **55**, 543–547.

Parsons, C. L. (1996) Infections of urologic prostheses. *Current Opinion in Urology*, **6**, 41–44.

Preston, C. A. K., Khoury, A. E., Reid, G. *et al.* (1996) Susceptibility of *Pseudomonas aeruginosa* in a biofilm to ciprofloxacin and tobramycin. *International Journal of Antimicrobial Agents*, **7**, 251–256.

Quirynen, M., van der Mei, H. C., Bollen, C. M. L. *et al.* (1994) The influence of surface free energy on supra- and subgingival plaque microbiology. *Journal of Periodontology*, **65**, 162–167.

Raad, I., Darouiche, R., Hachem, R. *et al.* (1995) Antibiotics and prevention of microbial colonization of catheters. *Antimicrobial Agents and Chemotherapy*, **39**, 2397–2400.

Ratner, B. D. (1980) Characterization of graft polymers for biomedical applications. *Journal of Biomedical Materials Research*, **14**, 665–687.

Reid, G. (1994) Microbial adhesion to biomaterials and infections of the urogenital tract. *Colloids and Surfactants B: Biointerfaces*, **2**, 377–385.

Reid, G., van der Mei, H. C. and Busscher, H. J. (1998) Microbial biofilms and urinary tract infections, in, *Urinary Tract Infections*, (eds W. Brumfitt, T. Hamilton-Miller and R. R. Bailey), Chapman & Hall, London, pp. 111–118.

Reid, G. and Tieszer, C. (1993) Preferential adhesion of bacteria from a mixed population to a urinary catheter. *Cells and Materials*, **3**, 171–176.

Reid, G. and Tieszer, C. (1995) Use of lactobacilli to reduce the adhesion of *Staphylococcus aureus* to catheters. *International Biodeterioration and Biodegradation*, **34**, 73–83.

Reid, G., Davidson, R. and Denstedt, J. D. (1994) XPS, SEM and EDX analysis of conditioning film deposition onto ureteral stents. *Surface Interface Analysis*, **21**, 581–586.

Reid, G., Tieszer, C. and Bailey, R. R. (1995) Bacterial biofilms on devices used in nephrology. *Nephrology*, **1**, 269–275.

Reid, G., Tieszer, C. and Lam, D. (1995) Influence of lactobacilli on the adhesion of *Staphylococcus aureus* and *Candida albicans* to diapers. *Journal of Industrial Microbiology*, **15**, 248–253.

Reid, G., Hawthorn, L. A., Mandatori, R. *et al.* (1988) Adhesion of lactobacillus to polymer surfaces *in vivo* and *in vitro*. *Microbial Ecology.* **16**(3), 241–251.

Reid, G., Beg, H. S., Preston, C. and Hawthorn, L. A. (1991) Effect of bacterial, urine and substratum surface tension properties on bacterial adhesion to biomaterials. *Biofouling*, **4**, 171–176.

Reid, G., Denstedt, J. D., Kang, Y. S. *et al.* (1992a) Microbial adhesion and biofilm formation on ureteral stents *in vitro* and *in vivo*. *Journal of Urology*, **148**, 1592–1594.

Reid, G., Tieszer, C., Foerch, R. *et al.* (1992b) The binding of urinary components and uropathogens to a silicone latex urethral catheter. *Cells and Materials*, **2**, 253–260.

Reid, G., Lam, D., Policova, Z. and Neumann, A. W. (1993a) Adhesion of two uropathogens to silicone and lubricious catheters: influence of pH, urea and creatinine. *Journal of Materials Science: Materials in Medicine*, **4**, 17–22.

Reid, G., Tieszer, C., Foerch, R. *et al.* (1993b) Adsorption of ciprofloxacin to urinary catheters and effect on subsequent bacterial adhesion and survival. *Colloids and Surfactants B: Biointerfaces*, **1**, 9–16.

Reid, G., Sharma, S., Advikolanu, K. *et al.* (1994) Effect of ciprofloxacin, norfloxacin and ofloxacin *in vitro* on the adhesion and survival of *Pseudomonas aeruginosa* on urinary catheters. *Antimicrobial Agents and Chemotherapy*, **38**, 1490–1495.

Reid, G., Busscher, H. J., Sharma, S. *et al.* (1995a) Surface properties of catheters, stents and bacteria associated with urinary tract infections. *Surface Science Reports*, **7**, 251–273.

Reid, G., Tieszer, C., Denstedt, J. and Kingston, D. (1995b) Examination of bacterial and encrustation deposition on ureteral stents of differing surface properties, after indwelling in humans. *Colloids and Surfactants B: Biointerfaces*, **5**, 171–179.

Reid, G., van der Mei, H. C., Tieszer, C. and Busscher, H. J. (1996) Uropathogenic *Escherichia coli* adhere to urinary catheters without using fimbriae. *FEMS Immunology and Medical Microbiology*, **16**, 159–162.

Rittle, K. H., Helmstetter, C. E., Meyer, A. E. and Baier, R. E. (1990) *Escherichia coli* retention on solid surfaces as functions of substratum surface energy and cell growth phase. *Biofouling*, **2**, 121–130.

Rupp, M. E., Sloot, N., Meyer, H. G. *et al.* (1995) Characterization of the hemagglutinin of *Staphylococcus epidermidis*. *Journal of Infectious Diseases*, **172**, 1509–1518.

Russell, A. D. (1995) Mechanisms of bacterial resistance to biocides. *International Biodeterioration and Biodegradation*, **36**, 247–265.

Savage, D. C. (1977) Microbial ecology of the gastrointestinal tract. *Annual Review of Microbiology*, **31**, 107–133.

Schakenraad, J. M., Noordmans, J., Wildevuur, C. R. H. *et al.* (1989) The effect of protein adsorption on substratum surface free energy, infrared absorption and cell spreading. *Biofouling*, **1**, 193–201.

Schrago, A. W., Chassy, B. M. and Dobrogosz, W. J. (1986) Conjugal plasmid transfer (pAM81) in *Lactobacillus plantarum*. *Applied and Environmental Microbiology*, **52**, 574–576.

Schulman, C. C. (1995) Non-autologous injected materials for the endoscopic treatment of vesico-ureteral reflux, in *Implanted and Injected Materials in Urology*, (ed. J. M. Buzelin), Isis Medical Media, Oxford, pp. 23–35.

Sjollema, J., van der Mei, H. C., Uyen, H. M. and Busscher, H. J. (1990) Direct observation of cooperative effects in oral streptococcal adhesion to glass by analysis of the spacial arrangement of adhering bacteria. *FEMS Microbiology Letters*, **69**, 263–270.

Stock, J. B., Stock, A. M. and Mottonen, J. M. (1990) Signal transduction in bacteria. *Nature*, **344**, 395–400.

Strachan, C. J. (1995) The prevention of orthopaedic implant and vascular graft infections. *Journal of Hospital Infection*, **30**(Suppl.), 54–63.

Tannock, G. W. (1987) Conjugal transfer of plasmid pAM81 in *Lactobacillus reuteri* and between lactobacilli and *Enterococcus faecalis*. *Applied and Environmental Microbiology*, **53**, 2693–2695.

Teichman, J. M., Abraham, V. E., Stein, P. C. and Parsons, C. L. (1994) Protamine sulfate and vancomycin are synergistic against *Staphylococcus epidermidis* prosthesis infection *in vivo*. *Journal of Urology*, **152**, 213–216.

Trieu-Cuot, P., Carlier, C., Martin, P. and Courvalin, P. (1987) Plasmid transfer by conjugation from *Escherichia coli* to Gram-positive bacteria. *FEMS Microbiology Letters*, **48**, 289–294.

Tuomanen, E. (1986) Phenotypic tolerance: the search for β-lactam antibiotics that kill non-growing bacteria. *Reviews of Infectious Diseases*, **8**(Suppl.): S279–S291.

van Loosdrecht, M. C. M., Norde, W. and Zehnder, A. J. B. (1987) Influence of cell surface characteristics on bacterial adhesion to solid surfaces, in *Proceedings of the 4th European Congress of Biotechnology*, pp. 575–580.

Vaudaux, P. E., Lew, D. P. and Waldvogel, F. A. (1989) Host factors predisposing to foreign body infections, in *Infections Associated with Indwelling Medical Devices*, (eds A. L. Bisno and F. A. Waldvogel), American Society for Microbiology, Washington, DC, pp. 3–26.

Wadström, T. (1989) Molecular aspects of bacterial adhesion, colonization, and development of infections associated with biomaterials. *Journal of Investigative Surgery*, **2**, 353–360.

Williams, G. (1995a) Surgical procedures, indications and results of injected materials for treating urinary incontinence (male and female), in *Implanted and Injected Materials in Urology*, (ed. J. M. Buzelin), Isis Medical Media, Oxford, pp. 36–47.

Williams, G. (1995b) Urethral stents: history, surgical procedure, indication and results for treating BPH, urethral stricture and neurological voiding dysfunction, in *Implanted and Injected Materials in Urology*, (ed. J. M. Buzelin), Isis Medical Media, Oxford, pp. 74–90.

Wilson, S. K. and Delk, J. R. II (1995) Inflatable penile implant infection: predisposing factors and treatment suggestions. *Journal of Urology*, **153**, 659–661.

Yasuda, H., Ajiki, Y., Koga, T. and Yokota, T. (1994) Interaction between clarithromycin and biofilms formed by *Staphylococcus epidermidis*. *Antimicrobial Agents and Chemotherapy*, **38**, 138–141.

17

Urogenital microflora and urinary tract infections

Gregor Reid and Marc Habash

17.1 INTRODUCTION

Urinary tract infections (UTI) constitute a worldwide problem affecting an estimated 150 million women per year. In Western countries such as Canada and the US, these infections are a common reason for adult females to visit their family practitioner and urologist. It has been estimated that each episode leads to 6 days of signs and symptoms (Foxman and Fredrichs, 1985).

The diseases involve inflammation and infection of the urinary bladder, ureters and kidneys (Kunin, 1987). Uncomplicated symptomatic infections of the lower tract (bladder) present with pain upon urination, frequency and urgency of voiding, suprapubic pain and often dysuria, pyuria and hematuria, along with bacterial presence in the bladder. Kidney infections also present with fever and lower back pain. Asymptomatic bacteriuria (ABU) occurs commonly, but no signs and symptoms of disease are present. Complicated UTI is a common condition where the patient's status is complicated by factors such as an indwelling catheter, urinary calculi, cancer, ureteric reflux or other ailments.

Many of the patients will experience a recurrence, particularly within the first year from the original infection. Of more concern are infections that are complicated by kidney involvement, as hospitalization is usually required, and kidney damage and even death can occur. Other serious complications can arise during pregnancy, sometimes leading to premature birth (Bergeron, 1972). All in all, there are enormous costs to the

G. W. Tannock (ed.), *Medical Importance of the Normal Microflora*, 423–440.
© 1999 *Kluwer Academic Publishers. Printed in Great Britain.*

healthcare system for treatment of UTI, and cost to the employer for time off work.

A clinical diagnosis is obtained primarily by culture of midstream urine and isolation of more than 1000–100 000 organisms per ml, depending upon the patient and the bacterium involved. In conjunction with culture, the presence of signs and symptoms is the key factor in diagnosis.

Most uncomplicated UTI cases are resolved by 1–7 days' antibiotic therapy. However, drug resistance is increasing among uropathogens against commonly used antibiotics (e.g. trimethoprim/sulfamethoxazole; Reid and Seidenfeld, 1997), and patients are experimenting more and more with alternative natural remedies such as cranberry juice. Currently, preventive therapies rely almost exclusively upon the use of antibiotics. In real terms, no true prophylaxis exists, because long-term, low-dose antibiotics are actually treating bacteria that enter the bladder.

17.2 DYNAMICS OF THE UROGENITAL MICROFLORA

Historical data indicate that the vast majority (around 80%) of UTI in a suburban, non-hospitalized community is caused by *Escherichia coli*, followed by other Enterobacteriaceae and *Staphylococcus saprophyticus*. However, a recent 7-year study by our group indicated a different result, whereby *E. coli* infections were less common and *Enterococcus faecalis* was the second most prevalent uropathogen (Reid and Seidenfeld, 1997). The latter result was also found in hospitalized patients (Preston, Bruce and Reid, 1992). Thus, remedies to treat and prevent UTI should target, at the very least, *E. coli*, *E. faecalis* and *S. saprophyticus*.

The urinary bladder is regarded as being normally a sterile conduit, while the urethral opening has organisms on the distal side. The origin of the uropathogens in uncomplicated UTI is the fecal microflora, whose path of colonization takes them to the vaginal introitus, urethra and then into the bladder and occasionally up to the kidney. For over 20 years, the key factor in pathogenesis has been regarded as the ability of the pathogens to attach to epithelial cells, thereby allowing them a niche to establish, multiply, spread and avoid host defenses (Svanborg Edén *et al.*, 1976; Reid and Sobel, 1987). Numerous studies have investigated and documented the adhesins and receptor sites involved in the attachment process. Factors such as hemolysins, aerobactin, capsular antigens and others play a role in *E. coli* pathogenesis (Brooks *et al.*, 1980; Hacker *et al.*, 1985; Hultgren *et al.*, 1985; Kallenius, Mollby and Winberg, 1979; Kist, Salit and Hofmann, 1990; Pere *et al.*, 1985; Rhen, Klemm and Korhonen, 1986; Vaisenen *et al.*, 1982). Furthermore, there are host susceptibility factors, including genetic determinants, age-related changes and mucosal differences, that influence the infection process (Perdigon *et al.*, 1995).

Recent advances have improved our understanding of the processes involved in the development of UTI. It is known that the most common bacterial cause of UTI, *E. coli*, produces two adhesins allowing it to adhere to receptor sites on uroepithelial cells. Type 1 fimbriae, which recognize mannose moieties on uroepithelial cells have been associated with the development of cystitis. P fimbriae bind to Galα(1–4)Gal receptors on uroepithelial cells and have been associated with the onset of pyelone-phritis. Recent studies have attempted to provide more information on type 1 fimbriae, including verification of their precise receptor on uro-epithelial cells and their role as a virulence factor. A possible receptor site for *E. coli* type 1 fimbriae involves uroplakins. These major glycoproteins of uroepithelial plaques have been identified as containing anchorage sites allowing for adhesion and colonization of the bladder by bacteria (Wu, Sun and Medina, 1996). Further, the presence of type 1 fimbriae has been shown to enhance the virulence of *E. coli*. The presence of type 1 fimbriae led to increased severity of infection in children when compared to strains of the same serotype minus type 1 fimbriae (Connell *et al.*, 1996).

Virulence factors found to occur frequently among urinary *E. coli* iso-lates include *pap*-related fimbriae (as described above), hemolysin, outer membrane protein T, S fimbrial adhesin, aerobactin and Dr binding adhesins. The prevalence of these factors in an initial UTI isolate ranged from 16% for Dr binding adhesins to 100% for *pap*-related fimbriae (Fox-man *et al.*, 1995). Further it was found that the presence of certain viru-lence factors in the initial infecting strain led to increased risk of acquiring a second UTI. Subjects with initial isolates containing aerobactin, outer membrane protein T, Dr binding protein, *pap*-related fimbriae and cap-sular proteins had the highest risk of obtaining a second UTI (Foxman *et al.*, 1995).

Other factors that may play a role in the establishment of a UTI include genes (*guaA* and *argC*) identified to be necessary for the growth of *E. coli* in urine and urovirulence. Random transposon mutants lacking the *guaA* or *argC* genes showed significantly diminished urovirulence in an *in vivo* mouse UTI model. Also, lipopolysaccharides containing the O4-specific antigen moiety have been found to have an important role in the patho-genesis of UTI. In a mouse model, isogenic strains of *E. coli* deficient in O4 antigen were less virulent than a parent strain, whereas group 2 capsule (K54)-deficient *E. coli* were as virulent as the parent strain (Russo *et al.*, 1996a).

While the adhesion phenomenon is unquestionably important, there appears to be another key element of pathogenesis that is less well under-stood, namely the ability of the pathogens to survive exposure to the microflora that exists on the external urogenitalia. It has been reported that over 50 species colonize the healthy vagina (Redondo-Lopez, Cook and Sobel, 1990). Studies have shown that urogenital cells are covered by dense bacterial biofilms (Sadhu *et al.*, 1989; Reid *et al.*, 1990b), whose composition changes constantly but in which lactobacilli predominate, at least until the menopause. A uropathogenic organism emerging from the intestine (Wold *et al.*, 1992) comes into contact with these biofilms on vaginal and urethral cells, yet little is known about what happens thereafter. It is presumed that the uropathogens can either bypass the microflora or successfully enter the biofilms, survive and continue their ascension into the bladder (Plate 7).

Biofilms have been described in the human host for over 20 years (Savage, 1977; Costerton, Irwin and Cheng, 1981), but their role in health has been less well studied than their association with disease. Current understanding of biofilms indicates that the structure and dynamics vary with the organisms, the surface to which they attach, the nutritional environment and shear forces present in any given case. The organisms appear to benefit from biofilm formation by gaining access to nutrients, escaping host immune cells and antimicrobial attack, and having an ability to better control their multiplication. A recurrent theme here is that various factors, including nutrients, antimicrobials and arriving bacteria, change the properties and the composition of the biofilms.

Studies have shown that exposure of a woman to antibiotics or spermicide can cause disruption of the urogenital microflora, and increase their risk of infection (Hooton *et al.*, 1991; Reid *et al.*, 1990a). Indeed, *in vitro* studies have shown that most lactobacilli are eradicated by exposure to low-dose nonoxynol-9, while uropathogens grow and prosper in high concentrations of this compound (McGroarty *et al.*, 1993). The fact that exposure to antimicrobials does not completely remove the urogenital flora is due not only to resistance of individual bacterial strains, but also to their existence in biofilms, which help confer resistance to attack (Costerton and Lappin-Scott, 1995). In the urinary tract, bacterial biofilms have been shown to exist when the voiding mechanism is deficient, as in spinal cord injury (Reid *et al.*, 1992a; Reid, 1994). Indeed, the dense biofilms are often not detected by traditional urine culture, nor are they easily eradicated by antibiotics (Reid and Howard, 1997).

Scanning confocal laser microscopy has shown that biofilms are mushroom-shaped and form water channels through which the organisms are nourished (James *et al.*, 1995; Sanford *et al.*, 1996; Potera, 1996).

Clearly, lactobacilli can grow using nutrients present in the urogenital tract (Fig. 17.1). Of these, it is not clear which have the most effect on lactobacillus numbers, but certainly estrogen has a great impact, because after menopause lactobacilli are scarce and they reappear with estrogen therapy (Raz and Stamm, 1993). Exposure of urogenital microflora to estrogen and other compounds under microscopic real-time analysis will allow us to determine how the normal biofilm changes. Furthermore, exposure of nutrients and lactobacillus byproducts (e.g. acid,

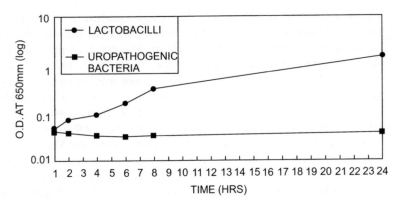

Figure 17.1 Results obtained when *Lactobacillus* (strains GR-1 and B-54) and uropathogens (five different species) were individually cultured in MRS broth plus 0.1% ascorbic acid. *Lactobacillus* growth was stimulated and the uropathogens did not grow. Controls without ascorbic acid showed no difference in growth patterns.

biosurfactant) to uropathogen-dominated biofilms will determine how easily a normal microflora can be restored.

17.3 WHY LACTOBACILLI WOULD MAKE A GOOD PROBIOTIC FOR PREVENTION OF UTI

The dominant presence of lactobacilli in the microflora of healthy women and their demise in patients who develop UTI (Stamey, 1973; Schaeffer and Stamey, 1977; Seddon *et al.*, 1976) and many other genital infections (Hillier *et al.*, 1993; Hay *et al.*, 1994) except candidiasis (Zdolsek *et al.*, 1995) has led us to focus our attention on these bacteria. *Lactobacillus* are Gram-positive rods, primarily facultative or strict anaerobes, which generally have a fastidious growth requirement. They prefer an acidic environment and help create such by production of lactic acid and other acids. In general, lactobacilli have not been associated with disease and for over 100 years they have been regarded as non-pathogenic members of the intestinal and urogenital microfloras.

Lactobacillus have long been of interest to the dairy and agriculture industries (Klaenhammer, 1982; Watkins, Miller and Neil, 1982), while over the past century studies of the impact of lactobacilli on human health have been sporadic and often inconclusive. Some examples can be found in which lactic acid bacteria have been used to treat or prevent infections of the intestinal and genital tracts, with varying degrees of success (Beck and Necheles, 1961; Alexander, 1967; Clements *et al.*, 1981; Hilton *et al.*, 1992). One study from 1915 suggested that lactic acid bacteria could clear bladder infections (Newman, 1915), however, the problem with that trial, as well as almost all of the others, has been a failure to identify the properties of lactobacilli required to prevent and treat disease, the optimal dosage and the duration and mode of delivery, and the failure to include a placebo in the study. None of the studies have recovered and documented lactobacilli properly, nor shown their mechanisms of action.

In recent years, the use of probiotics *per se*, and lactobacilli specifically, has received greater attention as an alternative, inexpensive, natural remedy to restore and maintain health (McGroarty, 1994; Chang, 1996; Elmer *et al.*, 1996; Reid, 1996; Shalev *et al.*, 1996; Hamilton-Miller, 1997). Two strains, *Lactobacillus* GG (ATCC 53103) and *L. rhamnosus* GR-1 (formerly *L. casei* var. *rhamnosus*; now ATCC 55826) appear to be effective at colonizing and protecting the intestine (Biller *et al.*, 1995; Gorbach, Chang and Goldin, 1987; Isolauri, Juntunen and Rautanen, 1991; Oksanen *et al.*, 1990; Siitonen *et al.*, 1990; Saxelin, Pessi and Salminen, 1995; Miller *et al.*, 1993) and urogenital tract (Bruce and Reid, 1988; Bruce *et al.*, 1992; Reid, Millsap and Bruce, 1994; Hilton, Rindos and Isenberg, 1995; Reid, Bruce and Taylor, 1995) respectively against microbial infection.

One obvious question is: what properties do strains possess that make them able to be effective probiotic agents against uropathogens? The answer is not fully known, but some common denominators appear to exist, namely the ability to adhere to and colonize tissues, and the capacity to inhibit the pathogenesis of disease-causing organisms. Another question can be raised: do we expect an exogenous probiotic strain to colonize the urogenital tract of a given person for a long time, and even become part of the microflora, replacing or coexisting with the endogenous lactobacilli? Or is the aim to replace the endogenous microflora only while it is altered (e.g. as a result of antibiotic therapy; Korshunov *et al.*, 1985)? Again, no clear answers are available. Indeed, it is not clear if probiotic therapy should simply be regarded as a means of temporary substitution while the indigenous microflora recovers and re-establishes itself. Obviously in the case of using oral ingestion to prevent UTI, fewer doses would be preferable, and this will only be possible if the lactobacilli colonize the urogenital tract quickly and well. The other consideration is if lactobacilli can eliminate or reduce uropathogen numbers in the intestine, thereby lowering the risk of them infecting the urogenital tract.

17.4 *IN VITRO* DATA ON LACTOBACILLI ACTIVE AGAINST UROPATHOGENS

In the first exploratory study of adherence of normal microflora *in vitro* (Chan, Bruce and Reid, 1984), it was found that:

• normal microflora (diphtheroids and lactobacilli) adhered to normal and infected uroepithelial cells;

- the menstrual cycle appeared to affect the adherence of lactobacilli in healthy women, in that days corresponding to high circulating concentrations of estrogens resulted in higher adherence;
- on day 10 of the menstrual cycle, lactobacilli adherence corresponded to a reduced adherence to cells by uropathogens;
- the strain that gave the best adherence results was a urethral strain isolated from a healthy women; later, this isolate was named *Lactobacillus casei* subsp. *rhamnosus* GR-1.

We showed the potential for whole viable and non-viable *Lactobacillus* cells and cell-wall fragments to significantly block the adherence of various uropathogens to urinary cell surfaces (Chan *et al.*, 1985). Lipoteichoic acid from the cell wall of *Lactobacillus* appeared to help to mediate adherence of the organisms to receptors on epithelial cells that were different from those used by uropathogens.

Lactobacilli have been found to produce substances that inhibit the growth of a range of uropathogens and also to coaggregate (adhere one species to another) with members of the urogenital microflora, including pathogens (Reid, Cook and Bruce, 1987; Reid *et al.*, 1988). A study of 25 healthy premenopausal women and nine women with a history of UTI confirmed that coaggregation did occur *in vivo* and *in vitro* (Reid *et al.*, 1990b). Competitive exclusion might be occurring on the urogenital epithelial mucosa in which lactobacilli bind to uropathogens, then inhibit their growth or kill them by producing antimicrobial substances. The lactobacilli, by binding to the pathogens, could also be preventing the latters' ascension into the bladder.

Hydrogen-peroxide-producing *Lactobacillus* strains are believed to be important in vaginal colonization (Eschenbach *et al.*, 1989), yet GR-1 is a non-producer, and perhaps in its case the resistance to spermicide is a more important attribute for its probiotic usage, as these agents can destroy the normal microflora (McGroarty *et al.*, 1990, 1992, 1993).

The inhibition of growth of uropathogenic *E. coli* by GR-1 and other strains was a killing effect that has not been explained by acid or hydrogen peroxide production. The substance produced by the lactobacilli was heat-labile, with a molecular weight greater than 12 000–14 000 (McGroarty and Reid, 1988a). The inhibitory product was found in urine, but was optimally produced after growth in MRS Lactobacilli medium at pH 4.3. Interestingly, an agar underlay that included skim milk showed the best degree of inhibition. It would later be shown *in vivo* that skim milk appeared to aid *Lactobacillus* competition against uropathogens. Lactobacilli inhibited the growth of enterococci, with *L. fermentum* B-54 and *L. casei* RC-15 doing so to 83% of the strains (McGroarty and Reid, 1988b). The inhibition was not due to low pH or acids, nor to hydrogen peroxide (aerobic incubation suppresses this, yet inhibition occurred). It was believed to be caused by bacteriocin-like substance(s).

In 1995, we discovered another characteristic of lactobacilli that appears to be important in conferring probiotic action against uropathogens. A total of 15 strains were found to produce biosurfactant (Velraeds *et al.*, 1996a). The substance(s) adsorbed to surfaces and inhibited by 70% the initial adhesion of *E. faecalis*. The crude substance was analysed and was found to contain proteins and carbohydrates. Amino-acid analysis of hydrolysed material from strain *L. acidophilus* RC-14 showed a high alanine content compared to less active strains. The activity is not due to lipoteichoic acid or glycosyldiglycerides or to factors such as acid or bacteriocins that inhibit bacterial growth. SDS-PAGE showed a variety of proteins from 14.4–140 kDa molecular weight (Velraeds *et al.*, 1996a, b). This biosurfactant substance produced by *L. acidophilus* RC-14 and previously termed surlactin has now been partially characterized (Reid *et al.*, in preparation). The crude surlactin was confirmed to contain protein, as shown by SDS-PAGE and carbohydrate, as determined by phenol–sulfuric acid analysis. The biosurfactant activity was resistant to trypsin and pepsin and sensitive to α-amylase and lysozyme, as well as being resistant to heating at 75°C.

The biosurfactant not only prevents uropathogens from adhering to surfaces, but may also cause them to detach, thereby providing a potentially important clinical use (Plate 8). In a study, using biosurfactant from RC-14, *L. fermentum* B-54 and *L. rhamnosus* GR-1, large numbers (65%, 70% and 43% respectively) of indigenous bacteria were detached within 24 hours from vaginal epithelial cells collected from a healthy female. The crude biosurfactant was also found to inhibit adhesion of a broad range of uropathogenic species (Velraeds *et al.*, 1998).

17.5 *IN VIVO* STUDIES WITH LACTOBACILLI

Using a rat model, we established chronic UTI in the bladder by instilling uropathogens within agar beads (Reid *et al.*, 1985). The infection remained for many weeks and the host was unable to flush it out. Pretreatment of the rats with *L. casei* subsp. *rhamnosus* GR-1 prevented UTI in 84% of the animals that were subsequently challenged with uropathogens.

Then, using an acute UTI mouse model, it was discovered that GR-1 did not infect the urinary tract of the mice and only adhered to kidney and bladder cells for 2–3 days (Reid *et al.*, 1989). Pretreatment of the mice with GR-1, followed 2 days later by challenge with *E. coli* Hu734 expressing type 1 and P fimbriae, and sacrifice the next day, showed that 99% of controls were infected with *E. coli* but only 66% of *Lactobacillus*-treated animals became infected; only 2% of kidneys that had more than 10 000 lactobacilli adherent to cells became infected with E. *coli*. This again indicated that lactobacilli could exclude adherence of E. *coli* to cells and prevent acute UTI. In those mice that became infected, fewer numbers of E. *coli* were recovered in *Lactobacillus*-treated animals. These findings have since been supported by de Ruiz, de Bocanera and de Macias (1996) and Demacias *et al.* (1996), who showed that L. *fermentum* could be used to treat and prevent UTI in mice.

A study of humans was undertaken in Sweden using lactobacilli inserted directly into the urinary bladder. It had been observed that avirulent E. *coli* introduced into the urinary tract of volunteers who had chronic UTI prevented or delayed infection by virulent uropathogens. The use of lactobacilli was preferred as they did not constitute a threat of acquiring virulence factors (such as drug-resistant plasmids, which were regularly passed between E. *coli*). The study was undertaken on nine patients (eight female, one male) with intractable UTI. The results showed that lactobacilli did not infect the patients and were in fact eliminated within 1 day of instillation into the bladder (Hagberg *et al.*, 1989). Avirulent E. *coli* persisted in the bladder for 4 months, but the patients reported having foul-smelling urine.

Further human studies were carried out with GR-1, selected because it adhered to epithelial cells, produced inhibitors against uropathogens and did not cause any side-effects when placed straight into the bladder. Five patients were recruited, including one child. They were all extremely problematic in that they suffered from many repeated episodes of acute UTI. Upon commencement, no UTI was present. The cells of strain GR-1 were suspended in buffer (10^{11} per dose) and instilled once or twice weekly at night into the vagina as a douche. A tampon was placed (except in the child) overnight to increase the likelihood that the solution stayed in the vagina. The results showed that vaginal colonization with GR-1 was achieved in all patients, without any side-effects (Bruce and Reid, 1988). Three of the five patients showed sustained clinical relief from UTI, and their urine was free of bacteria for periods of over 6 months. The other two patients had relief from UTI for only a short time. Two patients had vaginal and bladder enterococci that persisted regardless of GR-1 treatment. The patients had no complaints about the lactobacillus therapy but, when given the option of having a gelatin suppository rather than a douche in future trials they said they would prefer the pessary.

In a subsequent human trial, suppositories ($> 1.6 \times 10^9$ organisms per 0.5 g freeze dried in skim milk) were used containing two types of lactobacillus (GR-1 and B-54). Ten women, six pre- and four postmenopausal, took one suppository weekly for over a year. The follow-up included 2 days, 2 weeks, 5 weeks, 3 and 6 months. There were no side-effects (Bruce *et al.*, 1992). Only two patients withdrew, in both cases because we did not believe there was compliance. The UTI rate in the remaining eight patients dropped 66.3% over 1 year, compared to the previous year's UTI rate. The UTI rate dropped 77.3% with inclusion of data for two of the eight patients when they were given a suppository twice weekly. This verified our belief that one dose per week would probably be sufficient in most patients but higher doses, at least for a period of time, might be required to optimize *Lactobacillus* colonization. Strains GR-1 and B-54 were identified by colony morphology and Gram stain identification.

The largest study carried out to date was a clinical trial where 55 premenopausal women were given either one suppository of GR-1 and B-54 (0.5 g) or one suppository of a *Lactobacillus* growth factor (LGF) weekly for 1 year. The patients were followed up after 2 weeks, then monthly (Reid, Bruce and Taylor, 1995). Six patients were excluded in the first month for non-compliance, moving house, becoming pregnant, etc. and 11 patients did not complete the study in full but their data were included to examine infection rates. Again, there were no side-effects in the study. The UTI infection rate was reduced from 6.0 per previous year to 1.6 (79% drop) for those given lactobacilli and 1.3 (83% drop) for those given LGF. Time to first infection was the same in both groups, namely 21–22 weeks. Vaginal pH was the same for both groups, namely 4.6–5.0. The viable lactobacillus counts recovered from vaginal swabs increased with therapy, especially for months 7–12 for lactobacillus-treated patients, during which time lower UTI rates were seen.

It had been hypothesized by some researchers that low vaginal pH (<5) is sufficient to prevent infection, including UTI, sexually transmitted diseases and AIDS. We were not convinced that this was the case, so we analysed ten patient files to that point and found that lactobacillus therapy maintained the vaginal pH at acidic levels (4.8 ± 0.5) during the study (Reid and Bruce, 1995). UTI rates fell by 79.4% but UTI did occur despite vaginal pH being acidic at 4.8, and the causative agents were common uropathogens (six *E. coli*, three streptococci and three staphylococci). We conclude that vaginal pH below 5 is not sufficient alone to prevent UTI.

It could be argued that the clinical trials had not actually proved that GR-1 and/or B-54 had colonized the host. In 1993/94, having developed a molecular identification system (ribotyping and chromotyping) for lactobacilli, strain GR-1 was instilled in a suppository into the vagina of a woman with a history of recurrent urogenital infections and we took vaginal swabs for 7 weeks. Strain GR-1 was identified colonizing the vagina on each sampling up to 7 weeks (Reid, Millsap and Bruce, 1994). The molecular typing results were clear and conclusive. Coincidentally, the patient reported being very well with no urogenital infection signs or symptoms. The enterococci that had dominated her vagina prior to treatment were replaced (as far as being the dominant organism is concerned) by GR-1.

It is clear that lactobacilli express several factors that allow them to compete in a variety of microenvironments, and different sets of properties are useful to compete with other organisms, depending upon the nature and site of the interaction. For example, lactobacilli can use many mechanisms to adhere to surfaces, such as electrostatic, hydrophobic, hydrophilic, capsular and fimbrial mechanisms (Cook, Harris and Reid, 1988; Cuperus *et al.*, 1992, 1993; Reid *et al.*, 1992b; McGroarty, 1994), in the urogenital tract hydrophilic GR-1 and hydrophobic *L. fermentum* B-54 both colonize (Reid, Bruce and Taylor, 1995).

Once implanted, the ability of lactobacilli to alter their characteristics in response to nutritional changes is well known. We have illustrated this by freeze substitution and electron microscopic examination of cell structure and morphology (Cook, Harris and Reid, 1988), and by physicochemical analyses, which have shown that strain B-54 becomes more hydrophilic with passage (Cuperus *et al.*, 1992). A critical issue that still needs to be resolved is what happens to lactobacilli in the human vagina or intestine, with respect to their structure, morphology, expression of adhesins and production of antimicrobial substances. This remains an important area for future investigation.

Presumably lactobacilli continually enter the intestine from food sources. There are studies that indicate that certain species, such as *L. acidophilus*, are better able to survive stomach pH and bile salt exposure to pass through into the intestine (Sarra and Dellaglio, 1984). In other studies of biopsied intestinal mucosa, strains of *L. plantarum*, *L. casei* subsp. *rhamnosus*, *L. reuteri* and *L. agilis* have been recovered postimplantation (Johansson *et al.*, 1993). However, without the use of molecular typing and specific probes one cannot be certain that they are transmitted or that they subsequently colonize the female urogenital tract. One could argue that the strains that appear in stool and in vaginal mucosa throughout life are those that colonized shortly after birth, and that subsequent events such as nutritional and hormonal changes have led to the emergence or suppression of certain strains. A lengthy and large epidemiological study might be needed to better understand the succession of the normal microflora during different phases of life.

We assume that artificial colonization of the urogenital tract is possible, and that *in vitro* analyses will allow us to construct methods for selecting and delivering strains that are counterproductive to uropathogens. However, an adjunct to such studies must be an examination of the strains already colonizing healthy humans and a clarification of how these organisms are altered by food intake. Based upon our knowledge, strains with different origins and probiotic properties seem to exist in the vagina of healthy women (Reid *et al.*, 1996; Zhong *et al.*, 1998). The challenge will be to identify which strains are the most beneficial for health and why.

The microenvironment in which these organisms live is a vibrant, dynamic ecosystem under constant change. In a study carried out over 20 years ago, of the urogenital microflora of healthy nurses it was shown that the microflora changed in composition hourly (Seddon *et al.*, 1976). In the intestine, the peristalsis and incoming food substances constitute the major change components. In the urogenital tract, hormone levels, semen, urine, menstruation and exposure to spermicides are the main factors influencing the microflora. It has been shown that lactobacilli grow in the presence of estrogen (Reid, unpublished) and increase their adhesion to epithelial cells during the menstrual cycle when estrogen levels are highest (Chan, Bruce and Reid, 1984), and that estrogen replacement therapy causes re-emergence of lactobacilli and reduction in the incidence of UTI (Raz and Stamm, 1993) and bacterial vaginosis (Parent *et al.*, 1996). Lactobacilli create and thrive in an acid pH environment (Reid and Bruce, 1995). Thus, many environmental factors affect colonization.

Our studies have shown that the vaginal microflora can be altered with application of nutrients such as skim milk based compounds (Reid, Bruce and Taylor, 1995). More recently, we have found *in vitro* that certain compounds, such as ascorbic acid, can promote growth of lactobacilli over uropathogens (Fig. 17.1; Reid *et al.*, 1998).

Thus, there is potential to alter the composition of the microflora by using natural nutrients. This would certainly have appeal to patients and health professionals, but much research remains to be done. Another food substance that appears to have functional properties in disease prevention is cranberry juice, which creates an acidic environment in the bladder urine and interferes with uropathogen adhesion to surfaces (Murphy, Zelman and Mau, 1965; Zafriri *et al.*, 1989). The combination of cranberry juice with lactobacillus probiotics has appeal in theory but has not yet been tested. Vitamin B complex, taken orally twice daily for 1–2 weeks, has been reported to be important for artificial implantation of *L. acidophilus* into the vagina to treat infection (Friedlander, Druker and Schachter, 1986). Clearly, there is a potential role for prebiotics (nutrients which stimulate probiotic growth) in prevention of disease in the urogenital tract, but further investigations are essential.

In summary, no single method has been found to effectively prevent UTI, apart from antibiotics. The concept of altering the urogenital microflora appears to hold some promise. Although no *Lactobacillus* isolate has been found that contains all the protective characteristics that appear important to prevent UTI, certain strains (GR-1, B-54, RC-14) do appear to possess enough of the characteristics to make them effective *in vivo*. Time will tell if restoration of the urogenital microflora is feasible in a large population and if this reduces significantly the risk of urinary tract infections. Studies that examine how uropathogens overcome the microflora's defenses will also be helpful in better understanding the disease process and devising suitable ways of intervening.

17.6 REFERENCES

Alexander, J. G. (1967) Thrush bowel infection: existence, incidence, prevention and treatment, particularly by a *Lactobacillus acidophilus* preparation. *Current Medicinal Drugs*, **8**, 3–11.

Beck, C. and Necheles, H. (1961) Beneficial effects of administration of *Lactobacillus acidophilus* in diarrheal and other intestinal disorders. *American Journal of Gastroenterology*, **35**, 522–530.

Bergeron, M. G. (1972) Treatment of pyelonephritis in adults. *Medical Clinics of North America*, **79**, 619–649.

Biller, J. A., Katz, A. J., Flores, A. F. *et al.* (1995) Treatment of recurrent *Clostridium difficile* colitis with *Lactobacillus* GG. *Journal of Pediatric Gastroenterology and Nutrition*, **21**, 224–226.

Brooks, H. J. L., O'Grady, F., McSherry, M. A. and Cattell, M. R. (1980) Uropathogenic properties of *Escherichia coli* in recurrent urinary-tract infection. *Journal of Medical Microbiology*, **13**, 57–68.

Bruce, A. W. and Reid, G. (1988) Intravaginal instillation of lactobacilli for prevention of recurrent urinary tract infections. *Canadian Journal of Microbiology*, **34**, 339–343.

Bruce, A. W., Reid, G., McGroarty, J. A. *et al.* (1992) Preliminary study on the prevention of recurrent urinary tract infections in ten adult women using intravaginal lactobacilli. *International Urogynecology Journal*, **3**, 22–25.

Chan, R. C. Y., Bruce, A. W. and Reid, G. (1984) Adherence of cervical, vaginal and distal urethral normal microbial flora to human uroepithelial cells and the inhibition of adherence of uropathogens by competitive exclusion. *Journal of Urology*, **131**, 596–601.

Chan, R. C. Y., Reid, G., Irvin, R. T. *et al.* (1985) Competitive exclusion of uropathogens from uroepithelial cells by *Lactobacillus* whole cells and cell wall fragments. *Infection and Immunity*, **47**, 84–89.

Chang, H. (1996) Genetic engineering to enhance microbial interference and related therapeutic applications. *Nature Biotechnology*, **14**, 444–447.

Clements, M. L., Levin, M. M., Black, R. E. *et al.* (1981) *Lactobacillus* prophylaxis for diarrhea due to enterotoxigenic *Escherichia coli*. *Antimicrobial Agents and Chemotherapy*, **20**, 104–108.

Connell, H., Agace, W., Klemms, P. *et al.* (1996) Type 1 fimbrial expression enhances *Escherichia coli* virulence for the urinary tract. *Proceedings of the National Academy of Sciences of the USA*, **93**, 9827–9832.

Cook, R. L., Harris, R. J. and Reid, G. (1988) Effect of culture media and growth phase on the morphology of lactobacilli and on their ability to adhere to epithelial cells. *Current Microbiology*, **17**(3), 159–166.

Costerton, J. W. and Lappin-Scott, H. M. (1995) Introduction to microbial biofilms, in *Microbial Biofilms*, (eds H. M. Lappin-Scott and J. W. Costerton), Cambridge University Press, Cambridge, pp. 1–11.

Costerton, J. W., Irwin, R. T. and Cheng, K.-J. (1981) The bacterial glycocalyx in nature and disease. *Annual Review of Microbiology*, **35**, 399–424.

Cuperus, P., van der Mei, H. C., Reid, G. *et al.* (1992) The effect of serial passaging of lactobacilli in liquid medium on their physico-chemical and structural surface characteristics. *Cells and Materials*, **2**, 271–280.

Cuperus, P. L., van der Mei, H. C., Reid, G. *et al.* (1993) Physico-chemical surface characteristics of urogenital and poultry lactobacilli. *Journal of Colloids and Interface Science*, **156**, 319–324.

Demacias, M. E. N., Deruiz, C. S., Debocanera, M. E. L. and Holgado, A. A. P. D. (1996) Preventive and therapeutic effects of lactobacilli on urinary tract infections in mice. *Anaerobe*, **2**, 85–93.

de Ruiz, S. C., de Bocanera, L. M. E. and de Macias, N. M. E. (1996) Effect of lactobacilli and antibiotics on *E. coli* urinary infections in mice. *Biological and Pharmaceutical Bulletin*, **19**, 88–93.

Elmer, G. W., Surawicz, C. M. and McFarland, L. V. (1996) Biotherapeutic agents: a neglected modality for the treatment and prevention of selected intestinal and vaginal infections. *Journal of the American Medical Association*, **275**, 870–876.

Eschenbach, D. A., Davick, P. R., Williams, S. J. *et al.* (1989) Prevalence of hydrogen peroxide-producing *Lactobacillus* species in normal women and women with bacterial vaginosis. *Journal of Clinical Microbiology*, **27**, 251–256.

Foxman, B. and Fredrichs, R. R. (1985) Epidemiology of UTI. Diaphragm use and sexual intercourse. *American Journal of Public Health*, **75**, 1308–1313.

Foxman, B., Zhang, L., Tallman, P. *et al.* (1995) Virulence characteristics of *Escherichia coli* causing first urinary tract infection predict risk of second infection. *Journal of Infectious Diseases*, **172**, 1536–1541.

Friedlander, A., Druker, M. M. and Schachter, A. (1986) *Lactobacillus acidophilus* and vitamin B complex in the treatment of vaginal infection. *Panminerva Medica*, **28**, 51–53.

Gorbach, S. L., Chang, T. and Goldin, B. (1987) Successful treatment of relapsing *Clostridium difficile* colitis with *Lactobacillus* GG. *Lancet*, **ii**, 1519.

Hacker, J., Schmidt, G., Hughes, C. *et al.* (1985) Cloning and characterization of genes involved in production of mannose-resistant, neuraminidase-susceptible (X) fimbriae from a uropathogenic O6:K15:H31 *Escherichia coli* strain. *Infection and Immunity*, **47**, 434–440.

Hagberg, L., Bruce, A. W., Reid, G. *et al.* (1989) Colonization of the urinary tract with live bacteria from the normal fecal and urethral flora in patients with recurrent symptomatic urinary tract infections, in *Host–Parasite Interactions in Urinary Tract Infections*, (eds E. H. Kass and C. Svanborg Edén), University of Chicago Press, Chicago, IL, pp. 194–197.

Hamilton-Miller, J. M. T. (1997) Living in the 'post-antibiotic era': could the use of probiotics be an effective strategy? *Clinical Microbiology and Infection*, **3**, 2–3.

Hay, P. E., Lamont, R. F, Taylor-Robinson, D. *et al.* (1994) Abnormal bacterial colonisation of the genital tract and subsequent preterm delivery and late miscarriage. *British Medical Journal*, **308**, 295–298.

Hillier, S. L., Krohn, M. A., Rabe, L. K. *et al.* (1993) The normal vaginal flora, H_2O_2-producing lactobacilli, and bacterial vaginosis in pregnant women. *Clinical Infectious Diseases*, **16**(Suppl. 4), S273–S281.

Hilton, E., Rindos, P. and Isenberg, H. D. (1995) *Lactobacillus* GG vaginal suppositories and vaginitis. *Journal of Clinical Microbiology*, **33**, 1433.

Hilton, E., Isenberg, H. D., Alperstein, P. *et al.* (1992) Ingestion of yogurt containing *Lactobacillus acidophilus* as prophylaxis for Candidal vaginitis. *Annals of Internal Medicine*, **116**, 353–357.

Hooton, T. M., Hillier, S., Johnson, C. *et al.* (1991) *Escherichia coli* bacteriuria and contraceptive method. *Journal of the American Medical Association*, **265**, 64–69.

Hultgren, S. J., Porter, T. N., Schaeffer, A. J. and Duncan, J. L. (1985) Role of type 1 pili and effects of phase variation on lower urinary tract infections produced by *Escherichia coli*. *Infection and Immunity*, **50**, 370–377.

Isolauri, E., Juntunen, M. and Rautanen, T. (1991) A human *Lactobacillus* strain (*Lactobacillus* GG) promotes recovery from acute diarrhea in children. *Pediatrics,* **88**, 90–97.

James, G. A., Korber, D. R., Caldwell, D. E. and Costerton, J. W. (1995) Digital image analysis of growth and starvation responses of surface-colonizing *Actinobacter* sp. *Journal of Bacteriology,* **177**, 907–915.

Johansson, M.-L., Molin, G., Jeppsson, B. *et al.* (1993) Administration of different *Lactobacillus* strains in fermented oatmeal soup: *in vivo* colonization of human intestinal mucosa and effect on the indigenous flora. *Applied Environmental Microbiology,* **59**, 15–20.

Kallenius, G., Mollby, R. and Winberg, J. (1979) Adhesion of *Escherichia coli* to periurethral cells correlated to mannose-resistant agglutination of human erythrocytes. *FEMS Microbiology Letters,* **5**, 295–299.

Kist, M. L., Salit, I. E. and Hofmann, T. (1990) Purification and characterization of the Dr hemagglutinins expressed by two uropathogenic *Escherichia coli* strains. *Infection and Immunity,* **58**, 695–702.

Klaenhammer, T. R. (1982) Microbiological considerations in selection and preparation of *Lactobacillus* strains for use as dairy adjuncts. *Journal of Dairy Science,* **65**, 1339–1349.

Korshunov, V. M., Sinitsyna, N. A., Ginodman, G. A. and Pinegin, B. V. (1985) Correction of the intestinal microflora in cases of chemotherapeutic dysbacteriosis by means of bifidobacterial and lactobacterial autostrains. *Zhurnal Microbiologii Epidemiologii i Immunobiologii,* **9**, 20–25.

Kunin, C. M. (1987) *Detection, Prevention and Management of Urinary Tract Infections,* 4th edn, Lea & Febiger, Philadelphia, PA.

McGroarty, J. A. (1994) Cell surface appendages of lactobacilli. *FEMS Microbiology Letters,* **124**, 405–409.

McGroarty, J. A. and Reid, G. (1988a) Detection of a *Lactobacillus* substance which inhibits *Escherichia coli*. *Canadian Journal of Microbiology,* **34**, 974–978.

McGroarty, J. A. and Reid, G. (1988b) Inhibition of enterococci by *Lactobacillus* species *in vitro*. *Microbial Ecology in Health and Disease,* **1**, 215–219.

McGroarty, J. A., Chong, S., Reid, G. and Bruce, A. W. (1990) Influence of the spermicidal compound nonoxynol-9 on the growth and adhesion of urogenital bacteria *in vitro*. *Current Microbiology,* **21**(4), 219–223.

McGroarty, J. A., Tomeczek, L., Pond, D. G. *et al.* (1992) Hydrogen peroxide production by *Lactobacillus* species, correlation with susceptibility to the spermicidal compound nonoxynol-9. *Journal of Infectious Diseases,* **165**(6), 1142–1144.

McGroarty, J. A., Faguy, D., Bruce, A. W. and Reid, G. (1993) Influence of the spermicidal compound nonoxynol-9 on adhesion of *E. coli*. *International Urogynecology Journal,* **4**(4), 194–198.

Miller, M. R., Bacon, C., Smith, S. L. *et al.* (1993) Enteral feeding of premature infants with *Lactobacillus* GG. *Archives of Disease in Childhood,* **69**, 483–487.

Murphy, F. J., Zelman, S. and Mau, W. (1965) Ascorbic acid as a urinary acidifying agent: 2. Its adjunctive role in chronic urinary infection. *Journal of Urology,* **94**, 300–304.

Newman, D. (1915) The treatment of cystitis by intravesical injections of lactic *Bacillus* cultures. *Lancet,* **Aug 14**, 330–332.

Oksanen, P., Salminen, S., Saxelin, M. *et al.,* (1990) Prevention of traveler's diarrhea by *Lactobacillus* GG. *Annals of Medicine,* **22**, 53–56.

Parent, D., Bossens, M., Bayot, D. *et al.* (1996) Therapy of bacterial vaginosis using exogenously-applied lactobacilli acidophili and a low dose of estriol – a placebo-controlled multicentric clinical trial. *Arzneimittel-Forschung,* **46**, 68–73.

Perdigon, G., Alvarez, S., Rachid, M. *et al.* (1995) Immune system stimulation by probiotics. *Journal of Dairy Science*, **78**, 1597–1606.

Pere, A., Leinonen, M., Vaisanen-Rhen, V. *et al.* (1985) Occurrence of type-1C fimbriae on *Escherichia coli* strains isolated from human extraintestinal infections. *Journal of General Microbiology*, **131**, 1705–1711.

Potera, C. (1996) Biofilms invade microbiology. *Science*, **273**, 1795–1797.

Preston, C. A. K., Bruce, A. W. and Reid, G. (1992) Antibiotic resistance of urinary pathogens isolated from patients attending The Toronto Hospital between 1986 and 1990. *Journal of Hospital Infection*, **21**, 129–135.

Raz, R. and Stamm, W. E. (1993) A controlled trial of intravaginal estriol in postmenopausal women with recurrent urinary tract infections. *New England Journal of Medicine*, **329**, 753–756.

Redondo-Lopez, V., Cook, R. L. and Sobel, J. D. (1990) Emerging role of lactobacilli in the control and maintenance of the vaginal bacterial microflora. *Reviews of Infectious Diseases*, **12**, 856–872.

Reid, G. (1994) Do antibiotics clear bladder infections? *Journal of Urology*, **152**, 865–867.

Reid, G. (1996) Probiotics: how organisms compete. *Journal of the American Medical Association*, **276**, 29.

Reid, G. and Bruce, A. W. (1995) Low vaginal pH and urinary-tract infection. *Lancet*, **346**, 1704.

Reid, G. and Howard, L. (1997) Effect on uropathogens of prophylaxis for urinary tract infection in spinal cord injured patients: preliminary study. *Spinal Cord*, **35**, 605–607.

Reid, G. and Seidenfeld, A. (1997) Drug resistance among uropathogens isolated from women in a suburban population: laboratory findings over 7 years. *Canadian Journal of Urology*, **4**, 432–437.

Reid, G. and Sobel, J. D. (1987) Bacterial adherence in the pathogenesis of urinary tract infection: a review. *Reviews of Infectious Diseases*, **9**(3), 470–487.

Reid, G., Bruce, A. W. and Taylor, M. (1995) Instillation of *Lactobacillus* and stimulation of indigenous organisms to prevent recurrence of urinary tract infections. *Microecology Therapy*, **23**, 32–45.

Reid, G., Cook, R. L. and Bruce, A. W. (1987) Examination of strains of lactobacilli for properties which may influence bacterial interference in the urinary tract. *Journal of Urology*, **138**, 330–335.

Reid, G., Millsap, K. and Bruce, A. W. (1994) Implantation of *Lactobacillus casei* var *rhamnosus* into the vagina. *Lancet*, **344**, 1229.

Reid, G., Chan, R. C. Y., Bruce, A. W. and Costerton, J. W. (1985) Prevention of urinary tract infection in rats with an indigenous *Lactobacillus casei* strain. *Infection and Immunity*, **49**(2), 320–324.

Reid, G., McGroarty, J. A., Angotti, R. and Cook, R. L. (1988) *Lactobacillus* inhibitor production against *E. coli* and coaggregation ability with uropathogens. *Canadian Journal of Microbiology*, **34**, 344–351.

Reid, G., Cook, R. L., Hagberg, L. and Bruce, A. W. (1989) Lactobacilli as competitive colonizers of the urinary tract, in *Host–Parasite Interactions in Urinary Tract Infections*, (eds E. H. Kass and C. Svanborg Edén), University of Chicago Press, Chicago, IL, pp. 390–396.

Reid, G., Bruce, A. W., Cook, R. L. and Llano, M. (1990a) Effect on the urogenital flora of antibiotic therapy for urinary tract infection. *Scandinavian Journal of Infectious Diseases*, **22**, 43–47.

Reid, G., McGroarty, J. A., Domingue, P. A. G. *et al.* (1990b) Coaggregation of urogenital bacteria *in vitro* and *in vivo*. *Current Microbiology*, **20**, 47–52.

Reid, G., Charbonneau-Smith, R., Lam, D. *et al.* (1992a) Bacterial biofilm formation in the urinary bladder of spinal cord injured patients. *Paraplegia*, **30**, 711–717.

Reid, G., Cuperus, P. L., Bruce, A. W. *et al.* (1992b) Comparison of contact angles and adhesion to hexadecane of urogenital, dairy and poultry lactobacilli: effect of serial culture passages. *Applied Environmental Microbiology*, **58**(5), 1549–1553.

Reid, G., McGroarty, J. A., Tomeczek, L. and Bruce, A. W. (1996) Identification and plasmid profiles of *Lactobacillus* species from the vagina of 100 healthy women. *FEMS Immunology and Medical Microbiology*, **15**, 23–26.

Reid, G., Bruce, A. W., Soboh, F. and Mittel man (1998) Effect of nutrient on the in vitro growth of urogenital *Lactobacillus* and uropathogens. *Canadian Journal of Microbiology*, in press.

Rhen, M., Klemm, P. and Korhonen, T. K. (1986) Identification of two new hemagglutinins of *Escherichia coli*, N-acetyl-D-glucosamine-specific fimbriae and a blood group M-specific agglutinin, by cloning the corresponding genes in *Escherichia coli* K-12. *Journal of Bacteriology*, **168**, 1234–1242.

Russo, T. A., Brown, J. J., Jodush, S. T. and Johnson, J. R. (1996a) The O4 specific antigen moiety of lipopolysaccharide but not the K54 group 2 capsule is important for urovirulence of an extraintestinal isolate of *Escherichia coli*. *Infection and Immunity*, **64**, 2343–2348.

Russo, T. A., Jodush, S. T., Brown, J. J. and Johnson, J. R. (1996b) Identification of two previously unrecognized genes (*guaA* and *argC*) important for uropathogenesis. *Molecular Microbiology*, **22**, 217–229.

Sadhu, K., Domingue, P. A. G., Chow, A. W. *et al.* (1989) A morphological study of the *in situ* tissue-associated autochthonous microflora of the human vagina. *Microbial Ecology in Health and Disease*, **2**, 99–106.

Sanford, B. A., de Feijter, A. W., Wade, M. H. and Thomas, V. L. (1996) A dual fluorescence technique for visualization of *Staphylococcus epidermidis* biofilm using scanning confocal laser microscopy. *Journal of Industrial Microbiology*, **16**, 48–56.

Sarra, P. C. and Dellaglio, F. (1984) Colonization of a human intestine by four different genotypes of *Lactobacillus acidophilus*. *Microbiologia*, **7**, 331–339.

Savage, D. C. (1977) Microbial ecology of the gastrointestinal tract. *Annual Review of Microbiology*, **31**, 107–133.

Saxelin, M., Pessi, T. and Salminen, S. (1995) Fecal recovery following oral administration of *Lactobacillus* strain GG (ATCC 53103) in gelatine capsules to healthy volunteers. *International Journal of Food Microbiology*, **25**, 199–203.

Schaeffer, A. J. and Stamey, T. A. (1977) Studies of introital colonization in women with recurrent urinary infections. IX. The role of antimicrobial therapy. *Journal of Urology*, **118**, 221–224.

Seddon, J. M., Bruce, A. W., Chadwick, P. and Carter, D. (1976) Introital bacterial flora – effect of increased frequency of micturition. *British Journal of Urology*, **48**, 211–218.

Shalev, E., Battino, S., Weiner, E. *et al.* (1996) Ingestion of yogurt containing *Lactobacillus acidophilus* compared with pasteurized yogurt as prophylaxis for recurrent Candidal vaginitis and bacterial vaginosis. *Archives of Family Medicine*, **5**, 593–596.

Siitonen, S., Vapaatalo, H., Salminen, S. *et al.* (1990) Effect of *Lactobacillus* GG yogurt and prevention of antibiotic-associated diarrhea. *Annals of Medicine*, **22**, 57–59.

Stamey, T. A. (1973) The role of introital enterobacteria in recurrent urinary infections. *Journal of Urology*, **109**, 467–472.

Svanborg Edén, C., Hanson, L. A., Jodal, U. *et al.* (1976) Variable adherence to normal human urinary-tract epithelial cells of *Escherichia coli* strains associated with various forms of urinary-tract infection. *Lancet,* **ii**, 490–492.

Vaisanen, V., Korhonen, T. K., Jokinen, M. *et al.* (1982) Blood group M specific haemagglutinin in pyelonephritogenic *Escherichia coli. Lancet,* **i**, 1192.

Velraeds, M. C., van der Mei, H. C., Reid, G. and Busscher, H. J. (1996a) Inhibition of initial adhesion of uropathogenic *Enterococcus faecalis* by biosurfactants from *Lactobacillus* isolates. *Applied Environmental Microbiology,* **62**, 1958–1963.

Velraeds, M. C., van der Mei, H. C., Reid, G. and Busscher, H. J. (1996b) Physico-chemical and biochemical characterization of biosurfactants released from *Lactobacillus* strains. *Colloids and Surfactants B: Biointerfaces,* **8**, 51–61.

Velraeds, M. C., van der Belt, B., van der Mei, H. C. *et al.* (1998) Interference in initial adhesion of uropathogenic bacteria and yeasts to silicone rubber by a *Lactobacillus acidophilus* biosurfactant. *Journal of Medical Microbiology,* in press.

Watkins, B. A., Miller, B. F. and Neil, D. H. (1982) *In vivo* inhibitory effects of *Lactobacillus acidophilus* against pathogenic *Escherichia coli* in gnotobiotic chicks. *Poultry Science,* **61**, 1298–1308.

Wold, A. E., Caugant, D. A., Lidin-Janson, G. *et al.* (1992) Resident colonic *Escherichia coli* strains frequently display uropathogenic characteristics. *Journal of Infectious Diseases,* **165**, 46–52.

Wu, X., Sun, T. and Medina, J. J. (1996) In vitro binding of type 1-fimbriated *Escherichia coli* to uroplakins Ia and Ib: relation to urinary tract infections. *Proceedings of the National Academy of Sciences of the USA,* **93**, 9630–9635.

Zafriri, D., Ofek, I., Adar, R. *et al.* (1989) Inhibitory activity of cranberry juice on adherence of type 1 and type P fimbriated *Escherichia coli* to eucaryotic cells. *Antimicrobial Agents and Chemotherapy,* **33**, 92–98.

Zdolsek, B., Hellberg, D., Froman, G. *et al.* (1995) Vaginal microbiological flora and sexually transmitted diseases in women with recurrent or current vulvo-vaginal candidiasis. *Infection,* **23**, 81–84.

Zhong, W., Millsap, K., Bialkowska-Hubrzanska, H. and Reid, G. (1998) Differentiation of *Lactobacillus* species by molecular typing. *Applied and Environmental Microbiology,* **64**, 2418–2423.

18

Candida albicans: from commensal to pathogen

Paul L. Fidel Jr

18.1 INTRODUCTION

Candida albicans is a ubiquitous fungal organism that is part of the normal microflora of most, if not all, individuals. However, it is also an opportunistic pathogen and can quickly transform from a harmless inhabitant of mucocutaneous tissues to a highly pathogenic organism capable of killing its host under the appropriate conditions. *Candida* species are dimorphic organisms that can grow as a yeast or as a multi-nucleated hyphal form. Innate and acquired host defense mechanisms are responsible for keeping the organism in the commensal state. However, when the organism overwhelms these defenses or the defenses become deficient or lost, an infection will incur. Infections can be acute, chronic or recurrent and affect cutaneous or mucocutaneous tissues as well as systemic blood/organs. This chapter reviews the properties of *C. albicans* as a commensal organism and as a pathogen. The respective protective host defense mechanisms responsible for keeping the organism in the commensal state are included, together with immunological changes that occur, or are postulated to occur, to initiate an infectious condition. In some cases, current data are used to formulate hypotheses regarding the initiation of infection and potential immune-based therapies. Finally, the incidence and potentia mechanisms associated with candidiasis in the HIV patient are reviewed.

G. W. Tannock (ed.), *Medical Importance of the Normal Microflora*, 441–476.

18.2 THE ORGANISM

Candida albicans is one of more than 200 species in the genus *Candida* (Kreger-van Rij, 1984). There are currently 23 different strain types (karyotypes) of *C. albicans* detected by contour-clamped homogenous electric field electrophoresis (CHEF; Vazquez *et al.*, 1994). Isolates of *C. albicans* occur in two serotypes, A and B (Hasenclever, Mitchell and Loewe, 1961). In the USA, the frequency of serotypes A and B are approximately equal, whereas serotype A is more prevalent in Canada, France, the UK and north Africa. Serotype B appears more prevalent in cases of antimycotic resistance and in AIDS patients (Kwon-Chung and Bennett, 1992). Other *Candida* species that cause infections in humans include *C. tropicalis*, *C. glabrata* (formally known as *Torulopsis glabrata*), *C. parapsilosis*, *C. stellatoidea* and *C. krusei*. *C. albicans* accounts for approximately 60–75% of candidal infections, while *C. tropicalis* accounts for 15–25%, *C. glabrata* for 10–15%, *C. parapsilosis* for 5–15%, *C. krusei* for 2–5% and *C. stellatoidea* for 1–2% (Kwon-Chung and Bennett, 1992). Although there may be slight changes in the rates of infection depending on the tissue infected, it is quite clear that *C. albicans* causes the majority of infections. Therefore, the remainder of the chapter will be devoted to *C. albicans* and the relevant host responses important for protection.

As stated above, *C. albicans* is classified as a dimorphic fungus. It can grow as an oval-shaped budding yeast (blastospore or blastoconidia), as pseudohyphae or as true hyphae (Fig. 18.1).

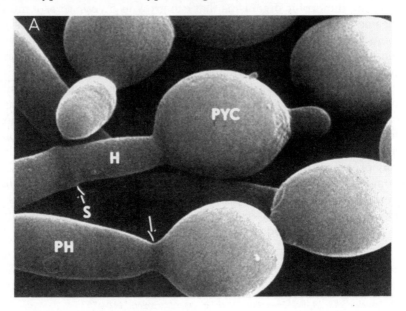

Figure 18.1 continued on facing page

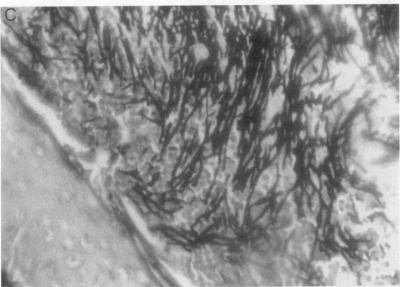

Figure 18.1 Forms of *C. albicans*. (A) Electron micrograph of *C. albicans* showing the parental yeast cell (PYC) (blastospore), pseudohyphae (PH), and hyphae (H) with septum (S). Arrow represents constriction at septum between the parent cell and pseudohyphae. Source: reproduced with permission from Cole, 1986. (B) Electron micrograph of mycelium. (C) Silver stain of hyphae associated with the vaginal mucosa.

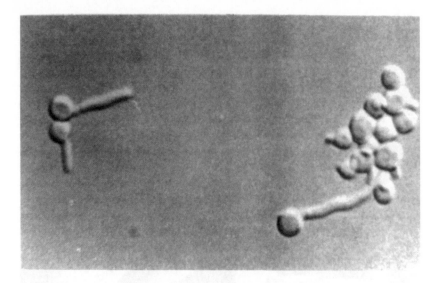

Figure 18.2 Wet mount preparation of *C. albicans* germ tubes. Source: reproduced with permission from Kwon-Chung, K. J. and Bennett, J. E. (1992) Candidiasis, in *Medical Mycology*, (ed. C. Cann), Lea & Febiger, Philadelphia, PA.

True hyphae ultimately form multinucleated masses called mycelium. Pseudohyphal and true hyphae are differentiated by the fact that pseudohyphae have invaginations at each septum whereas true hyphae do not. *C. albicans* and *C. stellatoidea* are unique in that they are the only *Candida* species that form an intermediate morphological structure *in vitro*, termed the germ tube, when incubated in several types of animal sera or commercial media designed to promote hyphal growth. Germ tubes are the tubular structures that initiate the hyphal stage of growth (Fig. 18.2).

When the tube has attained sufficient length a septum is produced dividing the cell into mononuclear compartments. This process of elongation continues in a similar manner, ultimately resulting in mycelium. A schematic diagram of the different morphologic transitions that *C. albicans* undergoes *in vitro* is shown in Fig. 18.3. *C. albicans* also uniquely forms a spherical, thick-walled chlamydospore that occurs along pseudohyphae or on the tip of hyphae (Fig. 18.4).

There are several environmental factors that allow *C. albicans* blastospores to transform into the hyphal form, the most common of which are temperature and pH. Acidic pH and temperatures below 35°C usually favor the blastospore form, whereas alkaline pH (> 6.5) and temperatures of 37°C or above favor hyphal formation. It is the hyphal form of *C. albicans* that is often observed in infected tissue, while the blastospore

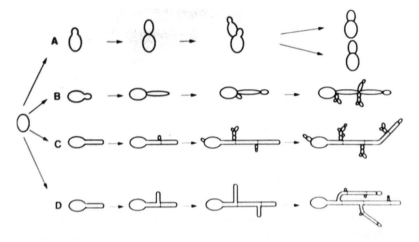

Figure 18.3 Schematic diagram of *C. albicans* morphogenesis. (A) Budding from blastospores. (B) Formation of pseudohyphae (invaginations at septa).
(C) Formation of hyphae from germ tubes (no invaginations at septa) under natural conditions (37°C). (D) Formation of hyphae under specific conditions of germ tube formation (serum at 37°C). Source: reproduced with permission from Kwon-Chung, K. J. and Bennett, J. E. (1992) Candidiasis, in *Medical Mycology*, (ed. C. Cann), Lea & Febiger, Philadelphia, PA.

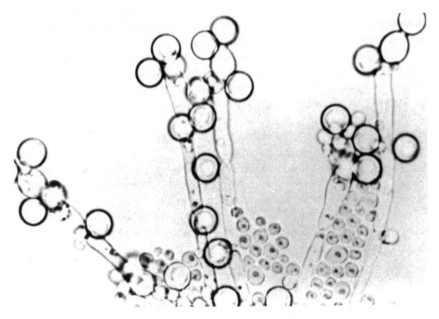

Figure 18.4 *C. albicans* chlamydospore. Source: reproduced with permission from Kwon-Chung and Bennett, 1992.

Table 18.1 Virulence factors of *C. albicans*

- Dimorphism
  - –Germination
  - –Hyphae/mycelium
  - –Antigenic variation
- Integrin analogs that promote adherence
- Phenotype switching
- Hydrolase secretion
- Heat-shock proteins
- Antimycotic drug resistance

is present in the commensal situation. The conditions promoting dimorphism of *C. albicans* are in sharp contrast to those that promote dimorphism in other fungi (e.g. *Blastomycoses dematiditis* and *Histoplasma capsulatum*), where the yeast form occurs at 37°C and the hyphal or mold form occurs at 25°C.

C. albicans has several virulence factors that promote adherence and ultimately disease. Table 18.1 contains a list of these virulence factors.

Dimorphism itself is considered a virulence factor of *C. albicans*. Hyphae appear to adhere and invade tissue by elaborating proteolytic enzymes. A secondary effect of dimorphism that can be associated with virulence is antigenic variation (DeBernardis *et al.*, 1994). Antigens present on the blastospore and hyphal surface may be different and thus evade immune surveillance. Another virulence factor is the presence of integrin analogs on *C. albicans* that enhance adhesion to epithelial cells (Bendel *et al.*, 1995). Interestingly, compared to non-*albicans* species, *C. albicans* has been reported to express the highest levels of the integrin analog. Phenotype switching, an *in vitro* phenomenon that involves phenotypic changes in the absence of genotypic changes, represents yet another possible virulence factor (Soll, 1992). If it occurs *in vivo*, it may be predicted to effect both adherence and immunological escape. Additional virulence factors are the secretion of hydrolases which enhance adherence to mucosal tissues (Ross *et al.*, 1990), and heat-shock proteins that bind to serum proteins and inhibit proper folding of the serum proteins as well as interactions with other proteins. Finally, antimycotic drug resistance of some *Candida* species allows them to escape therapeutic treatment regimes.

18.3 *CANDIDA ALBICANS*: THE COMMENSAL

As stated previously, *C. albicans* is a commensal organism of the gastrointestinal and reproductive tracts. In this colonized state, subjects are considered to be asymptomatic carriers. It is widely recognized that, as a commensal, *C. albicans* exists exclusively in the blastospore form, although it is rarely detectable in wet mount preparations of tissue scrapings. Instead, asymptomatic colonization is usually detected by swab culture. The most common sites showing asymptomatic colonization are the rectum, oral cavity and vagina. Although many *Candida* species are considered normal microflora, *C. albicans* is the most common species recovered. In the oral cavity, the approximate rate of yeast carriage is 25–30% but has been reported to be as low as 2% and as high as 69% (Odds, 1988). *C. albicans* was identified in approximately 75–80% of those with positive oral yeast carriage (Odds, 1988). The distribution of *C. albicans* in the oral cavity includes the tongue, palate and buccal mucosa. It can also be recovered from diverse areas of the gastrointestinal tract, including oropharynx, jejunum, ileum and colon. Rectal swabs were positive for *C. albicans* in 20–25% of healthy individuals (Odds, 1988). A similar proportion of normal healthy women (i.e. 20–25%) carry *C. albicans* in the vagina (Sobel, 1988), although one study reported carriage in 40% of women (Soll *et al.*, 1991). The recovery of yeast increases during pregnancy and during selected stages of the menstrual cycle, in particular when reproductive hormones are elevated (luteal phase). In contrast to colonization with *C. albicans* at mucosal sites, *C. albicans* is not considered normal microflora of the skin. *C. albicans* is also infrequently isolated from water sources, soil, air and plants, and those isolates recovered from food and beverages are considered contaminants from human and animal sources (Hasenclever, Mitchell and Loewe, 1961).

Colonization with *C. albicans* at mucosal tissues is considered to occur at an early age, the organism having been acquired either by traveling through the birth canal, by nursing or from food sources. Aside from direct culture of mucosal surfaces, long-term colonization with *C. albicans* is assumed to be responsible for acquired immune responsiveness. Anti-*Candida* antibodies, circulating IgG and mucosal IgA and IgG, can be detected in most healthy individuals. Moreover, more than 80% of healthy persons have positive cutaneous skin test reactivity to *Candida* skin test antigen, and peripheral blood lymphocytes from over 90% of healthy individuals proliferate *in vitro* to *Candida* antigen. It is presumed that these acquired host responses in conjunction with innate resistance (i.e. polymorphonuclear leukocytes and macrophages) play a significant role in restricting *C. albicans* to mucosal surfaces.

18.4 *CANDIDA ALBICANS*: THE PATHOGEN

There are several scenarios in which *C. albicans* can convert from the harmless commensal to a serious pathogen. These include direct inoculation (i.e. colonized indwelling catheter), a lack of protective host defenses, expression of virulence factors by the organism or environmental conditions that allow the yeast to overwhelm the protective host defenses.

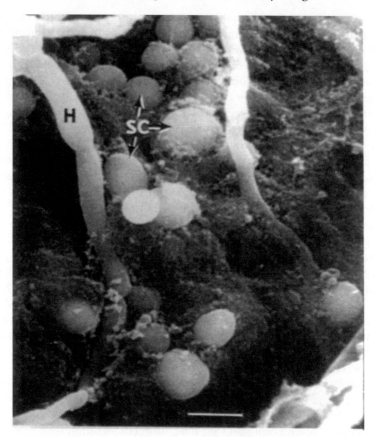

Figure 18.5 Superficial penetration of tissue by *C. albicans*. Figure shows an electron micrograph of *C. albicans* blastoconidia, pseudohyphae and hyphae superficially penetrating the gut mucosa. Source: reproduced with permission from Cole, 1986.

Tissues affected can include cutaneous, mucocutaneous and systemic blood/organs. Irrespective of the tissue infected, *C. albicans* can be observed in lesions as a mixture of pseudohyphae, hyphae and yeasts (Fig. 18.5).

Cutaneous and mucocutaneous infections are generally superficial, with invasion limited to outer layers of the tissue. Regardless though of the level of penetration, there can be large numbers of organisms associated with the tissue. It is generally considered that, as a result of some change in the host, the organism grows in large enough numbers to overcome the host defenses, penetrates the tissue and begins to cause the symptoms associated with the infection. It is unclear, however, at what point in the pathway the organism converts into the hyphal form. It is postulated that the conversion begins prior to penetration and that

Table 18.2 Clinical infections caused by *C. albicans*

Superficial candidiasis
- Cutaneous
- Mucocutaneous
    - –Oropharyngeal
    - –Gastrointestinal
    - –Vaginal
    - –Conjunctival
    - –Urinary tract

Disseminated candidiasis
- Candidemia
- Chronic disseminated
- Endocarditis
- Meningitis
- Phlebitis

---

the extension on the germ tube initiates the penetration into the tissue. Other proponents suggest that the yeast initially invades the tissue and the hyphal form occurs as a result of exposure to serum components as the immune cells infiltrate the tissue to eliminate the organism. The latter is supported by the high incidence of mucosal infections caused by *C. glabrata*, an organism that does not form hyphae.

Infections by *C. albicans* can be superficial or systemic (Table 18.2). The superficial candidal infections can involve cutaneous or mucocutaneous (i.e. oropharyngeal, gastrointestinal, vaginal, urinary tract and conjunctival) tissue. The systemic infections are disseminated and include candidemia, chronic disseminated infection, endocarditis, meningitis and phlebitis. Cutaneous candidiasis usually occurs as a diaper rash in infants or between folds of skin in obese individuals, seen as erythematous patches on the thigh and buttocks or as superficial infections of the hands in individuals who wear plastic gloves routinely. If left untreated, infections usually persist as long as the tissue remains moist. The major symptoms are burning and itching. Mucocutaneous candidiasis is a heterogenous description that applies to infections involving a number of mucosal membranes. Oropharyngeal candidiasis involves infections of the buccal mucosa, palate and tongue. The infections can be atrophic or pseudomembranous (thrush) and white lesions are seen. Chewing and swallowing can be difficult under these conditions. Infections can be acute or recurrent and are common in immunocompromised patients. Vaginitis involves infections of the vaginal lumen and often the vulva as well. Symptoms include burning, itching, soreness, an abnormal discharge and dysparunia. Signs include vaginal and vulvar erythema and edema. An estimated 75% of all women will experience an episode of acute vulvovaginitis (VVC) in their lifetime, with up to 5% succumbing to

recurrent VVC (RVVC) (Sobel, 1988). With respect to RVVC, it is postulated that recurrences are a result of relapse rather than reinfection, since
strain types of C. *albicans* tend to remain the same in women for multiple
recurrences over several years (Vazquez *et al.*, 1994). This suggests that
antifungal therapy does not completely eliminate the organisms, but
reduces their numbers to a level compatible with the commensal relationship. Conjunctivitis caused by C. *albicans* infects the mucous membranes
of the eye causing symptoms that include conjunctival erythema, cheesy
discharge and corneal ulceration. Esophagitis is the most common form
of gastrointestinal candidiasis. It is extremely painful and can promote
wasting because of the difficulty with swallowing. White plaques are
usually observed similar to thrush. Complications include bleeding and
perforation. The second most common site of gastrointestinal candidiasis
is the stomach, where symptoms of gastritis occur. Gastrointesinal candidiasis is sometimes difficult to diagnose as opposed to vaginal and oral
candidiasis due to the lack of specific symptoms and the lack of ability to
differentiate between infectious and colonized C. *albicans*. The presence of
Candida species in urine usually represents contamination of the urine
with vaginal flora. However, if a true urinary tract infection ensues, it is
most often associated with erythema and edema rather than lesions.
Unlike the more superficial forms of candidiasis, disseminated candidiasis can be fatal. It can involve multiple internal organs causing endocarditis, arthritis, osteomyelitis and meningitis.

18.4.1 Diagnosis

Together with compatible signs and symptoms, a definitive diagnosis of
cutaneous and mucocutaneous C. *albicans* infections involves the demonstration of hyphae detected by microscopic analysis of tissue scraping
(KOH wet mount preparation) or tissue invasion by histopathological
examination. A swab culture is used to isolate the organism. The organism is speciated in the laboratory using a variety of tests, such as germ
tube formation, chlamydospore development in cornmeal agar and an
assimilation of sugars, sugar alcohols and nitrate. Germ tubes can be
formed readily *in vitro* within 2 hours of culture in serum. Thus, a germ
tube test is a tremendous tool for identifying C. *albicans* or C. *stellatoidea* as
an etiological agent as opposed to one of the other *Candida* species. Since
C. *albicans* accounts for approximately 70% of candidal infections, while
C. *stellatoidea* accounts for less than 2%, a positive germ tube test on an
isolated yeast is considered diagnostic of C. *albicans*. However, to separate
C. *albicans* from C. *stellatoidea* only, one must incubate C. *albicans* under
conditions that lead to the production of the chlamydospore. Chlamydo-
spores are produced within 48–72 hours of culture when cornmeal agar is
inoculated appropriately. In cases of mucosal infections, because of the

high frequency of *C. albicans* infections, the causative agent is usually identified as *C. albicans* or non-*albicans* species. For isolates identified as *C. albicans*, CHEF karyotyping can be used for strain identification.

Systemic candidiasis is best diagnosed using the lysis centrifugation method with blood followed by specific biochemical tests described above. Pseudohyphae and hyphae can rarely be seen in blood specimens and is considered a false negative. It is difficult, however, to determine the significance of *Candida* in the blood. A catheter, for example, colonized with the organism could continually seed (inoculate) the blood. In such cases, removal of the catheter may 'cure' the septicemia.

18.4.2 Treatment

Nystatin, a topical polyene antifungal related chemically to amphotericin B, can be used for treatment of cutaneous candidiasis. Azoles (i.e. fluconazole, ketoconazole, butaconazole, itraconazole, etc.) are recommended for mucosal candidiasis. Amphotericin B is the drug of choice for systemic candidiasis. Antifungal drugs are highly effective for individual attacks or acute infections, but recurrence remains a problem. In fact, recurrent attacks can occur within days of cessation of treatment. Thus, in the appropriate clinical setting, maintenance therapy is encouraged.

18.4.3 Changes in the organism and/or host that promote infection

Since *C. albicans* is a commensal organism of the mucosa, and innate resistance and acquired protective host defenses are intact in the majority of individuals, certain criteria or events must occur with either the organism and/or the host for a mucosal infection to ensue. Mucosal *C. albicans* infections are categorized as primary or secondary. Secondary infections are defined as those candidal infections that occur as a result of some endogenous or exogenous known predisposing factor in the host (i.e. change in pH, normal microflora, hormonal influence, glycogen content, etc.). Primary infections, on the other hand are idiopathic. It is postulated that these idiopathic infections are caused by expression of virulence factors by the organism or the result of some as yet unknown abnormalities in the protective host defenses. More specific mechanisms of idiopathic candidiasis will be discussed later.

(a) Changes in the host

There are numerous host factors, both endogenous and exogenous, that are known to promote the mucosal conversion of *C. albicans* from commensal to pathogen. The most common exogenous factor is antibiotic usage. Antibiotics often reduce the normal bacterial microflora and thus influence the growth and expression of *C. albicans*. This may involve decreased competition for nutrients, but there may also be changes in the tissue, such as changes in pH, that allow *C. albicans* to overgrow and invade. It is also known that increases in glycogen (i.e. the result of diabetes) at mucosal surfaces can increase the carbon source and facilitate the growth of *C. albicans*. Endogenous factors that promote candidal infections include increased hormone production and altered host defenses. It is widely accepted that premenarchal and postmenopausal women do not develop vaginal yeast infections. In contrast, pregnant women, women taking oral contraceptives and women treated with hormone replacement therapy (HRT) are particularly susceptible (Sobel, 1988). Additionally, vaginitis occurs most often during the luteal phase of the menstrual cycle when both progesterone and estrogen are high. Thus, the expression of reproductive hormones appears to influence the incidence of vaginal candidiasis. Immunocompromising conditions (e.g. immunosuppressive therapy) have a tremendous impact on the incidence of mucosal candidiasis as well. Transplant recipients, lymphoma patients and individuals treated with corticosteroids are susceptible to many forms of mucosal candidiasis. Additionally, patients with immunodeficiency syndromes (i.e. AIDS, severe combined immunodeficiency, etc.) are highly susceptible to mucosal candidiasis. These observations stress the significance of the host defenses that help maintain *C. albicans* in the commensal state.

Systemic candidiasis usually occurs as a result of host factors or extenuating circumstances involving the host. Specific immunosuppressive therapies that depress both innate and acquired immune defenses will promote overgrowth and dissemination of *C. albicans*. Frequently, however, as discussed above, systemic candidiasis occurs as a result of direct inoculation. Because of the ubiquitous nature of *C. albicans*, venous or bladder catheters can become contaminated and seed *C. albicans* into the bloodstream.

(b) Changes in the organism

In addition to, or as a result of, changes in the host, *C. albicans* may express more virulence or more virulent organisms may emerge. Organisms that possess more virulence factors will obviously enhance the chances for infection. These virulence factors may be associated with the blastospore and/or hyphae. Germination is considered an important virulence factor, on the basis of the presence of hyphae in infected tissues. As discussed above, however, there has been no described cause-and-effect relationship for hyphae and infectivity. There are several factors that can be modified in the organism as a result of germination that potentially influence pathogenicity. One of these is antigenic modulation (DeBernardis et al., 1994). Deviation in antigen composition from those that induced protective responses would enhance the organism's ability to escape acquired immunological defenses. Related to this is the issue of whether the hyphae has the same or different antigenic determinants. If blastospores convert to hyphae, do host defenses recognize the hyphae, or are additional responses required to interact effectively with the new form? In addition to antigenic modulation, *C. albicans* strains have been known to switch phenotypes in the absence of an identifiable change in genotype (Soll, 1992). Although strictly an *in vitro* phenomena to date, which is detected in colony morphology, phenotypic switching may affect tissue adherence and hyphal formation. Virulence of *C. albicans* may also be enhanced by its products or secretions. The presence or emergence of *C. albicans* strains that secrete hydrolases such as aspartate proteinase, has been reported to facilitate enhanced mucosal invasion (Ross *et al.*, 1990). Finally, antifungal drug resistance has become an increasing problem, especially with azoles in immunocompromised patients (Bart-Delabesse *et al.*, 1993). It is unclear what the association is with the immunocompromised condition, but the incidence has increased sufficiently to deter the use of fluconazole as prophylactic therapy in AIDS patients.

18.5 HOST DEFENSE MECHANISMS

Defense mechanisms of the host against invasion by *C. albicans* include innate resistance and acquired immunity. The relative contributions of these two systems, however, appear to differ depending upon the site at which *C. albicans* enters the host. Innate host defenses (i.e. polymorphonuclear leukocytes, macrophages) are heavily involved in protecting the systemic circulation, antibodies (i.e. IgA, IgG) appear to play a role at both systemic and mucosal sites, and cell-mediated immunity (CMI; i.e. T cells, cytokines) protects primarily mucosal tissues. Some data support a role for CMI in protection against systemic candidiasis as well. A summary is provided in Table 18.3.

Table 18.3 Anti-*Candida* immunity and site-specific protection (PMNL = polymorphonuclear leukocyte; sIgA = secretory IgA)

Immune response	Site of protection	Immune mediators
Innate resistance	Systemic circulation	PMNLs, macrophages
Humoral	Systemic/mucosal	IgG, IgA, sIgA
Cell-mediated	Mucosal	T cells, cytokines
Oral cavity		Systemic + local
Gastrointestinal tract		Systemic + local
Vagina		Local only

Data from both clinical studies and experimental animal models that support or argue against each type of immune response for protection of various tissues are reviewed below. Particular attention is given to the role of systemic and local CMI against vaginal, gastrointestinal and oropharyngeal candidiasis. Finally, a biological perspective is given regarding mucosal immune mechanisms.

18.5.1 Innate resistance

Innate resistance mechanisms involve both cellular and acellular components of blood. The polymorphonuclear leukocyte (PMNL) is the first line of defense against *C. albicans*. The role of macrophages in defense is less clear. Soluble factors such as lactoferrin and transferrin, specific iron binding proteins in serum (Caroline *et al.*, 1964), are critical. For example, normal human serum inhibited the growth of *C. albicans in vitro*, but if iron was added the effect was reversed. Interestingly, an iron binding protein present in PMNL granules, apolactoferrin, has been shown to have candidastatic activity as well. The most impressive data supporting a major role for PMNL in protection against systemic candidiasis comes from clinical observations. Individuals with severe neutropenia (less than 500 cells/ml) are highly susceptible to systemic candidiasis. Transplantation recipients given immunotherapy that promotes severe neutropenia often succumb to systemic candidiasis (Clift, 1984). Although it is clear from animal studies that circulating PMNLs are responsible for clearance of *C. albicans* in the blood, it is unclear if PMNLs also function to protect against vascular entry of *C. albicans* from mucosal tissues. As will be discussed later, CMI is the dominant host defense mechanism at mucosal surfaces. However, individuals with T-cell deficiencies (e.g. AIDS) are generally not susceptible to systemic candidiasis, suggesting that the lack of T cells does not promote vascular entry of *C. albicans*. This has been an arduous field to study despite the availability of experimental animal models. Since conventional mice are not colonized with, nor sensitized to, *C. albicans*, it has been difficult to obtain colonization to assess dissemination. Balish and coworkers have attempted to address this with the use of germfree mice (no normal flora) (Cantorna and Balish, 1990). It was reported that, in order for systemic dissemination to occur following GI colonization with *C. albicans* under germfree conditions, mice had to be deficient in both phagocytic and T cells. In contrast, B-cell-deficient mice were resistant to dissemination (Wagner *et al.*, 1996), while mice deficient in class I MHC antigens, those antigens required for CD8 cell differentiation, were only slightly more susceptible to dissemination (Balish *et al.*, 1996). Taken together, the data suggest that both T cells and PMNLs play some role in prevention against dissemination, although clinical observations suggest a stronger role for PMNLs.

Innate immunity at mucosal sites has not received much attention, in part because of the difficulty in obtaining or extracting cells from the tissue. Thus, it is not clear how PMNLs, macrophages or natural killer (NK) cells function in the oral cavity, GI tract or vaginal mucosa. It is presumed that macrophages play at least a prominent role in CMI at mucosal sites, since delayed-type hypersensitivity (DTH) responses involve macrophages as final effector cells.

The mechanisms by which PMNLs inhibit and/or kill microbes are oxidative or non-oxidative. It is the oxidative mechanisms that appear most important against *C. albicans*. They involve the production of potent oxidants such as H_2O_2, hydroxyl radicals, hypochlorous acids and chloramines. These oxidative products produced by PMNLs can kill *C. albicans* in either the blastospore or hyphal form (Lehrer and Cline, 1969). Non-oxidative mechanisms of killing include the production of proteins/peptides with antifungal activity. Of these non-oxidative proteins (e.g. defensins, lysozyme and serine proteases) the defensins and some serine proteases in particular have anti-*Candida* activity. These proteins are present in a subset of cytoplasmic granules called 'primary' or 'azurophil' granules. Defensins are low-molecular-weight peptides with antimicrobial activity. Although multiple defensins have a high degree of homology to one another, only two defensins are known to kill *C. albicans* (i.e. HNP-1, HNP-2). The serine proteases are higher-molecular-weight, highly cationic proteins with chymotrypsin-like activity. Lysozyme, present in the PMNLs as well, also has killing activity, but only in solutions of low ionic strength. Thus, its role *in vivo* is unclear. While oxidative and non-oxidative mechanisms are distinct, both probably function for most efficient killing.

PMNLs are relatively effective, providing that organisms gain access to the vasculature or that the PMNLs attach to hyphae. However, PMNL function can be greatly enhanced with cytokines such as tumor necrosis factor, gamma-interferon, interleukin-2, interleukin-8 and colony stimulating factors (Djeu, 1993). Oxidative mechanisms are enhanced in the presence of serum as well. There are also both classical and alternate complement pathways that become activated by *C. albicans* which promote increased adherence to PMNL (Kozel, Brown and Pfrommer, 1987; Heidenreich and Dierich, 1985; Edwards *et al.*, 1986). Opsonization by complement components is important for innate resistance, particularly when they involve phagocytosis. The most prominent complement component involved in anti-*Candida* activity is C3. It was determined that C3b, iC3b and C3d are deposited on *C. albicans* as a result of alternate pathway activation and thus serve as opsonins for more efficient phagocytosis. In fact, *C. albicans* has receptors for these complement components (CR2 and CR3), which are similar to mammalian cells.

The functional mechanisms by which monocytes and macrophages affect *C. albicans* are less well defined. In general, once phagocytosed *in vitro*, killing is at least as effective as PMNLs and involves oxidative mechanisms and candidicidal molecules. Activation of macrophages with cytokines appears to increase the killing activity, although the effects appear short-lived at best. Killing of both blastospores and hyphae can be demonstrated, but the rate and efficiency depends on where the cells resided prior to removal for assay. For example, pulmonary macrophages appear particularly efficient against *C. albicans*. The variable killing efficiency of macrophages may in fact shed some light on the intracellular presence or existence of *C. albicans*. Under conditions that favor phagocytosis, the macrophages may ingest *Candida* and allow it to survive intracellularly for an extended period in the commensal state. This could contribute to keeping the organism 'in check' as a commensal.

The activity of NK cells against *C. albicans* has been controversial. Inhibition of growth by leukocyte suspensions has been demonstrated against *C. albicans* in animal models, but caution must be exercised in attributing the activity to NK cells as suspensions may easily have been contaminated with PMNLs and macrophages. Suffice it to say that although inhibition of *C. albicans* can be mediated by cells from lymphoid tissue, the primary and secondary effector cells are PMNLs and macrophages respectively. Effectors are not characteristically Asialo GM1[+] leukocytes (NK cells).

A final type of innate resistance effector cell with inhibitory activity against *C. albicans* is the CD8[+] T cells that are stimulated with IL-2 (Beno, Stover and Mathews, 1995). It has been demonstrated that T cells from naive mice stimulated with IL-2 had considerable anti-*Candida* activity. Enrichment for CD8[+] T cells showed that these naive T cells were the primary effector cell against *C. albicans*. This is the first evidence that naive T cells can be stimulated with cytokines in the absence of specific antigen and inhibit the growth of *C. albicans*, although the mechanisms of this unique interaction are not yet known.

18.5.2 Humoral immunity

Antigens of *C. albicans* readily induce the production of immunoglobulins as evidenced by the presence of *Candida*-specific IgG in serum and IgA in mucosal secretions. Thus it might be presumed that humoral immune mechanisms are important host defenses both in the systemic circulation and at mucosal sites. However, a role for antibody-mediated immunity either systemically or at mucosal sites has been quite controversial. A major argument against a role for humoral immunity involves clinical observations wherein individuals with congenital or acquired B-cell abnormalities are not more susceptible to mucosal or systemic candidiasis than immunocompetent individuals (Rogers and Balish, 1980). Additionally:

- levels of *Candida*-specific IgA were higher in the saliva of AIDS patients compared to HIV-negative controls;
- levels of *Candida*-specific IgA and IgG in women with acute VVC were similar to those in healthy control women;
- women with RVVC had normal levels of *Candida*-specific IgG in serum and IgA in vaginal secretions (Kirkpatrick, Rich and Bennett, 1971; Mathur *et al.*, 1977);
- *in vitro* studies have provided little evidence that serum antibody and complement can kill *C. albicans* (Rogers and Balish, 1980), despite increased phagocytosis of *C. albicans* in the presence of complement.

In contrast to the above, *C. albicans* has been shown, in fact, to produce proteases that degrade immunoglobulins and block phagocytosis (Diamond, Krzesicki and Wellington, 1978; LaForce *et al.*, 1975), and it has been postulated that *Candida*-specific Igs may bind to *Candida* species, blocking ligands required for adherence to mucosal surfaces (Liljemark and Gibbons, 1973). This could result in decreased invasion, but may not affect colonization.

There are data both in favor of and against a role for antibodies in protection against systemic candidiasis. Arguing against their role is the fact that *Candida*-specific antibodies are often detected in sera of patients with disseminated candidiasis (Lehner, 1997). Evidence supporting a protective role is the presence of antibodies specific for a 47 kDa breakdown product of *C. albicans* heat shock protein 90 (hsp90) in patients who have recovered from clinically active disease (Matthews, Burnie and Tabaqchali, 1984). Those who succumbed had low or undetectable titers. There is also evidence that these anti-47 kDa antibodies protected against disseminated disease in AIDS patients (Matthews *et al.*, 1988) and enhanced survival of infected mice (Matthews *et al.*, 1995). Other studies showed that mice with high *Candida*-specific antibody titers were resistant to hematogenous disseminated disease and that protection was transferable to naive recipients with serum (Mourad and Friedman, 1968). In yet another study, Domer and coworkers showed that experimental cutaneous candidiasis in mice induced *Candida*-specific antibodies and that these infected mice were less susceptible to disseminated disease (Giger *et al.*, 1978). Finally, it was recently demonstrated that antibodies directed against one specific cell surface adhesin antigen of *C. albicans* protected mice against disseminated candidiasis whereas antibodies directed against a different epitope had no effect (Han and Cutler, 1995), providing evidence that both protective and non-protective antibodies can be produced by different antigens.

Despite the lack of clinical evidence to support a protective role for antibodies at mucosal surfaces, including the vagina, there are considerable animal data that suggest a role for *Candida*-specific antibodies against vaginal candidiasis in rats. Polonelli *et al.* recently showed that intravaginal deposition with a monoclonal antibody specific for a yeast killer toxin elicited an anti-idiotypic secretory IgA (sIgA) response that protected rats against vaginal candidiasis (Polonelli *et al.*, 1994). In fact, the antibody was fungicidal presumably as a result of the mimicry of the yeast killer toxin. In another study, Cassone and coworkers showed that rats infected vaginally for a first time with *C. albicans* developed antibodies specific to aspartyl proteinase and mannan of *C. albicans,* and that protection against subsequent infections was evident in such animals. Moreover, the transfer of the immunoglobulin from such animals to naive rats provided protection against primary challenge (Cassone *et al.*, 1995).

Taking into account all the available data, an interesting concept to explain the contradictory role of antibodies against *C. albicans* was presented in a recent review by Casadevall (1995). In it, it was suggested that the controversies in the literature could be explained by the presence of immunoprotective, non-protective and deleterious antibodies in sera. These 'good', 'bad' or 'indifferent' antibodies may all exist together and in fact counter each other's actions. For instance, in the sera of one individual, immunoprotective antibodies in low concentration might be countered by deleterious antibodies and lead to enhanced pathogenesis, while another individual in whose sera immunoprotective antibodies predominate is protected against *C. albicans* infections.

18.5.3 Cell-mediated immunity

In contrast to the controversial issues regarding the role of humoral immunity against mucosal and systemic candidiasis, the role of CMI is somewhat better defined, at least as far as clinical observations are concerned. Clinical observations support an important role for CMI in protection of mucosal tissues against invasion by *C. albicans*. The strongest evidence is that individuals with reduced or absent CMI (e.g. AIDS patients, transplantation recipients, lymphoma patients treated with T cell immunosuppressive therapy and patients on corticosteroid therapy) have a high incidence of mucosal candidiasis (oropharyngeal, esophageal, mucocutaneous, vaginal, etc.; Odds, 1988), but not systemic candidiasis. In addition, there is a direct causal relationship between reduced *Candida*-specific CMI in the peripheral circulation and the incidence of individuals with chronic mucocutaneous candidiasis (CMC) (Odds, 1988). The role of T cells at each respective mucosal site, however, (i.e. oral mucosa, vaginal mucosa, gastrointestinal tract) is poorly understood, as is the relative role of cells from the systemic circulation *versus* those resident in the local tissue. There is also now some evidence from animal studies that T cells do in fact play some role in protection against systemic candidiasis (Romani, Puccetti and Bistoni 1996), but this is not supported clinically.

The mechanisms associated with CMI against *C. albicans* appear to be dominated by CD4+ T cells. Protection often correlates with delayed-type hypersensitivity (DTH), but a direct role has been difficult to assess. There have been reports where the presence of DTH did not correlate with protection (Fidel, Lynch and Sobel, 1994; Garner and Domer, 1994). Nevertheless, *Candida*-specific DTH develops *in vivo* in the immunocompetent individual. DTH is demonstrable at an early age and probably results from early exposure to *C. albicans*. Thus, a pool of *Candida*-specific memory T cells continue to circulate in the peripheral circulation. The response was presumably initiated at the mucosal site, where antigen presenting cells such as Langerhans cells or dendritic cells processed the antigen and presented it to locally associated T cells, or transported it to the draining lymph nodes where T cells were stimulated. Cytokine secretion by *Candida*-specific T cells following contact with processed candidal antigen can activate macrophages and signal them to infiltrate to the site of insult (mononuclear infiltrate), and attempt to eliminate or reduce the organism. This may in fact be a suitable mechanism whereby *C. albicans* is kept 'in check'.

In contrast to the role CD4+ cells play in protection from *C. albicans* infections at mucosal tissues, a role for CD8+ cells as an acquired host defense is unknown. To date, there has been no evidence that CD8+ cytotoxic T lymphocytes (CTL) are involved in host defense against *C. albicans*. CD8+ cells do, however, have a role in suppression of CMI in candidiasis. Although the whole field of T suppressor (Ts) cells has fallen out of favor in the scientific community, there are numerous reports showing that CD8+ cells specific for *C. albicans* cell wall antigens, both glycoprotein and mannan, suppress DTH and lymphocyte proliferation in infected mice (Garner *et al.*, 1990; Domer, Garner and Befidi-Mengue, 1989). These suppressor T cells also appear to hinder the ability to clear the infection (Carrow and Domer, 1985). The underlying mechanisms involved in this type of suppression, however, have eluded investigators for many years.

Recently CD4+ T cells have been subdivided by their cytokine profiles produced in response to antigen. A T helper (Th) 1-type response is characterized by the production of IL-2, gamma interferon (IFN-γ) and IL-12 by T cells and/or accessory cells in response to antigen. In contrast, a Th2-type response is characterized by the production of IL-4, IL-5 and IL-10 in response to antigen (Mosman and Coffman, 1989). Th1-type responses are involved in elicitation of DTH and provide help for production of opsonizing and complement-fixing antibodies (phagocytic-dependent), while Th2-type responses provide help for phagocytic-independent antibody production (e.g. IgE). Most importantly is the fact that the cytokines of each response cross-regulate those of the alternate response. For example, IL-4 and IL-10 inhibit Th1-type cytokines, while IFN-γ inhibits IL-4. Although the two types of response tend to be mutually exclusive in many experimental animal models, distinct profiles have not been as easily defined in humans. Nevertheless, the identification of Th1/Th2-type responses has shed considerable light on resistance and susceptibility to specific infectious agents, as well as on issues regarding regulation of immune responses. In fact, it is now considered that regulation of immune responses probably involves the action of these cross-regulatory cytokines, potentially elicited by CD4^{+} and/or CD8^{+} T cells.

In most infectious diseases where the role of Th responses has been investigated, one of the two Th responses is associated with resistance to infection while the alternate response is associated with susceptibility to infection. For intracellular pathogens (e.g. intracellular bacteria, viruses) and some fungi, the general rule is that Th1-type responses are associated with resistance to infection whereas Th2-type responses are associated with susceptibility to infection. In contrast, for extracellular pathogens, extracellular bacteria, parasitic helminths, etc., a Th2-type response is associated with protection and a Th1-type response is associated with susceptibility. The same is true for autoimmune diseases where inflammatory responses are deleterious. With *C. albicans*, although it is generally not considered an obligate intracellular organism, Th1-type responses have been associated with resistance to infection while Th2-type responses have been postulated to create a state of susceptibility to infection.

Animal models have been and continue to be extremely useful for dissecting CMI responses to *C. albicans* infections. In fact, the putative roles of Th1- and Th2-type responses against *C. albicans* have been identified by studies performed in a murine model of systemic candidiasis. Although clinical evidence does not suggest a role for T cells in protection against systemic disease, Bistoni and coworkers have eloquently identified and characterized a dichotomous Th cell response involved in resistance and susceptibility to experimental systemic infection, which has now been demonstrated as well for specific mucosal *C. albicans* infections (Romani, Puccetti and Bistoni 1996). Specifically, they showed that mice susceptible to systemic disease had a characteristic Th2-type response identified by the cytokines produced by splenocytes *in vitro*. In contrast, vaccination with live attenuated blastoconidia derived from an agerminative mutant strain of *C. albicans* protected mice from lethal challenge using the wild-type germinative strain. Protected animals had Th1-type responses, as evidenced by the production of Th1-type cytokines *in vitro* in response to *Candida* antigens. Additionally, mice given antibodies specific for IL-4 or IL-10 or soluble IL-4 receptor before or during a lethal challenge had an increased rate of survival that correlated with Th1-type responses. Alternatively, mice treated with anti-IFNγ or anti-IL-12 antibodies developed Th2-type responses and eventually succumbed to infection. These results provided the first evidence that CD4 T cell subsets influenced resistance (Th1) and susceptibility (Th2) to candidal infections, and that the responses could be converted from one to the other by manipulating the cytokine expression. There are many parallels between these data and those observed in leishmaniasis caused by the intracellular parasite *Leishmania donovani* (Boom *et al.*, 1990).

The role of CMI in protecting against gastrointestinal (GI) candidiasis has been studied primarily through the use of immunodeficient mice. Studies using both congenitally deficient and severe-combined-immuno-deficient (*scid*) mice have shown that T cells are critical for effective protection against GI candidiasis (Cantorna and Balish, 1990). *Scid* mice were shown to be more susceptible to GI tract colonization but were not susceptible to intravenous challenge or dissemination from the GI tract. Mice with congenital T cell (*nude/nude*), phagocytic cell (*beige/beige*) or both (*nu/nu, bg/bg*) deficiencies developed persistent GI candidiasis, while those with only phagocytic cell deficiencies (*nu/+, bg/bg*) effectively cleared *C. albicans* from the GI tract. The *nu/+* mice (wild-type) responded to *Candida* antigens with IL-2 production and DTH, while *nu/nu* mice did not, suggesting the correlation between a Th1-type response and clearance of the organisms. An important question, however, is: what role do circulating and mucosal T cells play in the protection? It has been shown that the mucosal tissue of the GI tract has intraepithelial lymphocytes (IELs), Peyer's patches (organized lymphoid tissue), T cells and antigen presenting cells that function in many types of GI infections. Much less is known, however, about the function of any of these cell populations against *C. albicans*, although Bistoni and coworkers have recently begun to address this issue. It was shown that Peyer's patch lymphocytes from colonized mice that went on to clear the fungus from the GI tract produced Th1- but not Th2-type cytokines in response to *Candida* antigen (Cenci *et al.*, 1994). On balance, GI candidiasis appears to conform to the Th1/Th2-type immune response pattern seen in systemic infection; i.e. Th1-type responses are associated with resistance against infection while Th2-type responses promote susceptibility.

There is much less known about the protective responses against *C. albicans* at the oral mucosa, although a recently developed murine model has been used to examine responses during experimental oral *C. albicans* infections. To date, *Candida* infections of the oral mucosa appear to be dominated by CMI (Chakir *et al.*, 1994). First, DTH appeared to correlate with resolution of the infection and resistance from reinfection. Second, both CD4$^+$ and CD8$^+$ T cells were recruited along with macrophages into the mucosal tissue and intraepithelial CD4$^+$ T cells persisted after resolution of the infection. Third, there was a time-dependent recruitment of γ/δ TCR$^+$ cells that correlated with the resolution of the infection. As with GI candidiasis, however, little is known concerning the specific roles of mucosal *versus* systemic T cells in protection against infection.

Until recently little was known regarding host defense mechanisms important at the vaginal mucosa against any vaginal pathogen, including *C. albicans*. Historically, animal models of vaginal candidiasis were used solely to evaluate the effects of new antifungal drugs. However, there are currently both rat and mouse models of vaginal candidiasis that have been used to investigate immunological issues. The rat model has been particularly instrumental in identifying the potential role of antibodies against *C. albicans* at the vaginal mucosa. The mouse model, on the other hand, has been used primarily to study the role of CMI. Regardless of the model, experimental vaginal infections in rodents are dependent on a state of pseudoestrus. It is assumed that this physiological state provides altered tissues and higher levels of glycogen so that *C. albicans* can proliferate, adhere and superficially invade the tissue. Based on the important role of CMI against *C. albicans* at mucosal surfaces suggested by clinical observations, and the lack of antibody deficiencies in women with RVVC, it has been postulated that women with RVVC suffer from deficiencies of CMI similar to patients with CMC.

Employing the mouse model in our laboratory, we first determined that mice inoculated under pseudoestrus conditions acquired a persistent vaginal infection for up to 10 weeks. The vaginal infection induced a systemic Th1-type response, as evidenced by Th1-type cytokine production in response to *Candida* antigen by lymph node cells draining the vaginal tissue and *Candida*-specific DTH (Fidel, Lynch and Sobel, 1993). Subsequent studies were directed toward examining the role of *Candida*-specific systemic CMI at the vaginal mucosa. The results of these studies showed that preinduced *Candida*-specific systemic Th1-type CMI did not protect mice against vaginal candidiasis (Fidel, Lynch and Sobel, 1994). Partial protection against vaginitis was achieved, however, in animals given a second inoculation following the spontaneous resolution of a primary infection in the absence of estrogen (Fidel *et al.*, 1995). Interestingly, although the vaginal infection had induced systemic *Candida*-specific Th1-type CMI, suppression of the systemic CMI, either by *Candida*-specific suppressor T cells or by depletion of all systemic CD4$^+$ cells, had no effect on the protection against vaginitis. These results suggested that systemic CMI had a limited role at the vaginal mucosa in protection against vaginal candidiasis, but it was unknown what effects the suppression or cellular depletion had on resident T cells in the vaginal mucosa.

As a result of subsequent studies that showed a virtual lack of effect on vaginal T cells by these inhibitory or depletion techniques, we postulated that some form of locally acquired mucosal immunity, T-cell and/or antibody-mediated, was responsible for protecting mice against vaginal *C. albicans* infection. This concept correlated with two clinical observations: (1) women with RVVC are not susceptible to chronic mucocutaneous candidiasis (CMC) or other forms of cutaneous candidiasis (Sobel, 1988); (2) women with CMC are generally not susceptible to RVVC (Odds, 1988); as well as with clinical studies showing that RVVC patients had normal levels of *Candida*-specific Th1-type CMI in the peripheral circulation during recurrent attacks (Fidel *et al.*, 1993). Thus the immune deficiency in RVVC patients is presumed to be localized to the vagina and not to exist at the systemic level. However, there are clinical reports that still promote a role for systemic immunity in RVVC (Witkin, 1991).

The data gathered in the murine model were the first evidence to suggest some level of immunological independence regarding CMI host defense mechanisms at the vaginal mucosa and that T cells in the vaginal mucosa should be examined for host defense mechanisms against *C. albicans*. Thus, it appears that vaginal protection from *C. albicans* infection by CMI is different from that associated with the mucosal tissue of the GI tract, at least on the basis of the role of systemic T cells.

Our current work, therefore, has focused on the presence of T cell subpopulations in the vaginal mucosa. It has recently been established that the vaginal mucosa is immunologically competent for both CMI and T-dependent antibody production since both T cells and MHC class II$^+$ cells are present. Interestingly, we and others have found that vaginal lymphocyte populations are phenotypically distinct from those in the periphery (Fidel, Wolf and KuKuruga, 1996; Nandi and Allison, 1991). To date, flow cytometric data suggests that although CD4$^+$, α/β TCR$^+$ cells predominate in the vaginal mucosa, as in lymph node cells, there is a fivefold higher percentage of γ/δ TCR$^+$ cells and few if any CD8$^+$ cells. Interestingly, through several independent assays, we additionally have evidence suggesting that vaginal CD4$^+$ cells express the CD4 protein in a different conformation from lymph node cells. This observation further supports the concept of immunological independence or compartmentalization of vaginal immunity.

Although data continue to accumulate on the presence of CMI at the vaginal mucosa, few studies have evaluated vaginal T cells during an infection. In fact, the only studies to date are in a murine model of a genital infection caused by *Chlamydia trachomatis* (Cain and Rank, 1995). In these studies, it was determined that vaginal lymphocytes were induced in infected mice and responded to *Chlamydia* antigens by proliferation. Moreover, cytokines were elicited by specific antigen *in vitro*.

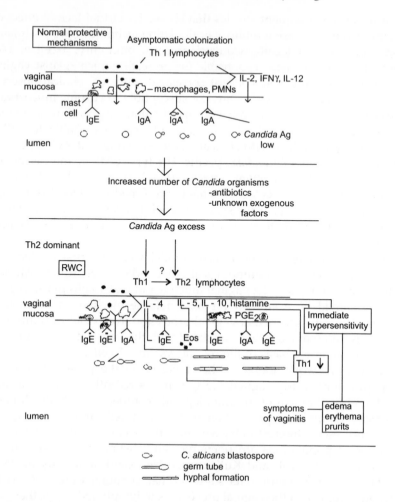

Figure 18.6 Proposed model of immunological susceptibility to RVVC. (See text for full explanation.) Briefly, a Th1-type response dominated by IL-2, IFNγ and IL-12 keeps *C. albicans* as a commensal organism as part of the normal protective host response mechanisms. The condition of RVVC occurs as a result of a rise in the population numbers of *C. albicans*, which leads to increased *Candida* antigen (Ag) expression. Under these conditions, a Th2-type response becomes dominant with the production of IL-4, IL-5 and IL-10. As a result of the Th2-type cytokines, immediate hypersensitivity increases and the protective Th1-type response is reduced or downregulated. The immediate hypersensitivity is responsible for the symptoms associated with vaginitis, while the lack of the protective Th1-type response allows *C. albicans* to convert to the pathogenic hyphal form and cause the superficial infection.

Studies on vaginal lymphocyte function during *C. albicans* infections have not yet been evaluated. We have, however, accumulated some preliminary clinical data suggesting that the intravaginal administration of *Candida* skin test antigen results in an increase of Th1-type cytokines in vaginal lavage fluid. Perhaps this is clinical evidence that expression of Th1-type cytokines contributes to protective responses against *C. albicans* infections in the vagina.

Based on all the available data, both at the vaginal mucosa and at other mucosal sites, we propose the following hypothesis to explain the incidence of RVVC (Fidel and Sobel, 1996). An illustration of the postulated mechanism(s) is provided in Fig. 18.6.

Under normal asymptomatic colonization with *C. albicans*, small numbers of ungerminated yeasts remain associated with the vaginal mucosa through a Th1-type dominant condition, including the expression of the Th1-type cytokines, anti-*Candida* IgA and IgG antibodies (top of figure). Increases in the number of *C. albicans* as a result of endogenous or exogenous factors (e.g. hormones, antibiotic usage, immune modulation) modify immunoprotective mechanisms such that a conversion to a Th2-type-dominant condition occurs. Under the influence of Th2-type reactivity (i.e. the presence of IL-4, IL-5 and IL-10), and anti-*Candida* IgE antibodies, histamine release and prostaglandin production, a down-regulation of Th1-type reactivity occurs. As a result, a conversion of *C. albicans* from the blastospore form to the more invasive hyphal form occurs and an immediate hypersensitivity reaction ensues. This immediate hypersensitivity leads to the initiation of symptoms associated with vaginitis (edema, erythema, pruritus, etc.; bottom of figure). Treatment with long- or short-term antimycotic therapy eventually decreases the numbers of *C. albicans* and *Candida* antigen returns to levels no longer capable of sustaining Th2-type and immediate hypersensitivity responses. Normal protective Th1-type CMI can then be re-expressed and proceeds to maintain *Candida* organisms in a commensal state until another modulation of protective immunity is triggered.

18.5.4 Speculations on mucosal immune mechanisms

Although many factors affect the ability of some organisms to maintain a commensal relationship with the host at mucosal surfaces, the host must also be capable of tolerating the presence of such organisms. In contrast to systemic immune mechanisms that provide an aseptic environment for internal organs and tissues, mucosal immunity appears designed to prevent tissue invasion and local disease while maintaining beneficial yet potentially pathogenic indigenous microflora. Since mucosal tissues are exposed to a large number of organisms, it is interesting to entertain the idea that immune reactivity at mucosal surfaces evolved to tolerate the presence of a variety of organisms in addition to defending against them. This theory includes non-pathogenic microflora and microflora that are potential opportunistic pathogens. *C. albicans* falls into the latter category. Thus, it is reasonable to predict that normal immunoprotective mechanisms acting at mucosal surfaces tolerate and fail to eliminate small numbers of *C. albicans* organisms, allowing them to colonize the tissue for an indeterminate amount of time. Therefore, immunoprotection against candidiasis at mucosal surfaces may be realized in the presence rather than the absence of *C. albicans*, and potentially in a manner that is separate and distinct for each tissue.

The current data for vaginal candidiasis support the concept that colonization promotes immunity. Considering the relative rates of asymptomatic vaginal colonization by *C. albicans* in healthy women (about 25%), and the receptive environment provided for *C. albicans* (e.g. pH, glucose content, hormones), protective immunity at the vaginal mucosa appears to be relatively effective. This immunity can be overwhelmed easily, however, in the presence of large numbers of organisms brought on by immune or hormonal changes that alter the vaginal environment. This same argument can be made for the other mucosal surfaces (e.g. gastrointestinal tract and oral mucosa), although the relative effectiveness of the protective immunity may vary for each site, depending on the types and amount of normal microflora and the relative colonization rates with *C. albicans* infections at those sites.

18.6 CANDIDIASIS AND THE HIV PATIENT

Mucosal candidiasis is a serious problem in HIV-infected individuals. Oropharyngeal candidiasis (OPC) is the most common oral manifestation of HIV-infected individuals, and the first clinical sign of immunosuppression during progression to AIDS. Esophageal candidiasis has been described as an AIDS-defining illness and has been considered an outward clinical sign of AIDS. As stated above, under immunocompromised situations, *C. albicans* can quickly become pathogenic, causing symptomatic infections in a variety of mucosal tissues. Any of these infections impact negatively on the quality of life. In fact, the high incidence of mucosal candidiasis in HIV-infected individuals has been an important clinical observation supporting a major role for CMI and T cell function in protection against *C. albicans* infections at mucosal surfaces. OPC has been the most dramatic form of candidiasis described for HIV-infected persons (Lamster *et al.*, 1994). While esophageal candidiasis is a specific clinical sign of AIDS, OPC correlates directly to the CD4$^+$ count during progression to AIDS. OPC does not usually occur in HIV-infected individuals when CD4$^+$ cell counts (cells/µl blood) are greater than 1000, although oral yeast colonization can be detected in up to 60% of normal or HIV-infected individuals. When CD4$^+$ cell counts drop to between 400 and 1000, OPC becomes more common. OPC in these individuals is atrophic and limited to the tongue and soft and hard palate. As CD4$^+$ cell numbers drop below 200, symptomatic disease is extremely common. The disease in these patients is 90% pseudomembranous with lesions present on the buccal mucosa, floor of the mouth, soft and hard palate and tongue. Interestingly, the national rates of OPC in HIV-infected individuals rose 23-fold between 1983 and 1989 (Fisher-Hoch and Hutwagner, 1995). Moreover, the greatest increase in OPC nationally occurred among individuals aged 14–44 years, the age range of most HIV-infected individuals. Thus, although a significant percentage of normal healthy individuals are colonized orally with *C. albicans*, the significant disease rates, excluding infant children, occur in the HIV population. Despite this positive correlation, there have been no identifiable characteristics of HIV that are known to promote *C. albicans* oral infections.

The majority of OPC cases are defined as primary and/or sporadic, which refers to patients without prior bouts of OPC or who have only experienced sporadic episodes (fewer than one or two per year). However, there is a subpopulation of HIV-infected individuals who succumb to recurrent OPC with rates of recurrence similar to women with RVVC. These individuals, like those with sporadic OPC, respond adequately to antifungal therapy but have intractable recurrences, usually within 1 month of cessation of therapy. It is unclear how these patients with recurrent OPC are different from those who acquire more sporadic episodes, although it is postulated that a greater magnitude of CMI deficiencies are present in those with recurrent OPC. Clearly, however, the incidence of OPC correlates with diminished systemic CMI function during progression to AIDS. It should be noted, however, that peripheral blood lymphocytes from many patients with decreased CD4$^+$ cell numbers and symptomatic OPC still have the ability to respond to *C. albicans in vitro*, although *Candida*-specific skin test reactivity is usually negative. Thus, it appears that both local and systemic immune factors play a role in protection against *C. albicans* in the oral mucosa, and decreases in local and/or systemic CMI can create susceptibilities to OPC. It is interesting to postulate that those patients with recurrent OPC have more significant immunological problems in the oral mucosa since there have been no detectable differences in the systemic *Candida*-specific CMI in either group of patients.

Vaginal candidiasis, unlike esophageal or oral candidiasis, has become a rather controversial issue in HIV-infected individuals over the past few years. Originally, it was reported that vaginal candidiasis occurred more frequently in HIV-infected women than in uninfected women (Rhoads *et al.*, 1987). In fact, RVVC in HIV-infected women was considered by some to be an AIDS-defining illness similar to OPC or esophageal candidiasis.

The fact is that VVC does occur in HIV-infected individuals and a specific subset of these women succumbs to RVVC similar to those who acquire recurrent OPC. However, new studies have shown no direct correlation between VVC or RVVC and $CD4^+$ counts. In fact, no specific systemic immune dysfunction in HIV-infected women has correlated with the incidence of VVC or RVVC. One problem associated with the previous studies was the comparison with HIV-negative women who were not matched for similar risk factors for HIV exposure, including intravenous drug use and multiple sexual partners. When studies were designed to include HIV-negative women matched to the socioeconomic risk for HIV exposure, there were no significant increases in the incidence of VVC or RVVC in HIV-infected women when compared to HIV-negative women matched to the same risk factors (White, 1996; Fidel *et al.*, 1996). Hence, VVC does not appear to be more common in HIV-infected individuals, although additional studies will be required to confirm this.

If HIV does not enhance susceptibility to VVC, the question then becomes: what is causing the increased rate of VVC? One explanation may be behavioral. The increased use of antibiotics or lack of hygiene under this socioeconomic classification might increase susceptibility to VVC or RVVC. Alternatively, there may also be a diminution of local host defense mechanisms as a result of this behavior, resulting in increased susceptibility to VVC and RVVC. A third explanation is that, although the incidence of VVC is increased under these socioeconomic conditions, these conditions might represent normal predisposing factors to secondary VVC or RVVC. One might reason then that, despite an increase in VVC or RVVC, vaginal host defenses remain intact in HIV-infected individuals longer than the systemic host defenses and simply become unavoidably overwhelmed.

In contrast to the high incidence of mucosal candidiasis in HIV-infected individuals, systemic candidiasis is relatively rare in such patients. However, if systemic candidiasis does occur, it is usually fatal. Systemic candidiasis will usually only occur during the very late stages of AIDS when the patient becomes severely neutropenic. In fact, in a recent study that evaluated PMNL function in HIV-infected individuals at late-stage AIDS, where the mean $CD4^+$ cell number was 27 and neutrophils numbers were subnormal, killing of *C. albicans* continued to be observed, albeit at a reduced rate (37%) compared to 60% in controls and early-stage HIV-infected patients (Tascini *et al.*, 1996; Vecchiarelli *et al.*, 1995). Thus, susceptibility to systemic candidiasis in HIV-infected individuals appears to require near-complete elimination of PMNLs. These data further support the role of PMNLs in host defense against *C. albicans* systemically.

Recently, a controversial mechanism to explain the progression to AIDS in HIV-infected individuals has been suggested, which may have some parallels to mucosal candidal infections. It was postulated, primarily from *in vitro* data, that in response to recall antigens (e.g. tetanus toxoid), alloantigens and/or mitogens a shift from the production of Th1-type cytokines to Th2-type cytokines occurred coincident with progression to AIDS (Clerici *et al.*, 1993). The levels of IL-2 and IFNγ in response to these antigens/mitogens shifted to the production of IL-4 and IL-10. The Th2-type response then inhibited any further Th1-type reactivity and promoted the production of antibodies instead of CMI. The lack of Th1-type immunity against the virus and the production of antibodies that were not protective led to the inability to control the virus and thus enhanced the progression to AIDS. There were also data suggesting that the addition of IL-12 into cultures of PBLs from HIV-infected individuals restored the production of IFNγ in response to recall antigens (Clerici and Shearer, 1993). However, case-controlled clinical studies as well as use of cloned T cells from HIV-infected individuals have not supported the Th1 to Th2 shift in immune reactivity (Romagnani and Maggi, 1994). It was shown, however, that HIV replicated much better in Th2-type and Th0-type clones than in Th1-type clones (Maggi *et al.*, 1994), suggesting that Th2-type cells may prolong the life of the virus. It has now been suggested that, instead of a switch to a Th2-type response, a lack of a Th1-type response or a shift to a Th0-type response occurs. A persistent presence of Th2-type cells might then be a reflection of their ability to support the replication of HIV virions better and thus decline slower than Th1-type cells during progression to AIDS. The lack of a Th1-type response may also involve a simple exhaustion of the protective response in its attempt to eliminate HIV. The recent finding that HIV has a very high rate of turnover in the human host supports this theory. It has also been suggested that the point of seroconversion might in fact be a critical point in the ability of the immune system to combat the HIV virus. Therefore, investigators have been encouraged to examine, by polymerase chain reaction (PCR), the presence of HIV in seronegative individuals who have been exposed to HIV and to evaluate CMI in those patients for HIV sensitization. If positive, such individuals might be candidates for specific immunotherapy, including efforts to keep Th1-type responses intact.

It is unknown whether any of these mechanisms also apply to *Candida*-specific reactivity during progression to AIDS, but studies are ongoing to examine both local and systemic *Candida*-specific CMI during different stages of HIV infection and AIDS. Interestingly, in one study, both IL-4 and IL-10 negatively affected the ability of PMNLs from HIV-infected and control individuals to kill *C. albicans*. These data suggested that a Th2-type response might increase the susceptibility of HIV-infected individuals to systemic candidiasis, particularly if their numbers of PMNLs are subnormal (Tascini *et al.*, 1996). With respect to CMI, there has been only one report describing a decrease in proliferation, IFNγ production and cytotoxic responses to *C. albicans* during progression to AIDS (Quinti *et al.*, 1991), a finding that requires confirmation. In fact, future studies should include both oral and vaginal immune reactivity to *C. albicans*, as well as *Candida*-specific systemic immunity in HIV-infected individuals with appropriate HIV-negative controls.

The final consideration regarding candidiasis in the HIV patient is the issue of candidal virulence and antifungal drug resistance. Several investigators have reported that *C. albicans* isolates recovered from the oral cavities of HIV-infected individuals have enhanced virulence. The increased virulence of these *Candida* isolates has been described at either the level of enhanced adherence to epithelial cells (Sweet, Cookson and Challacombe, 1995) or at the level of increased proteinase production (De Bernardis *et al.*, 1996). While the relationship between proteinase production and pathogenesis is somewhat controversial, some isolates with increased proteinase production have been shown to be more virulent in an experimental animal model of systemic candidiasis.

There have also been several reports describing resistance to fluconazole antifungal therapy in HIV-infected individuals (Bart-Delabesse *et al.*, 1993). Originally, the striking effectiveness of fluconazole for treatment of *Candida* infections prompted physicians to use the drug prophylactically in high-risk HIV-infected individuals. However, the appearance of fluconazole-resistant organisms has now discouraged this practice. It is unclear what mechanisms are responsible for this increased resistance to fluconazole and what, if any, significance there is to HIV infection. Another interesting clinical observation that was made recently is the emergence of multi-drug-resistant *Candida* isolates causing mucosal infections in HIV-infected individuals whose CD4$^+$ cell numbers are extremely low (< 10). Such infections were unresponsive to conventional antifungal treatment. Surprisingly, however, as the patients were treated with new antiviral agents (e.g. protease inhibitors), which quickly prompted enhanced immune function, the multi-drug-resistant *Candida* organisms disappeared. These preliminary observations suggest that the multi-drug-resistant organisms are commensal *Candida* species that are of low virulence and only emerge when protective host defenses are completely or almost completely obliterated. Although they are multi-drug-resistant, these organisms quickly revert to the commensal state when host defenses improve. This interesting hypothesis requires confirmation both by additional clinical studies and by testing the relative virulence of the multi-drug-resistant isolates in experimental animals. Preliminary experiments in our laboratory show that fluconazole-resistant *C. albicans* isolates from HIV-infected individuals are in fact, less pathogenic in the murine model of vaginal candidiasis.

18.7 CONCLUSIONS

The focus of the chapter has been the relationship of *Candida albicans* as a commensal and pathogen in the human host. Areas of concentration included a review of (1) characteristics of the organism itself, including its dimorphic properties and structural characteristics; (2) the characteristics of *C. albicans* as a commensal and pathogen, including host tissues affected; (3) host defense mechanisms important for protection of mucosal and systemic tissues against *C. albicans* infection; and (4) a review of host/pathogen issues related to candidiasis in HIV-infected persons. The most important issues to consider are that *C. albicans* exists as a commensal organism of mucosal tissues but can be triggered by a number of factors, both endogenous and exogenous, to become an opportunistic pathogen. In the commensal state, acquired immune sensitization can be demonstrated by cutaneous skin test reactivity and peripheral blood lymphocyte cytokine production and blastogenesis in response to *C. albicans* antigen, and by the presence of *Candida*-specific immunoglobulins in serum. Lapses or loss of host defense mechanisms permit the conversion of *C. albicans* from 'commensal' to 'pathogen'. CMI appears to be the predominant host defense mechanism against *C. albicans* at mucosal tissues. Therefore, conditions that limit or reduce CMI promote mucosal *C. albicans* infections. In contrast, PMNLs are the primary host defense against *C. albicans* in the peripheral circulation. Severe neutropenia, therefore, promotes systemic candidal infections. It is unclear what role immunoglobulins play, although arguments have been made for and against the role of antibodies in both mucosal and systemic infections. It is also unclear what role local *versus* systemic host defense mechanisms play for protection of mucosal tissues. There is good evidence that systemic CMI plays a prominent role in protection of the oral cavity and gastro-intestinal tract but is much less of a factor at the vaginal mucosa. At the vaginal mucosa, current data suggest that local rather than systemic host defense mechanisms are more important for protection against vaginal *C. albicans* infections.

Finally, mucosal candidiasis is a significant problem in HIV-infected individuals. Oral candidiasis appears to be the most common clinical form of the disease and correlates directly with the loss of CD4$^+$ cell numbers. Esophageal candidiasis occurs during very late stages of AIDS and is considered an AIDS-defining illness. Vaginal candidiasis is common but does not appear to be associated with HIV infection specifically. Instead, risk factors associated with exposure to HIV appear to predispose to acute or recurrent vaginal candidiasis. The mechanisms associated with the incidence of mucosal candidiasis in HIV-infected individuals are not known, except for the reduction in T cells, but current theories involve an exhaustion of protective host defense mechanisms

(Th1-type) by attempting to eliminate the virus, or a switch from the protective Th1-type to a Th0-type or the non-protective Th2-type immune response. There is also the issue of enhanced virulence of *C. albicans* as a cause of infection, and the emergence of multi-drug-resistant *C. albicans* that does not respond to conventional antimycotic therapy.

The problems, significance and prominence of *C. albicans* infections stress the urgency in understanding the host defense mechanisms important for protection against *C. albicans* 'the pathogen'. The fact that most individuals are sensitized to *C. albicans* should prove useful in understanding these mechanisms and what specific immunological mediators become modulated to promote susceptibility to infection. While antifungal therapy is usually effective for individual attacks of candidiasis, it certainly does not provide a cure. It is only through enhancing the immune status of individuals at risk, together with antimycotic therapy, that we can expect to impact the incidence of candidiasis. Currently, together with attempts to understand tissue-specific anti-*Candida* host defense mechanisms, efforts are being made to vaccinate or enhance anti-*Candida* immunity, as well as to identify new antifungal targets and develop new antifungal drugs. It is anticipated that significant progress will be made regarding prevention of this medically important disease caused by *Candida albicans* as we approach the new millennium.

18.8 ACKNOWLEDGMENTS

I wish to thank to Judith Domer and Chad Steele for help in constructing this manuscript and for critical review of its contents.

18.9 REFERENCES

Balish, E., Vazquez-Torres, A., Jones-Carson, J. *et al.* (1996) Importance of β2-microglobulin in murine resistance to mucosal and systemic candidiasis. *Infection and Immunity*, **64**, 5092–5097.

Bart-Delabesse, E., Boiron, P., Carlotti, A. and Dupont, B. (1993) *Candida albicans* genotyping in studies with patients with AIDS developing resistance to fluconazole. *Journal of Clinical Microbiology*, **32**, 2933–2937.

Bendel, C. M., St Sauver, J., Carlson, S. and Hostetter, M. K. (1995) Epithelial adhesion in yeast species: correlation with surface expression of the integrin analog. *Journal of Infectious Diseases*, **171**, 1660–1663.

Beno, D. W. A., Stover, A. G. and Mathews, H. L. (1995) Growth inhibition of *Candida albicans* hyphae by CD8+ lymphocytes. *Journal of Immunology*, **154**, 5273–5281.

Boom, W. H., Liebster, L., Abbas, A. K. and Titus, R. G. (1990) Patterns of cytokine secretion in murine leishmaniasis: correlation with disease progression or resolution. *Infection and Immunity*, **58**, 3863–3870.

Cain, T. K. and Rank, R. G. (1995) Local Th1-like responses are induced by intravaginal infection of mice with the mouse pneumonitis biovar of *Chlamydia trachomatis*. *Infection and Immunity*, **63**, 1784–1789.

Cantorna, M. T. and Balish, E. (1990) mucosal and systemic candidiasis in congenitally immunodeficient mice. *Infection and Immunity*, **58**, 1093–1100.

Caroline, L., Taschdjian, C. L., Kozinn, P. J. and Schade, A. L. (1964) Reversal of serum fungistasis by addition of iron. *Journal of Investigative Dermatology*, **42**, 415–419.

Carrow, E. W. and Domer, J. E. (1985) Immunoregulation in experimental murine candidiasis: specific suppression induced by *Candida albicans* cell wall glycoprotein. *Infection and Immunity*, **49**, 172–181.

Casadevall, A. (1995) Antibody immunity and invasive fungal infections. *Infection and Immunity*, **63**, 4211–4218.

Cassone, A., Boccanera, M., Adriani, D. A. *et al.* (1995) Rats clearing a vaginal infection by *Candida albicans* acquire specific, antibody-mediated resistance to vaginal infection. *Infection and Immunity*, **63**, 2619–2624.

Cenci, E., Mencacci, A., Spaccapelo, R. *et al.* (1994) T helper cell type 1 (Th1)- and Th2-like responses are present in mice with gastric candidiasis but protective immunity is associated with Th1 development. *Journal of Infectious Diseases*, **171**, 1279–1288.

Chakir, J., Cote, L., Coulombe, C. and Deslauriers, N. (1994) Differential pattern of infection and immune response during experimental oral candidiasis in BALB/c and DBA/2 (H-2d) mice. *Oral Microbiology and Immunology*, **9**, 88–94.

Clerici, M. and Shearer, G. M. (1993) A TH1–TH2 switch is a critical step in the etiology of HIV infection. *Immunology Today*, **14**, 107–111.

Clerici, M., Lucey, D. R., Berzofsky, J. A. *et al.* (1993) Restoration of HIV-specific cell mediated immune responses by interleukin-12 *in vitro*. *Science*, **262**, 1721–1724.

Clift, R. A. (1984) Candidiasis in the transplant patient. *American Journal of Medicine*, **77**(Suppl. 4D), 34–38.

Cole, G. T. (1986) Preparation of microfungi for scanning electron microscopy, in *Ultrastructure Techniques for Microorganisms*, Plenum Press, New York.

DeBernardis, F., Molinari, A., Boccanera, M. *et al.* (1994) Modulation of cell surface-associated mannoprotein antigen expression in experimental candidal vaginitis. *Infection and Immunity* **62**, 509–519.

De Bernardis, F., Chiani, P., Ciccozzi, M. *et al.* (1996) Elevated aspartic proteinase secretion and experimental pathogenicity of *Candida albicans* isolates from oral cavities of subjects infected with human immunodeficiency virus. *Infection and Immunity*, **64**, 466–471.

Diamond, R. D., Krzesicki, R. and Wellington, J. (1978) Damage to pseudohyphal forms of *Candida albicans* by neutrophils in the absence of serum *in vitro*. *Journal of Clinical Investigation*, **61**, 349–359.

Djeu, J. Y. (1993) Modulators of immune response to fungi, in *Fungal Infections and Immune Responses*, (eds J. W. Murphy, H. Friedman and M. Bendinelli), Plenum Press, New York, p. 521.

Domer, J. E., Garner, R. E. and Befidi-Mengue, R. N. (1989) Mannan as an antigen in cell-mediated immunity (CMI) assays and as a modulator of mannan-specific CMI. *Infection and Immunity*, **57**, 693–700.

Edwards, J. E. Jr, Gaither, T. A., O'Shea, J. J. *et al.* (1986) Expression of specific binding sites on *Candida* with functional and antigenic characteristics of human complement receptors. *Journal of Immunology*, **137**, 3577–3583.

Fidel P. L. Jr and Sobel, J. D. (1996) Immunopathogenesis of recurrent vulvovaginal candidiasis. *Clinical Microbiology Reviews*, **9**, 335–348.

Fidel, P. L. Jr, Lynch, M. E. and Sobel, J. D. (1993) *Candida*-specific Th1-type responsiveness in mice with experimental vaginal candidiasis. *Infection and Immunity*, **61**, 4202–4207.

Fidel, P. L. Jr, Lynch, M. E. and Sobel, J. D. (1994) Effects of preinduced *Candida*-specific systemic cell-mediated immunity on experimental vaginal candidiasis. *Infection and Immunity*, **62**, 1032–1038.

Fidel, P. L. Jr, Wolf, N. A. and KuKuruga, M. A. (1996) T lymphocytes in the murine vaginal mucosa are phenotypically distinct from those in the periphery. *Infection and Immunity*, **64**, 3793–3799.

Fidel, P. L., Lynch, M. E., Redondo-Lopez, V. *et al.* (1993) Systemic cell-mediated immune reactivity in women with recurrent vulvovaginal candidiasis (RVVC). *Journal of Infectious Diseases*, **168**, 1458–1465.

Fidel, P. L. Jr, Lynch, M. E., Conaway, D. H. *et al.* (1995) Mice immunized by primary vaginal *C. albicans* infection develop acquired vaginal mucosal immunity. *Infection and Immunity*, **63**, 547–553.

Fidel, P. L. Jr, Wolf, N. A., Cutright, J. L. *et al.* (1996) *Candida*-specific Th1/Th2 reactivity in HIV⁺ women. Paper given at the ASM Conference on *Candida* and Candidiasis: Biology, Pathogenesis, and Management (abstract).

Fisher-Hoch, S. P. and Hutwagner, L. (1995) Opportunistic candidiasis: an epidemic of the 1980s. *Clinical Infectious Diseases*, **21**, 897–904.

Garner, R. E. and Domer, J. E. (1994) Lack of effect of *Candida albicans* mannan on development of protective immune responses in experimental murine candidiasis. *Infection and Immunity*, **62**, 738–741.

Garner, R. E., Childress, A. M., Human, L. G. and Domer, J. E. (1990) Characterization of *Candida albicans* mannan-induced, mannan-specific delayed hypersensitivity suppressor cells. *Infection and Immunity*, **58**, 2613–2620.

Giger, D. K., Domer, J. E., Moser, S. A. and McQuitty, J. T. Jr (1978) Experimental murine candidiasis: pathological and immune responses in T-lymphocyte-depleted mice. *Infection and Immunity*, **21**, 729–737.

Han, Y. and Cutler, J. E. (1995) Antibody response that protects against disseminated candidiasis. *Infection and Immunity*, **63**, 2714–2719.

Hasenclever, H. F., Mitchell, W. O. and Loewe, J. (1961) Antigenic studies of *Candida*. II. Antigenic relation of *Candida albicans* Group A and Group B to *Candida stellatoidea* and *Candida tropicalis*. *Journal of Bacteriology*, **82**, 570–573.

Heidenreich, F. and Dierich, M. P. (1985) *Candida albicans* and *Candida stellatoidea*, in contrast to other *Candida* species, bind iC3b and C3d but not C3b. *Infection and Immunity*, **50**, 598–600.

Kirkpatrick, C. H., Rich, R. R. and Bennett, J. E. (1971) Chronic mucocutaneous candidiasis: model building in cellular immunity. *Annals of Internal Medicine*, **75**, 955–978.

Kozel, T. R., Brown, R. R. and Pfrommer, G. S. T. (1987) Activation and binding of C3 by *Candida albicans*. *Infection and Immunity*, **55**, 1890–1894.

Kreger-van Rij, N. J. W. (1984) A taxonomic study, in *The Yeasts*, Elsevier Science Publishers, Amsterdam, p. 584.

Kwon-Chung, K. J. and Bennett, J. E. (1992) Candidiasis, in *Medical Mycology*, (ed. C. Cann), Lea & Febiger, Philadelphia, PA, p. 280.

LaForce, F. M., Mills, D. M., Iverson, K. *et al.* (1975) Inhibition of leukocyte candidacidal activity by serum from patients with disseminated candidiasis. *Journal of Laboratory and Clinical Medicine*, **86**, 657–666.

Lamster, I. B., Begg, M. D., Mitchell-Lewis, D. *et al.* (1994) Oral manifestations of HIV infection in homosexual men and intravenous drug users. *Oral Surgery, Oral Medicine, and Oral Pathology*, **78**, 163–174.

Lehner, T. (1997) Serum fluorescent antibody and immunoglobulin estimations in candidosis. *Journal of Medical Microbiology*, **3**, 475–481.

Lehrer, R. I. and Cline, M. J. (1969) Interaction of *Candida albicans* with human leukocytes and serum. *Journal of Bacteriology*, **98**, 996–1004.

Liljemark, W. F. and Gibbons, R. J. (1973) Suppression of *Candida albicans* by human oral streptococci in gnotobiotic mice. *Infection and Immunity* **8**, 846–849.

Maggi, E., Mazzetti, M., Ravina, A. (1994) Ability of HIV to promote a Th1 to Th0 shift and to replicate preferentially in Th2 and Th0 cells. *Science*, **265**, 244–248.

Mathur, S., Virella, G., Koistinen, J. *et al.* (1977) Humoral immunity in vaginal candidiasis. *Infection and Immunity*, **15**, 287–294.

Matthews, R. C., Burnie, J. P. and Tabaqchali, S. (1984) Immunoblot analysis of the serological response in systemic candidosis. *Lancet*, **ii**, 1415–1418.

Matthews, R., Burnie, J., Smith, D. *et al.* (1988) *Candida* and AIDS: evidence for protective antibody. *Lancet*, **ii**, 263–265.

Matthews, R. C., Burnie, J. P., Howat, D. *et al.* (1995) Preliminary assessment of a human recombinant antibody fragment to hsp90 in murine invasive candidiasis. *Journal of Infectious Diseases*, **171**, 1668–1771.

Mosman, T. R. and Coffman, R. L. (1989) Th1 and Th2 cells: different patterns of lymphokine secretion lead to different functional properties. *Annual Reviews of Immunology*, **7**, 145–173.

Mourad, S. and Friedman, L. (1968) Passive immunization of mice against *Candida albicans*. *Sabouraudia*, **6**, 103–105.

Nandi, D. and Allison, J. P. (1991) Phenotypic analysis and gamma/delta-T cell receptor repertoire of murine T cells associated with the vaginal epithelium. *Journal of Immunology*, **147**, 1773–1778.

Odds, F. C. (1988) Chronic mucocutaneous candidiosis, in Candida *and Candidosis*, University Park Press, Baltimore, MD, p. 104.

Polonelli, L., De Bernardis, F., Conti, S. *et al.* (1994) Idiotypic intravaginal vaccination to protect against candidal vaginitis by secretory, yeast killer toxin-like anti-idiotypic antibodies. *Journal of Immunology*, **152**, 3175–3182.

Quinti, I., Palma, C., Guerra, E. C. *et al.* (1991) Proliferative and cytotoxic responses to mannoproteins of *Candida albicans* by peripheral blood lymphocytes of HIV-infected subjects. *Clinical and Experimental Immunology*, **85**, 1–8.

Rhoads, J. L., Wright, D. C., Redfield, R. R. and Burke, D. S. (1987) Chronic vaginal candidiasis in women with human immunodeficiency virus infection. *Journal of the American Medical Association*, **257**, 3105–3107.

Rogers, T. J. and Balish, E. (1980) Immunity to *Candida albicans*. *Microbiological Reviews*, **44**, 660–682.

Romagnani, S. and Maggi, E. (1994) Th1 versus Th2 responses in AIDS. *Current Opinions in Immunology*, **4**, 616–622.

Romani, L., Puccetti, P. and Bistoni, F. (1996) Biological role of Th cell subsets in candidiasis. *Chemical Immunology*, **63**, 115–137.

Ross, I. K., DeBernardis, F., Emerson, G. W. *et al.* (1990) The secreted aspartate proteinase of *Candida albicans*: physiology of secretion and virulence of a proteinase-deficient mutant. *Journal of General Microbiology*, **136**, 687–694.

Sobel, J. D. (1988) Pathogenesis and epidemiology of vulvovaginal candidiasis. *Annals of the New York Academy of Science*, **544**, 547–557.

Soll, D. R. (1992) High-frequency switching in *Candida albicans*. *Clinical Microbiology Reviews*, **5**, 183–203.

Soll, D. R., Galask, R., Schmid, J. *et al.* (1991) Genetic dissimilarity of commensal strains of *Candida* spp. carried in different anatomical locations of the same healthy women. *Journal of Clinical Microbiology*, **29**, 1702–1710.

Sweet, S. P., Cookson, S. and Challacombe, S. J. (1995) *Candida albicans* isolates from HIV-infected and AIDS patients exhibit enhanced adherence to epithelial cells. *Journal of Medical Microbiology*, **43**, 452–457.

Tascini, C., Baldelli, F., Monari, C. *et al.* (1996) Inhibition of fungicidal activity of polymorphonuclear leukocytes from HIV-infected patients by interleukin (IL)-4 and IL-10. *Acquired Immune Deficiency Syndrome*, **10**, 477–483.

Vazquez, J. A., Sobel, J. D., Demitriou, R. *et al.* (1994) Karyotyping of *Candida albicans* isolates obtained longitudinally in women with recurrent vulvovaginal candidiasis. *Journal of Infectious Diseases*, **170**, 1566–1569.

Vecchiarelli, A., Monari, C., Baldelli, F. *et al.* (1995) Beneficial effect of recombinant human granulocyte colony-stimulating factor on fungicidal activity of polymorphonuclear leukocytes from patients with AIDS. *Journal of Infectious Diseases*, **171**, 1448–1454.

Wagner, R. D., Vazquez-Torres, A., Jones-Carson, J. and Balish, E. (1996) B cell knockout mice are resistant to mucosal and systemic candidiasis of endogenous origin but susceptible to experimental systemic candidasis. *Journal of Infectious Diseases*, **174**, 589–597.

White, M. H. (1996) Is vulvovaginal candidiasis an AIDS-related illness. *Clinical Infectious Diseases*, **22**(Suppl. 2), S124–S127.

Witkin, S. S. (1991) Immunologic factors influencing susceptibility to recurrent candidal vaginitis. *Clinical Obstetrics and Gynecology*, **34**, 662–668.

19

The role of the microflora in bacterial vaginosis

Gregor Reid and Christine Heinemann

19.1 INTRODUCTION

Microbial diseases of the female genital tract are large in number and common in occurrence. They include organisms that are sexually transmitted, such as *Chlamydia trachomatis*, *Neisseria gonorrhoeae*, Herpes simplex virus, HIV, Papillomavirus, *Treponema pallidum* and *Trichomonas vaginalis* (Aynaud, Bijaoui and Huynh, 1993; Graves and Gardner, 1993; Biro, Rosenthal and Kiniyalocts, 1995; Nilsson *et al.*, 1997; Sewankambo *et al.*, 1997).

For the purposes of this report, only bacterial vaginosis (BV) will be discussed. The reason is that this disease is extremely common, yet the mechanisms whereby the normal microflora is altered are still uncertain. Furthermore, unlike sexually transmitted organisms, the causative organisms in BV are presumed to ascend from the intestine. For the host, the consequences are significant.

In North America each year, many millions of cases of BV are reported, and complications in pregnancy can lead to premature birth and the death of the fetus or newborn (Gibbs *et al.*, 1992; McGregor, French and Seo, 1993; Chaim, Mazor and Leiberman, 1997; Goldenberg *et al.*, 1997; Newton, Piper and Peairs, 1997). Premature rupture of the membranes (PROM) occurs in 5% to 10% of pregnancies and one third of preterm births, and the outcome is often maternal morbidity and newborn mortality (Alger and Pupkin, 1986). PROM can be associated with BV, urinary tract infections, group B streptococcal infections and the presence of

G. W. Tannock (ed.), Medical Importance of the Normal Microflora, 477–486.

organisms such as *Ureaplasma* and *Mycoplasma* in the urogenital tract (Kass, 1960; Kincaid-Smith and Bullen, 1965; Braun *et al.*, 1971; Brumfitt, 1975; Regan, Chao and James, 1981; Eschenbach *et al.*, 1991; Hay *et al.*, 1994; Holst, Goffeng and Andersch, 1994; Calleri *et al.*, 1997).

The common denominator in these conditions is the involvement of organisms within the urogenital microflora and infection of the expectant mother and/or fetus. While virulence factors play a role in pathogenesis, it is also likely that biofilm formation, overgrowth and sequential adherence and spread of bacteria in the vagina, cervix and uterus are important factors in onset of disease.

19.2 THE NORMAL GENITAL MICROFLORA IN ADULT FEMALES

Attempts to define and characterize the normal urogenital microflora date back very many years. The obvious difference in microbial inhabitants at different times in the female reproductive cycle (from prepuberty to postmenopause) makes generalizations impossible. However, studies focusing on specific populations have revealed important information. A study of 100 healthy, premenopausal women in 1977 (Pfau and Sacks, 1977) showed that *Lactobacillus* were the dominant organisms in the vaginal vestibule, mid-vagina and distal urethra, followed by *Staphylococcus*, diphtheroids and streptococci, while enterobacteria were rarely present. This work confirmed earlier findings (Fair *et al.*, 1970; Bruce *et al.*, 1973). The methodology used in these studies was aerobic culture, but similar results using anaerobic conditions have been reported (Corbishley, 1977; Bartlett *et al.*, 1978). Cervical anaerobic cultures also yielded lactobacilli as the dominant organism (Gorbach *et al.*, 1973). The microflora was reported to be stable and lactobacilli-dominated, even with use of oral contraception and tampons (Morris and Morris, 1967).

A recent paper by Priestley *et al.* (1997) is entitled 'What is normal vaginal flora?', and this title alone indicates how current the quest is for a better understanding of organisms within the microflora and their role in health and disease. The introduction of new microbiology detection systems, a better appreciation for growth conditions of microorganisms and changes in usage of contraceptives (such as spermicides, intrauterine devices, sponges, etc.) make it worthwhile to re-examine the composition of the urogenital microflora.

Some recent studies provide additional perspectives on the vaginal microflora. For example, a study carried out in seven urban medical centers in the USA comprising 13 747 women showed that the highest rates of potentially pathogenic colonization (BV organisms, *C. trachomatis*, *N. gonorrhoeae*) occurred in African-American women and the lowest in Asian-Pacific Islanders (Goldenberg *et al.*, 1996). There is no obvious explanation, such as the number of male partners in the past year, for the differences. It would be interesting to investigate black women living in Africa and Asian-Pacific Islanders living on Pacific islands to determine whether or not prevalence of BV and STD were high, and if the urogenital microflora was significantly different from the migrant population living in the USA. It could be hypothesized that nutrient intake influences intestinal microbial content, and in turn organisms that ascend into the genital area and, therefore, food could play a role in the susceptibility of African-American women to infection. Hopefully, studies will be designed to expand upon the findings of the Goldenberg paper.

With respect to the colonization process, the first stages of lactobacillus colonization involve electrostatic and hydrophobic interactions. These features can vary with nutrient availability and between strains. Urogenital lactobacilli have been found to have a range of hydrophobic to hydrophilic surface properties (Reid *et al.*, 1992), indicating that adherence to epithelial cell surfaces is probably mediated by several adhesins. To date, there is evidence that lactobacilli produce fimbriae (McGroarty, 1994), hemagglutinate sheep and human red blood cells (Andreu *et al.*, 1995), have lipoteichoic acid which can mediate adherence to epithelial cells (Chan *et al.*, 1985), and express proteinaceous and non-proteinaceous adhesins (Cook, Harris and Reid, 1988; Reid *et al.*, 1988, 1993; Cuperus *et al.*, 1992, 1993; Alderberth *et al.*, 1996). There is variation between strains in their ability to adhere to epithelial cells (Reid, Cook and Bruce, 1987), but it remains unclear as to how this reflects variations in expression of specific and non-specific adhesion mechanisms.

The later stages of colonization of lactobacilli and their formation of biofilms may well involve coaggregation and extracellular factors, such as fimbriae, but further studies are required to verify this hypothesis. A morphological and descriptive study of the human vaginal microflora has shown adhesion of lactobacilli to cells surrounded by connective glycocalyces (Sadhu *et al.*, 1989). In addition, lactobacilli have been shown to coaggregate *in vitro* and *in vivo* with members of the urogenital microflora (Reid *et al.*, 1988, 1990b), indicating an ability of the lactobacilli to become part of the urogenital microflora.

As shown in Chapters 16 and 17, it is proposed that lactobacilli can penetrate uropathogen biofilms. It is also possible that the reverse can occur, leading to the onset of diseases such as BV. However, the antilactobacillus factors involved in this activity have not yet been identified.

19.3 BACTERIAL VAGINOSIS

Bacterial vaginosis, formerly referred to as non-specific vaginitis (Gardner and Dukes, 1955), is defined as a mild infection of the lower female genital tract, characterized by the presence of three of five criteria:

- release of an amine (one or more of putrescine, cadaverine and tri-methylamine – fishy);
- release of an amine odor after addition of 10% potassium hydroxide;
- a vaginal pH greater than 4.5;
- clue cells in the vaginal fluid;
- a milky homogenous vaginal discharge (Amsel *et al.*, 1983; Hillier, 1993).

The examination of clue cells consists of scoring the cell population as to being normal (0–3) and dominated by Gram-positive bacilli resembling lactobacilli; intermediate (4–6) with colonization by small Gram-negative or Gram-variable rods (*Bacteroides* or *Gardnerella*) and curved Gram-variable rods (*Mobiluncus*); and BV (7–10) with domination by the pathogens (Nugent, Krohn and Hillier, 1991). The diagnosis has its flaws, ranging from low predictive values for Gram smears and culture findings to causes other than BV for amine production (Hillier, 1993; Hay and Taylor-Robinson, 1996). In addition, high vaginal and endocervical swabs and culture of fetal blood and amniotic fluid are not as sensitive and specific as preferred, making prediction of PROM based upon intrauterine culture and isolation of BV organisms poor in some centers (Carroll *et al.*, 1996). Future improvements using DNA probes, fast and accurate detection of amines, short-chain volatile acids and enzymes, should make diagnosis easier and more reliable.

The organisms associated with BV include a variety of anaerobic Gram-negative rod-shaped bacteria – *Gardnerella*, *Mobiluncus*, *Bacteroides* and possibly *Fusobacterium*, *Prevotella*, *Peptostreptococcus*, *Porphyromonas* and *Mycoplasma* species (Mardh and Taylor-Robinson, 1984; Van der Meijden, 1984; Westrom *et al.*, 1984; Spiegel *et al.*, 1980; Cook *et al.*, 1989; Hillier, 1993; Holst, Goffeng and Andersch, 1994). The mechanisms by which these bacteria displace and replace the normal *Lactobacillus*-predominant microflora remains unclear. It could be argued that the phenomenon involves a biofilm whose composition changes, perhaps through competition for nutrients and space and through the pathogens producing substances detrimental to the normal microflora. Restoration of the normal microflora is a complicated process, started by antibiotic therapy to remove pathogens (which is often not successful, perhaps because of biofilm formation by the organisms).

More information has been forthcoming with respect to the properties of the lactobacilli that appear to be important in the maintenance of the normal microflora. There is a strong case to be made that hydrogen-peroxide (H_2O_2) production is a key factor in resisting BV disease (Klebanoff *et al.*, 1991; Hillier *et al.*, 1992). H_2O_2-producing strains of lactobacilli have been found in 61% of pregnant women with normal microflora yet in only 5% of women with BV (Hillier *et al.*, 1993). The H_2O_2 has been shown to be toxic to BV causative organisms, i.e. *G. vaginalis* and *Prevotella bivia*. Therefore, it is argued that an advantage to the organisms would be gained by increased production of H_2O_2. In the presence of low iron, it has been shown that lactobacilli that grow aerobically can accumulate up to $7000\,\mu mol/l$ of H_2O_2 (Weinberg, 1997). These strains possess elevated manganese ions as a defense against the superoxide radical. Thus, the presence of iron-lowering agents such as human recombinant lactoferrin can complement the ability of lactobacilli to compete against pathogens.

It is possible that competing pathogens displace lactobacilli because of host factors such as stress (which is loosely defined and poorly understood in terms of outcome on normal microflora), and exposure to antimicrobials including antibiotics and spermicides. Both human and animal studies have shown that relatively small exposures to amoxycillin can severely deplete the normal vaginal microflora resulting in the dominant growth of pathogens (Herthelius *et al.*, 1989; Reid *et al.*, 1990a). Similarly, large proportions of urogenital lactobacilli are killed by exposure to low concentrations (1%) of the spermicidal agent nonoxynol-9 (N-9; McGroarty *et al.*, 1990; Klebanoff, 1992). Moreover, it is the H_2O_2-producing strains of lactobacillus that are particularly susceptible to N-9 (McGroarty *et al.*, 1992) and, given the increased use of these spermicidal compounds, it seems certain that this is one explanation for disruption of the host's normal microflora.

19.4 TREATMENT OF BV WITH PROBIOTICS

In selected clinical settings, antibiotic therapy to treat BV and other genital infections has been recommended to prolong pregnancy and prevent maternal–neonatal complications (Gibbs and Eschenbach, 1997). Treating women who are carrying group B streptococci is less clear as a way of predicting and preventing PROM and preterm labor (Chua *et al.*, 1995). The lack of predictable success with culture and antibiotic treatment for genital infections makes the search for alternatives at least worthy of investigation.

Having established that disruption of the lactobacilli-dominated vaginal microflora plays a role in onset of BV, it can be questioned whether the lactobacillus populations can be restored by encouraging endogenous growth or by exogenous application of probiotic strains. While the results of clinical trials currently under way in the USA, are pending and preliminary feedback is positive (Hillier, personal communication), there is a growing reason to believe that probiotics can help to treat and prevent recurrence of urogenital infections.

The evidence dates back many years (Mohler and Brown, 1932) and includes a study that showed that endocervical microflora could potentially prevent sexually transmitted disease, specifically gonorrhea (Saigh,

Sanders and Sanders, 1978). Many probiotic preparations are commercially available and, while not all appear to be reliable in their content (Hughes and Hillier, 1990), they are widely used, especially in Europe and Asia (O'Sullivan *et al.*, 1992). Indeed, there are an increasing number of proponents for probiotics in science and medicine (McGroarty, 1993; Conway and Henriksson, 1994; Fuller, 1994; Elmer, Surawicz and McFarland, 1996; Reid, 1996). It has been proposed that probiotic lactobacillus therapy could reduce the incidence of preterm delivery and late miscarriage (Hay *et al.*, 1994).

The nature of the therapy could involve viable lactobacilli, or perhaps even their byproducts, such as lipoteichoic acid. Animal studies using lipoteichoic acid have shown that group B streptococcal colonization can be significantly reduced (Cox, 1982). In human trials, the combined use of vaginal douches with *L. acidophilus* and supplementary oral vitamin B improved the well-being and condition of patients suffering from various vaginal infections (Friedlander, Druker and Schachter, 1986). The studies by Hillier currently under way should provide the definitive answer to the clinical question.

With respect to the mechanisms of action, should lactobacillus probiotics be successful, some questions will remain. The strain being used by Hillier is *L. crispatus* CTV05, a highly adherent, H_2O_2-producing strain representative of a species found in the vagina of healthy women (Reid *et al.*, 1996). To date, there are no reports of this strain producing substances, other than H_2O_2, inhibitory to the growth of genital pathogens (Skarin and Sylwan, 1986), nor of the ability of this strain to competitively exclude pathogenic adherence *in vitro*. Therefore, linking effective *in vivo* findings with mechanisms of action may not be possible. Nevertheless, if this organism or other lactobacilli provide an effective clinical outcome, it will represent an important breakthrough in the management of BV.

19.5 ACKNOWLEDGMENTS

The support provided by an NSERC Scholarship for Ms Heinemann and by Lawson Research Institute, London, Ontario is much appreciated.

19.6 REFERENCES

Alderberth, I., Ahrne, S., Johansson, M.-L. *et al.* (1996) A mannose-specific adherence mechanism in *Lactobacillus plantarum* conferring binding to the human colonic cell line HT-29. *Applied Environmental Microbiology*, **62**, 2244–2251.

Alger, L. and Pupkin, M. (1986) Etiology of preterm rupture of membranes. *Clinical Obstetrics and Gynecology*, **29**, 758–770.

Amsel, R., Totten, P. A., Spiegel, C. A. *et al.* (1983) Nonspecific vaginitis. Diagnostic criteria and microbial and epidemiologic associations. *American Journal of Medicine*, **74**, 14–22.

Andreu, A., Stapleton, A. E., Fennell, C. L. *et al.* (1995) Hemagglutination, adherence, and surface properties of vaginal *Lactobacillus* species. *Journal of Infectious Diseases*, **171**, 1237–1243.

Aynaud, O., Bijaoui, G. and Huynh, B. (1993) Genital bacterial infections associated with papillomavirus: value of screening and basis for treatment. *Contraception, Fertilité, Sexualité*, **21**, 149–152.

Bartlett, J. G., Moon, N. E., Goldstein, P. R. *et al.* (1978) Cervical and vaginal bacterial flora: ecological niches in the female lower genital tract. *American Journal of Obstetrics and Gynecology*, **130**, 658–661.

Biro, F. M., Rosenthal, S. L. and Kiniyalocts, M. (1995) Gonococcal and chlamydia genitourinary infections in symptomatic and asymptomatic adolescent women. *Clinical Pediatrics*, **34**, 419–423.

Braun, P., Yhu-Hsiung, L., Klein, J. O. *et al.*, (1971) Birth weight and genital mycoplasmas in pregnancy. *New England Journal of Medicine*, **284**, 167.

Bruce, A. W., Chadwick, P., Hassan, A. and VanCott, G. F. (1973) Recurrent urethritis in women. *Canadian Medical Association Journal*, **108**, 973–976.

Brumfitt, W. (1975) The effects of bacteriuria in pregnancy on maternal and fetal health. *Kidney International Supplement*, **8**, 113.

Calleri, L., Porcelli, A., Gallello, D. *et al.* (1997) Bacterial vaginosis and premature membrane rupture: an open study. *Minerva Ginecologica*, **49**, 19–23.

Carroll, S. G., Papioannou, S., Ntumazah, I. L. *et al.* (1996) Lower genital tract swabs in the prediction of intrauterine infection in preterm prelabour rupture of the membranes. *British Journal of Obstetrics and Gynaecology*, **103**, 54–59.

Chaim, W., Mazor, M. and Leiberman, J. R. (1997) The relationship between bacterial vaginosis and preterm birth. *Archives of Gynecology and Obstetrics*, **259**, 51–58.

Chan, R. C. Y., Reid, G., Irvin, R. T. *et al.* (1985) Competitive exclusion of uropathogens from uroepithelial cells by *Lactobacillus* whole cells and cell wall fragments. *Infection and Immunity*, **47**, 84–89.

Chua, S., Arulkumaran, S., Chow, C. *et al.* (1995) Genital Group B *Streptococcus* carriage in the antinatal period: its role in PROM and preterm labour. *Singapore Medical Journal*, **36**, 383–385.

Conway, P. L. and Henriksson, A. (1994) Strategies for the isolation and characterisation of functional probiotics, in *Human Health: the Contribution of Microorganisms*, (ed. S. A. W. Gibson), Springer-Verlag, London, pp. 75–93.

Cook, R. L., Harris, R. J. and Reid, G. (1988) Effect of culture media and growth phase on the morphology of lactobacilli and on their ability to adhere to epithelial cells. *Current Microbiology*, **17**(3), 159–166.

Cook, R. L., Reid, G., Pond, D. G. *et al.* (1989) Clue cells in bacterial vaginosis: immunofluorescent identification of the adherent Gram-negative bacteria as *Gardnerella vaginalis*. *Journal of Infectious Diseases*, **160**, 490–496.

Corbishley, C. M. (1977) Microbial flora of the vagina and cervix. *Journal of Clinical Pathology*, **30**, 745–748.

Cox F. (1982) Prevention of group B streptococcal colonization with topically applied lipoteichoic acid in a maternal-newborn mouse model. *Pediatric Research*, **16**, 816–819.

Cuperus, P., van der Mei, H. C., Reid, G. *et al.* (1992) The effect of serial passaging of lactobacilli in liquid medium on their physico-chemical and structural surface characteristics. *Cells and Materials*, **2**, 271–280.

Cuperus, P. L., van der Mei, H. C., Reid, G. *et al.* (1993) Physico-chemical surface characteristics of urogenital and poultry lactobacilli. *Journal of Colloids and Interface Science*, **156**, 319–324.

Elmer, G. W., Surawicz, C. M. and McFarland, L. V. (1996) Biotherapeutic agents: a neglected modality for the treatment and prevention of selected intestinal and vaginal infections. *Journal of the American Medical Association*, **275**, 870–876.

Eschenbach, D. A., Nugent, R. P., Rao, A. V. *et al.*, (1991) A randomized placebo-controlled trial of erythromycin for the treatment of *Ureaplasma urealyticum* to prevent premature delivery. *American Journal of Obstetrics and Gynecology*, **164**, 734–742.

Fair, W. R., Timothy, M. M., Millar, M. A. and Stamey, T. A. (1970) Bacteriologic and hormonal observations of the urethra and vaginal vestibule in normal, premenopausal women. *Journal of Urology*, **104**, 426–431.

Friedlander, A., Druker, M. M. and Schachter, A. (1986) *Lactobacillus acidophilus* and vitamin B complex in the treatment of vaginal infection. *Panminerva Medica*, **28**, 51–53.

Fuller, R. (1994) Probiotics: an overview, in *Human Health: the Contribution of Microorganisms*, (ed. S. A. W. Gibson), Springer-Verlag, London, pp. 63–74.

Gardner, H. L. and Dukes, C. D. (1955) *Haemophilus vaginalis* vaginitis: a newly defined specific infection previously classified non-specific vaginitis. *American Journal of Obstetrics and Gynecology*, **69**, 962–976.

Gibbs, R. S. and Eschenbach, D. A. (1997) Use of antibiotics to prevent preterm birth. *American Journal of Obstetrics and Gynecology*, **177**, 375–380.

Gibbs, R. S., Romero, R., Hillier, S. L. *et al.* (1992) A review of premature birth and subclinical infection. *American Journal of Obstetrics and Gynecology*, **166**, 1515–1528.

Goldenberg, R. L., Klebanoff, M. A., Nugent, R. *et al.* (1996) Bacterial colonization of the vagina during pregnancy in four ethnic groups. Vaginal infections and prematurity. *American Journal of Obstetrics and Gynecology*, **174**, 1618–1621.

Goldenberg, R. L., Andrews, W. W., Yuan, A. C. *et al.* (1997) Sexually transmitted diseases and adverse outcomes of pregnancy. *Clinical Perinatology*, **24**, 23–41.

Gorbach, S. L., Menda, K. B., Thadepalli, H. and Keith, L. (1973) Anaerobic microflora of the cervix in healthy women. *American Journal of Obstetrics and Gynecology*, **117**, 1053–1055.

Graves, A. and Gardner, W. A. Jr (1993) Pathogenicity of *Trichomonas vaginalis*. *Clinical Obstetrics and Gynecology*, **36**, 145–152.

Hay, P. E. and Taylor-Robinson, D. (1996) Defining bacterial vaginosis: to BV or not to BV, that is the question. *International Journal of STD and AIDS*, **7**, 233–235.

Hay, P. E., Lamont, R. F., Taylor-Robinson, D. *et al.* (1994) Abnormal bacterial colonisation of the genital tract and subsequent preterm delivery and late miscarriage. *British Medical Journal*, **308**, 295–298.

Herthelius, M., Gorbach, S. L., Mollby, R. *et al.* (1989) Elimination of vaginal colonization with *Escherichia coli* by administration of indigenous flora. *Infection and Immunity*, **57**, 2447–2451.

Hillier, S. L. (1993) Diagnostic microbiology of bacterial vaginosis. *American Journal of Obstetrics and Gynecology*, **169**, 455–459.

Hillier, S. L., Krohn, M. A., Klebanoff, S. J. and Eschenbach, D. A. (1992) The relationship of hydrogen peroxide-producing lactobacilli to bacterial vaginosis and genital microflora in pregnant women. *Obstetrics and Gynecology*, **79**, 369–373.

Hillier, S. L., Krohn, M. A., Rabe, L. K. *et al.* (1993) The normal vaginal flora, H_2O_2-producing lactobacilli, and bacterial vaginosis in pregnant women. *Clinical Infectious Diseases*, **16**(Suppl. 4), S273–S281.

Holst, E., Goffeng, A. R. and Andersch, B. (1994) Bacterial vaginosis and vaginal microorganisms in idiopathic premature labor and association with pregnancy outcome. *Journal of Clinical Microbiology*, **32**, 176–186.

Hughes, V. L. and Hillier, S. L. (1990) Microbiologic characteristics of *Lactobacillus* products used for colonization of the vagina. *Obstetrics and Gynecology*, **75**, 244–248.

Kass, E. H. (1960) Bacteriuria and pyelonephritis of pregnancy. *Archives of Internal Medicine*, **105**, 194.

Kincaid-Smith, P. and Bullen, M. (1965) Bacteriuria in pregnancy. *Lancet*, **i**, 395.

Klebanoff, S. J. (1992) Effects of the spermicidal agent nonoxynol-9 on vaginal microbial flora. *Journal of Infectious Diseases*, **165**, 19–25.

Klebanoff, S. J., Hillier, S. L., Eschenbach, D. A. and Waltersdorph, A. M. (1991) Control of the microbial flora of the vagina by H_2O_2-generating lactobacilli. *Journal of Infectious Diseases*, **164**, 94–100.

McGregor, J. A., French, J. I. and Seo, K. (1993) Premature rupture of membranes and bacterial vaginosis. *American Journal of Obstetrics and Gynecology*, **169**, 463–466.

McGroarty, J. A. (1993) Probiotic use of lactobacilli in the human female urogenital tract. *FEMS Immunology and Medical Microbiology*, **6**, 251–164.

McGroarty, J. A. (1994) Cell surface appendages of lactobacilli. *FEMS Microbiology Letters*, **124**, 405–410.

McGroarty, J. A., Soboh, F., Bruce, A. W. and Reid, G. (1990) The spermicidal compound nonoxynol-9 increases adhesion of *Candida* species to human epithelial cells *in vitro*. *Infection and Immunity*, **58**, 2005–2007.

McGroarty, J. A. Tomeczek, L., Pond, D. G. *et al.* (1992) Hydrogen peroxide production by *Lactobacillus* species, correlation with susceptibility to the spermicidal compound nonoxynol-9. *Journal of Infectious Diseases*, **165**(6), 1142–1144.

Mardh, P.-A. and Taylor-Robinson, D. (eds) (1984) *Bacterial Vaginosis*, Almqvist & Wiksell International, Stockholm.

Mohler, R. W. and Brown, C. P. (1932) Döderlein's bacillus in the treatment of vaginitis. *American Journal of Obstetrics and Gynecology*, 718–723.

Morris, C. A. and Morris, D. F. (1967) 'Normal' vaginal microbiology of women of childbearing age in relation to the use of oral contraceptives and vaginal tampons. *Journal of Clinical Pathology*, **20**, 636–640.

Newton, E. R., Piper, J. and Peairs, W. (1997) Bacterial vaginosis and intraamniotic infection. *American Journal of Obstetrics and Gynecology*, **176**, 672–677.

Nilsson, U., Hellberg, D., Shoubnikova, M. *et al.* (1997) Sexual behavior risk factors associated with bacterial vaginosis and *Chlamydia trachomatis* infection. *Sexually Transmitted Diseases*, **24**, 241–246.

Nugent, R. P., Krohn, M. A. and Hillier, S. L. (1991) Reliability of diagnosing bacterial vaginosis is improved by a standardized method of Gram stain interpretation. *Journal of Clinical Microbiology*, **29**, 297–301.

O'Sullivan, M. G., Thornton, G., O'Sullivan, G. C. and Collins, J. K. (1992) Probiotic bacteria: myth or reality? *Trends in Food Science and Technology*, **3**, 309–314.

Pfau, A. and Sacks, T. (1977) The bacterial flora of the vaginal vestibule, urethra and vagina in the normal premenopausal woman. *Journal of Urology*, **118**, 292–295.

Priestley, C. J., Jones, B. M., Dhar, J. and Goodwin, L. (1997) What is normal vaginal flora? *Genitourinary Medicine*, **73**, 23–28.

Regan, J. A., Chao, S. and James, L. S. (1981) Premature rupture of membranes, preterm delivery and group B streptococcal colonization of mothers. *American Journal of Obstetrics and Gynecology*, **141**, 184–186.

Reid, G. (1996) Probiotics: how microorganisms compete. *Journal of the American Medical Association*, **276**, 29–30.

Reid, G., Cook, R. L. and Bruce, A. W. (1987) Examination of strains of lactobacilli for properties which may influence bacterial interference in the urinary tract. *Journal of Urology*, **138**, 330–335.

Reid, G., McGroarty, J. A., Angotti, R. and Cook, R. L. (1988) *Lactobacillus* inhibitor production against E. coli and coaggregation ability with uropathogens. *Canadian Journal of Microbiology*, **34**, 344–351.

Reid, G., Bruce, A. W., Cook, R. L. and Llano, M. (1990a) Effect on the urogenital flora of antibiotic therapy for urinary tract infection. *Scandinavian Journal of Infectious Diseases*, **22**, 43–47.

Reid, G., McGroarty, J. A., Domingue, P. A. G. *et al.* (1990b) Coaggregation of urogenital bacteria *in vitro* and *in vivo*. *Current Microbiology*, **20**, 47–52.

Reid, G., Cuperus, P. L., Bruce, A. W. *et al.* (1992) Comparison of contact angles and adhesion to hexadecane of urogenital, dairy and poultry lactobacilli: effect of serial culture passages. *Applied and Environmental Microbiology*, **58**(5), 1549–1553.

Reid, G., Servin, A., Bruce, A. W. and Busscher, H. J. (1993) Adhesion of three *Lactobacillus* strains to human urinary and intestinal epithelial cells. *Microbios*, **75**, 57–65.

Reid, G., McGroarty, J. A., Tomeczek, L. and Bruce, A. W. (1996) Identification and plasmid profiles of *Lactobacillus* species from the vagina of 100 women. *FEMS Immunology and Medical Microbiology*, **15**, 23–26.

Sadhu, K., Domingue, P. A. G., Chow, A. W. *et al.* (1989) A morphological study of the *in situ* tissue-associated autochthonous microflora of the human vagina. *Microbial Ecology in Health and Disease*, **2**, 99–106.

Saigh, J. H., Sanders, C. C. and Sanders, W. E. (1978) Inhibition of *Neisseria gonorrhoeae* by aerobic and facultatively anaerobic components of the endocervical flora: evidence for a protective effect against infection. *Infection and Immunity*, **19**, 704–710.

Sewankambo, N., Gray, R. H., Wawer, M. J. *et al.* (1997) HIV-1 infection associated with abnormal vaginal flora morphology and bacterial vaginosis. *Lancet*, **350**, 530–531.

Skarin, A. and Sylwan, J. (1986) Vaginal lactobacilli inhibiting growth of *Gardnerella vaginalis*, *Mobiluncus* and other bacterial species cultured from vaginal content of women with bacterial vaginosis. *Acta Pathologica Microbiologica et Immunologica Scandinavica*, **94**, 399–403.

Spiegel, C. A., Amsel, R., Eschenbach, D. *et al.* (1980) Anaerobic bacteria in nonspecific vaginitis. *New England Journal of Medicine*, **303**, 601–607.

Van der Meijden, W. I. (1984) Clinical aspects of *Gardnerella vaginalis*-associated vaginitis. A review of the literature, in *Bacterial Vaginosis*, (eds P.-A. Mardh and D. Taylor-Robinson), Almqvist & Wiksell International, Stockholm, pp. 135–141.

Weinberg, E. D. (1997) The *Lactobacillus* anomaly: total iron abstinence. *Perspectives in Biology and Medicine*, **40**, 578–583.

Westrom, L., Evaldson, G., Holmes, K. K. *et al.* (1984) Taxonomy of vaginosis; bacterial vaginosis – a definition, in *Bacterial Vaginosis*, (eds P.-A. Mardh and D. Taylor-Robinson), Almqvist & Wiksell International, Stockholm, pp. 259–260.

20

Modification of the normal microflora

Gerald W. Tannock

The preceding chapters demonstrate that, far from being a benign collection of microbial associates, the normal microflora constitutes a considerable hazard to the well-being of the human host. Lurking on and within the human body are numerous microbial species that, although of low virulence, have the capacity to produce disease. These opportunistic pathogens require conditions to exist in the host that predispose to the abnormal spread and multiplication of the microbes before diseases become manifest. The predisposing factors upset the usual balance that exists between the host and the microflora (Chapter 1). Under these conditions, certain microbial species become the aetiological agents of diseases. These particular members of the normal microflora are a major source of morbidity in hospitals, resulting in increased suffering for patients and higher medical costs because of the necessity for additional patient treatment and care.

As described throughout this book, the predisposing conditions for diseases caused by members of the normal microflora are, by and large, well known. Therefore it should be possible to devise strategies that would prevent diseases due to opportunist members of the normal microflora from occurring. This is sometimes possible, as in the case of the rational choice of antimicrobial drugs. The effect of antimicrobial drugs on the composition of the normal microflora is important, for example, in the treatment of patients with granulocytic disease, or those who are receiving immunosuppressive therapy. Such patients have increased susceptibility to opportunistic infections. Various antimicrobial

G. W. Tannock (ed.), *Medical Importance of the Normal Microflora*, 487–506.
© 1999 *Kluwer Academic Publishers. Printed in Great Britain.*

regimens have been used to 'decontaminate' the oropharynx and gastro-intestinal tract of these patients because their infections are often due to microbes (*Escherichia coli*, yeasts, *Pseudomonas* species) that inhabit these regions (Cockerill *et al.*, 1992). Some broad-spectrum antimicrobial drugs may worsen the situation, however, because by altering the composition of the normal microflora they permit the replication of *Enterobacteriaceae*, *Pseudomonas* and *Candida*, which increase in numbers as the result of the suppression of obligately anaerobic bacteria whose activities are important in regulating the populations of these potential pathogens. Van der Waaij (1992) proposed that patients at risk from these opportunistic infections should be given antimicrobial drugs that do not affect the obligately anaerobic populations of the normal microflora and the Gram-positive bacteria (streptococci, staphylococci, enterococci) while inhibiting potentially pathogenic microbes. This approach has been referred to as 'selective decontamination'. Newly developed antimicrobial drugs should therefore be tested for their influence on the normal microflora in addition to their pharmacological properties. Finegold (1986) has described the influence of various antimicrobial drugs on the large bowel microflora and this information is summarized in Table 20.1.

Antimicrobial prophylaxis in elective surgery varies from country to country. Some surgeons favour the use of the oral administration of an antimicrobial drug that is not absorbed from the intestinal tract (e.g. neomycin) to reduce the number of opportunistic pathogens (e.g. *E. coli*) in the 24 hours prior to colonic surgery. Under this regimen, erythromycin is also administered as prophylaxis against staphylococcal infections. Currently, however, the use of a single antimicrobial drug, given in a single parenteral dose just before the surgical incision is made, is popular. The aim of modern, antimicrobial drug prophylaxis is to achieve, at the time of the surgical incision, maximal levels of antimicrobial drug in the patient's tissues and fluids, rather than to alter the composition of the normal microflora. A single dose given parenterally 30 minutes before surgery (90 minutes if given orally) satisfies these requirements and avoids altering the composition of the microflora in terms of obligately anaerobic species and of selecting for antimicrobial drug resistance through exposure of the intestinal microflora to the drugs. If surgery extends beyond a 2-hour period, further parenteral doses are administered at 2-hour intervals. The choice of antimicrobial agent used is governed by the bacterial species that commonly cause postsurgical sepsis for a particular anatomical region (e.g. colonic surgery: *E. coli* and *Bacteroides fragilis*, cefoxitin; gall bladder surgery: *E. coli* and enterococci, amoxycillin/clavulanic acid). In all anatomical locations, wound infections due to *Staphylococcus aureus* are likely, but prophylaxis is usually covered by the use of the cephalosporin or amoxycillin/clavulanic acid (Sanderson, 1993; Wittmann and Condon, 1991).

Table 20.1 Examples of influences of antibiotics on the large bowel microflora (Source: after Finegold, 1986)

Antibiotics having a major effect on the microflora
(Bacteroides, anaerobic cocci, bifidobacteria, lactobacilli eliminated or decreased in numbers; streptococci increased)

- Ampicillin
- Cefoperazone
- Clindamycin
- Neomycin + tetracycline, erythromycin or metronidazole
- Kanamycin + tetracycline, erythromycin or metronidazole

Antibiotics having a moderate effect on the microflora
(Coliforms and obligate anaerobes slightly decreased in numbers; streptococci slightly increased)

- Neomycin
- Kanamycin
- Co-trimoxazole
- Cephalin
- Cefoxitin
- Moxalactam
- Erythromycin

Antibiotics having minor effects on the microflora

- Sulphonamides
- Penicillins G and V
- Bacampicillin
- Dicloxacillin
- Cefataxime
- Imipenem
- Tetracyclines (major effect soon after introduction, but nowadays many bacteria are resistant)
- Chloramphenicol
- Colistin
- Metronidazole

In reality, prediction of the occurrence of opportunistic infections due to members of the normal microflora is not as simple as it is in theory because of the numerous factors that produce a predisposition to disease (e.g. in the case of urinary tract infections). Some of the diseases occur in the community rather than being nosocomial (e.g. in the case of oral diseases). There is little that can be done realistically to provide additional prophylaxis to that which already exists against these diseases. Nevertheless, it has been tempting through much of the history of medical microbiology to consider the possibility of altering the composition of the normal microflora by some sort of 'biological control', thus reducing exposure of the human host to certain microbes considered to be undesirable.

The attempted modification of the intestinal microflora provides a suitable example of this approach. Elie Metchnikoff was interested in the concept of the biological control of microbes in the large bowel. He believed that the large bowel was a non-essential organ and merely acted as a storage site for body waste. Arterial sclerosis in the elderly was, in Metchnikoff's time, attributed to the effects of alcohol, metals and syphilis. Many cases of sclerosis of organs, however, did not seem to be due to these causes. Metchnikoff proposed that these cases were due to poisoning of the body's organs by the metabolic products of the microflora inhabiting the large bowel. Although the microbes were confined within the intestinal tract, the soluble products of their metabolism were absorbed into the blood and lymph, and produced toxic effects on the body's tissues. The large intestine 'acting as an asylum of harmful microbes is a source of intoxication from within' (Metchnikoff, 1907, 1908). This was the theory of gastrointestinal autointoxication. The microbes inhabiting the large bowel, according to this theory, obtained their energy by means of the decomposition of proteins (putrefaction) or the fermentation of carbohydrates. Putrefaction occurred only slightly, or not at all, under conditions of good health. But conditions could arise whereby the balance between fermentative and putrefactive microbes could alter, the putrefactive microbes becoming numerically dominant and releasing toxic products in abundance and causing autointoxication of the host. The remedy or prevention of autointoxication was clearly surgical removal of the large bowel! A less radical approach, however, was to encourage the suppression of the putrefactive microbes in the large bowel by enhancing the activities of fermentative microbes.

Herter (1897) reported that cultures of 'colon bacillus' (indole-producing) injected into the jejunum of dogs resulted in strong indican reactions in the urine of the animals. 'Lactic acid bacillus' cultures injected into the jejunum reduced the intensity of indican reactions in the urine, which was said to indicate that the activities of indole-producing (putrefactive) bacteria in the intestine had been suppressed. It was also observed that milk soured by lactic-acid-producing bacteria was inhibitory to the growth of putrefactive microbes. Since the lactic fermentation served so well in preventing the putrefaction of milk, it should have the same effect in the digestive tract. The medicinal value of consuming fermented milk, and the lactic acid bacteria that it contained, thus became a popular topic of discussion and experimentation. Consumption of fermented milk products by humans, or ingestion of cultures of the 'Bulgarian bacillus' (*Lactobacillus delbrueckii* subsp. *bulgaricus*) isolated from yoghurt (eastern European peasants consumed large amounts of fermented milk and were apparently long-lived), resulted in a reduction in the concentration of microbial products in the urine. It was believed that the lactic-acid-producing bacteria established in the intestinal tract following daily consumption of the fermented milk products. The lactic acid bacteria then suppressed the multiplication of putrefactive microbes in the intestine by producing lactic acid and another, unidentified substance. Descriptions of the diagnosis and treatment of gastrointestinal autointoxication can be obtained by reference to the publication of Combe (1908).

Rettger and colleagues (1935) found that the ingestion of *L. delbrueckii* subsp. *bulgaricus* in milk did not lead to the appearance of viable cells of this bacterium in the faeces. Interest turned to another lactobacillus species, *Lactobacillus acidophilus*, as a useful lactic-acid-producing bacterium to use in the treatment of intestinal disease. *Lactobacillus acidophilus* was considered to be an important member of the intestinal microflora. It was reported that the administration of large numbers of *L. acidophilus* by mouth, together with the ingestion of large amounts of lactose (up to 90 g daily) or dextrin, resulted in the appearance of the lactobacillus in high numbers in the faeces. Best results were obtained when the bacteria were ingested in the form of 'acidophilus milk' or bouillon culture accompanied by a large amount of lactose. Acidophilus milk was used in the treatment of simple constipation, constipation with biliary symptoms, chronic diarrhoea following bacillary dysentery, colitis, sprue and eczema. Treatment with acidophilus milk, in the words of Rettger and colleagues, 'brought at least temporary relief to a large majority of these subjects'.

Acidophilus therapy was thought to produce beneficial effects through alteration of the composition of the intestinal microflora rather than by purely physical or chemical means. Torrey and Kahn (1923) reported that lactic acid production by *L. acidophilus* suppressed the proteolytic action of anaerobes *in vitro*. Smith (1924) found that the excretion of *E. coli* in the faeces was reduced by consumption of acidophilus milk, but the excretion of typhoid and paratyphoid bacilli by chronic carriers was not influenced by the treatment. Nissle (1916) compared the ability of different coliform strains to suppress the growth of typhoid bacilli *in vitro*. He found that some isolates had a greater ability to inhibit *Salmonella typhi* than did other strains. These differences were expressed as an 'antagonism index'. Nissle treated patients suffering from intestinal infections with capsules (*per os*) containing coliform bacteria. The health of some of his patients apparently improved following this treatment. It was concluded that the antagonistic power of coliforms harboured in the intestinal tract had a major influence on the resistance of the host to infection with a pathogen. Individuals harbouring coliforms of low antagonistic power contracted intestinal infections. Implanting coliforms of high antagonistic power in the intestine of such individuals, particularly chronic cases and 'carriers', suppressed the coliforms of low antagonistic ability along with the pathogen. A 'strong' strain of *E. coli* was marketed under the name of Mutaflor and was said to be beneficial in the treatment of intestinal infections and constipation.

The suppression of one microbial type by another was also reported for situations other than the gastrointestinal tract. Pasteur and Joubert (see Florey, 1946) observed an antagonistic effect between 'common bacteria' and anthrax bacilli cultured in urine. Cantani, according to Florey (1946), was the first investigator to record a treatment regimen in which one microbial type (but possibly a mixed culture) was used to suppress the multiplication of mycobacteria. Schitz (see Florey, 1946) observed that a patient who had a staphylococcal throat infection and had been wrongly placed in a diphtheria ward did not develop diphtheria. Schitz then sprayed suspensions of staphylococci into the throats of diphtheria carriers: good results in eliminating the carrier state were claimed. Similarly, colonization of the nasal passages by one *S. aureus* strain can prevent establishment of a different strain in that site. Shinefield and colleagues (1963) found that a nurse caring for babies in a nursery was a carrier of a strain of *S. aureus* (phage type 80/81) that was responsible for an outbreak of staphylococcal infection in the neonates. The investigators observed that only infants less than 24 hours of age were colonized by the 80/81 strain. Older infants were not susceptible because they had acquired a different phage type of *S. aureus* prior to being handled by the nurse. Shinefield and colleagues inoculated the umbilical stump and nasal mucosa of newborn infants with a strain of *S. aureus* (502A) isolated from another nurse. Strain 502A prevented colonization of the babies by strain 80/81 and eliminated the disease-producing strain from the nursery.

It is easy to understand, therefore, how attempts to alter the composition of the normal microflora by intentional exposure of subjects to particular strains of bacteria have continued to be made as a means of treatment or prophylaxis for diseases with a microbial aetiology. Interestingly, in the last decade, it has not been medical science that has promoted the development of microbe-based products supposed to improve the well-being of the consumer, but the dairy industry. The influence of the work of Metchnikoff and colleagues earlier this century in the marketing of these products is easily seen, and the desire to produce novel dairy products that could lead to an increased use of milk is apparent. These novel products, mostly yoghurts containing *Lactobacillus* species considered to be of intestinal origin, belong to the category of dietary supplements known as 'probiotics' (Table 20.2).

A probiotic is, by definition, a 'live microbial feed supplement which beneficially affects the host animal by improving its intestinal microbial balance' (Fuller, 1989). The marketing of probiotic products based on milk uses this definition because statements relating to the benefits to the consumer are generally restricted to 'helps maintain a healthy balance

Table 20.2 Examples of microbial species used in probiotic products

Products for humans	Products for farm animals
Lactobacillus acidophilus	Lactobacillus acidophilus
L. rhamnosus	L. casei
L. casei Shirota strain	L. delbrueckii subsp. bulgaricus
L. delbrueckii subsp. bulgaricus	L. plantarum
L. reuteri	L. reuteri
Bifidobacterium adolescentis	Bifidobacterium bifidum
B. bifidum	Bacillus subtilis
B. breve	Streptococcus thermophilus
B. longum	Pediococcus pentosaceus
B. infantis	Enterococcus faecium
Streptococcus thermophilus	Saccharomyces cerevisiae
Saccharomyces boulardii	Aspergillus oryzae
	Torulopsis spp.

of beneficial bacteria', 'stabilizes the intestinal microflora and modulates its function', 'promotes the positive balance of the intestinal flora' or similar. Implicit in the marketing of these products, therefore, is that consumption of the probiotic will impact in some way on the intestinal microflora.

In some ways, it might be better to use less specific phraseology in defining a probiotic. Alterations to the intestinal microflora by the consumption of a probiotic have not been demonstrated, except for the detection of the administered strain (Goldin *et al.*, 1992). Probiotics might best be referred to as 'living microbial cells administered as dietary supplements with the aim of improving health' or 'direct-fed microbials' (the term favoured in animal husbandry in the USA), since the proposed mechanisms by which these products promote health have been considerably broadened in recent years. Increasing non-specific resistance to disease by promoting microbial interference, stimulating the immunological system, modulating the metabolic activities of other members of the microflora, and supplying digestive enzymes that the host lacks have all been included as probiotic effects (Table 20.3).

Lactic-acid-producing bacteria are popular choices for use in probiotics (Table 20.2). This choice is partly due to the historical reasons noted in the preceding paragraphs, but also because lactobacilli have long been used in the production of dairy products without harm to the consumer and are 'generally regarded as safe'. Additionally, the lactic acid bacteria can be cultured on an industrial scale, as demonstrated by the dairy industry. Presumably, the choice of a particular bacterial strain incorporated into a probiotic product would be based on the desired effect of the probiotic (disease resistance, nutrition, etc.). The bacterial strains included in

Table 20.3 Health benefits ascribed to *Lactobacillus rhamnosus* GG (Valio, Finland); see Salminen and Saxelin, 1996

- Prevention of antibiotic-associated diarrhoea
- Treatment and prevention of Rotavirus diarrhoea
- Treatment of relapsing *Clostridium difficile* diarrhoea
- Prevention of acute diarrhoea
- Treatment of Crohn's disease and juvenile rheumatoid arthritis
- Antagonist against cariogenic bacteria
- Vaccine adjuvant
- Decrease in faecal ammonia and faecal pH
- Decrease in faecal enzyme activities: β-glucuronidase, nitroreductase and azoreductase
- Modifies immunoactivity to food antigens and promotes endogenous barrier mechanisms

Table 20.4 Examples of probiotic dairy products containing specific *Lactobacillus* strains

Company	Bacterial strain
Biogaia Biologics, Sweden	*Lactobacillus reuteri*
Nestlé, Switzerland	*Lactobacillus johnsonii* LC-1
Valio, Finland	*Lactobacillus rhamnosus* GG
Yakult, Japan	*Lactobacillus casei* Shirota

currently available 'generic' probiotics (e.g. *acidophilus–bifidus* yoghurts containing bacterial strains purchased from companies selling starter cultures for industrial use) have mainly been chosen for their amenability to industrial-scale culture and for their survival characteristics so as to ensure a finished product that contains viable bacterial cells. Products containing specific strains of lactic acid bacteria 'owned' by a particular food company are now becoming common with each strain apparently having special probiotic characteristics (Table 20.4).

The choice of bacterial strain for inclusion in a probiotic product must be extremely difficult to make considering the enormous number of strains from which to choose in nature (Chapter 1). Perhaps surprisingly, single strains used in probiotic products often have multiple influences on the consumer (Table 20.3). While the probiotic effects listed by manufacturers are clinically oriented, the vast proportion of retail sales of probiotic products are made to healthy humans.

Several *Lactobacillus* species have been included in probiotic products (Table 20.2). *Lactobacillus acidophilus*, until recently a catch-all species composed of disparate isolates and used widely to describe the lactobacillus components of probiotics, has been split into six new groups

(*L. acidophilus, L. amylovorus, L. crispatus, L. gallinarum, L. gasseri, L. johnsonii*). Consequently, the exact identity of the lactobacilli included in probiotics is often not known because their manufacturers have not kept step with changes in bacterial taxonomy, or because of a desire to continue to use 'acidophilus' as a generic term that is accepted by yoghurt consumers. *Lactobacillus reuteri* is considered by some researchers to be the most prevalent heterofermentative *Lactobacillus* species inhabiting the intestinal tract of humans and other animals. There is clearly extensive research that needs to be conducted in this area to determine the commonly occurring species of intestinal lactobacilli.

While the purported benefits of the consumption of probiotic products are several and often astounding, the relationship between microbes inhabiting the intestinal tract and the immune system of their host can be investigated scientifically. Interest in the ability of lactic-acid-producing bacteria to stimulate the defence mechanisms of the body originated in the use of extracts of fermented mistletoe to treat cancer patients in the 1920s. The extracts were demonstrated to stimulate the immune system of experimental animals; the stimulating components were the cells of lactobacilli (Bloksma *et al.*, 1979). This immunological effect, which principally involves the activation of macrophages, is due to constituents of the bacterial cell wall (Hashimoto *et al.*, 1984). Lactobacilli are Gram-positive bacteria, and thus their cell wall is composed mostly of peptidoglycan. Degradation products of peptidoglycan include muramyl peptides, which can be detected in the systemic tissues and which, as reviewed in Chapter 1, in addition to pharmacological activities affecting sleep patterns, body temperature and appetite, are also strong adjuvants, since they enhance immunological reactivity. The use of bacterial cells or their components to stimulate the immunological system must be approached with caution, since, in some subjects, components of the immune system that are activated and respond to bacterial substances may also react with host components and produce autoimmune disease. Cell wall material from *Bifidobacterium breve, Bifidobacterium adolescentis, Eubacterium aerofaciens* and *Lactobacillus casei*, for example, when injected intraperitoneally into rats, produces severe arthritis in the animals (Lehman *et al.*, 1987; Severijnen *et al.*, 1989).

Perhaps the most interesting application of members of the normal microflora in the promotion of health in terms of immunity is provided by investigations of the use of recombinant microbes as live vehicles for mucosal vaccines. Ideally, vaccines to stimulate mucosal immunity should be able to deliver non-denatured immunogen to an appropriate mucosal surface, ensure a sufficiently long period of exposure of the mucosa-associated lymphoid tissue to the immunogen for immunity to result (i.e. to avoid the need for multiple doses of vaccine), and ensure that the antibodies produced in response to the immunogens do not cross-react with animal tissues (i.e. to avoid the generation of auto-immune diseases). Recombinant DNA technology has potential in the development of such vaccines in the following ways. Strains of an aviru-lent bacterium could be genetically modified so that they synthesized a novel immunogen characteristic of a specific pathogen. The recombinant strain would be introduced into an appropriate body habitat where it would infect or colonize for a limited period but would not produce a symptomatology. During this period, however, the immunogen would be synthesized *in situ* by the recombinant bacteria in sufficient quantities over a sufficiently long period for an immunological response to be stimulated. Under experimental conditions, this approach has achieved success on numerous occasions using avirulent *Salmonella* strains produc-ing various immunogens as a result of genetic modification (Cardenas and Clements, 1992). Avirulence is achieved by causing mutations in genes whose products are involved in metabolic pathways. Only DNA encoding amino acid sequences that contain epitopes essential for stimu-lating immunity to a specific virulence factor need be incorporated into the recombinant strain (Newton *et al.*, 1991). The use of avirulent *Salmo-nella* strains as immunizing strains may, however, meet with some resist-ance from the community and public health authorities since there will always be the fear that the vaccine strains could revert to virulence through recombinational events with the DNA of closely related mem-bers of the *Enterobacteriaceae* that normally inhabit the intestinal tract. Further, dissemination of the vaccine strains in the general population might complicate the diagnosis of infections caused by virulent *Salmonella* species. The derivation of genetically modified strains of bacteria that normally inhabit the human body, members of the normal microflora, may prove more acceptable in the development of living vaccine strains. Intestinal lactobacilli, which are generally regarded as safe (GRAS) bac-teria, provide excellent candidates for this role.

A suitable *Lactobacillus* strain would be genetically modified to synthes-ize a novel immunogen. The recombinant lactobacillus cells would be ingested, as is the case with the current orally administered poliovirus vaccine, and the bacterial cells would colonize the intestinal tract, synthesizing the immunogen *in situ* and providing a sufficiently long exposure of the mucosal tissues to an immunogen for immunization to occur. Should the recombinant strains be incapable of colonizing the intestinal tract, ingestion of the strains in a food on a regular basis could be considered as an alternative procedure. The development of oral tolerance is sometimes a problem with administration of protein antigens administered at mucosal surfaces. Whether it will preclude the use of genetically modified bacteria as immunogen delivery systems can only be decided after appropriate experiments have been carried out. Antibodies reactive with *Lactobacillus* antigens are detectable in the sera of

Table 20.5 Examples of genetic modification of bacterial strains for experimental use as vaccine delivery vehicles

Bacterial species	Antigen produced as the result of genetic modification	Reference
Lactobacillus plantarum	N-terminal part of M6 fibrillar protein of *Streptococcus pyogenes* fused with gp41E protein of HIV	Hols *et al.*, 1997
Lactococcus lactis	Tetanus toxin fragment C	Wells, Norton and Le Page, 1995
Streptococcus gordonii	E7 protein of Human papillomavirus type 16	Medaglini *et al.*, 1997
Streptococcus gordonii	C-terminal part of M6 fibrillar protein of *S. pyogenes* fused with hornet venom allergen Ag5.2	Medaglini *et al.*, 1995
Staphylococcus xylosus	Fusion of cell wall-anchoring parts of *S. aureus* protein A and LTB subunit of cholera toxin	Liljeqvist *et al.* 1997
Salmonella dublin (avirulent)	*S. pyogenes* M protein 5 sequence inserted into flagellin gene	Newton *et al.*, 1991
Salmonella enteritidis (avirulent)	LTB subunit of heat-labile toxin of enterotoxigenic *E. coli*	Clements *et al.*, 1986
Vibrio cholerae (avirulent)	Fusion of *E. coli* haemolysin A secretion signal and C-terminus of toxin A of *Clostridium difficile*	Ryan *et al.*, 1997

healthy humans (Chapter 1); thus stimulation of the immune system by immunogens produced by recombinant lactobacilli is feasible. The pioneering work using recombinant lactobacilli has been carried out by Mercenier and colleagues (see Hols *et al.*, 1997). Other normal microflora members are under investigation as vaccine vectors: genetically modified *Streptococcus gordonii* which colonizes the oral cavity and murine vagina, and *Staphylococcus xylosus* which is an inhabitant of skin. This is undoubtedly one of the more worthy research topics in the field of probiotics (Table 20.5).

Another potentially interesting area of probiotics concerns the derivation of fermented milk products for lactose-intolerant subjects. The disaccharide lactose, which constitutes about 5% of cow's milk, is hydrolysed to glucose and galactose by lactase (a β-galactosidase) associated with the brush border of enterocytes lining the small intestinal tract of children. These monosaccharides can be absorbed and metabolized. Congenital deficiency in lactase is extremely rare, but late-onset hypolactasia, developing during childhood, is common in many regions of the

Table 20.6 Incidence of lactose-intolerance in various human populations

Ethnic group	Proportion of population that are lactose intolerant (%)
Northern European	1–5
Middle European	10–20
Mediterranean	60–90
African	70–100
Mexican Americans	80–100
Nomadic Bedouins	24
New Zealand Maori	64
New Zealand European	9
New Zealand Samoan	54

world, including Asian, African and Arab populations (80–100% of individuals) and also occurs in caucasians, but at a lower incidence (Table 20.6).

Reduction in the production of lactase in the intestinal tract (typically about 10% of childhood values) results in the passage of dietary lactose in an unaltered state to the lower small bowel and colon. Several members of the intestinal microflora produce β-galactosidases that enable them to hydrolyse the lactose. The products of the hydrolysis are fermented by the bacteria, with the production of gases (including hydrogen) and organic acids which give rise to the symptoms of abdominal discomfort and osmotic diarrhoea that are experienced by lactose-intolerant individuals (Gracey, 1991).

Clinically, lactose-intolerance is defined as the occurrence of gastrointestinal symptoms after the administration of a single test dose of 50 g of lactose in aqueous solution. More specifically, lactose-intolerant individuals have a breath hydrogen concentration of over 20 ppm during an 8-hour period following ingestion of 20 g of lactose (Kolars *et al.*, 1984). This test provides a standard, internationally recognized, non-invasive clinical assay by which lactose intolerance can be diagnosed and the effects of fermented milk products on lactose-intolerant individuals can be evaluated under conditions that exclude a placebo effect. Studies have demonstrated that lactose-intolerant individuals can tolerate fermented milk (yoghurt) better than base milk containing the same amount of lactose (Pochart *et al.*, 1989). Patient perceptions correlate with hydrogen breath test results. It appears that, although the lactic acid starter bacteria in the yoghurt cannot replicate in the intestinal tract, β-galactosidases produced by *Streptococcus salivarius* subsp. *thermophilus* and *Lactobacillus delbrueckii* subsp. *bulgaricus* during production of the yoghurt can pass through the stomach (the enzyme is intracellular). The β-galactosidases are inactive at the acid pH of the final yoghurt product and do not hydrolyse the lactose in the yoghurt until the digesta reaches the lower regions of the small bowel. By this time, the pH of the digesta, even containing acidic substances such as yoghurt, has reached near-neutrality, which is optimal for the activity of β-galactosidases produced by streptococci and lactobacilli (Pochart *et al.*, 1989). Additionally, the bacterial cells may lyse as they pass through the small bowel.

On the basis of these studies, it can be proposed that yoghurts prepared using lactic acid bacteria that produce β-galactosidases that are active at acid pH would enable lactose-intolerant individuals to ingest more milk-based food. With the associated importance in nutrition, relating to protein and calcium requirements, it could thus be an aid to the dietary supplementation of malnourished populations in Africa and other countries where lactose intolerance is common. β-Galactosidases with these properties would pass through the stomach in the yoghurt and begin to hydrolyse lactose as soon as the duodenal contents became slightly more alkaline than the yoghurt. Lactose hydrolysis would therefore occur in the proximal small bowel rather than in the distal bowel and would mean that the products of the hydrolysis would be absorbed, assuming that the bacterial cells lysed, from the intestine before they reached regions of the intestinal tract colonized by the normal microflora. Burton and Tannock (1997) have reported a study in which strains of lactobacilli isolated from the gastrointestinal tract of piglets were screened for β-glactosidases active at an acid pH. A total of 35 isolates of lactobacilli obtained from gastric samples were examined with respect to lactose fermentation, growth in milk and β-galactosidase activity. One of the isolates had β-galactosidase activity at acid pH (5.5): at pH 5.5 35% of its β-galactosidase activity at pH 7.0 remained. In comparison, yoghurt starter strains of lactobacilli had less than 5% activity remaining at pH 5.5. *Lactobacillus* strains of this phenotype might be useful, therefore, in the derivation of novel fermented milk products suitable for consumption by lactose-intolerant individuals. Lactobacilli isolated from the gastrointestinal tract may not have qualities necessary for the production of a quality yoghurt, but molecular biological methods could be used to incorporate the appropriate β-galactosidase-encoding gene into a strain of lactic acid bacterium already used as a starter culture.

A major problem with the probiotic concept is that the ingested bacterial strain does not colonize the body in the accepted ecological sense of the term: 'to become established in a new environment'. Maintenance of the probiotic strain requires daily consumption of the probiotic and, for the majority of consumers, the bacteria are eliminated relatively rapidly from the body once inoculation ceases. This inability of probiotic strains to persist in the ecosystem is understandable when it is considered that unique collections of microbial strains are harboured by each individual (Chapter 1). This is true of the intestinal tract at least and possibly applies to the normal microflora of other body habitats. The well-adapted indigenous strains prevent the persistence of probiotic strains through the phenomenon of microbial interference (Chapter 1).

Rather than to attempt to modify the composition of the normal microflora of the intestinal tract through the addition of bacterial strains, biological control might be better attempted by adding substances produced by the microbes in the intestine to the diet. This might lead to the suppression of the allegedly undesirable members of the microflora. Lactic acid has been added to the feed of farm animals and produced growth-promoting results similar to the use of subtherapeutic levels of antimicrobial drugs (Burnett and Hanna, 1963). Inorganic acids, propionate and acrylate have been used in the same way. The addition of such substances to the diet is not an economic proposition on the large scale, however, and the additives often make the food unpalatable. The additives, moreover, may not reach the distal intestinal tract where most members of the digestive tract microflora reside, but instead are likely to be degraded or absorbed in the small bowel.

Oligosaccharides that are not hydrolysed in the small bowel of humans have been proposed as agents that, upon ingestion, could modify the composition of the large bowel microflora. Occurring naturally in foods (fruits, vegetables, milk and honey), these non-digestible oligosaccharides can also be derived by enzyme-catalysed processes from various carbohydrates (Table 20.7).

Table 20.7 Oligosaccharides derived from various carbohydrates; see Crittenden and Playne, 1996 for manufacturing details

Carbohydrate	Oligosaccharide
Lactose	Galacto-oligosaccharides
	Lactulose
	Lactosucrose
Sucrose	Fructo-oligosaccharides
	Lactosucrose
	Glycerol sucrose
	Palatinose oligosaccharides
Inulin	Fructo-oligosaccharides
Starch	Cyclodextrins
	Malto-oligosaccharides
	Isomalto-oligosaccharides
	Gentio-oligosaccharides
Xylan	Xylo-oligosaccharides
Soybean whey	Soybean oligosaccharides

They have become more common as additives in foods during the past decade, especially in Japan where they are considered to confer health benefits on the consumer. Generally, oligosaccharides added to foods are not pure preparations but contain a mixture of oligosaccharides of different degrees of polymerization, the parent carbohydrate that was used as substrate and monosaccharides. Non-hydrolysed oligosaccharides pass to the distal intestinal tract of humans where some members of the normal microflora are able to utilize them as fermentable energy sources (Table 20.8).

Only a limited number of bacterial species that comprise the intestinal microflora have been screened for their ability to utilize the oligosaccharides. Currently, it appears that bifidobacteria, in general, are able to ferment the oligosaccharides (Tables 20.8, 20.9). They have therefore been described as bifidogenic factors.

A relatively small amount of information concerning the influence of ingested oligosaccharides on the intestinal microflora has been published. Lactulose (β-D-galactosyl-(1–4)-β-D-fructose) has been used in the treatment of hepatic coma. Lactulose, when fermented by the microflora in the distal intestinal tract, lowers colonic pH. The dose of lactulose is adjusted to produce the excretion of two soft stools per day, under which conditions faecal pH is 4–5. At this acidic pH, much of the ammonia produced in the large bowel is in the form of ammonium ions. In this form, ammonia is less likely to be absorbed from the bowel contents into the blood circulation. The concentration of ammonia in the blood is reduced, thereby avoiding mental or neurological deterioration in the patient (Drasar and Hill, 1974).

Other studies have focussed on the effect of incorporating oligosaccharides in infant formulas. According to the Snow Brand Company, Japan, consumption of a formula containing a galactosyl-lactose preparation increases the proportion of bifidobacteria in the faeces of babies (breast milk: 92.7% of microflora was bifidobacteria; oligosaccharide-containing formula: 69.3%; cows' milk: 61.1%) and increases the faecal concentration of acetic and lactic acids (breast milk: acetic 53.4%, lactic 20.0%; formula: acetic 39.2%, lactic 19.4%; cows' milk: acetic 24.8%, lactic 11.1%). The addition of oligosaccharide to the formula also lowered the faecal pH and altered the odour, colour and consistency (softer) of the faeces. According to Gibson and Roberfroid (1995), the consumption of fructo-oligosaccharides (15 g/d) by human subjects over a 2-week period resulted in an increase in the proportion of bifidobacteria in faeces from 6–22%, whereas bacteroides, clostridial and fusobacterial populations decreased.

While oligosaccharides are generally considered to be bifidogenic, most investigators have reported less than, or little more than, a tenfold increase in the numbers of bifidobacteria in the faeces of ex-germfree

Table 20.8 Utilization of galacto-oligosaccharides by species of intestinal bacteria (Source: after Tanaka *et al.*, 1983)

Bacterial species	Number of strains tested	Utilization*
Bifidobacterium bifidum	1	+++
B. infantis	1	+++
B. lactocentis	1	+++
B. liberorum	1	+++
B. breve	1	+++
B. longum	1	+++
B. adolescentis	1	+++
Bacteroides fragilis	2	+++
B. ovatus	1	+
B. vulgatus	1	+
B. distasonis	1	+
Fusobacteria	3	−
Eubacteria	6	−
Clostridia	7	−
Clostridium difficile	1	+
Propionibacterium acnes	1	−
Lactobacillus plantarum	3	++ / ++++†
L. casei	1	+++
L. fermentum	3	−/++
L. acidophilus	3	−/+++
L. salivarius	1	+++
L. helveticus	1	−
L. delbrueckii subsp. *bulgaricus*	1	−
L. brevis	1	−
Enterococcus faecium	3	−/+
E. faecalis	2	−
Streptococcus bovis	2	−
S. salivarius	2	−/++
S. mitis	1	++
Escherichia coli	3	++ / +++
Klebsiella pneumoniae	1	+++

* Relative rate of acidification of a culture medium containing 1% galacto-oligosaccharide
† Variation in results between bacterial strains

rats colonized by a 'human microflora', or humans whose diet has been supplemented with these substances (Bouhnik *et al.*, 1996). It is probable that a less than one log difference in viable counts determined by conventional culture methods is within experimental error. Gibson and colleagues, however, have reported about 1000-fold increases in the number of bifidobacteria in batch cultures containing palatinose-oligosaccharide or soya-oligosaccharide and inoculated with the faeces of humans.

Table 20.9 Utilization of oligosaccharides by bifidobacteria (see Mitsuoka, 1996)

Bifidobacterial species	Oligosaccharides			
	Soybean	Raffinose	Stachyose	Fructo-
B. bifidum	−*	−	+/−	−
B. longum	+++	++	+++	++
B. breve	+++	+++	+++	+
B. infantis	+++	+++	+++	++
B. adolescentis	++	++	++	++

* Relative amount of growth in presence of stipulated substrate

Oddly, although enterobacteria can ferment oligosaccharides *in vitro*, their populations in faeces have been reported to decrease or remain unaltered when the oligosaccharides are consumed by humans (Tanaka *et al.*, 1983). More convincingly, microflora-associated enzyme activities have been observed to be altered when oligosaccharides were consumed. β-fructosidase activity in human faeces was increased above basal level when fructo-oligosaccharide was ingested (9.6–13.8 IU/g dry weight, $p < 0.01$; Bouhnik *et al.*, 1996), and β-glucuronidase activity was reduced in the caecal contents of rats fed galactosyl-oligosaccharide (4.08–2.53 mmol/h/mg of protein, $p < 0.001$; Rowland and Tanaka, 1993). The impact of oligosaccharides on the intestinal microflora is clearly worthy of further investigation, especially since the administration of fructo-oligosaccharides to poultry has been reported to reduce colonization of the intestinal tract by *Salmonella* and *Campylobacter* (Schoeni and Wong, 1994; Oyarzabal and Conner, 1996).

Bifidobacterium species are now nearly as common as lactobacilli in probiotic yoghurts, and oligosaccharides are additives in probiotic drinks in Japan. It is curious that so much emphasis has been placed on boosting the populations of bifidobacteria in the intestinal tract even though the functions that bifidobacteria perform in the intestinal ecosystem remain unknown. How do bifidobacterial activities influence the well-being of the consumer and, therefore, what is the scientific rationale for including them in dairy-based products? Are intestinal populations of these bacteria in need of stimulation when all reliable published reports show that bifidobacteria are consistently one of the most numerous members of the human intestinal microflora?

The innovations concerning oligosaccharides in the food industry have resulted in the coinage of new terms: 'prebiotic', a non-digestible food ingredient that beneficially affects the host by selectively stimulating the growth and/or activity of one or a limited number of bacterial species already resident in the colon, and thus improves host health (Gibson and Roberfroid, 1995); 'synbiotic', a food product that contains both probiotic and prebiotic ingredients (Gibson and Roberfroid, 1995). This terminology belongs to the brave new world of 'functional foods'. It will be interesting to observe the impact of pre-, pro- and synbiotic philosophies, if any, on the practice of medicine.

The study of the normal microflora flourished in the 1970s and early 1980s but has received relatively little attention during the last decade. Technical difficulties associated with the investigation of such a large and complex microbial collection caused interest in the topic to diminish in both microbiological and medical circles. New molecular technologies are likely to rejuvenate research concerning the normal microflora, especially where they can be applied to medical conditions in which the indigenous microbes are aetiological agents. Studies of the maintenance or controlled modification of the normal microflora will benefit from these technical advances, and these studies will in turn be of benefit in the prophylaxis and treatment of diseases related to the activities of the normal microflora.

20.1 REFERENCES

Bloksma, N., De Heer, E., Van Dijk, H. and Willers, J. M. (1979) Adjuvancy of lactobacilli. I. Differential effects of viable and killed bacteria. *Clinical and Experimental Immunology*, **37**, 367–375.

Bouhnik, Y., Flourie, B., Riottot, M. *et al.* (1996) Effects of fructo-oligosaccharides ingestion on fecal bifidobacteria and selected metabolic indexes of colon carcinogenesis in healthy humans. *Nutrition and Cancer*, **26**, 21–29.

Burnett, G. S. and Hanna, J. (1963) Effect of dietary calcium lactate and lactic acid on faecal *Escherichia coli* counts in pigs. *Nature (London)*, **197**, 815.

Burton, J. P. and Tannock, G. W. (1997) Properties of porcine and yogurt lactobacilli in relation to lactose intolerance. *Journal of Dairy Science*, **80**, 2318–2324.

Cardenas, L. and Clements, J. D. (1992) Oral immunization using live attenuated *Salmonella* spp. as carriers of foreign antigens. *Clinical Microbiology Reviews*, **5**, 328–342.

Clements, J. D., Lyon, F. L., Lowe, K. L. *et al.* (1986) Oral immunization of mice with attenuated *Salmonella* enteritidis containing a recombinant plasmid which codes for production of the B subunit of heat-labile *Escherichia coli* enterotoxin. *Infection and Immunity*, **53**, 685–692.

Cockerill, F. R., Muller, S. R., Anhalt, J. P. *et al.* (1992) Prevention of infection in critically ill patients by selective decontamination of the digestive tract. *Annals of Internal Medicine*, **117**, 545–553.

Combe, A. (1908) *Intestinal Auto-Intoxication*, Rebman Company, New York.

Crittenden, R. G. and Playne, M. J. (1996) Production, properties and applications of food-grade oligosaccharides. *Trends in Food Science and Technology*, **7**, 353–361.

Drasar, B. S. and Hill, M. J. (1974) *Human Intestinal Flora*, Academic Press, London.

Finegold, S. M. (1986) Intestinal microbial changes and disease as a result of antimicrobial use. *Pediatric Infectious Diseases*, **5**, S88–S90.

Florey, H. (1946) The use of micro-organisms for therapeutic purposes. *Yale Journal of Biology and Medicine*, **19**, 101–117.

Fuller, R. (1989) Probiotics in man and animals. *Journal of Applied Bacteriology*, **66**, 365–378.

Gibson, G. R. and Roberfroid, M. B. (1995) Dietary modulation of the human colonic microbiota: introducing the concept of prebiotics. *Journal of Nutrition*, **125**, 1401–1412.

Goldin, B. R., Gorbach, S. L., Saxelin, M. *et al.* (1992) Survival of *Lactobacillus* species (strain GG) in human gastrointestinal tract. *Digestive Diseases and Sciences*, **37**, 121–128.

Gracey, M. (1991) Sugar intolerance, in *Diarrhoea*, (ed. M. Gracey), CRC Press, Boca Raton, FL, pp. 283–297.

Hashimoto, S., Nomoto, K., Matsuzaki, T. *et al.* (1984) Oxygen radical production by peritoneal macrophages and Kupffer cells elicited with *Lactobacillus casei*. *Infection and Immunity*, **44**, 61–67.

Herter, C. A. (1897) On certain relations between bacterial activity in the intestine and indican in the urine. *British Medical Journal*, Dec. 25, 1847–1849.

Hols, P., Slos, P., Dutot, P. *et al.* (1997) Efficient secretion of the model antigen M6-gp41E in *Lactobacillus plantarum* NCIMB 8826. *Microbiology*, **143**, 2733–2741.

Kolars, J. C., Levitt, M. D., Aouji, M. and Savaiano, D. A. (1984) Yogurt – an autodigesting source of lactose. *New England Journal of Medicine*, **310**, 1–3.

Lehman, T. J., Cremer, M. A., Walker, S. M. and Dillon, A. M. (1987) The role of humoral immunity in *Lactobacillus casei* cell wall induced arthritis. *Journal of Rheumatology*, **14**, 415–419.

Liljeqvist, S., Samuelson, P., Hansson, M. *et al.* (1997) Surface display of the cholera toxin B subunit on *Staphylococcus xylosus* and *Staphylococcus carnosus*. *Applied and Environmental Microbiology*, **63**, 2481–2488.

Medaglini, D., Pozzi, G., King, T. P. and Fischetti, V. A. (1995) Mucosal and systemic immune responses to a recombinant protein expressed on the surface of the oral commensal bacterium *Streptococcus gordonii* after oral colonization. *Proceedings of the National Academy of Sciences of the USA*, **92**, 6868–6872.

Medaglini, D., Rush, C. M., Sestini, P. and Pozzi, G. (1997) Commensal bacteria as vectors for mucosal vaccines against sexually transmitted diseases: vaginal colonization with recombinant streptococci induces local and systemic antibodies in mice. *Vaccine*, **15**, 1330–1337.

Metchnikoff, E. (1907) *The Prolongation of Life. Optimistic Studies*, William Heinemann, London.

Metchnikoff, E. (1908) *The Nature of Man. Studies in Optimistic Philosophy*, William Heinemann, London.

Mitsuoka, T. (1996) Intestinal flora and human health. *Asia Pacific Journal of Clinical Nutrition*, **5**, 2–9.

Newton, S. M., Kotb, M., Poirier, T. P. *et al.* (1991) Expression and immunogenicity of a streptococcal M protein epitope inserted in *Salmonella* flagellin. *Infection and Immunity*, **59**, 2158–2165.

Nissle (1916) Ueber die Grundlagen einer neuen ursachlichen Bekampfung der pathologischen Darmflora. *Deutsche Medizinisches Wochenschrift*, **42**, 1181–1184.

Oyarzabal, O. A. and Connor, D. E. (1996) Application of direct-fed microbial bacteria and fructooligosaccharides for *Salmonella* control in broilers during feed withdrawal. *Poultry Science*, **75**, 186–190.

Pochart, P., Dewit, O., Desjeux, J.-F. and Bourlioux, P. (1989) Viable starter culture, β-galactosidase activity, and lactose in duodenum after yogurt ingestion in lactase-deficient humans. *American Journal of Clinical Nutrition*, **49**, 828–831.

Rettger, L. F., Levy, M. N., Weinstein, L. and Weiss, J. E. (1935) Lactobacillus Acidophilus *and its Therapeutic Application*, Yale University Press, New Haven, CT.

Rowland, I. R. and Tanaka, R. (1993) The effects of transgalactosylated oligosac-charides on gut flora metabolism in rats associated with a human faecal micro-flora. *Journal of Applied Bacteriology*, **74**, 667–674.

Ryan, E. T., Butterton, J. R., Smith, R. N. *et al.* (1997) Protective immunity against *Clostridium difficile* toxin A induced by oral immunization with a live, attenu-ated *Vibrio cholerae* vector strain. *Infection and Immunity*, **65**, 2941–2949.

Salminen, S. J. and Saxelin, M. (1996) Comparison of successful probiotic strains. *Nutrition Today Supplement*, **31**, 32S–34S.

Sanderson, P. J. (1993) Antimicrobial prophylaxis in surgery: microbiological factors. *Journal of Antimicrobial Chemotherapy*, **31**(Suppl. B), 1–9.

Schoeni, J. L and Wong, A. C. L. (1994) Inhibition of *Campylobacter jejuni* coloniza-tion in chicks by defined competitive exclusion bacteria. *Applied and Environ-mental Microbiology*, **60**, 1191–1197.

Severijnen, A. J., van Kleef, R., Hazenberg, M. P. and van der Merwe, J. (1989) Cell wall fragments from major residents of the human intestinal flora induce chronic arthritis in rats. *Journal of Rheumatology*, **16**, 1061–1068.

Shinefield, H. R., Ribble, J. C., Boris, M. and Eichenwald, H. F. (1963) Bacterial interference: its effect on nursery-acquired infection with *Staphylococcus aureus*. I. preliminary observations on artificial colonization of newborns. *American Journal of Diseases of Children*, **105**, 646–654.

Smith, R. P. (1924) The implantation or enrichment of Bacillus acidophilus and other organisms in the intestine. *British Medical Journal*, **ii**, 948–950.

Tanaka, R., Takayama, H., Morotomi, M. *et al.* (1983) Effects of administration of TOS and *Bifidobacterium breve* 4006 on the human fecal flora. *Bifidobacteria Microflora*, **2**, 17–24.

Torrey, J. C. and Kahn, M. C. (1923) The inhibition of putrefactive spore-bearing anaerobes by *Bacterium acidophilus*. *Journal of Infectious Diseases*, **33**, 482–497.

Van der Waaij, D. (1992) Selective gastrointestinal decontamination. Epidemio-logy and Infection, **109**, 315–326.

Wells, J. M., Norton, P. M. and Le Page, R. W. F. (1995) Progress in the develop-ment of mucosal vaccines based on *Lactococcus lactis*. *International Dairy Journal*, **5**, 1071–1079.

Wittmann, D. H. and Condon, R. E. (1991) Prophylaxis of postoperative infections. *Infection*, **19**, S337–S344.

Index

Note: page numbers in *italics* refer to tables, those in **bold** refer to figures.

Index